Photovoltaic Thermal Passive House System

Photovoltaic Thermal Passive House System

Basic Principle, Modeling, Energy and Exergy Analysis

Gopal Nath Tiwari
and
Neha Gupta

CRC Press is an imprint of the
Taylor & Francis Group, an **informa** business

First edition published 2023
by CRC Press
6000 Broken Sound Parkway NW, Suite 300, Boca Raton, FL 33487-2742

and by CRC Press
2 Park Square, Milton Park, Abingdon, Oxon, OX14 4RN

© 2023 Taylor & Francis Group, LLC
CRC Press is an imprint of Taylor & Francis Group, LLC

Reasonable efforts have been made to publish reliable data and information, but the author and publisher cannot assume responsibility for the validity of all materials or the consequences of their use. The authors and publishers have attempted to trace the copyright holders of all material reproduced in this publication and apologize to copyright holders if permission to publish in this form has not been obtained. If any copyright material has not been acknowledged please write and let us know so we may rectify in any future reprint.

Except as permitted under U.S. Copyright Law, no part of this book may be reprinted, reproduced, transmitted, or utilized in any form by any electronic, mechanical, or other means, now known or hereafter invented, including photocopying, micro-filming, and recording, or in any information storage or retrieval system, without written permission from the publishers.

For permission to photocopy or use material electronically from this work, access www.copyright.com or contact the Copyright Clearance Center, Inc. (CCC), 222 Rosewood Drive, Danvers, MA 01923, 978-750-8400. For works that are not available on CCC please contact mpkbookspermissions@tandf.co.uk

Trademark notice: Product or corporate names may be trademarks or registered trademarks and are used only for identification and explanation without intent to infringe.

Library of Congress Cataloging-in-Publication Data
Names: Tiwari, G. N., author. | Gupta, Neha, author.
Title: Photovoltaic thermal passive house system : basic principle,
modeling, energy and exergy analysis / Gopal Nath Tiwari and Neha Gupta.
Description: First edition. | Boca Raton, FL : CRC Press, 2021. | Includes
bibliographical references and index.
Identifiers: LCCN 2021008913 | ISBN 9781138333550 (hardback) | ISBN
9781032047010 (paperback) | ISBN 9780429445903 (ebook)
Subjects: LCSH: Solar houses. | Solar energy--Passive systems. |
Building-integrated photovoltaic systems.
Classification: LCC TH7414 .T59 2021 | DDC 697/.78--dc23
LC record available at https://lccn.loc.gov/2021008913

ISBN: 978-1-138-33355-0 (hbk)
ISBN: 978-1-032-04701-0 (pbk)
ISBN: 978-0-429-44590-3 (ebk)

DOI: 10.1201/9780429445903

Typeset in Times
by SPi Technologies India Pvt Ltd (Straive)

It is our deepest gratitude and warmest affection
that we dedicate this book
to our respected teacher and Guru ji
**Padmashri Professor M.S. Sodha, FNA on his 91st birthday,
February 8, 2023**

Contents

Preface..xvii

Authors..xix

Chapter 1   General Introduction..1

    1.1   Zero Energy Buildings...1

    1.2   The Sun and the Earth...2

          1.2.1   The Sun...2

          1.2.2   The Earth..4

          1.2.3   Sun–Earth Angles...5

          1.2.4   Solar Radiation...9

    1.3   Climate...13

          1.3.1   Climatic Conditions...13

          1.3.2   Weather Conditions..15

          1.3.3   Macro- and Microclimate..15

                1.3.3.1   Macroclimate..15

                1.3.3.2   Microclimate..15

    1.4   Passive Houses..16

          1.4.1   Strategies for Passive Design...16

                1.4.1.1   Solar Access..16

                1.4.1.2   Wind Control..16

    1.5   Architectural Design of Passive Buildings...17

          1.5.1   Site Planning..17

                1.5.1.1   Building Location and Orientation................................17

                1.5.1.2   Building Orientation..18

                1.5.1.3   Clustering...18

           1.5.2   Envelope Design or Building Envelope..18

                1.5.2.1   Building Shape..18

                1.5.2.2   Entrances and Windows...19

                1.5.2.3   Solar Shading Techniques...19

                1.5.2.4   Insulation...20

                1.5.2.5   Infiltration Reduction...22

           1.5.3   Interior Design...22

    1.6   Bioclimatic Design..23

          1.6.1   History..23

          1.6.2   Building Design Strategies Depending on Climatic Conditions.......25

    1.7   Energy Conservation...26

          1.7.1   Introduction...26

          1.7.2   Energy Consumption..26

          1.7.3   Energy Efficiency...26

    1.8   Design Approach to ZEB...28

          1.8.1   Stage 0: Research...28

          1.8.2   Stage 1: Reduce...29

           1.8.3   Stage 2: Reuse...31

           1.8.4   Stage 3: Produce..32

    1.9   Case Study of an Energy Neutral Building...34

    Objective Questions...35

vii

viii   Contents

Answers 36
Problems 37
References 37

Chapter 2   Basic Heat Transfer 41

   2.1   Introduction 41
   2.2   Conduction 41
         2.2.1   Temperature Field 42
         2.2.2   Fourier's Heat Conduction Equation 42
         2.2.3   Thermal Conductivity 43
         2.2.4   Thermal Diffusivity 43
         2.2.5   Conductive Heat Transfer Coefficient 44
         2.2.6   Dimensionless Heat Conduction Parameters 44
                 2.2.6.1   Biot Number (Bi) 44
                 2.2.6.2   Fourier Number 45
   2.3   Convection 46
         2.3.1   Dimensionless Heat Convection Parameters 46
                 2.3.1.1   Nusselt Number (Nu) 47
                 2.3.1.2   Reynolds Number (Re) 47
                 2.3.1.3   Prandtl Number (Pr) 47
                 2.3.1.4   Grashof Number (Gr) 48
                 2.3.1.5   Rayleigh Number (Ra) 48
         2.3.2   Types of Convection 56
                 2.3.2.1   Free Convection 56
                 2.3.2.2   Forced Convection 56
                 2.3.2.3   Mixed-Mode Convection 59
   2.4   Convective Heat Transfer Coefficient 59
   2.5   Radiation 60
         2.5.1   Radiation Involving Real Surfaces 60
         2.5.2   Kirchhoff's Law 61
         2.5.3   Laws of Thermal Radiation 61
                 2.5.3.1   Planck's Law 61
                 2.5.3.2   Wien's Displacement Law 62
                 2.5.3.3   Stefan–Boltzmann Law 62
                 2.5.3.4   Sky Radiation 62
         2.5.4   Radiative Heat Transfer Coefficient 64
   2.6   Evaporation (Mass Transfer) 66
   2.7   Total Heat Transfer Coefficient 68
   2.8   Overall Heat Transfer Coefficient 69
         2.8.1   Parallel Slabs 70
         2.8.2   Parallel Slabs with Air Cavity 71
   2.9   Thermal Circuit Analysis 73
         2.9.1   Composite Wall 73
         2.9.2   Composite Roof 74
   2.10 Energy Balance 75
         2.10.1   Energy Balance for Winter's Day 75
         2.10.2   Energy Balance on a Cloudy Day 75
         2.10.3   Energy Balance on a Summer's Day in an
                  Air-Conditioned Building 77
         2.10.4   Energy Balance for Intermediate Season Like Spring and Autumn 78
   Objective Questions 79

Contents   ix

Answers 81
Problems 81
References 82

Chapter 3 Thermal Comfort 85
    3.1   Introduction 85
    3.2   Physical Aspects 86
        3.2.1   Air Temperature 86
        3.2.2   Relative Humidity 86
        3.2.3   Air Movement 87
        3.2.4   Mean Radiant Temperature 88
        3.2.5   Air Pressure 90
        3.2.6   Air Ingredients 90
        3.2.7   Air Electricity 90
        3.2.8   Acoustics 90
        3.2.9   Daylighting 90
            3.2.9.1   Windows and Fenestrations 91
            3.2.9.2   Skylights 92
            3.2.9.3   Solar Tubes 92
            3.2.9.4   Semi-Transparent Solar Photovoltaic Lighting
                       System (SSPLS) and Transparent Facades 92
            3.2.9.5   Light Shelves 93
            3.2.9.6   Sawtooth Roofs 94
            3.2.9.7   Heliostats 94
            3.2.9.8   Smart Glass Windows 94
            3.2.9.9   Hybrid Solar Lighting (HSL) 95
    3.3   Physiological Aspects 95
        3.3.1   Nutritional Intake 95
        3.3.2   Age 95
        3.3.3   Ethnic Influences 95
        3.3.4   Gender Differences 95
        3.3.5   Constitution 95
    3.4   Behavioral Aspects 96
        3.4.1   Clothing 96
        3.4.2   Activity Level 96
        3.4.3   Adaptation and Acclimatization 97
        3.4.4   Time of the Day/Season 98
        3.4.5   Occupancy 98
        3.4.6   Psychological Factors 98
    3.5   The Comfort Equation 99
        3.5.1   Conduction 99
        3.5.2   Convection 100
        3.5.3   Radiation 101
        3.5.4   Evaporation 102
        3.5.5   Respiration 103
    3.6   Thermal Comfort Indices 103
        3.6.1   Predicted Mean Vote (PMV) Index 103
        3.6.2   Predicted Percentage Dissatisfied (PPD) Index 104
        3.6.3   Adaptive Comfort Standard 106
            3.6.3.1   Field Studies and Rational Indices 106
            3.6.3.2   Rational Approach 106

x   Contents

	3.6.4	Visual Comfort 106
3.7		Building Performance Parameters 106
	3.7.1	Thermal Load Leveling (TLL) 106
	3.7.2	Decrement Factor 107
3.8		Related Standards 107

Objective Questions.. 107

Answers 108

References 109

Chapter 4   Energy and Exergy Analysis 111

4.1		Introduction 111
	4.1.1	Brief History of Thermodynamics 111
4.2		Laws of Thermodynamics 111
	4.2.1	The Zeroth Law of Thermodynamics 112
	4.2.2	The First Law of Thermodynamics 112
	4.2.3	The Second Law of Thermodynamics 112
	4.2.4	The Third Law of Thermodynamics 114
4.3		Energy Analysis 114
	4.3.1	Introduction 114
	4.3.2	Energy Matrices 115
	4.3.3	Embodied Energy Analysis 115
	4.3.4	Energy Density Analysis 115
		4.3.4.1 Process Analysis 116
		4.3.4.2 Input-Output Analysis 116
		4.3.4.3 Hybrid Analysis 116
	4.3.5	An Overall Thermal Energy 116
		4.3.5.1 Energy Payback Time (EPBT) 116
	4.3.6	Energy Production Factor (EPF) 118
	4.3.7	Life Cycle Conversion Efficiency (LCCE) 118
	4.3.8	Energy Matrices of Photovoltaic (PV) Module 121
4.4		Exergy Analysis 122
	4.4.1	Low-Grade and High-Grade Energy 124
		4.4.1.1 Exergy as a Process 124
	4.4.2	Exergy Efficiency 125
	4.4.3	Solar Radiation Exergy 126
		4.4.3.1 Exergy Analysis Methods 127
	4.4.4	Exergy Analysis of Photovoltaic Thermal (PVT) Systems 129
4.5		Case Study with Roof-Mounted BiPVT System 130
	4.5.1	Description 130
	4.5.2	Overall Embodied Energy, EPBT, EPF 131

Objective Questions.. 133

Answers 134

Problems... 135

References 135

Chapter 5   Solar Cell Materials, PV Modules and Arrays 139

5.1		Introduction 139
5.2		Basics of Semiconductors and Solar Cells 139
	5.2.1	Intrinsic Semiconductor 141
	5.2.2	Non-Intrinsic Semiconductor 142

Contents   xi

5.2.3   Fermi Level in Semiconductor 142
5.2.4   p-n Junction 143
5.2.5   Photovoltaic Effect 143
5.2.6   Solar Cell (Photovoltaic) Materials 144
     5.2.6.1   Silicon (Si) 145
     5.2.6.2   Single-Crystal Solar Cell 145
5.2.7   Basic Parameters of Solar Cells 147
5.3   Photovoltaic (PV) Modules and PV Arrays... 151
   5.3.1   Single-Crystal Solar Cells PV Module.. 151
   5.3.2   Thin-Film PV Modules 151
   5.3.3   Packing Factor (β_c) of PV Module 152
   5.3.4   Efficiency of PV Modules 152
   5.3.5   Energy Balance Equations for PV Modules..................................... 153
     5.3.5.1   For Opaque (Glass to Tedlar) PV Modules..................... 153
     5.3.5.2   For Semi-Transparent (Glass-to-Glass) PV Modules 154
   5.3.6   Series and Parallel Combination of PV Modules............................. 156
   5.3.7   Degradation of Solar Cell Materials... 156
     5.3.7.1   Dust Effect 156
     5.3.7.2   Aging Effect.. 156
Objective Questions.. 157
Answers 158
References 159

Chapter 6   Static Design Concept for a Light-Structured Building for Cold Climatic
Conditions 161

6.1   Introduction 161
6.2   Sol-Air Temperature 161
   6.2.1   Bare Surface 161
   6.2.2   Wetted Surface.. 163
   6.2.3   Blackened and Glazed Surface.. 163
6.3   Thermal Gain.. 164
   6.3.1   Direct Gain 164
     6.3.1.1   Direct Gain through Semi-Transparent
         Photovoltaic (SPV) System... 165
     6.3.1.2   Direct Gain through Glazed Windows 167
     6.3.1.3   Net Thermal Energy Gains 168
   6.3.2   Indirect Gains 170
     6.3.2.1   Thermal Storage Wall/Roofs.. 171
     6.3.2.2   Trombe Walls 171
     6.3.2.3   Waterwalls.. 174
     6.3.2.4   Trans Walls 174
     6.3.2.5   Solariums 174
   6.3.3   Isolated Gain... 175
   6.3.4   Direct and Indirect Gain through Photovoltaic Thermal
      (PVT) Systems Integrated with Building 176
     6.3.4.1   Semi-Transparent Photovoltaic (SPV) Roof
         Integrated with Building's Rooftop 176
     6.3.4.2   Photovoltaic Thermal (PVT) Trombe Walls 177
     6.3.4.3   Integration of Roof (with Vent) with
         Semi-Transparent Photovoltaic Modules....................... 177
     6.3.4.4   Integration of Roof with Opaque Photovoltaic Modules 179

xii   Contents

| | | 6.3.4.5 | PVT Solariums | 180 |

Objective Questions 180
Answers 181
References 181

Chapter 7   Dynamic Design Concepts for Hot Climatic Conditions 183

7.1   Introduction 183
7.2   Phase Change Materials (PCMs) 183
7.3   Infiltration/Natural Ventilation 184
     7.3.1   Smart Windows 186
     7.3.2   Literature Study: Infiltration/Natural Ventilation 186
     7.3.3   Shading 186
     7.3.4   Windows 187
          7.3.4.1   Self-Inflating Curtains 187
          7.3.4.2   Window Quilt Shade 187
          7.3.4.3   Venetian Blind between the Glasses 188
          7.3.4.4   Transparent Heat Mirrors 188
          7.3.4.5   Solar Shading Devices 188
          7.3.4.6   Roofs 188
     7.3.5   Walls 188
          7.3.5.1   Heat Trap 188
          7.3.5.2   Optical Shutter 188
          7.3.5.3   Shading by Textured Surface 189
          7.3.5.4   Trees and Vegetation 189
7.4   Literature Study: Shading 189
7.5   Thermotropic and Thermochromic Coatings 190
7.6   Courtyards 190
7.7   Air Cavities 191
     7.7.1   Literature Study: Air Cavity 191
7.8   Green Roofs/Cool Roofs 192
     7.8.1   Literature Study: Cool Roof 193
     7.8.2   Evaporative Cooling 193
     7.8.3   Literature Study: Evaporative Cooling 195
7.9   Radiative Cooling 195
     7.9.1   Literature Study: Radiative Cooling 196
7.10   Movable Insulation 197
7.11   Dynamic Insulation Walls 197
     7.11.1 Exterior Insulation 197
     7.11.2 Interior Insulation 197
7.12   Wind Towers 197
     7.12.1 Literature Study: Wind Towers 198
7.13   Air Vents 198
7.14   Rock Bed Regenerative Cooler 199
7.15   Earth Coupling 199
     7.15.1 Earth-Air Heat Exchanger (EAHE) 199
          7.15.1.1 Literature Study: EAHE 202
7.16   Roof Pond 203
     7.16.1 Literature Study: Roof Pond-Passive Cooling 203
     7.16.2 Trombe Walls 204
7.17   Different Compositions of Trombe Wall 205

Contents

xiii

	7.17.1	Vented Trombe Wall	205
	7.17.2	Phase Change Material (PCM) Trombe Wall	205
	7.17.3	Photovoltaic Integrated Phase Change Materials (PV-PCM) Wall	207
	7.17.4	Heat Transfer in Trombe Walls	207
		7.17.4.1 U-Value	207
		7.17.4.2 Rate of Heat Transfer	209
	7.17.5	Efficiency Analysis of Trombe Wall	209
		7.17.5.1 Vent	209
		7.17.5.2 Size	209
		7.17.5.3 Fan	210
		7.17.5.4 Material and Color	210
		7.17.5.5 Insulation	210
7.18	Solar Cooling		211
	7.18.1	Solar Photovoltaic Cooling	211
Objective Questions			211
Answers			212
Problems			213
References			213

Chapter 8 Building Integrated Photovoltaic Thermal System (BiPVT) 219

8.1	Introduction		219
8.2	Literature Review of BiPV/ BiPVT Systems		220
8.3	Types of PV Integrations with Buildings		220
	8.3.1	Rooftop	220
	8.3.2	Façade	221
	8.3.3	Other Applications	221
8.4	Building Integrated Opaque Photovoltaic Systems (BiOPV)		223
	8.4.1	Opaque Photovoltaic System Integrated with Rooftop	223
	8.4.2	Opaque Photovoltaic System Integrated with Façade	224
8.5	Building Integrated Semi-Transparent Photovoltaic (BiSPVT) System		225
	8.5.1	Semi-Transparent Photovoltaic System Integrated with Rooftop	225
	8.5.2	Facade-Building Integrated Semi-Transparent Photovoltaic (BiSPVT) System	226
8.6	BiOPVT and BiSPVT System on Rooftop and Façade		227
8.7	Use of PV Modules in an Urban Settings		227
8.8	Energy and Exergy Analysis of BiSPVT System		228
	8.8.1	Working Principle	229
	8.8.2	Thermal Modeling	229
	8.8.3	Basic Energy Balance Equations	230
	8.8.4	Comparative Statement of Proposed Cases (a–d)	237
8.9	Performance Evaluation of the Proposed Systems		241
	8.9.1	For BiSPVT System (Case a)	241
	8.9.2	For BiSPVT System with Water Flow (Case b)	241
	8.9.3	For BiSPVT System with Heat Capacity (Case c)	241
	8.9.4	For BiSPVT System with Heat Capacity and Water Flow (Case d)	243
	8.9.5	BiSPVT System Heat Capacity with Movable Insulation and South-Facing Window (Case e)	243

8.10		Input Variables of BiSPVT System: Case Studies	246
	8.10.1	Number of Air Changes	246
	8.10.2	Velocity of the System	248
	8.10.3	Packing Factor	248
	8.10.4	Relative Humidity	248
	8.10.5	Transmissivity of Glass	248
	8.10.6	Mass of Water	248
	8.10.7	Mass Flow Rate	249
8.11		BiSPVT System Based on the PV Cell Type	249
	8.11.1	Literature Study	249
	8.11.2	Performance of BiSPVT System Based on PV Types: A Case Study	250

Objective Questions 252
Answers 253
References 254

Chapter 9   Environmental Aspects 259

9.1		Introduction	259
9.2		Life Cycle Assessment	259
	9.2.1	Basic Definitions of Life Cycle Assessment	261
	9.2.2	The Main Stages of Life Cycle Assessment	261
9.3		Embodied Energy	261
	9.3.1	Embodied Energy of Different Materials	262
	9.3.2	Embodied Energy of Different Construction Materials	262
	9.3.3	Embodied Energy in Floor/Roofing Systems	263
	9.3.4	Embodied Energy in Transportation of Building Materials	264
	9.3.5	Embodied Energy of PV Module	264
		9.3.5.1 Energy for Non-Silicon PV Modules	266
		9.3.5.2 Energy for Balance of System (BOS)	266
	9.3.6	Embodied Energy and Annual Output of Renewable Energy Technologies	267
	9.3.7	Guidelines for Reducing Embodied Energy	267
9.4		Modeling of Embodied Energy for BiPVT Systems	268
	9.4.1	Masonry Building	268
	9.4.2	Photovoltaic Thermal (PVT) System	268
	9.4.3	Balance of System (BOS)	269
9.5		Embodied Carbon	270
	9.5.1	Example of Estimation of Embodied Carbon Dioxide for Concret	270
9.6		Carbon Dioxide Emissions	271
9.7		Earned Carbon Credits and Carbon Dioxide Mitigation	274
	9.7.1	Formulation	275
9.8		Case Study with the BiPVT System	276
9.9		Kyoto Protocol and the United Nations Framework Convention on Climate Change	276
	9.9.1	The Protocol and the Green Growth	278
9.10		Carbon Dioxide Mitigation with Use of Photovoltaics	278

Objective Questions 279
Answers 280
References 280

Contents   **xv**

Chapter 10 Life Cycle Analysis 283

    10.1   Introduction 283
    10.2   Cash Flow Diagram 284
    10.3   Cost Analysis 285
          10.3.1   Capital Recovery Factor 285
          10.3.2   Uniform Annual Cost 297
          10.3.3   Sinking Fund Factor 299
          10.3.4   Linear Gradient Series Present Value Factor................304
          10.3.5   Gradient to Equal Payment Series Conversion Factor 306
          10.3.6   Linear Gradient Series Future Value Factor 307
    10.4   Capitalized Cost...308
    10.5   Cost Comparisons with Equal Duration 309
    10.6   Net Present Value (NPV) 311
          10.6.1   Limitations of the NPV Method....................................312
    10.7   Cost Comparisons with Unequal Duration..................................312
          10.7.1   Single Present Value Method (Method I) 313
          10.7.2   Annual Cost Method (Method II)..................................314
          10.7.3   Capitalized Cost Method (Method III)..........................315
          10.7.4   Method IV 315
    10.8   Payback Time 316
          10.8.1   Analytical Expression for Payback Time 316
          10.8.2   Payback Period without Interest....................................318
          10.8.3   Payback Period with Interest...318
    10.9   Benefit–Cost Analysis 320
          10.9.1   Types of Benefit–Cost Analysis 322
                   10.9.1.1   Aggregate B/C Ratio....................................322
                   10.9.1.2   Net B/C Ratio..322
          10.9.2   Advantages and Disadvantages of B/C Ratio................323
    10.10 Internal Rate of Return 327
          10.10.1 Iterative Method to Compute IRR.................................328
          10.10.2 Multiple Values of IRR...331
    10.11 Effect of Depreciation 332
          10.11.1 Expression for Book Value..334
          10.11.2 Straight-Line Depreciation..334
          10.11.3 Sinking Fund Depreciation..335
          10.11.4 Accelerated Depreciation 336
    10.12 Cost Comparison after Taxes...338
          10.12.1 Without Depreciation 338
          10.12.2 With Depreciation 339
    10.13 Estimating Cost of a Project..342
          10.13.1 Capital Cost..342
          10.13.2 Variable Cost 342
          10.13.3 Step-Variable Cost..342
          10.13.4 Non-Product Cost 342
    10.14 A Case Study of Building Integrated Photovoltaic
          Thermal (BiPVT) Systems 343
          10.14.1 Cost Estimation 343
          10.14.2 Modeling of Annualized Uniform Cost.........................344
          10.14.3 Methodology 344
          10.14.4 Results and Discussions 345

xvi Contents

Objective Questions..347
Answers 349
Problems...349
References 350

Chapter 11 Photovoltaic Application in Architecture 351

11.1 Introduction 351
11.2 Implementation of PV Systems around the World 351
 11.2.1 China 352
 11.2.2 United States of America..............................355
 11.2.3 Japan 357
 11.2.4 Germany 360
 11.2.5 India..363
 11.2.6 Spain...364
11.3 Case Study: BiSPVT System Installed at Sodha Bers Complex,
 Varanasi, India 366
 11.3.1 Introduction and Planning 367
 11.3.1.1 Basement......................................367
 11.3.1.2 Ground Floor................................369
 11.3.1.3 First and Second Floor 369
 11.3.1.4 Terrace Floor Integrated with Semi-Transparent
 Photovoltaic (SPV) System....................370
 11.3.2 Zones 370
 11.3.3 Construction Details and Materials Used.......371
 11.3.4 Thermal Heat Gains..372
 11.3.5 Electrical Power (E_p) 375
 11.3.6 Daylight Energy Savings 375
 11.3.7 Total Energy Savings......................................376
 11.3.8 Embodied Energy 376
 11.3.9 Energy Payback Time (EPBT) 377
 11.3.10 Energy Production Factor (EPF) 378
 11.3.11 Life Cycle Conversion Efficiency 378
 11.3.12 Carbon Dioxide Emission 379
 11.3.13 Net Carbon Dioxide Mitigation......................379
 11.3.14 Earned Carbon Credits 380
Objective Questions..381
Answers 381
References 381

Appendix A 385

Appendix B 391

Appendix C 393

Appendix D 405

Appendix E 407

Appendix F..419

Appendix G 423

Index...427

Preface

Energy conservation is equivalent to energy production, and thus, money saving. Use of solar energy as an alternative to conventional fossil fuels is the most economical solution to energy crisis. It has been estimated that about 6.7% of the global energy is used in the building sector for its thermal management (precise Arab data not available). About 35% of the total building's energy demands may be satisfied via alternate resources. Using solar energy to satisfy the building's energy requirement is the most cost-effective method, and thus, the concept of solar house came into existence. In order to reduce the dependence on artificial heating and cooling of buildings, the passive concepts must be integrated with the buildings from the very beginning, i.e., at the stage of conceptualization. Passive solar architecture is a cost- and resource-efficient concept to achieve the natural harmony between the environment, architecture, and people. The buildings should be self-sustaining, i.e., energy production for its own consumption.

The objective of this book is to provide a platform to disseminate the knowledge regarding fundamentals of solar energy, heat transfers, and solar house. The book aims at:

- Knowledge of solar radiation,
- Energy matrices (embodied energy, energy pay-back time, energy production factor, life cycle conversion efficiency)
- Solar passive heating and cooling
- Architecture design
- Low-cost building
- Energy and exergy analysis
- Building integrated photovoltaic thermal system
- Economic analysis, and
- Energy conservation.

The book is appropriate for undergraduate and post graduate students, learners, researchers, professionals, practitioners, designers, and architects.

The authors have drawn the materials for inclusion in the book with references at appropriate places that need to be mentioned. These include *Solar Energy* by G.N. Tiwari; *Building Integrated Photovoltaic Thermal Systems: For Sustainable Developments* by Basant Agrawal and G.N. Tiwari; *Solar Engineering of Thermal Processes* by John A. Duffie and William A. Beckman; and *Solar Passive Building* by M.S. Sodha, N.K. Bansal, P.K. Bansal, *P.K.,* A. Kumar, and M.A.S. Malik; *Advanced Renewable Energy Sources* by G.N. Tiwari and R.K. Mishra; *Handbook of Solar Energy: Theory, Analysis and Applications* by G.N. Tiwari, A. Tiwari, and Shyam.

The current book has been divided into 11 chapters. The basic concepts of solar radiation, passive house, heating, and cooling have been broadly considered in this book. Basic concepts of solar radiation, solar chart, climatology with respect to different climatic zones, micro and macro climate, thermal heating and cooling concepts, orientation, and architectural design of a passive house have been discussed in Chapter 1. In Chapter 2, total heat transfer and overall heat transfer coefficients have been highlighted. Chapter 3 describes various physical, physiological, psychological, and behavioral aspects related to thermal comfort. Thermal comfort indices have also been covered in this chapter. In Chapter 4, energy and exergy analysis have been discussed in detail. Chapter 5 gives us details about photovoltaic modules and its applications. Static design concept for a light structures building have been covered in Chapter 6 whereas the dynamic design concepts have been discussed in Chapter 7. In Chapter 8, integration of photovoltaic modules with façade and rooftop and its thermal modelling have been highlighted in detail. Chapter 9 discusses the environments aspects like embodied energy, carbon credits, carbon dioxide mitigation. Life cycle cost and energy

xvii

analysis has been covered in Chapter 10. Chapter 11 gives various case studies related to building integrated photovoltaic thermal systems.

To understand the basic parameters, SI units have been used throughout the length of the book. Appendix has been given at the end of the book. Appendix includes (a) conversion of units, (b) physical properties of metals and non-metals, (c) thermo-physical properties of air and saturated water, (d) absorptivity of various surfaces for sun rays, and (e) Fourier coefficients of periodic functions. The appendix has been frequently used in the text, especially in the solved examples.

This book has been aimed to provide a great insight in the subject, particularly to the learning students and professionals doing self-study. In spite of the best efforts, some errors might have been crept in the text. The authors welcome suggestions and comments, if any, from all readers for further improvement of the book in the next edition.

The authors are very grateful and thank Prof. A.S.K. Sinha, Director, Rajiv Gandhi Institute of Petroleum Technology (RGIPT), Jais, Amethi (UP), India, for his kind moral support during preparation of book from time to time.

Last but not least, we express our deep gratitude to Late Smt. Bhagirathi Tiwari; Late Shree Bashisht Tiwari; Late Shree R.P. Gupta; Smt. Sharda Mishra; Shree Shashi Bhusan Mishra; Smt. Bimla Gupta for their blessing during writing of the book. Further, we also thank Smt. Kamalawati Tiwari; Smt Kamini Gupta; Smt Rekha Gupta; Shree S.K. Gupta; Shree Atam Gupta, Shree Sri Vats Tiwari, Shree Ganeshu Tiwari, Ms. Shrivani Tiwari, and Kumari Hera Gupta for keeping our moral high during this work.

Gopal Nath Tiwari

Neha Gupta

Authors

Professor Gopal Nath Tiwari, was born on July 1, 1951 at Adarsh Nagar, Sagerpali, Ballia (UP), India. He received postgraduate and doctoral degrees in 1972 and 1976, respectively, from Banaras Hindu University (BHU). Over several years since 1977, he has been actively involved in the teaching program at the Centre for Energy Studies, IIT Delhi. His research interest in the field of solar energy applications are solar distillation, water/air heating system, greenhouse technology for agriculture, as well as for aquaculture, earth-to-air heat exchangers, passive building design and hybrid photovoltaic thermal (HPVT) systems, climate change, energy security, etc. He has one patent on solar still dated 1983. He has guided more than 100 PhD students and published over 700 research papers in journals of repute with h-index as 86 and citations more than 30,000. He has authored more than 20 books associated with reputed publishers, namely Pergamon Press (UK), CRC Press (US), Narosa Publishing House, Ahsan (UK), Alpha-science (UK), Royal Society of Chemistry (UK), etc. He was a co-recipient of the "Hariom Ashram Prerit S.S. Bhatnagar" Award in 1982. Professor Tiwari has been recognized both at national and international levels. He was the Energy and Environment Expert at the University of Papua, New Guinea from 1987 to 1989. He was also a recipient of European Fellow in 1997 and been to the University of Ulster (U.K.) in 1993. He has also been nominated for an IDEA award in the past. He is responsible for development of "Solar Energy Park" at IIT Delhi and Energy Laboratory at University of Papua New Guinea, Port Moresby. Dr. Tiwari has visited many countries, namely Italy, Canada, the United States, the United Kingdom, Australia, Sweden, Germany, Greece, France, Thailand, Singapore, Papua New Guinea, Hong Kong and Taiwan, etc. for invited talks, chairing international conferences, as an expert in renewable energy, presenting research papers, etc. He has successfully co-coordinated various research projects on solar distillation, water heating systems, greenhouse technology, hybrid photovoltaic thermal (HPVT), etc., funded by the government of India in the recent past.

Dr. Tiwari was editor of *International Journal of Agricultural Engineering* for three years (2006–2008). He is associate editor for *Solar Energy Journal* (SEJ) in the area of solar distillation since 2007. He is also editor of *International Journal of Energy Research*. Recently, he has also been appointed senior editor of Electrical Engineering, Electronic and Energy, Elsevier, for a period of three years.

He is a member of various scholarly societies and editorial boards for various journals. Professor Tiwari has received various recognitions like best paper on jaggery drying (2014) from ICR and energy conservation in passive solar buildings for hot and arid zones in India (1995) from NBCC. He has also won the best book (Greenhouse Technology) award in 2002. Professor Tiwari has been conferred "Vigyan Ratna" by the government of UP, India, on March 26, 2008 and the Valued Associated Editor award by the *Journal of Solar Energy*. He organized SOLARIS 2007, as well as the third international conference on "Solar Radiation and Daylighting" held at IIT Delhi, New Delhi, India in 2017. He is also president of Bag Energy Research Society (BERS:www.bers.in), which is responsible for energy education in rural India.

At present he is actively looking (i) the ongoing project on GiSPVT funded by DST at his home town Ballia (UP) at Margupur, Chilkhar Ballia and (ii) an international joint project with King Saudi University (KSU), Saudi Arabia under banner of Bag Energy Research Society (BERS) for which he is founder President.

Dr. Tiwari has been world ranked at 449 with C-score of 3.76131963 for Sriramswoop Memorial University (SRMU), Dew-Road, Lucknow and at 1603 with C-score of 3.34672147 for IIT Delhi among top 2% world scientists from India by Stanford University, UK.

Dr. Neha Gupta obtained her PhD from IIT Delhi in the field of semi-transparent photovoltaic thermal system after completing her M. Ekistics (Hons.) and B. Architecture from Jamia Millia Islamia University. She has also completed courses such as "application of renewable energy sources", "Non-conventional sources of energy" and "Solar architecture" from IIT Delhi. She is an active researcher in the area of building integrated semi-transparent photovoltaic thermal systems having several publications under her name with an h-index of 8 and is also an active reviewer for various international journals of repute. She has also presented papers at various national and international conferences. Her research work in this field has been recognized internationally and has the potential to help achieve self-sustainable energy-efficient buildings in rural areas for both hot and cold climatic conditions and enable them to meet energy security at an affordable price.

She became the first architect in 3 years to receive the prestigious "Young Engineers Award" by Institute of Engineers, India, for her contribution in research and design of sustainable mass rapid transport systems. For her research work in the field she has also been twice acknowledged with the professional membership of International Solar Energy Society (ISES) and accreditation of World Research Council (WRC). She was also awarded with the Research Ratna Award as a distinguished researcher in building integrated semi-transparent photovoltaic systems.

She is currently working with Delhi Metro Rail Corporation (DMRC), the premier metro rail organization in India, where she has undertaken several projects on sustainable mass rapid transport systems and green homes including the decorated Noida–Greater Noida metro corridor where her work received special appreciation from the Managing Director, Noida Metro Rail Corporation. Prior to this, she has also worked with UP Public Works Department as an architect (gazetted officer) and as an assistant professor with Jamia Millia Islamia. She was conferred with the prestigious "Director Award" at DMRC and was invited as a speaker in the field of passive design and solar architecture by the organization. Her work has also been regularly acknowledged with special mentions in several DMRC newsletters and she has been bestowed with multiple rewards for her work in the field of green design.

Her core motivation for research in this field is driven by her passion to design self-sustainable energy efficient buildings for the rural areas of India. The concept has already started being explored in the developed nations around the world but is still at a very nascent stage in a country like India where it can help provide energy security at an affordable price.

1 General Introduction

1.1 ZERO ENERGY BUILDINGS

The zero energy building (ZEB) concept is gaining momentum across the globe to meet energy savings. It is important for us to know about the need for zero energy buildings. This is because we live on Earth—the only planet that is inhabitable. Yet we constantly disturb our planet, and today we face significant sustainability issues. The first issue is climatic change, which is already happening, and we need to be prepared for extreme weather conditions and become climate adaptive. The aggravated climate change is caused by greenhouse gases, mostly emitted through human processes. Estimates vary, but the temperature on Earth will most probably rise between 1.5°C and 5°C within this century, depending on actions taken or not [1]. The second issue is the depletion of fossil fuels. In order to avoid these climatic changes, we need to shift to renewable sources of energy. The third issue is vulnerability. This means that urban areas have been experiencing disturbances in the supply of energy over the last few decades. Therefore, in order to create a resilient society, our cities have to become more self-sufficient.

The only way to solve this is to become entirely circular in our material use. Carbon emissions must be reduced, necessitating a rapid energy transition to clean renewables.

The built environment constitutes an important factor because it involves the use of 30%–40% of all the energy, thus there is a need for it to become carbon neutral. To achieve this, there is a need to design zero energy buildings and also to make existing structures energy neutral [2].

The zero energy building (ZEB) is a special relationship between the design of a building and natural processes that offer the potential for an inexhaustible source of energy. The concept is not new, and various examples from the vernacular architecture can be traced. The definition of zero energy building should be flexible and include assessing the quality of energy used for heating/cooling of buildings, different climatic conditions, types of buildings and existing building masses.

Zero energy building (ZEB) may also be known as energy neutral, zero energy or net zero energy. In the Netherlands the term "zero on the meter" is also well-known, indicating that over a year's time there are no energy costs.

The ZEB may be defined as one that does not consume any fossil fuels during its construction. In Indian history, architecture has evolved over time, depending on the traditional building materials, technologies and craftsmanship available. An example of constructing a ZEB depending on the technology and materials available can be referred to as *kachcha* building (mud house), which is constructed using building materials like mud, grass, bamboo, thatch or sticks. Then came buildings known as *gatiya*. These are constructed using cow dung, bricks, mud plaster, etc. These raw materials are weak durable materials but better than the ones used in kachcha construction. The gatiya buildings were a step towards net zero energy building (NZEB). Embodied energy in ZEB is zero, and when this embodied energy tends to increase, ZEB moves towards the NZEB concept. Embodied energy may be defined as the quantity of energy required by all the activities, i.e., the overall expenditure of energy required to construct and maintain the building over the whole life cycle [3].

The concept of ZEB has been defined by various researchers on the basis on energy use.

- As suggested by Buildings Performance Institute Europe (BPIE), "nearly zero energy building should include a threshold for household electricity used for integrated building equipment (like lifts, etc.) going beyond the building services (heating, cooling, ventilation and lighting)" [4].

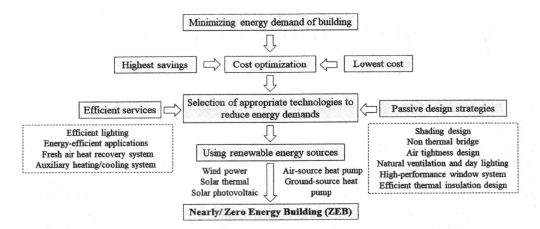

FIGURE 1.1 Design concept of Nearly ZEBs [5].

- According to Gilijamse [6] a ZEB is "a house in which no fossil fuels are consumed, and the annual electricity consumption equals annual electricity production. Unlike the autarkic situation, the electricity grid acts as a virtual buffer with annually balanced delivers and returns".
- It has also been defined by Torcellini et al. [7] as "a residential or commercial building with greatly reduced energy needs through efficiency gains such that the balance of energy needs can be supplied with renewable technologies".
- With reference to the above definition made by Torcellini et al. [7], Kilkis [8] came out with the concept of NZEB and defined it as "a building which has a total annual sum of zero exergy transfer across the building-district boundary in a district energy system, during all electric and any other transfer that is taking place in a certain period of time".
- Another definition for ZEB was given in the report by the International Energy Agency (IEA), a report by Laustsen [9] which states: "Zero energy buildings do not use fossil fuels but only get all their required energy from solar energy and other renewable sources". Marszal and Heiselberg [10] have also indicated the key pointers to define a ZEB.
- According to [5] ZEB consists of two strategies—minimizing demand through energy-efficient measures (passive design) and using renewable energy and other technologies to meet remaining energy requirements (active design), as shown in Figure 1.1. The first term includes the passive strategies to reduce the energy demand of the ZEB as much as possible. Passive design strategies include shading, thermal bridges, air tightness design, natural ventilation and lighting, and high-performance envelopes. The term "active design" means using high-efficient active technologies that use renewable energy. Efficient lighting, energy-efficient appliances and fresh air heat recovery systems are included in the active design. The cost-optimal model of ZEBs based on passive and active technologies should be evaluated carefully. The technical guidelines stipulate the minimum requirements for ZEBs to meet the relevant energy consumption indicators, air tightness indicators and indoor environmental parameters.

1.2 THE SUN AND THE EARTH

1.2.1 The Sun

The sun is the largest member of the solar system and is about 1.5×10^{11} m away from the earth. It is spherical and made up of intense hot gaseous matter with a diameter of about 1.39×10^9 m [11]. The structure of the sun is shown in Figure 1.2. The sun rotates on its axis about once every four weeks

General Introduction

and does not rotate as a solid body. The time taken for each rotation by the equator and polar regions are 27 and 30 days, respectively. The effective black body temperature (T_s) of the sun is 5777 K. The sun is a continuous fusion reactor; the most important fusion reaction to supply the energy radiated by the sun is given in Equation (1.1). The hydrogen (i.e., four protons) combines together to form a helium nucleus. The mass of the helium formed is less than the mass of four protons due to the loss in mass during the reaction process and conversion to energy.

$$4(_1H^1) \rightarrow {}_2H^4 + 26.7 \text{ MeV} \tag{1.1}$$

This energy is produced in the core of the solar sphere with temperature equal to some millions of degrees. This energy is then transferred out to the surface and radiated to the space as given in Equation (1.2).

$$E = \varepsilon \sigma T_s^4 \tag{1.2}$$

where ε and σ are, respectively, the emissivity of surface and Stefan-Boltzmann constant.

This energy is produced in the core of the solar sphere with temperature equal to some millions of degrees. This energy is then transferred out to the surface and radiated to the space as given in Equation (1.2) (Figure 1.2).

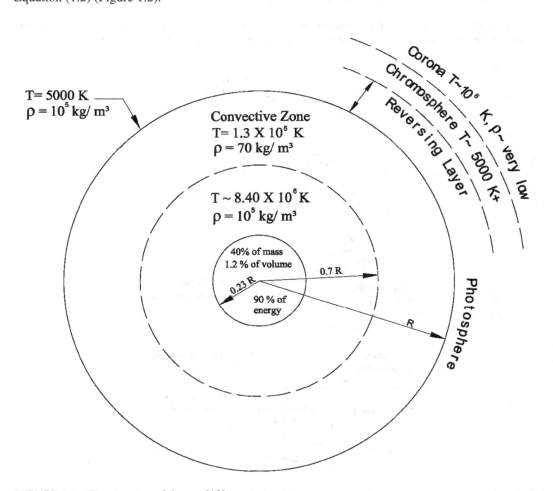

FIGURE 1.2 The structure of the sun [12].

1.2.2 The Earth

About 4.6×10^9 years ago, Earth came into existence. The inner core of the earth is solid, whereas the outer core is a melted state composed of iron and nickel. The layer covering the outer core is known as the mantle, which is made up of rocks. The outermost crust that covers the mantle is made up of rock too. The oldest rocks of sedimentary origin appear to be about 3.7×10^9 years old.

The earth is elliptical in shape with diameter of about 13,000 km, revolving around the sun about once a year. Nearly 70% of the earth is covered by water and the remaining 30% is land. At a time, only half of the earth is covered by solar radiation and reflects about one third of the radiation falling onto it. This is called Earth's albedo. Earth spins about its axis constantly, and the axis is inclined at an angle of 23.5°; hence, the length of the days and night changes accordingly.

The primary source of energy for Earth is solar radiation. Earth receives solar energy at the rate of 5.4×10^{24} J/year. Blue-green algae mark the beginnings of photosynthesis, as a result of which the level of O_2 and O_3 has increased in the atmosphere.

Unique properties of atmosphere are listed below:

- Short-wavelength radiation (0.23–2.26 µm) coming from the sun is transmitted by atmosphere consisting of greenhouse gases.
- It acts as an opaque surface for long-wavelength radiation greater than 2.26 µm.

The region between the atmosphere and the sun is known as terrestrial region, and the one between the sun and earth's atmosphere is known as the extraterrestrial region as given in Figure 1.3.

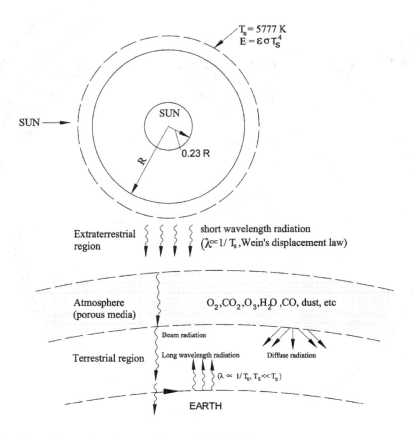

FIGURE 1.3 View of atmosphere between Sun and Earth [12].

General Introduction

1.2.3 Sun–Earth Angles

To estimate the availability of solar intensity throughout the year for any surface at any place with desired inclination and orientation, knowledge of Sun–Earth angles becomes necessary.

Latitude (ϕ): The angle between the radial line, joining the location (observer) to the Earth's center, and its projection on the equatorial plane is the latitude of that location (observer) as shown in Figure 1.4a. In the northern hemisphere, the latitude is positive, whereas it is negative for the southern hemisphere. Table 1.1 [11] gives the latitude for some places in the world.

Declination (δ): The angle between the line joining the centers of the sun and the earth, which also has the direction of direct sun's rays coming from the sun and its projection on the equatorial plane, is known as the angle of declination as shown in Figure 1.4a. Declination is due to the rotation of the earth about its axis, which makes an angle of 66.5° with the plane of its rotation around the sun refer Figure 1.4b. The declination varies from a maximum value of 23.45° on June 21 to a minimum value of −23.45° on December 21. Variation of declination angle with nth day of year is given in Figure 1.5 [12, 13]. Celik and Muneer [14] gave the following relation:

$$\delta = 23.45 \, sin\left[\frac{360}{365}\left(284 + n\right)\right] \tag{1.3a}$$

Hour angle (ω): The earth must be rotated at a particular angle to bring the meridian of the plane directly under the sun. This angle of rotation is termed as hour angle (Figures 1.4a and 1.4b). In other words, it is the angular displacement of sun from the local meridian due to Earth's rotation about its own axis at 15° per hour.

Table 1.2 [11] gives the values of the hour angle for the northern hemisphere. It can be seen that it is zero at noon hours, negative in the daytime and positive in the afternoon for the northern hemisphere and vice versa for the southern hemisphere. Hour angle can be expressed as:

$$\omega = \left(solar\ time - 12\right) \times 15° \tag{1.3b}$$

The total hour angle from sunrise to sunset is ($2\omega_s$). The $\pm\omega_s$ correspond to hour angle with reference to sunrise and sunset, respectively. The apparent solar time varies from 0 to 24 hours. At noon it is 12 hours.

TABLE 1.1
Latitude, Longitude and Elevation for Different Places in World [11]

Sr. No.	Place	Latitude (ϕ)	Longitude (L_{loc})	Elevation (E_0)
1	Beijing	39°54′ N	116°23′ E	50 m above msl
2	Berlin	52°31′ N	13°23′ E	34 m above msl
3	Chennai	13°00′ N	80°11′ E	16 m above msl
4	Kolkata	22°32′ N	88°20′ E	6 m above msl
5	London	51°30′ N	00°07′ W	35 m above msl
6	Moscow	55°45′ N	37°37′ E	156 m above msl
7	Mumbai	18°54′ N	72°49′ E	11 m above msl
8	New Delhi	28°35′ N	77°12′ E	216 m above msl
9	New York	40°42′ N	74°00′ W	10 m above msl
10	Paris	48°51′ N	02°21′ E	35 m above msl
11	Singapore	01°17′ N	103°50′ E	6 m above msl

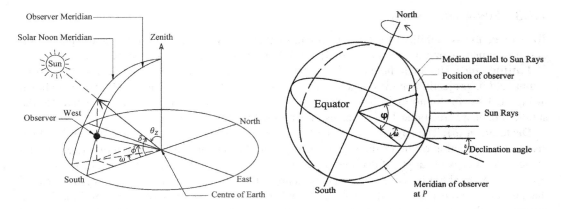

FIGURE 1.4 (a) View of different Sun–Earth angles [13]. (b) Earth always inclined at 23.5° rotating about self-axis [12].

TABLE 1.2
The Value of Hour Angle with Time of the Day (for Northern Hemisphere) [11]

Time of the Day	Hour Angle	Time of the Day	Hour Angle
6:00 am	−90°	1:00 pm	+15°
7:00 am	−75°	2:00 pm	+30°
8:00 am	−60°	3:00 pm	+45°
9:00 am	−45°	4:00 pm	+60°
10:00 am	−30°	5:00 pm	+75°
11:00 am	−15°	6:00 pm	+90°
12:00 pm	0°		

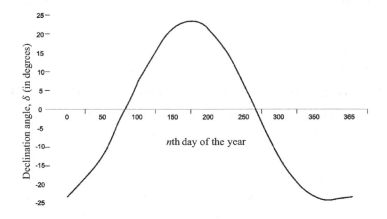

FIGURE 1.5 Variation of declination angle with nth day of year [11].

Zenith (θ_z): The angle between the center of Sun's disk and a perpendicular line to the horizontal plane is referred to as the zenith angle (Figure 1.4a). The value varies from 0° to 90° with movement

General Introduction 7

of the sun all through the day. The value corresponds closer to 90° (maximum) while the sun is rising or setting. However, it is near to 0° at the solar noon. It may be calculated as:

$$\cos\theta_z = \cos\varnothing\cos\delta\cos\omega + \sin\delta\sin\varnothing \tag{1.3c}$$

Altitude or solar altitude angle (α): The angle between center of Sun's disk and a horizontal plane is referred to as a solar altitude angle. It may be expressed as:

$$\alpha = 90 - \theta_z \tag{1.3d}$$

Slope (β): The angle between the horizontal and plane surface (under consideration) is referred to as slope (Figure 1.6). The numerical value is negative for surfaces sloping towards north and positive for surfaces sloping towards south.

Surface azimuth angle (γ): This is the angle in the horizontal plane, between the line due south and the projection of the normal to the surface (inclined plane) on the horizontal plane (Figure 1.6). By convention, the numerical value is negative for the northern hemisphere and positive for the southern hemisphere, if the projection is east of south. If the projection is west of south, the numerical value is positive in nature. The values of surface azimuth angle for some orientations are tabulated in Table 1.3.

Solar azimuth angle (γ_s): The angle between the line due south and the projection of beam radiation on the horizontal plane is referred to as the solar azimuth angle (Figure 1.6). By convention, the angle is taken to be positive if the projection is east of south and negative if the projection is west of south for the northern hemisphere and vice versa for the southern hemisphere. For example, due east would be 90° and due west would be −90°. (For details refer to Duffie and Beckman [15].)

Angle of incidence (θ_i): The angle between beam radiation on a surface and the normal to that surface is referred to as angle of incidence (Figure 1.6). The expression for angle of incidence (θ_i) is given below:

$$\begin{aligned}\cos\theta_i = &\left(\cos\varphi\cos\beta + \sin\varphi\sin\beta\cos\gamma\right)\cos\delta\cos\omega \\ &+ \cos\delta\sin\omega\sin\beta\sin\gamma + \sin\varepsilon\left(\sin\varphi\cos\beta - \cos\varphi\sin\beta\cos\gamma\right)\end{aligned} \tag{1.3e}$$

For a horizontal plane facing south, $\gamma = 0$, $\beta = 0$, $\theta_i = \theta_z$ (zenith angle)

$$\cos\theta_z = \cos\varphi\cos\delta\cos\omega + \sin\delta\sin\varphi \tag{1.3f}$$

TABLE 1.3

Surface Azimuth Angle (γ) for Various Orientations in Northern Hemisphere [11]

Surface Orientation with Slope Towards	Surface Azimuth Angle (γ)
South	0°
North	180°
East	−90°
West	+90°
South-East	−45°
South-West	+45°

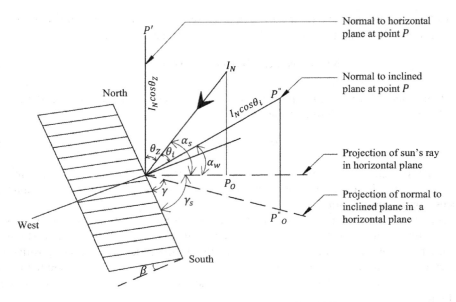

FIGURE 1.6 View of various Sun–Earth angles for an inclined surface [11].

EXAMPLE 1.1

Calculate the declination angle on September 23, 1995.

Solution

For September 23, $n = 266$. Using Equation (1.3a), we get:

$$\delta = 23.45 \sin\left[\frac{360}{365}(284 + 266)\right] = -1.01°.$$

EXAMPLE 1.2

Evaluate the angle of incidence of direct irradiance/solar radiation on an inclined surface at 45° from the horizontal and has orientation of 30° west of south and located at New Delhi at 1:30 (solar time) on February 16, 2013.

Solution

For the present case, the value of n is 47 and $\delta = -13.0°$ (Equation 1.3a); $\omega = +22.5°$ (Equation 1.3b):

$$\gamma = 30°; \beta = 45°; \varphi = +28.58° (\text{New Delhi})$$

Now, the angle of incidence of direct irradiance/solar radiation on an inclined surface can be calculated by using Equation (1.3e) as follows:

General Introduction

$$\cos\theta_i = \sin(-13°)\sin(28.58°)\cos(45°) - \sin(-13°)\cos(28.58°)\sin(45°)\cos(30°)$$
$$+ \cos(-13°)\cos(28.58°)\cos(45°)\cos(22.5°)$$
$$+ \cos(-13°)\sin(28.58°)\sin(45°)\cos(30°)\cos(22.5°)$$
$$+ \cos(-13°)\sin(45°)\sin(30°)\sin(22.5°) = 0.999$$

$$\theta_i = \cos^{-1}(0.999) = 2.56°$$

EXAMPLE 1.3

Evaluate the zenith angle of the sun at New Delhi at 2:30 pm on February 20, 2013.

Solution

For the present case, $n = 51$; $\phi = 28.35'$ (New Delhi, Table 1.1); $\delta = -11.58°$ (Equation 1.3a); $\omega = 37.5°$ (Equation 1.3b)
From Equation (1.3f), we have,

$$\cos\theta_z = \cos(28.58°)\cos(-11.58°)\cos(37.5°) + \sin(-11.58°)\sin(28.58°) = 0.587$$
$$\theta_z = \cos^{-1}(0.587) = 54.03°$$

1.2.4 SOLAR RADIATION

The orientation of the earth's orbit around the sun is such that the Sun–Earth distance varies only by 1.7%. Since the solar radiation outside the earth's atmosphere is nearly of fixed intensities, the radiant energy flux received per second by a surface of unit area held normal to the direction of Sun's rays at the mean Sun–Earth distance, outside the atmosphere, is practically constant throughout the year. This is termed the solar constant I_{sc}, and its value is now adopted to be 1367 W/m². However, this extraterrestrial radiation suffers variation due to the fact that the earth revolves around the sun not in a circular orbit but follows an elliptic path, with Sun at one of the foci. The intensity of extra-terrestrial radiation measured on a plane normal to the radiation on the nth day of the year is given in terms of solar constant (I_{sc}) as follows [15]:

$$I_{ext} = I_{sc}\left[1.0 + 0.033\cos\left(\frac{360n}{365}\right)\right] \tag{1.4a}$$

Various losses due to scattering and atmospheric absorption are incurred by the solar radiation while traveling from the extraterrestrial to terrestrial region via Earth's atmosphere. Thus, the rate of normal solar flux, i.e., normal solar radiation/irradiance reaching Earth's surface is given by:

$$I_N = I_{ext} \times \exp\left[-(m.\varepsilon.T_R + \alpha)\right] \tag{1.4b}$$

where, m is the air mass,

$$m = \left[\cos\theta_Z + 0.15 \times (93.885 - \theta_Z)^{-1.253}\right]^{-1} \tag{1.4c}$$

At noon, $\theta_Z = 0$, $m = 1$; for $\theta_Z = 60°$, $m = 2$ and $m = 0$ for outside Earth's atmosphere.

T_R is the turbidity factor and is defined as the cloudiness/haziness factor for the lumped atmosphere. The values of T_R and α for different weather and flatland conditions (cases a, b, c and d as defined earlier) are given in Appendix A. These values were calculated from the 10 years' average data obtained from the Indian Metrological Department (IMD), Government of India, Pune, India. These values have been calculated for Delhi but are valid for latitude after meeting weather conditions defined earlier. The turbidity factor (T_R) for different months is given in Appendix B.

The expression for ε, known as integrated optical-thickness of the terrestrial clear and dry atmosphere/Rayleigh atmosphere (dimensionless), is given by

$$\varepsilon = 4.529 \times 10^{-4} \times m^2 - 9.66865 \times 10^{-3} \times m + 0.108014 \qquad (1.4d)$$

The α in Equation (1.4b) is known as lumped atmospheric parameters, which accommodate the further attenuation of direct normal irradiance in the terrestrial zone because of the cloudiness/haziness level, anisotropic behavior and unpredictable changes in atmospheric conditions while reaching the earth's surface.

Direct normal irradiance (i.e., solar radiation) for a terrestrial region is also given by Perez et al. [16]

$$I_N = I_{ext} \times \exp\left[\frac{-T_R}{(0.9 + 9.4\cos\theta_Z)}\right] \qquad (1.5)$$

In the terrestrial region, the wave-length range of solar radiation (I_N) shall be considered between 0.23–2.26 µm. About 30% of solar radiation reaching the earth's surface is reflected back to the atmosphere without any change in its wavelength range and is further allowed to travel from the terrestrial to extraterrestrial region. The remaining solar irradiance is absorbed by the earth's surface. The terrestrial and extraterrestrial regions are shown in Figure 1.3. X-rays and very small ultraviolet radiations are absorbed by the gases present in the ionosphere. The ozone layer absorbs the ultraviolet radiation of $\lambda < 0.40$ µm, and water vapors absorb the infrared radiations of $\lambda > 2.3$ µm. Thus, solar radiation will be completely absorbed as it passes through the atmosphere ($\lambda < 0.29$ µm and $\lambda > 2.3$ µm). The solar radiation, as mentioned earlier, goes through scattering losses due to the presence of air molecules, water vapor and other particles, solar irradiance is further reduced. This scattered radiation is known as diffuse radiation (refer Figure 1.7). The available range of wavelength of solar radiation in the terrestrial region for utilization of solar energy applications lies between 0.29–2.3 µm. Solar radiation reaching Earth's surface through the atmosphere (terrestrial region) has two components, namely diffuse and beam radiation due to the scattering (Figure 1.3).

Beam radiation (I_b) or direct radiation: It is a normal component of the solar radiation in W/m² with a definite direction propagating along the line joining the receiving surface and the sun on a horizontal/inclined surface.

Diffuse radiation (I_d): The component of solar radiation in W/m², which got scattered due to dust, molecules, aerosols and other particles without any definite direction.

The total radiation (I) or global radiation: It is the sum of the beam and diffuse radiation measured in W/m².

The following expression can be used for beam and diffuse radiation on the horizontal surface:

$$I_b = I_N \cos\theta_z \qquad (1.6a)$$

$$I_d = \left(\frac{1}{3}\right)\left[I_{ext} - I_N\right]\cos\theta_z \qquad (1.6b)$$

General Introduction

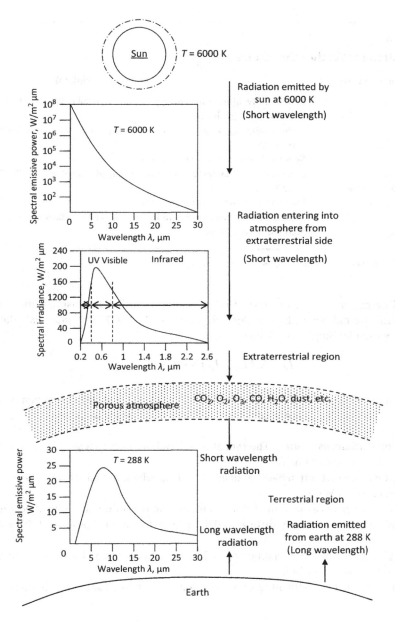

FIGURE 1.7 Propagation of solar radiation to the earth through atmosphere [17].

The following are a few terms commonly used in solar energy applications:

Irradiance: It is the rate at which radiant energy is incident on the per unit area of surface measured in W/m^2.

Irradiation or radiant exposure: The incident energy (irradiance) per unit area of surface in J/m^2 over a specified time (an hour or a day) obtained by integrating beam/diffuse/total radiation. The term insolation is specifically used for solar energy irradiation. Insolation per day (H) and insolation per hour (I) can represent beam/diffuse/total radiation for surface of any orientation.

TABLE 1.4

Classification of Weather Conditions [18]

Weather Condition (Type)	Weather Condition (Description)
a (Clear days)	Ratio of daily diffuse radiation in J/m^2 to daily global radiation in $J/m^2 \leq 0.25$; Sunshine hours ≥ 9 hours
b (Hazy days)	Ratio of daily diffuse radiation in J/m^2 to daily global radiation in J/m^2 between 0.25–0.50; Sunshine hours are between 7–9 hours
c (Hazy and cloudy days)	Ratio of daily diffuse radiation in J/m^2 to daily global radiation in J/m^2 between 0.50–0.75; Sunshine hours are between 5–7 hours
d (Cloudy days)	Ratio of daily diffuse radiation in J/m^2 to daily global radiation in $J/m^2 \geq 0.75$; Sunshine hours ≤ 5 hours

Reproduced from [18], with the permission of AIP Publishing.

Magnitude of normal irradiance in a terrestrial region can be calculated using Equation (1.4b). The magnitude of diffuse radiation (directionless) depends on $(I_{ext} - I_N)$ and can be calculated from the expression proposed by Singh and Tiwari [19].

$$I_d = K_1 \left(I_{ext} - I_N \right) \cos \theta_Z + K_2 \tag{1.7}$$

where the numerical values for K_1 and K_2 for different weather conditions are given in Appendix A. Equation (1.7) is applicable to horizontal surface (tangential surface to the observer at outer surface of Earth).

Radiosity or radiant existence: The rate at which radiant energy leaves a surface per unit area by reflection and transmission in W/m^2.

Emissive power or radiant self-existence: The rate at which radiant energy leaves a surface per unit area by emission in W/m^2.

Further, the weather classifications for a given climatic condition are defined according to sunshine hours (N) and ratios of daily diffuse to daily global radiation. These are briefly described as in Table 1.4

Appendix C gives the hourly variation of ambient air temperature and solar intensity for different months for various stations in India.

A program developed for estimating solar radiation at an inclined angle is offered in Appendix D.

EXAMPLE 1.4

Evaluate air mass of normal direct irradiance coming from sun at New Delhi at 2.30 pm on February 20, 2013.

Solution

Here, $\cos \theta_Z = 0.587°$ and $\theta_Z = 54.03°$ at 2.30 pm on February 20, 2013, New Delhi.
After substituting the above values in Equation (1.4b), we have

$$m = \left[0.587 + 0.15 \left(\left(93.885 - 54.03 \right) \right)^{-1.253} \right]^{-1} = 1.699$$

General Introduction

1.3 CLIMATE

1.3.1 CLIMATIC CONDITIONS

Typically, climate is classified into six climatic conditions namely (a) hot and dry, (b) warm and humid, (c) moderate, (d) cold and cloudy, (e) cold and sunny and (f) composite climatic conditions. These classifications are based upon mean monthly temperature, relative humidity, precipitation and number of clear days as given in Table 1.5. Temperature of any macro space is determined by the amount of solar radiation falling upon that area from one season to another. Regions that are completely exposed to the solar insolation for a large part of the year are hot; those that receive solar radiation at low angles and for smaller portions of the year are cold.

A land-based classification of climatic zones, the Köppen classification system divides the earth into five major types, represented with the letters A, B, C, D and E. The system utilizes both precipitation and temperature, as well as corresponding vegetation, to categorize biomes across the world. All zones except Zone B can be defined by temperature because the determining criteria for the vegetation in this zone is dryness, which falls under precipitation. The five zones are: (A) tropical moist climate, (B) dry climates, (C) moist mid-latitude climates with mild winters, (D) moist mid-latitude climates with cold winters and (E) polar climates.

Further, these zones are subdivided and briefly explained as follows:

A. **Tropical moist climates:** They are located around 15°–25° latitude northwards and southwards of the equator. A temperature of about 18°C is found year-round with annual precipitation of about 1500 mm. They are subcategorized as:

- Tropical wet climate: In this type, precipitation occurs year-round. Variations in temperature are less than 3°C with extremely high humidity, and the surface temperature results in the formation of cumulus and cumulonimbus clouds in the early afternoon time, resulting in high daily rainfall.
- Tropical monsoon climate: Here, the annual rainfall is similar to that of a tropical wet climate, but precipitation usually occurs within the seven to nine of the warmest months of the year.
- Tropical wet and dry climate is also known as the savanna climate: In this type of climate, there is an extended dry season during the winter. During the wet season, rainfall is less than 1000 mm, occurring mainly in the summertime.

TABLE 1.5
Criteria for the Classification of Climates [20]

Climate	Mean Monthly Temperature	Relative Humidity (%)	Precipitation (mm)	Number of Clear Days	Example (India)
Hot and dry (HD)	Greater than 30°C	Lesser than 55%	Lesser than 5 mm	Greater than 20	Jodhpur, Jaipur
Warm and humid (WH)	Greater than 30°C	Greater than 55	Greater than 5 mm	Lesser than 20	Mumbai, Pondicherry
Moderate (MO)	Between 25°C–30°C	Lesser than 75%	Lesser than 5 mm	Lesser than 20	Bangalore
Cold and cloudy (CC)	Lesser than 25°C	Greater than 55	Greater than 5 mm	Lesser than 20	Srinagar, Shimla
Cold and sunny (CS)	Lesser than 25°C	Lesser than 55%	Lesser than 5 mm	Greater than 20	Leh
Composite (CO)	This applies, when six months or more do not fall within any of the above categories				New Delhi, Dehradun

B. **Dry climates:** Evaporation and transpiration play a greater role in shaping the vegetative state than temperature, here. The regions extend 20°–35° latitude northwards and southwards from the equator. There are four subdivisions within this region:

- Hot, arid climate or true desert climate spans about 12% of Earth's total land. Xerophytic vegetation usually grows in this climatic zone.
- Hot, semi-arid climate also referred to as steppe climate spans about 14% of Earth's land and makes up grassland-type climate. These regions receive more rainfall than their true desert climate counterparts because of mid-latitude cyclones and areas of inter-tropical convergence.
- Cold desert climates is typically found at higher altitudes than hot desert climates and are usually drier than hot desert climates.
- Cold, semi-arid climates tend to be located in elevated portions of temperate zones, typically bordering a humid continental climate or a Mediterranean climate.

C. **Moist mid-latitude climates with mild winters:** These zones typically face hot and humid summers and mild winters. Extending between 30°–50° latitude northwards and southwards from the equator, these regions are typically the eastern and western extremes of each continent. Sometimes, summer months may feature convective thunderstorms, and winter months may feature mid-latitudinal cyclones. The climatic classification is further broken into three types:

- Humid subtropical climate: In this type of climate, summers are sweltering and humid, with frequent thunderstorms. Winters are milder in comparison, and precipitation occurs due to mid-latitude cyclones.
- Marine climates are usually on the western coasts of each continent. These areas are generally humid, with a hot and dry summer. While winters are milder, they come with heavy rainfall due to mid-latitude cyclones.
- Mediterranean climatic zones receive heaviest precipitation during the winters due to mid-latitude cyclones. There is hardly any rainfall during the summer. Examples include Portland, Oregon, or California.

D. **Moist mid-latitude climates with cold winters:** Summers are typically warm but can also be cool, while winters are cold. During the summer months, average temperatures climb above 10°C while in the colder months it can be less than 3°C. Winter in this region is usually bitingly cold, with strong winds and the possibility of snowstorms coming in from the continental polar and the arctic air masses. This Köppen climate classification is further divided into three subsections:

- *Dw*: Denotes dry winters.
- *Ds*: Denotes dry summers.
- *Df*: Year-round rainfall.

E. **Polar climates:** Temperatures are usually low all year round in the polar regions. The warmest months see temperatures less than 10°C. Usually occurring in the northern coastal regions of North America, Asia, Europe, and in Greenland and Antarctica, these are classified as:

- Polar tundra, where soil remains permanently frozen as permafrost and can be hundreds of meters deep. The only vegetation in this region consists of lichen, mosses, dwarf trees and woody shrubs.

General Introduction

- Polar ice caps is the second category, in which the surface is permanently covered in ice or snow.

1.3.2 Weather Conditions

Weather conditions are classified as (a) clear days, (b) hazy days, (c) hazy and cloudy days and (d) cloudy days. The classification is based upon the ratio of daily diffuse to daily global radiation and the number of sunshine hours (N). Table 1.4 gives the description of weather conditions.

1.3.3 Macro- and Microclimate

The knowledge of macro- and microclimate is very important to understand the energy performance and environmental performance of the buildings for both heating and cooling seasons.

1.3.3.1 Macroclimate

Macroclimate is the climate of a larger region/area. Any change in design cannot affect the macroclimate around the building, however the building design can be developed based on the macroclimate prevalent in that region. The parameters considered to develop a passive house for macroclimate are seasonal accumulated temperature differences, typical wind speeds and direction, annual global horizontal solar radiation, the driving rain index, etc.

1.3.3.2 Microclimate

The building site may have various microclimates due to the presence of hills, slopes, streams, water bodies, trees and other buildings.

Local terrain has an effect on the microclimate, e.g., the surrounding slopes affect the air movement of the area. This is the reason why valley floors are significantly colder than locations way up the slope; the cooler air drifts down the slope and settles, replacing the warm air. The crests of hills and ridges have unfavorable wind profiles as the wind flow is compressed, thus leading to high wind velocities.

The nearby buildings also have a profound impact on the microclimate by shading the ground or changing the localized wind direction. Heat island effect is one of the examples wherein large cities have higher average temperature as compared to the surrounding areas. The solar radiation is absorbed and radiated back to the ambient by the building surfaces, pavements, etc., thus raising the surrounding temperature. The building also acts as a hindrance in the wind flow, thus leading to lower wind speeds and causing warm air to stay around.

Presence of a water body (lakes, ponds etc.) also affects the microclimate to a large extent. The air that is in contact with the dry earth will have higher temperature than the air in contact with wet ground/water body.

With passive building design, microclimate may be improved, which has various advantages. This reduces winter heating costs, as well as summer overheating (reduced cooling loads) and maximizes the outdoor comfort. Durability of building material is improved, tree growth encouraged, and better visual connection occurs in spaces around the building. This also discourages growth of mosses and algae, facilitates open-air drying of clothes, etc.

Means of enhancing of microclimate around a building are listed below:

- Maximum solar insolation and daylight to penetrate into the buildings.
- Solar shading devices to be installed to protect from prolonged exposure to summer sun and protect the interiors from glare.
- The space shall be protected against the prevailing cold winds, rain and snow.
- Provision of thermal mass to moderate extreme temperature fluctuations.

- Trees/vegetation to be used as buffering against solar gains and wind. Transmission helps to moderate the high temperature (Section 7.4).
- Provision of water features like pools and fountains for cooling via evaporation.

1.4 PASSIVE HOUSES

Passive design is the designing strategy that does not require auxiliary technology to be effective. Also, this is how smart bioclimatic design is defined (Section 1.6). A passive solar house refers to conceptualizing and designing a building that deploys local characteristics intelligently in the sustainable design and respecting the site, climate, local building materials, geomorphology, wind direction and the sun for thermal comfort. Local characteristics may also refer to manmade interventions like the landscape, built surroundings, etc. [21].

In modern-day building architecture, solar housing is an important concept that incorporates the motion of the sun, location of building (latitude), climatic conditions and locally available materials required for building construction. The concept of a solar house also includes the strong correlation between available renewable sources of energy and building design. It ensures the utilization of natural sources as much as possible for the design, construction of building, as well as daylighting and thermal comfort in building. Solar houses are not an entirely new concept. Socrates (400 B.C.) stated: "Now in houses with a southern orientation, the sun's rays penetrate into porticos in winter, but in summer the path of the sun is right over our heads and above the roof. If then, this is the best arrangement, one should build the south side loftier to keep out the cold winds to reflects this" [11].

Examples are given in Section 1.6.1.

1.4.1 STRATEGIES FOR PASSIVE DESIGN

The building concept and the site have a profound impact upon the interaction between the building and its surroundings. The site and the topography affect the building's exposure to the prevailing wind direction, solar gains, precipitation levels, etc. The orientation, form and geometry of the building affect the wind capture and the solar gains. The trees provide shade, affect the wind flow patterns and also protect the building against the rain. As discussed previously, microclimate may be influenced by solar access and wind control. The strategies for the same are listed as follows.

1.4.1.1 Solar Access

It is important to understand solar access in order to minimize solar overheating during the summer and maximize solar gains during the winter. Buildings with heating requirements shall be oriented north-south with maximum glazing towards the south facade. An excellent means of site shading in summer months may be provided by deciduous trees, which is reduced during the winter months as the trees shed their leaves and allow the solar radiation to percolate. Grass planted outside buildings reduce ground-reflected solar. Courtyard planning and introduction of water features or vegetation also helps to moderate the effects of high temperature in the summer as seen in Figure 1.8. Further, the color of the nearby surfaces also has a pronounced effect on the availability of solar radiation on the building. As an example, light-color pavements increase the reflection of solar radiation from the ground to the surroundings. Paver blocks may act as external thermal mass, thus moderating the temperature fluctuations immediately adjacent to the building.

1.4.1.2 Wind Control

Buildings should be oriented in the prevailing wind direction so as to capture the maximum wind. Also, building form greatly impacts the effect of wind. The following should be kept under consideration:

General Introduction

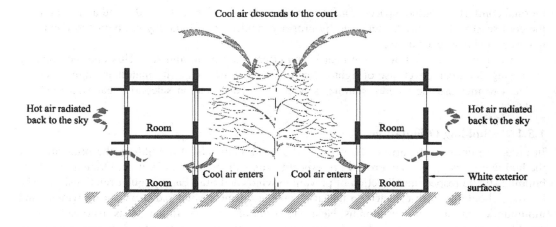

FIGURE 1.8 Courtyard planning [22].

- Building flank should be avoided to face the wind.
- Funnel-like gaps between buildings should be avoided.
- Cubical forms and flat roof structures should be avoided. Pitched roof and stepped forms are preferable for high rises.
- The long axis of the building should be oriented towards the prevailing wind direction.
- Abrupt changes in the building height shall be avoided.
- Wind tunneling shall be avoided, arranging the buildings in irregular fashion.
- Podiums shall be constructed to limit the draught at ground level.
- Coniferous trees, fencing and other landscape elements like earth mounds and hedges can also reduce the impact of wind and driving rain on buildings.

1.5 ARCHITECTURAL DESIGN OF PASSIVE BUILDINGS

The design of a passive building, whether small or big, is based upon certain thumb rules. These are listed under categories:

- Site planning
- Envelope design
- Interior design

1.5.1 Site Planning

Site planning is a major design guideline for any passive building design. Site planning includes building location, orientation and clustering.

1.5.1.1 Building Location and Orientation

Ideally, passive solar considerations and microclimate should be the prime factor of choosing a site. Knowledge of microclimate helps planners formulate a strategy to improve local site climatic conditions and take up a decision of how much the solar gains are to be exploited. For cold climates, the building should be located where it is able to receive maximum solar gains during the sun time. The building shall be located towards the northern portion of the sunny area. This strategy will ensure that the open areas/gardens receive maximum winter sun. For cool climates, the heat island effect is generally beneficial since it can reduce the building artificial heating costs and also improve the thermal comfort in outdoor spaces. The main concern is to moderate the cold, wind and wet during the cool season. This may be achieved by proper consideration of site layout, built form, exterior materials and landscape design.

In warm climatic conditions, the heat island effect shall be minimized. This may be done by increasing the green cover, use of light-color facades, low heat-capacity materials, high-reflective surfaces to increase the albedo, summer shading by trees, location selection near water bodies, parks, etc.

1.5.1.2 Building Orientation

In order to minimize the artificial heating and cooling demands of a building, the orientation of the building mass plays a vital role with respect to sun path and wind direction. Orientation of the building should respect the direction of prevailing winds and Sun–Earth angular relationships. The building should be such oriented so that it receives maximum solar radiation during winters and minimum during the summer months. Near equatorial regions, where the sun is overhead, simple overhangs can be used to shade the solar radiation on south and north facades. However, significant solar gains will be received on east and west facades, which cannot be minimized by sunshades due to low sun angles. The west elevation will receive maximum sun during the hottest time of the day (afternoon). This may cause overheating of the building. Thus, a linear building with east-west orientation (i.e., minimum exposure of east and west facades) and minimum openings is optimum in terms of solar orientation. This building elongation along the east-west axis helps in maximum surface area exposure towards the south direction. Therefore, this allows maximum solar gain during winter months while receiving only marginal solar insolation during summer months. A south-facing building facade receives maximum insolation during the heating season and is easier to shade in summer months when excessive solar gains are unwanted.

Another consideration for building orientation should be predominant wind direction. It is not necessary for the wind to be perpendicular or even near perpendicular to the façade to encourage the air flow within the building. In order to reduce air infiltration and heat loss, long faces of building are advised to be oriented parallel to the wind direction. Maximum air infiltration occurs when the wind is at a 45° angle to the building face [23].

1.5.1.3 Clustering

Building clusters help in mutual shading of the buildings, thus minimizing the incident solar radiation on the building envelope. This mutual shading is beneficial when the streets between the buildings are narrow or built around the courtyards. Building layouts have a huge impact on daylight, solar gains and ventilation of the mass and spaces around them.

1.5.2 Envelope Design or Building Envelope

Envelope design is a moderate guideline for any passive design. Building envelope includes building shape, protected entrances, window locations, shading, insulation and infiltration reduction. With appropriate selection, sizing and installation of these concepts, building envelope acts as an energy-efficient filter between the outside and inside environments.

1.5.2.1 Building Shape

The overall proportion of the building, surface area to volume aspect ratio is the most important consideration when finalizing/conceptualizing a floor layout. For direct gain concepts, the depth of the building shall not exceed 2% times the height of the window from the floor area for an effective solar gain and daylight savings. For spaces with larger depth, large south-facing windows are not preferable. In fact, for such spaces, south-facing clerestory windows or skylights are advisable (Figure 3.4). Multistory buildings have significant surface area savings when compared to single-story buildings [24]. Building heights and widths directly affect the wind flow patterns. For buildings in

General Introduction

cold climatic conditions, high surface-to-volume ratio renders the building more susceptible to environmental stress resulting in a poorer thermal performance. Various ways have been listed by [23] to reduce wind sensitivity of buildings and their immediate surroundings as listed here:

- Long axis to orient in wind direction.
- Avoid flat roof building and cubical forms. Stepped or pitched roofs are preferable as the latter reduce the solar access to adjoining buildings. Pyramid-shaped buildings allow daylight and solar gains on upper floors.
- If the north side slopes towards the grounds, this is considered to be the best shape since this minimizes the exposed surface in the north direction and reduces the shadow casting [24].

1.5.2.2   Entrances and Windows

Infiltration of air through cracks, frames or through the door each time it is opened is the major source of heat gain/heat loss. Thus, the main entrance should be designed with a vestibule/foyer (an enclosed space). This space will provide an airlock between the building and the exteriors. This air lock will prevent the inside air (warm or cool) to move outside the building every time the door is opened, thus limiting the infiltration losses. The orientation of the entrance should be away from the prevailing wind direction.

Window location and size should be the most important consideration when designing a passive building as they may act as sources of heat loss if not planned properly. Window openings shall be located on the southeast, south or southwards direction to allow winter sun. to minimize the heat losses; smaller-sized windows shall be placed in east, west or north directions. Recessed windows further reduce the heat losses. Daylight also depends on the window orientation and size as discussed in Section 3.2.9.

Natural ventilation is one of the major passive cooling techniques and can be done through the openings in the building façade. Ventilation can be established by over- and under-pressure. Another way is through thermal stacking. Let us study the following example:

Suppose a dark wall is constructed of stone and oriented south. This surface will be heated during the day and consequently will heat the air above it. This air becomes lighter and rises, creating a thermal draft. This principle can be used where enforced ventilation has to be created, such as with solar chimneys.

1.5.2.3   Solar Shading Techniques

Solar shading devices reduce the heat gains and add a degree of thermal and solar control. Thus, they reduce the cooling costs. These devices, if planned as an integral part of the building, may act as an aesthetic element and should not act as a hindrance for natural daylight.

The design of effective shading devices depends upon the solar orientation of a building façade. For example, simple fixed overhangs are very effective at shading south-facing windows in the summer when sun angles are high. However, the same horizontal device is ineffective at blocking low afternoon sun from entering west-facing windows during peak heat gain periods in the summer. Shading devices can be classified as follows:

- Movable opaque: Roller blind curtains, awnings, etc. reduce solar gains but impede air movement and block the view.
- Louvers: They are adjustable or can be fixed. To a certain extent, they impede air movement and provide shade to the building from solar radiation. Shading devices can be classified as vertical (vertical louvers, projecting fins), horizontal devices (canopies, awnings, horizontal

louvers, overhangs), egg-crate devices (concrete grill blocks, metal grills) and screenings (venetian blinds, double-glass windows, window quilt shades, movable insulation curtains, natural vegetation, etc.). Horizontal shading devices are best suited for south-oriented openings, whereas vertical shading devices are better for east and west-facing facades [22].
- Fixed: In Indian architecture, overhangs of chajjas protect the wall and openings against sun and rain. Shading devices have been used since ancient times. Mughal architecture (India) has used deep-inclined shades so that they cover a larger surface area. Heavy carvings were done on the exterior facade to allow mutual shading of the space; extended surface helps increase the heat transfer due to convection, thus reducing the heat flux entering into the building. In the evening, when the ambient conditions are cool, increased surface area helps in cooling faster. This can be seen in the fort city of Jaisalmer, India.

Various roof shading techniques are shown in Figure 1.9 and explained in Section 7.4.

1.5.2.4 Insulation

The most efficient and cost-effective energy design strategy is addition of insulation in building elements like walls, floors, ceilings, foundations, etc. For locations where solar radiation is not available during the heating season, higher levels of insulation are preferable. Table 1.6 gives the insulation values for different levels of thermal integrity.

This is an effective way to reduce the energy demand of a building because of the temperature difference between the inside and the outside, which can cause heat transmission. Each part of the building contributes to this heat flow depending on the size and thermal properties of the elements. The thermal transmittance, known as the U-value (refer to Section 2.8), is the most important

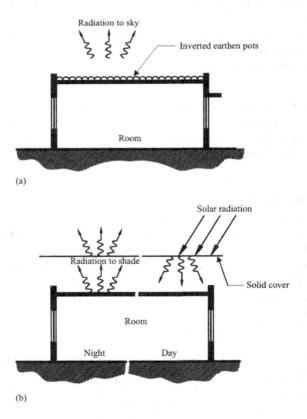

FIGURE 1.9 Roof shading by (a) Earthen pots, (b) solid cover.

(Continued)

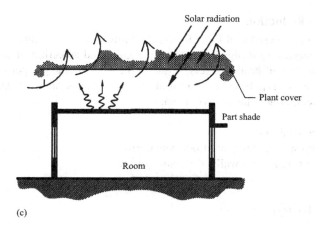

(c)

FIGURE 1.9 *(Continued)* Roof shading by (c) plant cover [22].

TABLE 1.6
Insulation Level Selection [24]

	U Values (W/m² °C)				
			Floor		
Thermal Integrity	Walls	Ceiling	Crawl Space	Slab	Basement
Level 1	0.515	0.299	0.515	1.136	1.136
Level 2	0.299	0.189	0.515	0.568	0.568
Level 3	0.236	0.149	0.299	0.568	0.568

thermal property of the building envelope. It gives an idea of how big the transmission loss per square meter is per degree temperature difference between indoors and outdoors.

Let us understand from the following examples.

CASE 1

U-value of an uninsulated masonry wall (210 mm thickness) = 2.7 W/m²K

Thermal resistance = $1/U$-value = 0.37 m²K/W

CASE 2

An addition of 50 mm layer of mineral wool insulation to the above uninsulated masonry wall is done with thermal resistance of 1.39 m²K/W, therefore the total thermal resistance of the setup becomes = 1.39 + 0.37 = 1.76 m²K/W. The U-value therefore becomes 0.568 W/m²K.

Therefore, with the addition of insulation material, there is a reduction in heat transmission loss by a factor of almost 5.

The impact of the addition of thermal insulation on the energy balance is given in Section 2.10

1.5.2.5 Infiltration Reduction

Energy consumption may be reduced by reducing the infiltration levels, thus comfort levels can be increased. Careful attention to design, construction details and selection of products needs to be done in order to reduce the infiltration levels. A rate of 0.5–0.7 air changes per hour (ACH) is found in energy-efficient homes [24]. To achieve infiltration levels within range of 0.6–0.7 ACH, the following construction techniques shall be followed:

- Low infiltration windows
- Sealed/minimum wall, ceiling and floor penetrations
- Continuous vapor barrier on walls, floor and ceiling
- Airlock entries

The rate of heat transfer $(\dot{Q})$ is given by:

$$\dot{Q} = 0.33NV\left(T_r - T_a\right) \tag{1.8}$$

where N is the number of air changes per hour, V is the volume of room air in m³, and T_r and T_a are room and ambient air temperature in °C, respectively. When the value of $N < 10$, it is referred to as infiltration, and for ventilation $N > 10$.

1.5.3 INTERIOR DESIGN

Interior design is a minor strategy and includes choosing the layout, appropriate materials and solar passive systems, movable insulation, thermal mass, reflectors, shading devices, thermal heating and cooling concepts etc. An effective interior design results in better utilization of passive solar energy to create better levels of thermal comfort.

Orienting the indoor spaces towards south of the building will allow maximum sun to enter. Thus, rooms should be oriented southeast, south, southwest as per the requirement of sunlight. Spaces that require minimum lighting and heating like garage, corridors, closets, etc., should be placed along the north of the building. These spaces will act as a buffer zone between the heated space and a colder north zone.

Systems choice should be done carefully. For passive heating, direct gain through windows, thermal storage walls and sunspaces should be created. Direct gain should achieve efficiency of about 30%–75% depending upon transmissivity. Windows facing 30° south utilize direct gain passive solar energy. Windows shall have an area of 0.19–0.38 m² of south-facing glass for each square meter of the floor space area for cold climate (av. winter temperature ranges between −6°C to 10°C). For temperate climate (av. winter temperature ranges between −1.5°C to 7°C), window area should be 0.11–0.25 m² of south-facing glass for each square meter of the floor space. These windows/ glass shall be sufficient enough to keep the interior spaces within a comfortable range of temperature (18°C–20°C). Thermal storage walls can be of various types like masonry, water walls or double-glazed walls. For same area and thermal storage capacity, a water wall is more efficient than masonry wall. Movable insulations should be used over entire glazed areas during off-sunshine hours. A single-glazed system with night insulation is more effective than a double-glazed system without night insulation. The insulation should be tightly sealed for the glazed opening. Heat losses for a single-glazed system is 100% (without night insulation) and 40% (with night insulation). Heat losses for a double-glazed system are 57% (without night insulation) and 20% (with night insulation). Reflectors should be used where large solar collector area is not feasible. These should be used for vertical glazing, equal to the width of glazing but 1–2 times higher than the glazing opening in length. The average solar radiation incident on vertical openings may be increased by 30%–40% during the winter months. The reflectors may be adjusted for summer months and may act as solar

General Introduction

shading devices. Horizontal overhang/shading devices are very important for south-facing openings. Providing ample daylighting is also very important, which is often overlooked in buildings. Chapter 3 illustrates various daylighting techniques.

1.6 BIOCLIMATIC DESIGN

1.6.1 HISTORY [25]

The term *bioclimatic design* has been coined by architect Ken Yeang as the passive low-energy approach that makes use of the ambient energies of the climate of the locality to create conditions of comfort for the users of the building.

Humans used to shelter in caves in the northern hemisphere for safety, and the caves are relatively stable in temperature, hovering around the annual mean temperature. These were more comfortable than the harsh cold winters and extremely hot summers outside. Therefore, humans started making cave dwellings in climates with huge temperature differences between summer and winters, day and night like deserts, or even climates with a benign average temperature. In colder climatic conditions, homes made from materials started to be constructed for stability with thatched roofs to protect from cold and rain. In regions with wind, buildings were designed to respond to the predominant wind direction. In areas with snow and rain, roofs were constructed in such a way that they didn't leak and the water could run off easily with construction of overhangs. These overhangs also helped to protect the underneath structure particularly in timber construction. Today, extreme protective measures against precipitation are found by complete umbrella structures. With water, another challenge is the humidity for which materials like loam have been used.

Loam, or adobe, has been a building material for ages, for typical wattle-and-daub or half-timbered buildings. Loam is capable of absorbing water vapor from the air in cases of high relative humidity, and if the air is relatively dry, it will release moist from the material to the air. Thus, loam can regulate the humidity making it suitable for indoor stucco plasterwork.

Similar to cave dwellings, they were comfortable because of their heavy mass, which stabilizes the indoor temperature. This is the reason old churches, mosques etc. were constructed with heavy stone that tempers the temperature differences between summer and winter. But in colder climatic conditions, this mass will not create comfortable conditions inside, thus additional heating is required. In medieval times, this thermal heat was provided by hearths inside the building. Other methods to achieve warmth included (i) the kitchen, around which other spaces were organized, (ii) compartmentalization of spatial zoning; (iii) on farms, thermal insulation was sometimes provided by a buffer zone of stables with cattle, and by a hay storage in the attic.

Romans invented a central heating system called hypocaust, which used in Roman bathhouses. Hot water from a stove was allowed to run through the ceramic pipes in the wall and under the floors. Romans also mastered construction and invented concrete, and immense structures like Pantheon were constructed in Rome. The Pantheon also demonstrated the climate aspect of daylight. The oculus in the roof allowed sunlight to illuminate the space sufficiently. Also, it was established that the higher the window, the more daylight is allowed.

This principle was explored during the Gothic architecture period after the 12th century. In the cathedral of Bayeux in France, high windows as large as the façade of a house bring daylight into the otherwise dark church interior.

In modern times, buildings have become transparent due to easy production of glass. The Post Tower in Bonn offers an extreme example with glass floors. The ventilation was done through façade to cool the interiors. Thus, ventilation is an important aspect as no building is inhabitable without proper air refreshment, which can also be achieved using mechanical means. With wind coming from one direction, there will always be a façade with over-pressure and one with under-pressure. If the temperature is not too low, windows or grills can be opened to create an air current through the building. Wind-driven ventilation can be seen in a building in Melbourne, Australia, where wind drives a fan on the roof. Creating an under-pressure in the building, sucking out exhaust air, thus induces ventilation.

Today, mechanical ventilation and air conditioning is very common to create comfortable thermal conditions, though increasing energy demand significantly. Therefore, understanding bioclimatic design is very important.

Another term that should be understood is *smart design,* which modern sources see as design that interacts intelligently with the environment. The concept of smart bioclimatic design is an approach that uses local characteristics intelligently in the sustainable design of buildings and urban plans.

The following is an example of a bioclimatic building located in Singapore [21] (Figure 1.10a). The Esplanade Theatres on the Bay, Singapore, completely respects the course of the sun without resulting to undesired overheating of the building. The steel flaps are added to avoid direct solar radiation but allow the daylight through the glass skin placed behind the flaps. Based on the location,

(a)

(b)

FIGURE 1.10 (a) Esplanade, Singapore [26]. (b) Sydney Opera House (by author).

General Introduction

with the sun being overhead, the flaps are smartly placed and the concept of the flaps is similar to overhangs in the Sydney Opera House, Australia (Figure 1.10b). The protruding, overhanging roof avoids direct insolation of the interior, thus reducing the cooling demand [21].

In old settlements, the urban planning used to be dense and compact, allowing sunlight onto the streets and facades. Light-reflecting surfaces were used to avoid the absorption of solar gain in the building envelope. The typical example of whitewashed walls and roofs can be seen in Greece. This reflectance factor of surfaces is referred to as albedo.

1.6.2 BUILDING DESIGN STRATEGIES DEPENDING ON CLIMATIC CONDITIONS

Classification of climatic conditions has been tabulated in Table 1.5. Based on which, building design typology has been defined as the following:

a. **Hot, dry climatic conditions**

In such climatic conditions, there is huge temperature variation during daytime and night-time. For passive design of buildings in hot, dry climatic conditions, the advantages of this temperature difference should be explored. This may be done by creating a time lag of solar gains so that it enters the building at nighttime when it is much needed. Materials like adobe bricks, mud walls, etc., which have great thermal inertia, or concepts like Trombe walls should be incorporated in the design. The buildings are arranged in compact patterns to reduce the exposed surface area to the solar radiation and increase the built weight per unit of volume. This increases the thermal storage. Other solutions for such climatic conditions are keeping the windows airtight during day and completely opened during night hours, underground construction, courtyard planning and double roofs or double walls with ventilated inner space.

b. **Warm, humid climatic conditions**

For hot, humid climatic conditions, thermal storage plays no important role for ensuring passive thermal comfort. The solar radiation is very intense in such conditions. Thus, it is important to curb the direct or diffuse solar gains. In addition to provision of solar shading devices, natural ventilation is equally important to dissipate the heat in the interiors and reduce the humidity. Steep roofs are preferable to drain off rainwater, and these also favor thermal stratification of hotter air at top where openings are made to vent out the hot air.

c. **Moderate climatic conditions**

The moderate climate has mild to warm summers and cool winters. The need for winter home heating is greater than the need for summer cooling. Passive features required can be listed as reduction of solar heat gain by orientation of the bedrooms towards north and by shading of east and west walls by neighboring buildings. Majorly, cross-ventilation is needed, and such measures for thermal heating and cooling are required.

d. **Cold and cloudy climatic conditions**

In cold climatic conditions, the buildings shall be designed in order to trap the heat inside, thus compact building forms are preferable. These spaces shall have few exposed surfaces to the outside to reduce the heat loss. In extreme cases, the building forms become semi-spherical, since this form ensures maximum volume for minimum shell surface. As discussed earlier, the architect should attempt to have maximum possible insulation along with a high level of airtightness. In earlier times, granaries and lofts storing straw were used to increase their insulating power. The buildings can be constructed in groups facing the sun, forming a compact layout to obtain mutual protection against the cold winds. Few examples of buildings are given below:

- The Hodges Residence [27] was the first underground (earth-sheltered) passive solar house located in Ames, Iowa. It was designed to incorporate the solar energy collection, storage and distribution with direct gain concept. The house successfully met about 85% of the heating requirements by means of solar energy only.
- The Benedictine Warehouse [28] was built on the concepts of direct gain and water wall storage concepts at New Mexico. It was designed to maintain temperature of 7.2°C and successfully attained the temperature with range of 8°C–10°C with ambient temperature of −7°C.

e. **Cold and sunny climatic conditions**

Passive design criteria for cold and sunny climatic conditions includes resisting the heat losses and promoting the heat gains. This may be achieved by decreasing the exposed surface area, increasing thermal resistance and thermal capacity (time lag), increasing the buffer spaces, decreasing the air exchange rate, and reducing shading. Passive features include direct gain; double-glazed windows on east, west and south sides of the building; high thermal insulation of roof and walls; night insulation; and creating a balance of temperature fluctuation by massive wall construction .

f. **Composite climatic conditions**

In such climatic conditions, combinations of various elements of building design are used that can easily change depending on the climatic condition. Flexible systems like mobile shading devices, movable insulation and intermediate spaces between inside and outside are proposed.

1.7 ENERGY CONSERVATION

1.7.1 INTRODUCTION

Reducing consumption of conventional energy is termed energy conservation. Due to reduction in the generation of conventional power by fossil fuels, there is reduction in the production of carbon dioxide, thus minimizing the climatic changes due to emission of carbon dioxide gas. Energy efficiency may be achieved by accommodating the following:

- Minimizing the use of fossil fuel based energy
- Using energy-efficient equipment
- Using renewable energy

1.7.2 ENERGY CONSUMPTION [29, 30]

It is important to analyze the energy usage of a building to make a ZEB. In a building, the energy use can be in form of heating, cooling, domestic hot water, ventilation, lighting, cooking, washing, refrigerating, electronic equipment, etc. The energy for all the purposes is provided by energy carriers. Energy carriers refer to the substance that contains the energy that can be converted into heat or used for any other energy consumption purpose. The most common and versatile example of an energy carrier is electricity. Other examples include fossil fuels, natural gas, coal, wood, oil, etc. After analyzing the energy carrier, one must figure out the energy consumption. The easiest way is to go through energy bills [2].

1.7.3 ENERGY EFFICIENCY

Efficient use of energy shall be the aim by using the minimum amount of required energy in our daily activities. According to the International Energy Agency (IEA), energy efficiency in buildings,

General Introduction

industrial processes and transportation could reduce the world's energy needs by one-third by 2050, which can help control global emissions of greenhouse gases [31].

To mitigate environmental abuses and to conserve natural resources, efforts that can be adopted in buildings to achieve energy efficiency are described under this section. The major building elements responsible for maximum energy consumption are structural (i.e., building envelope), heating, ventilation and air conditioning (HVAC), devices and appliances and lighting systems. The efficiency of these systems may be improved by installing or implementing energy-efficient equipment/methods. Following are a few examples to be implemented in order to achieve energy efficiency in the building sector.

- Insulating the interior walls, floor or ceiling in an air-conditioned room will lessen the use of heating/cooling by fossil fuel–based energy in order to maintain the thermal comfort. Use of synthetic fibers as insulating material reduces the heating/cooling loads by restricting the heat transfer through walls, floor or ceilings. Higher R-value indicates better insulating properties. The appropriate insulating material is based on type of climate, building and recommended R-values.
- The openings may be treated further to control the winter heat losses and summer heat gains, thus reducing the space conditioning loads. This may be done by using the interior thermal shades, solar shading devices. The most effective method is to seal the edge joints to minimize the convection effects. Reversible thermal curtains may also be used, one side made of vinyl, to insulate the openings in winter and the other side of aluminized polyester to reflect the summer sun. Window blinds/curtains with low-emissivity selective coatings, solar screens, exterior films etc. may also be used. The solar-load ratio (SLR) method is widely used to design glass windows (direct-gain), collector-cum-storage walls (indirect gain) and sunspace systems to save fossil fuel for thermal heating for cold climatic condition.
- Infiltration is basically the leakage of air through the cracks, joints, ceilings, walls, floors etc. that is not under control. This is a result of differences in the pressure and temperature between the inside and outside of the building caused by wind movements, natural convection and other forces. Measures suggested to control these include sealing the structural joints (like door and window frames, intersections of floor and walls) and cracks, weather stripping openings (windows and doors), installing gaskets, taping the leaky ducts, sealing holes made for utilities, etc.
- Use of skylights allows natural illumination for the interiors, thus eliminating or restricting the need for artificial lighting. Smart sensors may be installed to monitor the illuminance level during the sunshine hours with natural light, and artificial means may be switched on/off as per the requirement.
- Passive solar design: Assembly of nonmechanical architectural elements that convert the solar energy into usable energy is referred to as solar passive design. In summers, sunshade devices, landscape features and openings reduce the heat gains, and vents allow the unwanted heat to dissipate. In winters, the solar gains through the openings are stored in the masonry walls or water for release during the off-sunshine hours. Interiors of the buildings should be clean and light in colors to enhance the reflection of light. Low partition walls shall be used in office areas to share the light.

The following elements may be used in various combinations depending on the site and climatic conditions.

- Double-glazed south-facing openings
- Masonry/water heat storage mediums
- Sunspaces
- Solar shading devices

28   Photovoltaic Thermal Passive House System

- Earth berms
- Movable insulations
- Vents
- Exhaust fans (occasionally)

To ensure the effectiveness of passive solar systems, high levels of weatherization are necessary.

- Sensible use of electronic devices: One of the best ways to conserve energy is to turn off electronic devices when not in use. Occupancy sensors may be installed to ensure the same. Sensor systems may also be installed to control air conditioning systems with closed doors/vents. Use of energy-efficient equipment is the finest approach towards energy savings. Devices like home electronics (television, computers, mobiles, fans, etc.), kitchen equipment, laundry appliances, refrigeration, space conditioning (like air conditioners, geysers, coolers, heaters etc.) shall have at least five-star ratings. Use of compact fluorescent lights (CFL) instead of conventional incandescent light bulbs can further reduce the amount of energy required while maintaining the same level of illumination. These CFLs consume one-third the energy of conventional lights with a longer life span (about 6–10 times the conventional ones). Lighting dimmers are very effective energy savers and also enhance the fixture's life.
- Photovoltaic (PV) systems are considered to be most economical and environmentally friendly when compared to conventional sources of electricity generation.

1.8 DESIGN APPROACH TO ZEB [32]

Zero energy building (ZEB) can be defined as a building that generates all energy it uses during one year's time fully from renewable sources. The ZEB concept can be explained as a scheme that explains the energy and climate system of the structure. The design approach is divided in various stages:

Stage 0: Research: We start with studying the local circumstances.
Stage 1: Reduce: In this stage, the demand is to be reduced. Use of passive and energy-efficient appliances are the solution to this stage. Generally, there is lot of residual energy in air, water and material that can be recovered and used in the building. This brings us to the next stage.
Stage 2: Reuse: This stage focusses on reusing the waste heat, wastewater, waste material.
Stage 3: Produce: This is the final stage to generate renewable energy.

1.8.1 STAGE 0: RESEARCH [33, 34]

This step on the current energy demands and the local climate conditions has to be focused.

Therefore, the study of the climate where the building is located should be considered for any building design. We have already discussed the climate zone classifications by Köppen-Geiger in Section 1.3.1. Building design should be for local circumstances wherein the building fits the characteristics and microclimate of the site. This is very well evident from the vernacular architecture. For example, in the tropics where the temperatures are always high, optimal use of the cooling capacity of the winds was considered by creating air-permeable facades and floors. At the same time, roofs were designed to withstand heavy precipitation. In the deserts, narrow streets were planned for shading, and houses were built with a lot of mass that tempered diurnal differences. White-plastered surfaces were used to reflect the sun, and flat roof structures that cool down under a clear night sky were

General Introduction

29

planned. In cold climatic conditions, well-insulated buildings were planned to preserve the heat with steep pitch roof structures to keep the rain flowing down and away and carry the weight of snow. Therefore, it is clear that a building should be designed as per prevailing local climatic conditions. Temperature (mean annual and diurnal differences), humidity (absolute and relative), sun (course of the sun and solar intensity), wind (predominant winds), precipitation (annual values and seasonal difference), soil conditions (underground and geology gives options to use the soil energy) and the surroundings (like trees, buildings, etc. that can shade the structure in question) shall be considered in particular to analyze the local energy potentials and their usage.

1.8.2 STAGE 1: REDUCE

Here, the focus should be on passive reduction. The following measures are restricted to passive reduction only.

Sr. No	Measure	Principle
Thermal insulation		
1	Thermal insulation	Insulation with low U-value (or high R-value) helps to resist the heat transfer (or cold) from outside to the inside of the structure and thus reduces the energy losses.
2	Airtightness	Air infiltration due to cracks in the building skin or when the building is not properly sealed leads to undesired heat gains or losses.
3	Type of glazing	In any building envelope, windows serve as a weakest thermal link. The amount of thermal energy that is transferred through the glass is expressed in the U-value. Low U-value means low heating or cooling losses. This low value can be attained by using double or triple glazing, or adding a gas-filled or vacuum void between the glass panes.
		Other characteristics for choosing the glazing type are the SHGCs (solar heat gain coefficients), which reflect solar radiation instead of absorbing it.
4	Frames of doors and windows	Heat transfers depends on the frames of doors and windows. Wooden material with a thermal bridge interruption and insulation can be one method to reduce the U-value and thus reduce the losses.
Bioclimatic design		
5	Building orientation and envelope design	The building should be oriented to maximize the wind breezes through the building (or minimize) as per the thermal comfort required. The building envelope should be designed to support the choices.
6	Shading (Section 7.4)	Solar shading is used to prevent the heat gains and can be used in form of blinds, louvers, vegetation, overhanging facades, cantilevered floors, etc. The type of shading depends on the climate, facade, orientation and the building design. For temperate climates, movable shading is preferred in order to remove it during the heating season to make optimal use of solar gains.
7	Organization of floor plans	To reduce the transmission and ventilation losses and to optimize the solar gains, partitioning and zoning of the dwellings must be done. Portioning in certain compartments can avoid unnecessary heating and ventilation of certain rooms. Zoning refers to organizing rooms close to each other that have similar thermal condition requirements. The areas that need solar gains or daylight shall be placed on the sunny side of the building.
8	Double-skin facades	This refers to extra glazing, which acts as a buffer between the exteriors and the interiors, thus improving the energy performance of the building and adding a comfort to the inhabitants. This is suitable for moderate and cold climatic conditions. An additional advantage of this is that they also act as acoustic barrier between the outside and the inside.

Sr. No	Measure	Principle
9	Atrium	An atrium is a large roof, mostly glazed, that reduces the thermal energy losses through the façade. This is less suited to the hot climatic zones because there is a chance of overheating. In the rest of the climatic zones, overheating can be avoided by adding a provision of openable parts in the atrium and also by solar shading devices.
10	Greenhouse	A greenhouse is an enclosed space usually made of glass to benefit from solar heat. When a conservatory/greenhouse is not climatized, it can benefit the energy performance of the building by pre-heating the ventilation air. This is also less advisable in hot climatic zones because, like the atrium, of the chances of overheating. The best-suited orientation is towards south.
11	Façade air collector	These are made of glass, where a glass pane is placed in front of the absorber with a cavity in between. The glass acts as an insulator and also allows solar insolation to heat the absorber. A fan is used in the cavity to allow the air flow or this can also be done by stack effect of the heated air. Multiple other modes can also be employed in which the façade collector can be use, making it suitable for different climatic conditions. 1. Fresh outside air moves into the cavity, and gets heated and introduced to the interiors. Thus, this incoming air is preheated, which helps to increase the indoor comfort and also decrease the energy demand. 2. Circulating indoor air through the cavity > passive heating. 3. Inside air is forced to the outside through the cavity, resulting in passive cooling.
12	Light tube (Figure 3.5)	This is internally equipped with reflective material and is connected to the outside through the building roof (Section 3.2.9).
13	Building shape	The compact building (i.e., smaller outer surface compared to its volume) means lower transmission losses through the building envelope. With this, the building will take time to heat up or cool down. The compactness also reduces the number of structural connections resulting in a lower chance for thermal bridges and air leakages.
14	Building mass	The heavier the building structure is the more energy is stored in building mass; thus the buffer is greater. This type of structure takes time to heat up or cool down. In contrast, a lightweight structure has low thermal mass and heats up or cools down quickly. Heavy structure is suitable to conditions where the temperature difference between day and night is relatively high.
15	Trombe walls (Section 6.3.2.2)	These are thermal storage, heat-absorbing, sun-facing walls, mostly made of dark, heat absorbent material.
Passive cooling		
16	Concrete core activation	This methods implies the use of thermal mass made of concrete (ceilings, floors, walls) to passively cool and contribute to the heating of the building.
17	Earth tube ventilation (Section 7.15)	Earth tubes use the relatively constant temperature of the ground to preheat or cool the incoming ventilation air. An air duct is placed about 1.5 m depth, and fresh air is supplied through it which heats or cools down by the ground temperature. During summertime, the heat from the warm intake air is absorbed by the ground earth, resulting in drop of temperature in intake air by about 10°C. During the winter months, this method uses the relatively constant temperature of the ground to preheat or cool the incoming ventilation air.
18	Night ventilation	During the hot season, the cool outside air (at off-sunshine hours) removes the heat from the building that was accumulated during the sunshine hours through cross-ventilation and thermal drafts (stack effect) that are enabled by openable windows or other openable façade elements.

General Introduction

Sr. No	Measure	Principle
19	Green roofs (Section 7.8)	Green roofs provide additional cooling by reducing the amount of solar insolation on the building roof. This reduces the amount of solar gains from the roof to the building. Green roofs also helps to reduce the urban heat island. They do not provide additional insulation to the roof
20	Cool roof (Section 7.8)	Roofs made of light colors are advisable in hot climatic zones since they reflect the sunlight and result in reduction in roof temperature. This reduces the heat transfer from roof to inside of the building.
21	Double roof	Double roof means placing a roof over an existing roof to shade it. It is opens on the sides, which allows the warms air between both the roofs to escape. This composition reduces the cooling demands in hot climates. Extending the roof surface will shade the exterior walls creating shaded outdoor spaces.
22	Evaporative cooling (Section 7.8.2)	The incoming air is allowed to pass over a damp surface before it enters the interiors thus cooling the incoming air stream, thus cooling the building with natural ventilation. Also, this concept can be used as an evaporative cooling tower in which the water is evaporated at the top, creating a downdraft of cooled air into the building. This means combining the evaporative cooling with wind towers. Evaporative cooling reduces the demands for cooling and can (partly) replace mechanical cooling. This principle is mainly suitable for hot and dry climates (deserts) because of their low humidity.
23	Vegetation (Section 7.4)	Shading by trees and vegetation reduces the surface temperature of the shaded area and reduces the reflected sunlight. The temperature of the surrounding air reduces by evaporation of water through the leaves.

1.8.3 STAGE 2: REUSE

Let us understand various heat losses in a building, i.e., transmission losses and ventilation (along with infiltration) losses. Also, domestic hot water, such as water from the shower, is wasted. Therefore, these losses should be recovered. Here are a few methods for heat recovery.

Sr. No	Measure	Principle
1	Greenhouse	This measure is used to recover the transmission loss as the losses are captured in the greenhouse. The greenhouse gets warmer when compared to the outside ambient temperature. This warm air can be used as the incoming air for the ventilation, thus a part of the losses can be regained and reused. Also, this greenhouse measure benefits from the passive solar gains. In the sunshine hours, it heats up, and this reduces the heat losses in winter months. During summers and the intermediate seasons (autumn and spring), the inside air of the greenhouse becomes too warm, and this heat can be captured to reuse it instead of opening the vents/windows to release this heat. This can be done by using a heat exchanger/heat collector [30].
2	Recovery ventilation by heat exchangers	Mechanical exhaust ventilation can be used to recover the thermal energy used for heating/cooling of the building before it is allowed to exit from the building. This recovered energy is used to preheat/cool the incoming ventilation air. The most common type used is the counter flow heat exchanger. A convector or radiator for heating is also a heat exchanger. Sometimes, an additional heat pump is required to boost the temperature to a useful temperature [30].

Sr. No	Measure	Principle
3	Heat pumps	Heat pumps covert heat sources at low temperatures to high temperatures; they are used for space heating and domestic hot water.
4	Mechanical ventilation system with mechanical air inlet and outlet	Mechanical ventilation boxes having four connections can be installed to recover the ventilation losses. The connections can be described as (i) fresh cold air coming in, (ii) warm waste air coming from the house, (iii) preheated fresh air supplied to the rooms and (iv) cooled-down waste air blown out. A heat exchanger is placed inside the box where the cold incoming air is preheated by the warm exhaust air, thus decreasing the ventilation losses by about 90%.
5	Ground duct ventilation	The incoming air for ventilation is led through a duct buried under the ground which is preheated or precooled to the ground temperature before it is released into the interiors. It is most effective in moderate climate with cold winters and warm summers. It is also suitable in places where the difference between day and night temperature is huge. Places with a high outdoor temperature are not suitable because the temperature difference between the air and the soil is too small [30].
6	Energy storage	Heat cold storage, i.e., seasonal storage of thermal energy underground; this principle can also be used for cold storage. Systems commonly used are aquifer thermal energy storage or borehole thermal energy storage. Short-term storage water tanks are usually comprised of a barrel for storing hot water to 80°C–90°C for several days. This is combined with solar collectors Home storage battery: when combined with photovoltaic cells, the battery will increase the self-sufficiency and the amount of self-consumption of renewable energy for a household.
7	Energy programming	Energetic synergy through exchange of heat and cold between functions: the use of waste heat/cold from buildings to heat/cool other buildings. Buildings with, for example, year-round cooling demand, like a data centers or supermarkets, have an abundance of heat produced by the cooling installations. This heat can be extracted and used for low temperature heating of other functions.

1.8.4 STAGE 3: PRODUCE

To generate renewable energy, this can be done by

1. Solar energy:
Solar energy is the most commonly available source of energy for buildings as it is everywhere, and the building's skin is exposed to it. Therefore, this solar energy can be converted into useful energy, to produce heat and electricity. Following are the methods to utilize this abundant source of energy.

- With use of photovoltaic (PV) panels that can convert solar energy into electrical energy. These panels can be placed on the building roof or integrated with the roof or façade of the building. The DC electrical energy received from the PV needs to be converted into AC that can be used by the electricity grid and thereby by the building's installations. The PV orientation should be south, and the output depends on the seasonal and daily weather conditions. They produce only electricity.
- Photovoltaic thermal systems (PVT) are the integration of solar thermal collectors and PV panels in one system, which generates both thermal and electrical energy.

The thermal collector is integrated at the rear side of the panel. The fluid is running through the collector to cool down the PV panel and thereby increase its output. These PVT panels are more expensive than standalone PV or solar collectors. They produce both electricity and heat.

- Solar collectors: These collectors absorb the solar radiation and convert it into the heat. A fluid water-glycerol mixture (antifreeze) or water (depending on the climate) is running through pipes and absorbs the heat of the sun. The solar collector is often combined with the storage tank. They produce higher temperatures than the PVT modules.

The application of solar energy products on buildings is done by mounting/integrating them with the building elements. This contributes to sustainability and, if planned properly, also enhances the building's appearance. To enhance the building's appearance there are different type of fully integrated PVT roof tiles, colored PV modules, thin films etc., available in the market. The amount of solar radiation is different at different locations. It is more at the locations that are near to equator. The highest annual production of solar energy can be reached when solar products are maximally oriented towards the sun. The orientation of the modules also influences the amount of solar radiation. This is often on pitched roofs facing the equator, south-oriented in the northern hemisphere and north-oriented in the southern hemisphere or on tilted solar modules on flat roofs. The further from the equator, the larger the tilt should be for a maximized production.

Energy Academy, Groningen, Netherlands (Figure 1.11) demonstrates an example of utilization of solar energy by a building structure. The electrical power produced by the roof has been maximized by orienting the PV modules east and west on a south-facing roof. By doing this, space left on the roof for daylighting although the yield per module is around 20% less compared to optimally positioned modules; the total yield of the roof is 30% more this way. The positioning of the modules also results in more evenly distributed energy production during the day and during the year [35]. Chapter 8 gives few more examples of building integrated/mounted with photovoltaic panels.

2. Small-scale wind turbines, biomass, environmental energy (like ground heat exchange, surface water cooling, balanced heat cold storage) etc. are other measures apart from solar energy.

FIGURE 1.11 Energy Academy, Groningen, Netherlands [36].

1.9 CASE STUDY OF AN ENERGY NEUTRAL BUILDING [37, 38]

The Pulse is an educational building located on the campus of TU Delft University (Figure 1.12a), Netherlands. The system emphasized on active learning, so there are not long lectures but based on group learning. Few of the measures are taken for their design processes which are listed below:

The building layout/zoning is one of the important aspects of a sustainable building. Initially, the classrooms were planned in the south-west side of the building, but later, it was decided to create open work spots. This decision was influenced by the solar path analysis according to which this side had the receiving sun from 11:00 a.m. until the sunset, keeping it warm. Therefore, the educational spaces were planned in the north-east to be occupied by large groups of students for longer durations of time on the second floor. For this reason, it made sense to organize them to a relatively cool orientation.

On the second floor, the infrequently used open spaces were placed in the south-west. Therefore, daylight is desired, but a bit of sunshine also does not matter. In the middle, utility spaces like lifts, stairs, toilets and technical plant rooms were planned, along with large lecture halls that require mechanical air treatment and where too much daylight is not desirable. The catering facilities were planned on the ground floor adjacent to the square on the north side. This is locally climatized so that it does not impact the educational activities.

Open spaces were planned in the west side of the building façade. The façade had shading screens to avoid overheating. On the northern façade side, north indirect sunlight let in through the design made it suitable for graphical work requiring better light. A large open space/multifunctional space for various activities, lectures and presentations was oriented towards the north side. Figure 1.12(b) gives a 3D sectional view of the building.

The interiors of the classrooms reflected three elements of the climate design. (i) heating and cooling, (ii) lighting and (iii) ventilation. The heating/cooling was planned through the under-floor system. Ventilation has been managed through carbon dioxide measurement. That means the more students present, the greater the ventilation. Lastly, the lighting is directly connected to the direct current coming from the photovoltaic panels. Thus, there is no need to convert it to alternating current, which would otherwise cause the loss of s significant amount of energy. Daylight access has been done through the ceiling, which has a shed shape and allows the north light, while the tilted side is very important for the building. This is because the power station of the building (photovoltaic panels) is located on the roof. The sheds have south-oriented photovoltaic panels and from the north side they capture the daylight. These daylight catchers can be seen from within the building.

The openings on the façade have been designed to make optimum use of daylight, thus decreasing the demand on artificial means of lighting. An educational building requires a lot of electricity, and this demand is catered by solar power by using around 490 solar panels on an area of 750 m^2 with an annual yield of 150,000 kWh [39].

In the climate of the Netherlands, heating and cooling are required in winter and summer, respectively, which has been established through aquifer thermal heat storage in the underground. The aquifer is a sand layer in-between clay layers, which carries water. They take the heat out from the building in summers and store it in a well to reuse it in winter months. That means warm water from summer is extracted from the warm well, it passes the heat exchanger, a heat pump boosts the temperature to the desired level, and warm water runs through the underfloor heating to heat the building. Cold remaining water goes into the cold well. This basically sums up the main energy system. During the summers, the cold water that has been stored in winters is extracted from the cold well in the aquifer and reused. It goes through the heat exchanger, and the building is cooled through the underfloor system and remaining warm water is pumped into the warm well. Thus, this system makes the building sustainable in its heating and cooling system. The building also has a green inner wall with acoustical measures taken up.

General Introduction

(a)

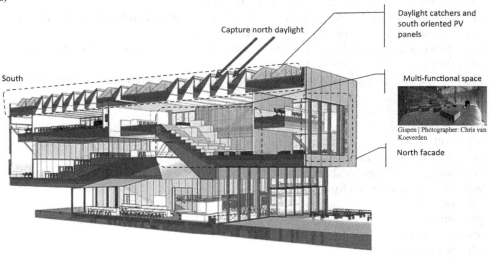

(b)

FIGURE 1.12 Energy neutral education building—TU Delft, The Netherlands. (a) Pulse, Netherlands [39]. (b) 3D sectional view of Pulse [39, 40].

OBJECTIVE QUESTIONS

1.1 In the terrestrial region, the wavelength range of solar radiation (I_N) shall be considered between
   (a) 0.23–2.26 μm
   (b) 3–30 μm
   (c) 0.3–3 μm
   (d) none of these

1.2 The relation between zenith (ϕ_z) and solar altitude $E\ a$) angles is
    (a) $\theta_z + \alpha = 60$
    (b) $\theta_z + \alpha = 90$
    (c) $\theta_z + \alpha = -90$
    (d) $\theta_z + \alpha = 0$

1.3 The energy generated at the core of the sun is due to
    (a) fission reaction
    (b) fusion reaction
    (c) conduction
    (d) radiation

1.4 Solar radiation is measured in
    (a) extraterrestrial region
    (b) terrestrial region
    (c) atmosphere
    (d) none of the above

1.5 Which climatic classification belongs to mean monthly temperature $< 25°C$, RH $>55\%$, precipitation >5 and number of clear days <20.
    (a) hot and dry
    (b) warm and humid
    (c) moderate
    (d) cold and cloudy

1.6 The classifications of weather conditions are based upon
    (a) the ratio of daily diffuse to daily global radiation and the number of sunshine hours
    (b) the ratio of daily diffuse to daily global radiation
    (c) the ratio of daily global to daily diffuse radiation and the number of sunshine hours
    (d) the ratio of daily global to daily diffuse radiation

1.7 To conserve the energy in a building for summer climate, exterior (exposed) walls should be insulated from
    (a) outside
    (b) inside
    (c) both sides
    (d) none of the above

1.8 To conserve fossil fuel for thermal heating of a building for winter climate, exterior (exposed) walls should be insulated from
    (a) outside
    (b) inside
    (c) both sides
    (d) none of the above

ANSWERS

1.1 (a)
1.2 (c)
1.3 (b)
1.4 (a)
1.5 (d)
1.6 (a)
1.7 (a)
1.8 (b)

General Introduction

PROBLEMS

1.1 Evaluate the declination angle (δ) for March 31, 2014.
Hint: Use Equation (1.3a).

1.2 Evaluate the hour angle (ω) at 2.30 pm.
Hint: Use Equation (1.3b).

1.3 Find out the hourly direct radiation (I_b) on (i) a horizontal surface and (ii) on an inclined surface with an inclination of $45°$ on January 15th, 2014 in the terrestrial region.
Hint: $\alpha = 90 - \theta_z$; for horizontal surface, $I_b = I_N \cos \theta_z$; for inclined surface, $I'_b = I_N \cos \theta_i$.

1.4 Derive an expression for the number of sunshine hours (N).
Hint: Use $\theta_z = 90°$ (sunset sunrise) in Equation (1.3f) with $\omega = \omega_s$; and 1 hour = $15°$, $N = 2\omega_s$

1.5 Give a few examples of places with hot and dry climatic conditions
Hint: Table 1.5.

1.6 Identify the energy demand of a building and draw the energy balance of the same.
Hint:

- Choose a building you want to make zero energy: it should be existing, with a heating/cooling system; the building should not be too big or complicated.
- Determine the functions that require energy in your building and the energy carriers that are used to supply the energy needed.

1.7 Analyze your own local climate and choose passive energy-saving design strategies that are appropriate for the climate and your building.

1.8 Analyze examples of vernacular architecture of your region.

1.9 Select the most suitable energy reuse measures for your building.

REFERENCES

[1] IPCC, "AR5 synthesis report: climate change 2014," 2014.
[2] A. V. D. Dobbelsteen, "Introduction," Dezign Ark, 2020a. [Online]. Accessed May 2021: https://www.youtube.com/watch?v=WVXfBO2CrAA&list=PLuhw3IJ5k9rjwAka_pI5Y-1wJlPcS8yad.
[3] N. Gupta and G. N. Tiwari, "Energy matrices of building integrated photovoltaic thermal systems: a case study," *Journal of Architectural Engineering*, 2017.
[4] Ecofys, "Towards nearly zero-energy buildings," Am Wassermann 36: ECOFYS Germany, 2013.
[5] Z. Liu, Q. Zhou, Z. Tian and B.-j. He, "A comprehensive analysis on definitions, development, and policies of nearly zero energy buildings in China," *Renewable and Sustainable Energy Reviews*, vol. 114, p. 109314, 2019.
[6] W. Gilijamse, *Zero-Energy Houses in the Netherlands*, Madison, WI, 1995.
[7] P. Torcellini, S. Pless and M. Deru, "Zero energy buildings: A critical look at the definition," National Renewable Energy Laboratory, 2006.
[8] S. Kilkis, "A new metric for net-zero carbon buildings," *ASME 2007 Energy Sustainability Conference*, Long Beach, CA, 2007.

[9] J. Laustsen, "Energy efficiency requirements in building codes, energy efficiency policies for new buildings," International Energy Agency (IEA), 2008.

[10] A. Marszal and P. Heiselberg, "A literature review of zero energy building (ZEB) definitions," Department of Civil Engineering, Aalborg University, DCE Technical reports No. 78 Denmark, 2009.

[11] G. N. Tiwari, A. Tiwari and Shyam, *Handbook of Solar Energy*, Springer, 2016.

[12] G. N. Tiwari, *Solar Energy: Fundamentals, Design, Modelling and Applications*, New Delhi: Narosa, 2002.

[13] G. N. Tiwari and R. Mishra, *Advanced Renewable Energy Sources*, UK: RSC Publishing, 2012.

[14] A. N. Celik and T. Muneer, "Neutral network based method for conversion of solar radiation data," *Energy Conversion and Management*, vol. 67, pp. 117–124, 2013.

[15] J. A. Duffie and W. A. Beckman, *Solar Engineering of Thermal Processes*, 2nd ed., New York: John Wiley & Sons, 1991, pp. 119–121.

[16] P. Ineichen, R. R. Perez, R. D. Seal, E. L. Maxwell and A. Zalenka, "Dynamic global-to-direct irradiance conversion models (RP-644)," *ASHRAE Transactions*, vol. 98, pp. 346–353, 1992.

[17] G. N. Tiwari and S. Dubey, *Fundamentals of Photovoltaic Modules and Their Applications*, U.K., Royal Society of Chemistry, 2010.

[18] N. Gupta and G. N. Tiwari, "Effect of heat capacity on monthly and yearly energy performance of building integrated semitransparent photovoltaic thermal system," *Journal of Renewable and Sustainable Energy*, vol. 9, 1589–1601, 2017.

[19] H. Singh and G. N. Tiwari, "Evaluation of cloudiness' haziness factor for composite climate," *Energy*, pp. 1589–1601, 2005.

[20] B. Agrawal and G. N. Tiwari, *Building Integrated Photovoltaic Thermal Systems: For Sustainable Developments*, UK: RSC Publishing, 2010.

[21] A. V. D. Dobbelsteen, "Passive and smart bioclimatic design," Dezign Ark, 2020b [Online]. Accessed May 2021: https://www.youtube.com/watch?v=fVI4ELvtdcg.

[22] N. Gupta and G. N. Tiwari, "Review of passive heating/cooling systems of buildings," *Energy Science and Engineering*, vol. 4, pp. 305–333, 2016.

[23] P. Littlefair, M. Santamuris, S. Alvarez, A. Dupagne, D. Hall, J. Teller, J.-F. Coronel and N. Papanikolaou, *Environmental Site Layout Planning: Solar Access, Micro Climate and Passive Cooling in Urban Areas* BRE Press, 2000.

[24] M. Sodha, N. Bansal, A. Kumar and M. Malik, *Solar Passive Building*, UK: Pergamon Press, 1985.

[25] A. V. D. Dobbelsteen, "History and principles of bioclimatic design," Dezign Ark, 2020c. [Online]. Accessed December 2020: https://www.youtube.com/watch?v=w0T2xeWRQJE&list=PLuhw3IJ5k9rjw Aka_pI5Y-1wJlPcS8yad&index=8.

[26] Wikipedia, 2020. [Online]. Accessed 2020: https://upload.wikimedia.org/wikipedia/commons/b/be/The_ Esplanade_4%2C_Singapore%2C_Dec_05.JPG.

[27] L. Hodges, The Hodges residence: performance of a direct gain passive solar home in Iowa, Private communication, 1983.

[28] J. K. Paul, *Passive Solar Energy Design and Materials*, Park Ridge, NJ: Noyes Data Corporation, 1979.

[29] E. V. D. Ham, "Analysing the current energy consumption," in *Zero-Energy Design an Approach to Make Your Building Sustainable*, EDX, 2020a.

[30] E. V. D. Ham, "Heat recovery," Dezign Ark, 2020b. [Online]. Available: https://www.youtube.com/watc h?v=LU64E8yw6JY&list=PLuhw3IJ5k9rjwAka_pI5Y-1wJlPcS8yad&index=18.

[31] F. Kreith and D. Y. Goswami, *Energy Management and Conservation Handbook*, CRC Press, Taylor & Francis, 2008. https://www.taylorfrancis.com/books/edit/10.1201/9781315374178/energy-management-conservation-handbook-frank-kreith-yogi-goswami

[32] EDX, "Zero-energy design: An approach to make your building design sustainable," DelftX: ZEBD01x, 2020.

[33] A. V. D. Dobbelsteen, "The climate," in *Zero-Energy Design: An Approach to Make Your Building Sustainable*, EDX, 2020d.

[34] A. V. D. Dobbelsteen, "History and principles of bioclimatic design," in *Zero-Energy Design: An Approach to Make your Building Sustainable*, EDX, 2020e.

[35] S. Broersma, "Produce-Sun," in *Zero-Energy Design: An Approach to Make Your Building Sustainable*, EDX, 2020.

General Introduction

[36] R. Zijlstra, "ArchDaily," 2016. [Online]. Available: https://www.archdaily.com/891205/energy-academy-europe-broekbakema-plus-de-unie-architecten/5ab3a185f197ccab090001f4-energy-academy-europe-broekbakema-plus-de-unie-architecten-photo.

[37] A. V. D. Dobbelsteen, "Pulse," in *Zero-Energy Design: An Approach to Make Your Building Sustainable*, EDX, 2020f.

[38] A. V. D. Dobbelsteen, "Zero-energy design: An approach to make your building sustainable," in *The Zero-Energy Design of Pulse*, EDX, 2020g.

[39] TUDellft, "Space for future science," [Online]. Accessed December 2020: https://campusdevelopment.tudelft.nl/en/project/pulse/.

[40] Gispen, 2020. [Online]. Available: https://www.gispen.com/en/projects/tu-delft-pulse-page.

2 Basic Heat Transfer

2.1 INTRODUCTION

Thermal energy is freely available from the sun (Chapter 1). The energy transfer takes place due to the temperature differences (ΔT). The science that deals with determining the rates of such energy transfer is known as heat transfer. The rate of heat transfer depends on the magnitude of temperature gradient in a certain direction. The higher the temperature difference, the higher the rate of heat transfer will be. The heat/thermal energy transfer takes place from the surface with the higher temperature to the surface with lower temperature (first and second laws of thermodynamics). The heat transfers occur through mechanical and electrical processes, and it is very important to understand the basic principles involved. In this chapter, fundamentals of basic heat transfers have been reviewed. Heat may be transferred by (a) conduction, (b) convection and (c) radiation. All aforementioned modes are involved in an overall heat transfer. These heat transfers are different in nature and are governed by different laws. It is also important to note the following:

- Each heat transfer mode acts independently, and
- Only one heat transfer mode behaves strongly in comparison to others.

Usually, all three methods are involved in an overall heat transfer problems.

2.2 CONDUCTION

Heat conduction is a process of energy transfer between the particles of a body in direct contact with each other (solids, liquids or gases) and that have a temperature difference. It is a molecular process, i.e., heat is transferred from one molecule to the other, and there is negligible movement of particles of the body. In gases and liquids, there is a collision of molecules due to their random motion. However, in solids, conduction is due to the combination of molecular vibrations in a lattice and the transport by free electrons. For example, if a cold canned drink is placed inside a warm room, its temperature eventually is raised to the room air temperature because of the heat transfer from the room to the drink through the can material via conduction. The rate of heat conduction through a medium depends on the following:

- Thickness of the medium,
- Material of the medium and
- The temperature difference across the medium.

For example, the rate of heat loss from an insulated hot water tank is reduced. The thicker the insulation, the smaller the heat loss will be. The rate of heat transfer can be increased by increasing the temperature gradient between the tank and the surrounding air. Also, the rate of heat loss will be more for the tank with the larger surface area.

DOI: 10.1201/9780429445903-2

2.2.1 Temperature Field

As discussed, it is clear that the heat conduction process takes place when there is a temperature difference. Analytical examination of heat conduction amounts to the study of space-time variations of temperature (T) namely, the determination of the temperature field, which is given by:

$$T = f\left(x,y,z,t\right) \tag{2.1}$$

In Equation (2.1), the temperature varies in time and space, which is a characteristic of the transient condition known as transient temperature field. However, if the temperature of the body does not vary with time and is a function if spatial coordinates only, the temperature field is known as steady state condition and may be expressed as:

$$T = f\left(x,y,z,t\right) \quad \text{or} \quad \frac{\partial T}{\partial t} = 0 \tag{2.2}$$

2.2.2 Fourier's Heat Conduction Equation

The basic equation for steady state heat conduction is known as Fourier's equation. Let us consider steady state heat transfer through a large vertical plane wall with the following design parameters:

- Change in thickness, $\Delta x = L$ in m
- Surface area, A in m^2
- Thermal conductivity of material, K in W/mK
- Temperature difference, $\Delta T = (T_2 - T_1)$ in °C

The direction of the heat flux is normal to the surface and is positive in the direction of decreasing temperature. The rate of heat conduction through a plane layer is proportional to temperature difference across the layer and the heat transfer area, but it is inversely proportional to the thickness of the layer (L) and may be expressed as:

$$Rate\ of\ heat\ conduction \propto \frac{\left(Cross\text{-}sectional\ area\right)\left(Temperature\ difference\right)}{Change\ in\ thickness\ of\ slab} \tag{2.3}$$

Equation (2.3) may be rewritten as:

$$\dot{Q} = -KA\frac{\Delta T}{\Delta x} \tag{2.3a}$$

where the constant K is the proportionality constant or thermal conductivity of the materials (Appendix E). It is the ability to transport thermal energy from a higher temperature to a lower temperature through conduction. In the limiting case of $\Delta x \to 0$, Equation (2.3a) reduces to the differential form and is referred as Fourier's law of heat conduction. The heat flux (Q) may be expressed as:

$$\dot{Q} = -KA\frac{dT}{dx} \tag{2.3b}$$

where $\frac{dT}{dx}$ is slope (temperature gradient) of the temperature curve on a (T–x) diagram at point x.

It may be seen from Equation (2.3b) that the rate of heat conduction in a direction is proportional

Basic Heat Transfer

to the temperature difference in that direction. Equation (2.3b) has a negative sign to make the heat transfer in positive direction. Heat is conducted in the direction of decreasing temperature, and the temperature gradient becomes negative when temperature decreases with increasing x.

2.2.3 THERMAL CONDUCTIVITY

As discussed, thermal conductivity is a physical property of a material and is given in Appendix E. The rate of heat transfer due to conduction under steady conditions (Equation 2.3a) may also be viewed as the defining equation for thermal conductivity. The thermal conductivity of a material (K) may be defined as the rate of heat transfer per unit surface area ($A = 1$ m^2) per unit thickness ($\Delta x = 1$ m) between two points having a unit temperature difference ($\Delta T = 1°C$). It is the measure of how fast heat travels through material. The higher the thermal conductivity, the higher the rate of heat transfer through the material will be. A material is a good heat conductor if it has a higher value of thermal conductivity; it is a poor heat conductor (or insulator) if the value of K is low. Thermal conductivity depends on the temperature, and this may be assumed to be linear as expressed:

$$K = K_0 \left[1 + \beta \left(T - T_0 \right) \right]$$ (2.4)

where

K_0 is the thermal conductivity at temperature T_0 and
β is a constant for the material.

The values of K will increase for $T > T_0$ and decrease for $T < T_0$. However, the value of K is unaffected during the medium temperature range of renewable energy technologies. In general, the conductivity of a gas increases (positive β) with increase in the temperature, however, it decreases for solid or liquid (negative β) with increase in the temperature. But there are a few exceptions to this generalization. Thermal conductivity is significantly influenced by moisture content of the material. Experimental studies have shown that it increases considerably with an increase in moisture content.

2.2.4 THERMAL DIFFUSIVITY

Thermal diffusivity (α) in m^2/s may be defined as the property of a material that represents how fast heat diffuses through a material. It is the thermal conductivity divided by density and specific heat capacity at constant pressure as given in Equation (2.5).

$$\alpha = \frac{Heat\ conducted}{Heat\ stored} = \frac{K}{\rho C_p}$$ (2.5)

where

Heat storage (or heat capacity, ρC_p) gives the energy stored by the material per unit volume
C_p is the specific heat of the material in J/kgK and
ρ is the density of the material in kg/m^3.

Any material will have high thermal diffusivity if it has a high thermal conductivity or a low heat capacity. High thermal diffusivity of the material means faster propagation of heat into the medium, whereas if the value is small, it means that most of the heat is absorbed by the material and only a small amount of heat will be further conducted. Appendix E gives thermal diffusivity (α) of various materials.

2.2.5 Conductive Heat Transfer Coefficient

Equation (2.3b) for the rate of heat transfer through conductance ($\dot{Q}_k$) in W may be rewritten as:

$$\dot{Q}_k = A\frac{K}{L}(T_2 - T_1) \tag{2.6}$$

In the above equation, the negative sign has been removed because the length has no negative sign. Equation (2.6) may be written for the rate of heat transfer per unit area in W/m² due to conduction as below:

$$\dot{q}_k = \frac{\dot{Q}}{A} = h_k(T_2 - T_1) \tag{2.6a}$$

where h_k is heat transfer coefficient due to conduction in W/m²°C.

Heat transfer coefficient through conduction (h_k) may be defined as the rate of heat transfer due to conduction for unit area and unit temperature difference in W/m²°C and may be expressed as:

$$h_k = \frac{K}{L} = \frac{\dot{q}_k}{(T_2 - T_1)} \tag{2.6b}$$

EXAMPLE 2.1

Evaluate the rate of heat transfer ($\dot{Q}_k$) through a rectangular plane wall 0.1524 m thick with thermal conductivity (K) of 0.432 W/mK in a steady state for uniform surface temperatures of $T_1 = 21.1°C$ and $T_2 = 71.1°C$.

Solution

From Equation (2.6), we have the following expression for $\dot{Q}_k$

$$\frac{\dot{Q}_k}{A} = \frac{K(T_2 - T_1)}{L} = \frac{0.432 \times 50}{0.1524} = 141.73\,\text{W/m}^2$$

The value of conductive heat transfer coefficient (h_k) is given by

$$h_k = \frac{K}{L} = \frac{0.432}{0.1542} = 2.80\,\text{W/m}^2°C$$

Here, it is important to mention that the unit of heat transfer coefficient can be expressed either in W/m²°C or in W/m²K due to cancellation of 273 in temperature difference.

2.2.6 Dimensionless Heat Conduction Parameters

2.2.6.1 Biot Number (Bi)

A Biot number is a unitless parameter used in heat conduction problems and is expressed as:

$$\text{Bi} = \frac{hL}{K} = \frac{h}{K/L} \tag{2.7}$$

where K is the thermal conductivity of the solid (W/mK)

Basic Heat Transfer   45

or

$$\mathrm{Bi} = \frac{Heat\ transfer\ coefficient\ at\ the\ surface\ of\ the\ solid}{Internal\ conductance\ of\ solid\ across\ length\ L} \tag{2.7a}$$

From the Equation (2.7a), the Biot number may be defined as the ratio of the convective heat transfer coefficient at the surface of a solid body to the conductive heat transfer coefficient within the body. This parameter is important in cases where a solid body is immersed in hot fluid for heating. Initially, the outer surface of the solid is heated via convection, and then the heat is transferred to the inner parts of the body through conduction. In terms of thermal resistance, it may also be defined as the ratio of thermal resistance faced by conductive heat transfer to the thermal resistance faced by the convective heat transfer. This means if the value of Biot number is small, then there will be lesser resistance to heat conduction, and thus, there will be small temperature gradients within the body.

2.2.6.2 Fourier Number

The Fourier number may be expressed as:

$$\mathrm{Fo}\frac{\alpha t}{L^2} = \frac{K\left(\dfrac{1}{L}\right)L^2}{\rho S L^3 / t} \tag{2.8}$$

or,

$$\mathrm{Fo} = \frac{Rate\ of\ heat\ conduction\ across\ L\ in\ volume\ L^3, \mathrm{W}/{}^{\circ}\mathrm{C}}{Rate\ of\ heat\ storage\ in\ volume\ L^3, \mathrm{W}/{}^{\circ}\mathrm{C}} = \frac{h_c}{K/L} \tag{2.8a}$$

Thus, the Fourier number may be defined as a measure of the rate of heat conduction in comparison with the rate of heat storage in a given volume (L^3). Therefore, the larger the Fourier number, the deeper the penetration of heat into a solid over a given period of time.

EXAMPLE 2.2

Estimate Biot number (Bi) for heat transfer as 3473 W/m²°C and 2.83 W/m²°C for water and air as medium with $L = 0.9$ m for the following conditions:
 (i)   Insulating solid surface ($K = 0.04$ W/m°C)
 (ii) Concrete solid surface ($K = 1.279$ W/m°C)
 (iii) Metallic solid surface ($K = 386$ W/m°C)

Solution

The expression for Biot number (Bi), Equation (2.7 and 2.7a), is given by

For water as a fluid:

$$\mathrm{Bi} = \begin{cases} \dfrac{3473 \times 0.9}{0.04} = 78{,}007.7 & \textit{for insulation} \\[2ex] \dfrac{3473 \times 0.9}{1.279} = 2445.97 & \textit{for concrete} \\[2ex] \dfrac{3473 \times 0.9}{386} = 8.1 & \textit{for copper} \end{cases}$$

For air as a fluid:

$$\text{Bi} = \begin{cases} \dfrac{2.83 \times 0.9}{0.04} = 63.67 & \textit{for insulation} \\[2ex] \dfrac{2.83 \times 0.9}{1.279} = 1.99 & \textit{for concrete} \\[2ex] \dfrac{2.83 \times 0.9}{386} = 6.60 \times 10^{-3} & \textit{for copper} \end{cases}$$

2.3 CONVECTION

The heat transfer due to convection occurs only in fluids, i.e., gases and liquids. Heat transfer (at a low temperature) from one place to another place by the movement of fluids is known as thermal convection. This mode of heat transfer occurs when the entire mass of a non-uniformly heated fluid (liquid or gas) is displaced and mixed. The higher the velocity of motion of the fluid, the higher the heat transfer through conduction due to high displacement of the particles of fluid per unit time will be. Since the particles at a different temperature are always in direct contact during the convection process, convection is accompanied by the process of conduction in solar energy applications.

The rate of heat transfers due to convection ($\dot{Q}_c$) between the fluid and the boundary surface in watts may be expressed as:

$$\dot{Q}_c = h_c A \Delta T \tag{2.9}$$

where,

h_c is the convective heat transfer coefficient (W/m²°C), and

ΔT is the temperature difference between the body surface and its surroundings. The rate of heat transfer per unit area ($\dot{q}_c$) in W/m²°C due to convection is given by:

$$\dot{q}_c = \frac{\dot{Q}_c}{A} = h_c \left(T_2 - T_1 \right) \tag{2.9a}$$

Also, the expression for h_c can be expressed as:

$$h_c = \frac{\dot{q}_c}{\left(T_2 - T_1 \right)} \tag{2.9b}$$

2.3.1 DIMENSIONLESS HEAT CONVECTION PARAMETERS

The value of a convective heat transfer coefficient (h_c) may be evaluated using the following dimensionless parameters depending on physical properties of fluid above the solid surface due to boundary problems.

There is a variation in density due to the temperature variation between the liquid and the surface in contact resulting in buoyancy. The fluid motion, thus produced is free convection (or natural).

However, if the fluid motion is due to the external force like fans, compressors, etc., and is independent of the temperature difference in the fluid, it is referred to as forced convection.

Basic Heat Transfer

2.3.1.1 Nusselt Number (Nu)

The ratio of the convective heat transfer coefficient to the conductive heat transfer coefficient for fluids is defined as the Nusselt number as given in Equation (2.10). It is required in order to understand the dominance of heat transfer (convective or conductive) for fluids, thus it is an important parameter for convective heat transfers. It characterizes the process of heat transfer at the wall–fluid boundary and may be expressed as:

$$\mathrm{Nu} = \frac{h_c X}{K} \qquad (2.10)$$

where,

K is the thermal conductivity of fluid over the solid surface (W/mK),
h_c is convective heat transfer coefficient between solid and fluid (W/m²K), and
X is characteristic dimension of the system.

Nusselt number (Equation 2.10) is different from the Biot number (Equation 2.7) as the latter includes thermal conductivity of solid, whereas the Nusselt number considers thermal conductivity of fluid.

2.3.1.2 Reynolds Number (Re)

The Reynolds number may be defined as the ratio of the fluid dynamic force (ρu_0^2) to the viscous drag force ($\mu u_0/X$) and is applicable for forced convection mode. It is necessary to evaluate the Reynolds number since it indicates the type of flow of fluid, which may be either laminar flow or turbulent flow. It characterizes the relation between the forces of inertia and viscosity.

Mathematically it may be expressed as:

$$\mathrm{Re} = \frac{\rho u_0^2}{\mu u_0 / X} = \frac{\rho u_0 X}{\mu} = \frac{u_0 X}{v} \qquad (2.11)$$

where,

ρ is the density in kg/m³,
μ is the dynamic viscosity in Pa. s,
$v = \dfrac{\mu}{\rho}$ is the kinematic viscosity in m²/s, and
X is characteristic length for the system of interest and given in Table 2.1.

2.3.1.3 Prandtl Number (Pr)

Prandtl may be defined as the ratio of momentum diffusivity (μ/ρ) to the thermal diffusivity ($K/\rho C_p$).·It relates fluid motion and heat transfer to the fluid and may be expressed as:

$$\mathrm{Pr} = \frac{\mu/\rho}{K/\rho C_p} = \frac{\mu C_p}{K} \qquad (2.12)$$

where

μ is the dynamic viscosity in Pa. s,
C_p is the specific heat at constant pressure (J/kgK), and
K is the thermal conductivity of fluid over the solid surface (W/mK).

2.3.1.4 Grashof Number (Gr)

The ratio of the buoyancy force to the viscous force is defined as the Grashof number. It represents the effects of hydrostatic lift force and viscous force of the fluid in free convection. It may be expressed as:

$$\text{Gr} = \frac{g\beta'\rho^2 X^3 \Delta T}{\mu^2} = \frac{g\beta' X^3 \Delta T}{v^2} \tag{2.13}$$

where,

β is the coefficient of volumetric thermal expansion in $m^3/m^3{}^\circ C$,
g is the acceleration due to gravitation in m/s^2,
ΔT is the operating temperature difference between the surface and the fluid (in $^\circ C$), and
$v = \dfrac{\mu}{\rho}$ is the kinematic viscosity in m^2/s.

2.3.1.5 Rayleigh Number (Ra)

The Rayleigh number may be defined as the ratio of the thermal buoyancy to viscous inertia. It is given as:

$$\text{Ra} = \text{GrPr} = \frac{g\beta'\rho^2 X^3 C_p \Delta T}{\mu K} \tag{2.14}$$

All the parameters defined in this section may be obtained using the properties of air and water given in Appendix E.

The Reynolds (Re), Prandtl (Pr) and Grashof (Gr) numbers have been calculated by using the physical properties of fluid at the average temperatures (T_f) of the hot surface (T_1) and surrounding air (T_2) given as following:

$$T_f = \frac{T_1 + T_2}{2} \tag{2.15}$$

The thermal expansion coefficient (β') at surrounding air (T_2) for exposed surface and fluid temperature (T_1) for parallel plate may be calculated as below:

$$\beta' = \frac{1}{(T_2 + 273)} \tag{2.16a}$$

$$\beta' = \frac{1}{(T_f + 273)} \tag{2.16b}$$

The characteristic dimension (X) for various shapes (irregular shapes) is calculated as following:

$$X = \frac{A\left(Area\right)}{P\left(Perimeter\right)} \tag{2.16c}$$

The characteristic dimension for a rectangular horizontal surface with dimensions ($L_0 \times B_0$) are determined using the relation given below:

$$X = \left(\frac{L_0 + B_0}{2}\right) \tag{2.16d}$$

TABLE 2.1
Free Convective Heat Transfer of Various Systems [1] (After [2])

System	Schematic Diagram	C'	n	Correlation Factor, K'	Operating Conditions
Horizontal Cylinder		0.47	0.25	1	Laminar flow
		0.1	0.33	1	Turbulent flow
Vertical cylinder with small diameter		0.686	0.25	$[\mathrm{Pr}/(1 + 1.05\,\mathrm{Pr})]^{1/4}$	Laminar flow $\overline{\mathrm{Nu}}_{\mathrm{local}} = \overline{\mathrm{Nu}} + 0.52\left(L/D\right)$
Vertical plate and vertical cylinder with large diameter		0.8	0.25	$\left[1 + \left(1 + \dfrac{1}{\sqrt{\mathrm{Pr}}}\right)\right]^{-1/4}$ $\left[\mathrm{Pr}^{\frac{1}{6}}/(1 + 0.496\,\mathrm{Pr}^{2/3})\right]^{2/3}$	Laminar flow to obtain local Nu, use $C' = 0.6$, $X = x$, formula applicable to vertical cylinder when $D/L > > 38\,\mathrm{Gr}^{-1/4}$ Turbulent flow; to obtain local Nu use $C' = 0.0296$, $X = x$

(*Continued*)

TABLE 2.1 (Continued)
Free Convective Heat Transfer of Various Systems [1] (After [2])

System	Schematic Diagram	C'	n	Correlation Factor, K'	Operating Conditions
Heated horizontal plate facing upward		0.54	0.25	1	Laminar flow $(10^5 < GrPr < 2 \times 10^7)$, $X = (L_0 + B_0)/2$ Laminar flow $(10^7 < GrPr < 10^{11})$, $X = A/P$ for circular disc of diameter D, use $X = 0.9D$
	$L_1 = L_0$ $L_2 = B_0$	0.14	0.33	1	Turbulent flow $(2 \times 10^7 < GrPr < 3 \times 10^{10})$, $X = (L_0 + B_0)/2$
		0.15	0.33	1	Turbulent flow $(10^7 < GrPr < 10^{11})$, $X = A/P$
Heated horizontal plate facing downward		0.27	0.25	1	Laminar flow only
Moderately inclined plane		0.8	0.25	$\left[\dfrac{\cos\theta}{1 + \left(1 + \dfrac{1}{\sqrt{Pr}}\right)^2} \right]^{1/4}$	Laminar flow (multiply Gr by $\cos\theta$ in the formula for vertical plate)

Basic Heat Transfer

Two vertical parallel plates at the same temperature 	0.04	1	$(d/L)^3$	Air layer
Hollow vertical cylinder with open ends 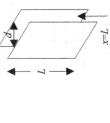	0.01	1	$(d/L)^3$	Air column
Two horizontal parallel plates hot plate uppermost 	0.27	0.25	1	Pure conduction $\dot{q} = K(T_h - T_c)/d$ Laminar flow (air) $(3 \times 10^5 < \text{GrPr} < 3 \times 10^{10})$
Two concentric cylinders 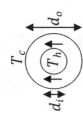	0.317	0.25	$\left[X^3 \left(\dfrac{1}{d_i^{3/5}} + \dfrac{1}{d_o^{3/5}} \right)^5 \right]^{-1/4}$	Laminar flow

$X = \frac{1}{2}(d_o - d_i)$
$A = 2\pi X L$

(*Continued*)

TABLE 2.1 (Continued)
Free Convective Heat Transfer of Various Systems [1] (After [2])

System	Schematic Diagram	C'	n	Correlation Factor, K'	Operating Conditions
Two vertical parallel plates of different temperatures (h for both surfaces)	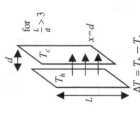 $\Delta T = T_h - T_c$	0.18	0.25	$\left(\dfrac{L}{d}\right)^{-1/9} (\mathrm{Pr})^{-1/4}$	Laminar flow (air) ($2 \times 10^4 < \mathrm{Gr} < 2 \times 10^5$)
		0.065	—	$\left(\dfrac{L}{d}\right)^{-1/9} (\mathrm{Pr})^{-1/3}$	Turbulent flow (air) ($2 \times 10^5 < \mathrm{Gr} < 2 \times 10^7$)
Two inclined parallel plates	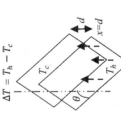 $\Delta T = T_h - T_c$			$\overline{\mathrm{Nu}} = \left[\dfrac{\overline{\mathrm{Nu}}_{\mathrm{vert}}\cos\theta + \overline{\mathrm{Nu}}_{\mathrm{horz}}\sin\theta}{2}\right]$	
Two horizontal parallel plates cold plate uppermost	 $\Delta T = T_h - T_c$	0.195	0.25	$\mathrm{Pr}^{-1/4}$	Laminar flow (air) ($10^4 < \mathrm{Gr} < 4 \times 10^5$)
		0.068		$\mathrm{Pr}^{-1/3}$	Turbulent flow (air) $\mathrm{Gr} > 4 \times 10^5$

Basic Heat Transfer

EXAMPLE 2.3

Estimate the convective heat transfer coefficient for a horizontal rectangular surface (1.0 m × 0.8 m), which is maintained at 134°C. The hot surface is exposed to (a) water and (b) air at 20°C.

Solution

For this exercise, in both cases, the average temperature, Equation (2.15), $T_f = (134 + 20)/2 = 77°C$ and the characteristic dimension $(L = X) = \dfrac{1.0 + 0.8}{2} = 0.90$ m will be same either for water or air as a fluid.

(a) For water

From Appendix E, the water thermal properties at $T_f = 77°C$ will be

$$\mu = 3.72 \times 10^{-4}\,kg\,/\,ms;\ K = 0.668\,W\,/\,m.K;\ \frac{\rho = 973.7\,kg}{m^3};\ Pr = 2.33\ and$$

$$\beta' = 1/(77 + 273) = 2.857 \times 10^{-3}\,K^{-1}$$

From Equation (2.13), the Grashof number can be calculated as

$$Gr_L = \frac{g\beta'\rho^2(\Delta T)X^3}{\mu^2} = \frac{9.8 \times 2.857 \times 10^{-3}(973.7)^2 \times 114 \times (0.9)^3}{(3.72 \times 10^{-4})^2} = 1.594 \times 10^{13}$$

This is a turbulent flow and for heated plate facing upward, the values of $C = 0.14$ and $n = 1/3$ (Table 2.1). Now, a convective heat transfer coefficient can be calculated as

$$h_c = \frac{K}{L}(0.14)(Gr_L\,Pr)^{1/3} = \frac{0.668}{0.9}(0.14)(1.594 \times 10^{13} \times 2.33)^{1/3} = 3467\,W\,/\,m^2 K$$

(b) For surrounding air

By using the physical properties of air at $T_f = 77°C$ (Appendix E) and $L = 0.90$ m

$$Gr_L.Pr = \frac{(9.8) \times (134 - 20) \times (0.90)^3 \times (0.697)}{(293) \times (2.08 \times 10^{-5})^2} = 4.51 \times 10^9$$

Using Table 2.1 for hot surface facing upward and turbulent flow condition, the heat transfer coefficient can be calculated as

$$h_c = \left(\frac{K}{L}\right) \times 0.14 \times (Gr_L\,Pr)^{0.333} = \left(\frac{0.03}{0.9}\right) \times (0.14) \times (4.91 \times 10^9)^{0.333} = 2.83\,W/m^2°C$$

It is important to note that the convective heat transfer coefficient changes from 3467 $W/m^2°C$ to 2.83 $W/m^2°C$ with the change of fluid from water to air for given parameters because it depends on physical properties of fluid.

EXAMPLE 2.4

Estimate the rate of heat loss from a horizontal rectangular surface (1.0 m × 0.8 m), which is maintained at 134°C. The hot surface is exposed to a plate placed at a distance of 0.10 m above it and temperature of plate is maintained at 20°C.

Solution

The average film temperature, $T_f = (134 + 20)/2 = 77°C$

From Appendix E, at $T_f = 77°C$; $v = 20.8 \times 10^{-6}\,m^2/s$; $K = 0.30\,W/mK$

Pr = 0.697 and $\beta' = 1/(77 + 273) = 2.857 \times 10^{-3}\,K^{-1}$ due to exposure to another plate and the characteristic dimension $(X = d) = 0.10\,m$

$$Gr = \frac{9.8 \times \left(2.857 \times 10^{-3}\right)\left(134 - 20\right)\left(0.1\right)^3}{\left(20.8 \times 10^{-6}\right)^2} = 7.377 \times 10^6$$

Using Table 2.1 for turbulent flow condition for two horizontal parallel plates (cold plate uppermost), the heat transfer coefficient can be calculated as

$$h = \frac{K}{L}(0.068)\left(Gr_L\,Pr\right)^{1/3} Pr^{-1/3}$$
$$= \frac{0.030}{0.10}(0.068)\left(7.377 \times 10^6\right)^{1/3} = 3.97\,W/m^2K$$

Hence, the rate of heat loss from a vertical wall is,

$$\dot{Q} = hA\left(T_s - T_\infty\right) = (3.97)(1.0 \times 0.8)(134 - 20) = 362.14\,W$$

From this example, one can observe that there is a change in the numerical value of the convective heat transfer coefficient from 7.47 W/m²K to 3.97 W/m²K due to a change in condition from open to close.

EXAMPLE 2.5

Estimate the rate of heat transfer from a horizontal rectangular surface (1.0 m × 0.8 m), which is maintained at 134°C. The hot surface is exposed to water at 20°C.

Solution

The average film temperature, $T_f = (134 + 20)/2 = 77°C$
From Appendix E, water thermal properties are at $T_f = 77°C$;

$$\mu = 3.72 \times 10^{-4}\,kg/ms; K = 0.668\,W/mK; \rho = 973.7\,kg/m^3; Pr = 2.33$$

and

$$\beta' = \frac{1}{77 + 273} = 2.857 \times 10^{-3}\,K^{-1}$$

Basic Heat Transfer

Consider the characteristic dimension $(X = d) = (1.0 + 0.8)/2 = 0.90$ m. The Grashof number is

$$\text{Gr} = \frac{g\beta'\rho^2 X^3 \Delta T}{\mu^2} = 1.594 \times 10^{13}$$

This is a turbulent flow, hence using Table 2.1 for a heated plate facing upward for $X = L = (L_0 + B_0)/2$, consider $C = 0.14$ and $n = 1/3$. Now, a convective heat transfer coefficient can be calculated as

$$h = \frac{K}{L}(0.14)(\text{Gr}_L \text{Pr})^{\frac{1}{3}} = \frac{0.668}{0.9}(0.14)(1.594 \times 10^{13} \times 2.33)^{\frac{1}{3}} = 3467 \text{ W/m}^2\text{K}$$

Hence, the rate of heat loss from a vertical wall is

$$\dot{Q} = hA(T_s - T_\infty) = (3467)(1.0 \times 0.8)(134 - 20) = 316,190.4 \text{ W} = 316.19 \text{ kW}$$

EXAMPLE 2.6

Water at an average bulk temperature of 80°C flows inside a 0.020 m ID circular tube with average bulk velocity 2.0 m/s. The pipe wall temperature is at each point 40°C below the local value of T_b (= 80°C), and the cooling takes place at constant heat flux. Estimate the rate of energy loss from a 4 m length of tubing.

Solution

The fluid properties at $T_b = 80$°C and $T_s = 40$°C are

$$v_b = 3.64 \times 10^{-7} \text{ m}^2/\text{s}, Pr_b = 2.22, K_b = 0.6676 \text{ W/mK}$$

$v_s = 6.58 \times 10^{-7} \text{ m}^2/\text{s}, \rho_b = 974 \text{ kg/m}^2, \rho_s = 995 \text{ kg/m}^3$ (Refer Appendix E)

The Reynolds number using Equation (2.11) and using $x = D$ is:

$$\text{Re} = \frac{Dv}{v_b} = \frac{(0.020)(2.0)}{3.64 \times 10^{-7}} = 1.10 \times 10^5, \text{highly turbulent}$$

Since, $T_s - T_b = 40$°C, and nothing that for $L/D = 4/0.02 = 200 > 60$, we have,

$$\bar{h} = \frac{K_b}{D}(0.027)(\text{Re})^{0.8}(\text{Pr})^{1/3}\left(\frac{\mu_b}{\mu_s}\right)^{0.14} \text{ (From Table 2.4)}$$

$$= (0.6676/0.02)(0.027)(1.10 \times 10^5)^{0.8}(2.22)^{\frac{1}{3}}\left[\left(\frac{3.64}{6.58}\right)\left(\frac{974}{995}\right)\right]^{0.14} = 11.65 \text{ W/m}^2\text{K}$$

Now using Equation (2.9), $\dot{Q}_{conv} = \bar{h}\pi DL(T_s - T_b) = -117.06 \text{ kW}$.

2.3.2 TYPES OF CONVECTION

2.3.2.1 Free Convection

The free/natural flow in the fluid is due to the heterogeneity of the mass forces acting upon the volume. When this natural flow is not limited within a boundary, it is known as free convection. For free convection, therefore, the terrestrial gravitational field, acting on the fluid with a non-uniform density distribution owing to the temperature difference between the fluid and the surface in contact, causes the fluid motion. Other types of body forces, such as centrifugal forces and Coriolis forces, also have influence on free convective heat transfer, particularly in rotating systems. The coefficient of heat transfer (h), usually combined with Nusselt number as given in Equation (2.10), depends on the type of flow (laminar or turbulent, free or forced).

$$\mathrm{Nu} = C'\left(\mathrm{GrPr}cos\beta\right)^{n} K' \tag{2.17a}$$

K' governs the entire physical behavior of the problem [3].

Equation (2.17) is obtained using dimensional analysis at the boundary layer. Values of C' and n have been assessed by using experimental data for systems with the same geometrical shapes and sizes. Few geometrical shapes used in solar thermal technology are given in Table 2.1. K' directs the entire physical behavior of the problem [3]. Some empirical relations used for free convection are also given in Table 2.1.

It has been found that average free-convection heat transfer coefficients can be represented in the following functional form for a variety of configurations:

$$\overline{\mathrm{Nu}}_f = C\left(\mathrm{Gr}_f\,\mathrm{Pr}_f\right)^{m} \tag{2.17b}$$

Table 2.2 gives the values of constants C and m to be used in Equation (2.17b) for various conditions.

Some empirical relations used for free convention are also given in Table 2.3 [1].

2.3.2.2 Forced Convection [4]

As discussed, when the fluid motion is due to artificially induced force and is independent of the temperature difference in the fluid, it is referred to as "forced convection". The rate of forced convection may be increased by flowing the fluid over the hot surface using an external source of energy

TABLE 2.2
The Value of C and m for Different Geometries [5]

Geometry	$\mathrm{Gr}_f\mathrm{Pr}_f$	C	m	Reference
Vertical planes and cylinders	10^4–10^9	0.59	0.25	[3]
	10^9–10^{13}	0.10	0.33	[6]
Horizontal cylinder	10^4–10^9	0.53	0.25	[3]
	10^9–10^{12}	0.13	0.33	[3]
	10^{-10}–10^{-2}	0.675	0.058	[7]
	10^{-2}–10^2	1.02	0.148	[7]
	10^2–10^4	0.85	0.188	[7]
Upper surface of heated plates or	2×10^4–8×10^6	0.54	0.25	[8]
lower surface of cooled plates	8×10^6–10^{11}	0.15	0.33	[8]
Lower surface of heated plates or	10^5–10^{11}	0.27	0.25	[8]
upper surface of cooled plates				[9]
Vertical cylinder, height = diameter	10^4–10^6	0.775	0.21	[10]

Basic Heat Transfer

TABLE 2.3
Simplified Equations for Free Convection from Various Surfaces to Air at Atmospheric Pressure [1, 4]

Cases	Surface	Schematic diagram	Laminar $10^4 < Gr_f Pr_f < 10^9$	Turbulent $Gr_f Pr_f > 10^9$
1	Horizontal hot plate facing upward or cool plate facing downward		$h = 1.32(\Delta T/L)^{1/4}$	$h = 1.52(\Delta T)^{1/3}$
2	Hot plate facing downward or cool plate facing upward		$h = 0.59(\Delta T/L)^{1/4}$	
3	Vertical plane and cylinder		$h = 1.42(\Delta T/L)^{1/4}$	$h = 1.31(\Delta T)^{1/3}$
4	Horizontal cylinder		$h = 1.32(\Delta T/d)^{1/4}$	$h = 1.24(\Delta T)^{1/3}$

such as a pump (liquid) or fan (air/gas). The external energy is supplied to maintain the process in which there are two types of forces (a) the fluid pressure related to flow velocity $\left(\frac{1}{2}\rho v^2\right)$ and (b) the frictional force produced by viscosity (μ. dv/dy) This type of heat transfer is affected by the type of flow, whether laminar or turbulent, and motion of the fluid, which are related to the Reynolds number and Prandtl number, respectively. Nusselt number under forced convection mode may be expressed as:

$$\text{Nu} = C\left(\text{RePr}\right)^n K \tag{2.18a}$$

where,

C and n are constants for a given type of flow and geometry, and
K is a correction factor (shape factor) added, to obtain a greater accuracy.

The empirical relation for forced convective heat transfer through cylindrical tubes may be expressed as:

$$\bar{\text{Nu}} = \frac{hD}{K_{th}} = C\,\text{Re}^m\,\text{Pr}^n\,K \tag{2.18b}$$

TABLE 2.4
The Value of Constants for Forced Convection [5]

Cross-section	D	C	m	n	Correlation Factor, K'	Flow	Operating Conditions
circle, diameter d	d	3.66	0	0	1	Laminar flow	Long tube $Re < 2000, Gz < 10$
		1.86	0.33	0.33	$(d/l)^{1/3}(\mu/\mu_w)^{0.14}$	Laminar flow	Short tube $Re < 2000, Gz > 10$
		0.027	0.8	0.33	$(\mu/\mu_w)^{0.14}$	Turbulent flow	Highly viscous liquids $0.6 < Pr < 100$
		0.023	0.8	0.4	1	Turbulent flow	Gases $Re > 2000$

where,

$D = 4\,A/P$, is the hydraulic diameter in m,
P is the perimeter of the section in m, and
K_{th} is the thermal conductivity in W/mK.

The values of C, m, n and K for various conditions are given in Table 2.4 after Yong [11]

Fully developed laminar flow in tubes at constant wall temperature give the following relation:

$$\mathrm{Nu_d} = 3.66 + \frac{0.0668\left(d/L\right)\mathrm{RePr}}{1 + 0.04\left[\left(d/L\right)\mathrm{RePr}\right]^{2/3}} \tag{2.19a}$$

The heat transfer coefficient, thus calculated from Equation (2.19a) is the average value over the entire length of the tube. When the tube is sufficiently long the Nusselt number approaches a constant value of 3.66.

For the plate heated over its entire length, average Nusselt number is given by:

$$\overline{\mathrm{Nu}}_L = 0.664\,\mathrm{Re}_L^{1/2}\,\mathrm{Pr}^{1/3} \tag{2.20b}$$

The thermo-physical properties of water (or any base fluid) can be improved for higher values of heat transfer coefficient by mixing nanoparticles in water (or base fluid).

EXAMPLE 2.7

Calculate an average convective heat transfer coefficient and the rate of heat transfer per m^2 from hot plate of 1 m length to the flowing water (0.20 m/s) if the hot plate temperature is 27.8°C above the flowing water temperature. The water physical properties at the fluid film temperature are: $\nu = 7.66 \times 10^{-7}\,\mathrm{m^2/s}$, $K = 0.621$ W/mK, $Pr = 5.13$.

Solution

The Reynolds number can be calculated from Equation (2.11) as,

$$\mathrm{Re_L} = \frac{(0.20)(1.0)}{7.66 \times 10^{-7}} = 261{,}096$$

Basic Heat Transfer

Since the flow is laminar, average convective heat transfer coefficient can be evaluated from Equation (2.20b) as

$$\bar{h} = \frac{K}{L}(0.664)\text{Re}_L^{1/2}\text{Pr}^{1/3}$$

$$= \frac{0.621}{1.0}(0.664)(261{,}096)^{1/2}(5.13)^{1/3} = 363.45\ \text{W/m}^2\text{K}$$

Using Equation (2.9), the average rate of heat transfer per m² to the water is given by:

$$\frac{\dot{Q}}{A} = \bar{h}\Delta T = (363.45)(27.8) = 10.104\ \text{kW/m}^2.$$

2.3.2.3 Mixed-Mode Convection

There may be many situations where the convective heat transfer is neither free nor forced in nature. This may occur when a fluid is forced to flow over a hot surface at low velocity. Coupled with the forced-flow velocity is a convective velocity, which is generated by the buoyant forces resulting from a reduction in fluid density near the heated surface. A high value of the Reynolds number means a large forced-flow velocity; thus free convection currents are of low influence, whereas a high value of the Grashof-Prandtl number product implies a larger influence of the free convection mode.

A correlation for the mixed convection, laminar flow region of flow through horizontal tubes was developed by Brown and Gauvin [11] and is expressed as:

$$\text{Nu} = 1.75\left(\frac{\mu_b}{\mu_w}\right)^{0.14}\left[\text{Gz} + 0.012\left(\text{GzGr}^{1/3}\right)^{4/3}\right]^{1/3} \tag{2.21}$$

where,

Gz = RePrd/L, is the Graetz number (dimensionless),
μ_b is the fluid viscosity at bulk temperature in (Pa.s), and
μ_w is the fluid viscosity at wall temperature in (Pa.s).

The general concept, which is applied in combined-convection analysis, is that the predominance of a heat-transfer mode is governed by the fluid velocity associated with that mode. A forced-convection situation involving a fluid velocity of 30 m/s, for example, would be expected to overshadow most free convection effects encountered in ordinary gravitational fields because the velocities of the free-convection currents are small compared to 30 m/s. On the other hand, a forced-flow situation at very low velocities (~0.3 m/s) might be influenced appreciably by free-convection currents. A general criterion is that when $\dfrac{\text{Gr}}{\text{Re}^2} > 10$, free convection is of primary importance.

2.4 CONVECTIVE HEAT TRANSFER COEFFICIENT

Various researchers have analyzed the heat transfer from a plate that is exposed to the outside winds. The convective heat transfer coefficient [3] may be expressed as:

$$h_c = 5.7 + 3.8V \text{ for } 0 \le V \le 5\,\text{m/s} \tag{2.22}$$

where V is the wind speed (m/s).

Heat loss by natural convection may be calculated by using Equation (2.22) with wind velocity equal to zero. The entire process is not as simple as it appears to be since the wind may not always be parallel to the surface. Probably [12] the effect of free convection and radiation are included in the above equation, thus Watmuff et al. [13] have proposed the following equation:

$$h_c = 2.8 + 3.0V \text{ for } 0 \le V \le 7\,\text{m/s} \tag{2.22a}$$

The sensibility of these parameters is also demonstrated through a comparison, and another relation for convective heat transfer coefficient is given by

$$h_c = 7.2 + 3.8V \tag{2.22b}$$

Several other correlations are also available in the literature and generally, it is determined from an expression expressed as:

$$h_c = a + vV_c^b \tag{2.22c}$$

where $a = 2.8$, $b = 3$ and $n = 1$ for $V_a < 5$ m/s and $a = 0$, $b = 6.15$ and $n = 0.8$ for >5 m/s.
Note: The source and reference of Equations (2.22) may be obtained from Tiwari [5].

2.5 RADIATION

Thermal radiation involves the transfer of heat through space/vacuum from a body at a higher temperature to another at a lower temperature by electromagnetic waves (0.1 to 100 μm). This process takes place in three stages: (a) a fraction of internal energy of one of the body gets converted into the energy of electromagnetic waves, (b) propagation of these waves in space and (c) absorption of radiant energy by the other body. Since thermal radiation lies in the range of infrared, thus it follows all the rules as that of light. This means it travels in straight line through a homogenous medium, is converted into heat when it strikes anybody which can absorb it and is reflected and refracted according to the same rule as that of light.

2.5.1 RADIATION INVOLVING REAL SURFACES

When the radiant energy (let us consider it to be solar radiation) is allowed to fall on the solid surface, some part of this energy is reflected, another is absorbed and the rest is transmitted through it in case it is a transparent solid body. Thus, the total energy (i.e., reflected, absorbed and transmitted) should be equal to the incident radiation as per the law of conservation of energy as given in Equation (2.23a):

$$I_r + I_a + I_t = I_T \tag{2.23a}$$

On dividing Equation (2.23a) by I_T, it may also be expressed as:

$$\rho' + \alpha' + \tau' = 1 \tag{2.23b}$$

where,

$\rho' = \dfrac{I_r}{I_T}$ is the reflectivity of the intercepting body,

$\alpha' = \dfrac{I_a}{I_T}$ is the absorptivity of the intercepting body, and

$\tau' = \dfrac{I_t}{I_T}$ is the transmittance of the intercepting body.

Reflectivity may be defined as the ratio of the solar energy reflected to the incident solar energy. Absorptivity refers to the ratio of the solar energy absorbed to the incident solar energy. The ratio of the solar energy transmitted to the incident solar energy and is referred as the transmittance of the intercepting body.

For an opaque surface, $\tau' = 0$, therefore from Equation (2.23b) $\rho' + \alpha' = 1$. When the substance absorbs the entire solar energy falling on it, then $\rho' = \tau' = 0$; $\alpha' = 1$ and the substance is known as a black body. Whereas a white body refers to the substance that reflects the entire radiation incident on it, and in this case $\alpha' = \tau' = 0$, $\rho' = 1$.

The energy that is absorbed is converted into heat, and this heated body, by virtue of its temperature, emits radiation according to Stefan-Boltzmann law. The radiant energy emitted per unit area of a surface in unit time is referred to as the emissive power (E_λ). However, this energy is defined as the amount of energy emitted per second per unit area perpendicular to the radiating surface in a cone formed by a unit solid angle between the wavelengths lying in the range $d\lambda$; it is called spectral emissive power (E_λ). Further, emissivity, defined as the ratio of the emissive power of a surface to the emissive power of a black body of some temperature, is the fundamental property of a surface.

2.5.2 Kirchhoff's Law

According to this law, the ratio of emissive power of a body (in thermal equilibrium) to that of a black body at the same temperature is equal to its absorptivity and is expressed as:

$$\frac{E_{b\lambda}}{E_b} = \alpha \ \ or \ \varepsilon = \alpha \tag{2.24}$$

At a given temperature, a body is able to absorb as much incident radiation as it is able to emit. However, this statement does not hold true in the case of incident radiation coming from a source at a different temperature. Further, it applies to surfaces bearing the grey surface characteristics viz. radiation intensity is taken to be a constant proportional to that of a black body. The radiative properties α_λ, ε_λ and ρ_λ are assumed to be uniform over the entire wavelength spectrum.

2.5.3 Laws of Thermal Radiation

Laws of thermal radiation have been obtained for black bodies and conditions of thermodynamics equilibrium.

2.5.3.1 Planck's Law

The emission of energy is dependent on the temperature and is non-uniform in nature. Planck's law establishes a relation between spectral emissive power, wavelength and temperature for the black body at temperature (T) in Kelvin, the magnitude of solar spectrum with function of wavelength (μm) is given by:

$$E_{b\lambda} = \frac{C_1}{\lambda^5 \left[\exp\{C_2 / (\lambda T)\} - 1 \right]} \tag{2.25}$$

where,

$E_{b\lambda}$ represents the energy emitted per unit area per unit time per unit wavelength (μm) interval at a given wavelength,
$C_1 = 3.742 \times 10^8 \ W\mu m^4 / m^2 \ (= 3.7405 \times 10^{-16} \ Wm^2)$ is the Planck's first radiation constant, and
$C_2 = 1.4387 \times 10^4 \ \mu mK \ (= 0.01439 \ mK)$ is the Planck's second radiation constant.

Planck's law has two limiting cases depending on the relative value of C_2 and λT:
Case (a) For $\lambda T \gg C_2$

$$E_{b\lambda} = \frac{C_1}{\lambda^5} \frac{\lambda T}{C_2} \tag{2.25a}$$

Equation (2.25a) is known as Rayleigh-Jeans law
Case (b) For $\lambda T \ll C_2$

$$E_{b\lambda} = \frac{C_1}{\lambda^5} \exp\left(\frac{-C_2}{\lambda T}\right) \tag{2.25b}$$

The variation of $E_{b\lambda}$ for temperature of 6000 K (sun's temperature) and 288 K (Earth's temperature) with wavelength in μm has been shown in Figures 2.1(a) and 2.2(b), respectively. Further the comparison of this radiation from the sun and the earth has been shown in Figure 2.1(c).

2.5.3.2 Wien's Displacement Law

The wavelength corresponding to the maximum intensity of black body radiation for a given temperature (T) is given by this law:

$$\lambda_{\max} T = C_3 \tag{2.26}$$

where $C_3 = 2897.6 \, \mu\text{m K}$

With increase in the temperature, a shift in the maximum black body radiation intensity may be seen towards the shorter wavelength.

2.5.3.3 Stefan–Boltzmann Law

This law relates the hemispherical total emissive power, viz. total energy and temperature. The total emitted radiation, i.e., emissive power from zero to any wavelength (λ) from the sun is expressed as:

$$E_{0-b,\lambda} = \int_0^\lambda E_{b\lambda} d\lambda \tag{2.27}$$

By integrating Planck's law over all wavelengths (between 0 to ∞), the total energy emitted by a black body is found to be:

$$E_{0-b,\lambda} = \int_0^\lambda E_{b\lambda} d\lambda = \sigma T^4 \tag{2.27a}$$

Equation (2.27a) is known as Stefan-Boltzmann law and $\sigma = 5.6697 \times 10^{-8}$ W/m²K⁴ is the Stefan-Boltzmann constant.

2.5.3.4 Sky Radiation

An equivalent black body temperature is defined to evaluate the radiation exchange from a horizontal surface directly exposed to the sky. This is due to non-uniform temperature of the atmosphere, and it radiates only in certain wavelength regions. Thus, the net radiation exchange $\left(\dot{Q}\right)$ in watts between horizontal surface (T_1) with emittance (ε), area (A) and sky radiation (T_{sky}) is expressed as:

$$\dot{Q} = A\varepsilon\sigma\left(T_1^4 - T_{sky}^4\right) \tag{2.28}$$

Basic Heat Transfer

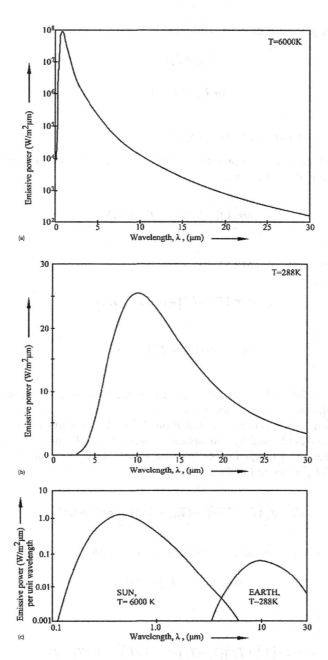

FIGURE 2.1 Effect of temperature of black body on emissive power [1].

The equivalent sky temperature (T_{sky}) has been expressed in terms of ambient air temperature in Equation (2.28a) given by Swinbank [14]. These expressions are only approximations.

$$T_{sky} = 0.0552 T_a^{1.5} \tag{2.28a}$$

where T_{sky} and T_a are both in degrees Kelvin.

Another commonly used relation by Whillier [15] is given as

$$T_{sky} = T_a - 6 \tag{2.28b}$$

$$\text{or } T_{sky} = T_a - 12 \tag{2.28c}$$

2.5.4 RADIATIVE HEAT TRANSFER COEFFICIENT

The radiant heat exchange between two infinite parallel surfaces per m^2 at temperatures T_1 and T_{sky} may be given from Equation (2.29) as:

$$\dot{q}_r = \varepsilon\sigma\left[\left(T_1 + 273\right)^4 - \left(T_{sky} + 273\right)^4\right] \tag{2.29}$$

The above equation may be rewritten as:

$$\dot{q}_r = \varepsilon\sigma\left(T_1^4 - T_a^4\right) + \varepsilon\sigma\left(T_a^4 - T_{sky}^4\right) \tag{2.29a}$$

$$\text{or } \dot{q}_r = \varepsilon\sigma\left(T_1^4 - T_a^4\right) + \varepsilon\Delta R \tag{2.29b}$$

where $\Delta R = \sigma[(T_a + 273)^4 - (T_{sky} + 273)^4]$ is the difference between the long wavelength radiation exchange between the horizontal surface at temperature T_a and the sky temperature at T_{sky}.

According to Wien's displacement law, Equation (2.26), the emitted radiation will be long wavelength radiation, which is blocked by atmosphere because T_a and T_{sky} are at low temperatures.

The value of values of ΔR is equal to 60 W/m^2 for $T_a = 25°C$ and by using the expression for relation between T_a and T_{sky} given in Equations (2.28a–c).

$$\Delta R = \sigma\left[\left(T_a + 273\right)^4 - \left(T_{sky} + 273\right)^4\right] = 60 \text{ W/m}^2 \tag{2.29c}$$

Further, after linearization of the first term of Equation (2.28),

$$\dot{q}_r = h_r\left(T_1 - T_a\right) + \varepsilon\Delta R \tag{2.30}$$

where,

$$h_r = \varepsilon\sigma\left(T_1^2 + T_2^2\right)\left(T_1 + T_2\right) = \varepsilon\left(4\sigma\overline{T}\right)^3 \quad for\,\overline{T}_1 \cong \overline{T}_2 \tag{2.30a}$$

and, $\varepsilon = \dfrac{1}{\varepsilon_1} + \dfrac{1}{\varepsilon_2} - 1$, for two parallel surfaces

$= \varepsilon$, for surface exposed to atmosphere

ε_1 and ε_2 are the emissivity of the two surfaces with one of the surface as sky. The numerical value of ΔR becomes zero for the surfaces not directly exposed to sky condition. The reduction of $\dot{q}_r$ in the form of Equation (2.30) will enable one to find an exact closed-form solution for T_1. It may be noted here that this solution is based on the assumption that T_a and T_{sky} are constant.

Basic Heat Transfer

65

EXAMPLE 2.8

Determine the rate of long wavelength radiation exchange (ΔR) between the ambient air ($T_a = 15°C$) and sky temperature.

Solution

$$\text{Using,} \Delta R = \sigma\left[\left(T_a + 273\right)^4 - \left(T_{sky} + 273\right)^4\right]$$

From Equations (2.28a–c), we have

$$T_{sky} = 0.0552(15)^{1.5} = 3.2°C$$

$$= 15 - 6 = 9°C$$

$$= 15 - 12 = 3°C$$

$$\text{Now,} \Delta R = 5.67\times10^{-8}\left[\left(15+273\right)^4 - \left(3.2+273\right)^4\right] = 60.11\,\text{W/m}^2$$

$$= 5.67\times10^{-8}\left[\left(15+273\right)^4 - \left(9.0+273\right)^4\right] = 31.50\,\text{W/m}^2$$

$$= 5.67\times10^{-8}\left[\left(15+273\right)^4 - \left(3.0+273\right)^4\right] = 61.06\,\text{W/m}^2$$

Here, it is important to mention that the value of T_{sky} is nearly the same for two cases, hence the numerical value of ΔR should be considered as 60 W/m².

EXAMPLE 2.9

Find out the radiative heat transfer coefficient between the surface of a wall at 25°C and room air temperature at $T_a = 24°C$.

Solution

Since the temperatures are approximately the same, from Equation (2.30a), we have,

$$h_r = 4\varepsilon\sigma T^3 = 4\times5.64\times10^{-8}\times\left(25+273\right)^3$$

$$= 6\,\text{W/m}^2°C$$

EXAMPLE 2.10

Evaluate the maximum monochromatic emissive power at 288 K.

Solution

From Equation (2.26), one gets, $\lambda_{max} = 2897.8/288 = 10.06$ µm.
Now by using Equation (2.25), the maximum monochromatic power can be obtained as follows:

$$E_{b\lambda} = \frac{3.742 \times 10^8}{(10.06)^5 \left[e^{(14,387/2897.6)} - 1 \right]} = 25.53 \text{ W/m}^2\mu\text{m}$$

2.6 EVAPORATION (MASS TRANSFER)

This is different from the cases discussed earlier since one of the surfaces is wetted for mass transfer to either surroundings or any other cooler surface. The rate of heat transfer for evaporation is given as:

$$\dot{Q} = h_{cw} \left(T_w - T_a \right) \tag{2.31}$$

where

T_w is fluid temperature, and
T_a is air temperature.

In the process of evaporation, transfer of mass takes place from one location of the system to another location, and the rate of mass transfer may be expressed as:

$$\dot{m} = h_D \left(\rho_w^0 - \rho_a^0 \right) \tag{2.32}$$

where,

$\dot{m}$ is the rate of mass flow per unit area in (kg/m²s),
h_D the mass transfer coefficient in [(kg/s)(m²/kg/m³)],
ρ_w^0 is the partial mass density of water vapor in (kg/m³), and
ρ_a^0 is the partial mass density of air in (kg/m³).

According to the Lewis relation for air and water vapor mixture, we know

$$\frac{h_{cw}}{h_D} = \rho^0 C_{pa} \tag{2.33}$$

By assuming $T_w = T_a = T$ at water-air interface, using Equation (2.33), Equation (2.32) becomes

$$\dot{m} = \frac{h_{cw}}{\rho^0 C_{pa}} \frac{M_w}{RT} \left(P_w - P_a \right) \tag{2.34}$$

The rate of heat transfer on account of mass transfer of water vapor is given by:

$$\dot{Q}_{ew} = \dot{m} L \tag{2.35}$$

Basic Heat Transfer

where

L is the latent heat of vaporization in (kJ/kg), and
P_w and P_a are partial pressure of water vapor and air, respectively.

From Equation (2.34) and (2.35), we have

$$\dot{Q}_{ew} = \frac{Lh_{cw}}{\rho^0 C_{pa}} \frac{M_w}{RT} \left(P_w - P_a \right) \tag{2.36}$$

let, $H_0 = \frac{Lh_{cw}}{\rho^0 C_{pa}} \frac{M_w}{RT}$ \hfill (2.36a)

Then Equation (2.36) becomes,

$$\dot{Q}_{ew} = H_0 \left(P_w - P_a \right) \tag{2.36b}$$

Using perfect gas equation $\rho^0 = \dfrac{P_a M_a}{RT}$ for air (for 1 mole of air) and by substituting in Equation (2.36a), we get

$$\frac{H_0}{h_{cw}} = \frac{L}{C_{pa}} \frac{M_w}{M_a} \frac{1}{P_a} \tag{2.36c}$$

For small values of P_w, $P_T = P_a$, Equation (2.36c) becomes

$$\frac{H_0}{h_{cw}} = \frac{L}{C_{pa}} \frac{M_w}{M_a} \frac{1}{P_T} \tag{2.36d}$$

where

L is the latent heat of vaporization, $L = 2200$ kJ/kg,
C_{pa} is the specific heat of air, $C_{pa} = 1.005$ kJ/kg°C,
M_w is the molar mass of water, $M_w = 18$ kg/mol, and
P_T is the total pressure of air-vapor mixture $= 1$ atm $= 101,325$ N/m².

Substituting these values in Equation (2.36d), we get

$$\frac{H_0}{h_{cw}} = 0.013 \tag{2.36e}$$

The best representation of heat and mass transfer phenomenon is obtained if the values of $\dfrac{H_0}{h_{cw}}$ is taken to be 16.276×10^{-3} instead of 0.013 [1]. Thus, the rate of heat transfer on account of mass transfer is expressed as

$$\dot{q}_{ew} = 16.276 \times 10^{-3} \times h_{cw} \times \left(P_w - P_a \right) \tag{2.37}$$

In case the surface is exposed to atmosphere, Equation (2.37) may be written as:

$$\dot{q}_{ew} = 16.276 \times 10^{-3} \times h_{cw} \times \left(P_w - \gamma P_a \right) \tag{2.37a}$$

where γ is the relative humidity of air in fraction.

Further, after linearization of Equation (2.37a), one can have

$$\dot{q}_{ew} = h_{ew}\left(T_w - T_a\right) \tag{2.37b}$$

where the evaporation heat transfer coefficient can be given as:

$$h_{ew} = \frac{16.276 \times 10^{-3} \times h_{cw} \times \left(\bar{P}_w - \gamma \bar{P}_a\right)}{\left(T_w - T_a\right)} \tag{2.38}$$

For wetted surface that is exposed to ambient moving air with velocity (V) then the numerical value of h_{cw} can be considered as given by Equation (2.22a).

The values of partial vapor pressure at temperature T for the ranges of temperature (10°C–90°C) can be obtained from the following expression [16, 17]

$$P(T) = \exp\left[25.317 - \frac{5114}{T + 273}\right] \tag{2.39}$$

EXAMPLE 2.11

Evaluate the rate of evaporative heat transfer coefficient in W/m² °C from wetted surface (35°C) to an ambient air temperature (15°C) with a relative humidity of 50%.

Solution

The vapor pressure $P(T)$ in N/m² at any temperature T in°C can be calculated from Equation (2.39).

Thus, the vapor pressures at wetted and ambient air temperatures can be calculated as

$$P(T_w) = \exp\left(25.317 - \frac{5144}{273 + 35}\right) = 5517.6 \, \text{N/m}^2$$

$$\text{and } P(T_a) = \exp\left(25.317 - \frac{5144}{273 + 15}\right) = 1730 \, \text{N/m}^2$$

Using h_c = 2.8 W/m²°C [Equation (2.22a) for V = 0] and substituting the values in Equation (2.37a), we have the rate of evaporation in W/m² as:

$$\dot{q}_{ew} = 16.273 \times 10^{-3} \times 2.8 \times (5517.6 - 0.5 \times 1730) = 211.99 \, \text{W/m}^2$$

The evaporative heat transfer coefficient can be calculated from Equation (3.47) as

$$h_{ew} = \frac{\dot{q}_{ew}}{(T_w - T_a)} = \frac{211.99}{35 - 15} = 10.60 \, \text{W/m}^2\text{°C}$$

2.7 TOTAL HEAT TRANSFER COEFFICIENT

The sum of heat transfer by conduction (Equation 2.6a), convection (Equation 2.9a), radiation (Equation 2.30) and evaporation (Equation 2.37b) is known as the total heat transfer per m² ($\dot{q}_k$)

Basic Heat Transfer

from any solid surface to surrounding (air). Heat transfer due to conduction may be neglected due to a very small value of thermal conductivity, and total heat transfer may be given as:

$$\dot{q}_T = \dot{q}_c + \dot{q}_r + \dot{q}_{ew} \tag{2.40}$$

Equation (2.40) may also be written in terms of respective heat transfer coefficient as:

$$\dot{q}_T = \left(h_c + h_r + h_{ew} \right) \left[T_s - T_a \right] \tag{2.41}$$

where h_c, h_r and h_{ew} are given in Equations (2.9b), (2.30a) and (2.38), respectively. It is important to mention here that T_s represents the temperature of solid dry/wetted surface as required in the analysis.

In Equation (2.41), the sum of convective (h_c) and radiative (h_r) heat transfer coefficients is also considered as total heat transfer coefficient without the effect of evaporation and is given as:

$$h = h_0 = h_c + h_r = 5.7 + 3.8V \tag{2.42}$$

With effect of evaporation, total heat transfer (i.e., convective + radiative + evaporative) from the solid surface to surrounding may be given as:

$$h = h_c + h_r + h_{ew} \tag{2.43}$$

2.8 OVERALL HEAT TRANSFER COEFFICIENT

The passive house concept is to use the materials as a way of storing and delivering the energy and also by moving within the house air and renewing it when necessary in order not to add more energy for air which is leaking, going outside or suffering without any need. This has been done since centuries by controlling the fenestrations and inside atmosphere either by high insolation or air tightness. Other measures as already discussed in previous chapters, orientation, thermal mass and natural ventilation are important passive controls. All these measures can be handled through the materials for comfort and passive house design. The U-value should be the first thing that should be delivered by the materials. As mentioned in Chapter 1, the overall heat transfer coefficient is the most important thermal property of a building element. This is the minimum analysis that is required for energy efficiency and is very significant to a passive house design.

Heat transfer from one medium at higher temperature to another medium at lower temperature may take place between many layers of different thermo-physical properties and thickness with more than one mode of heat transfer (conduction, convection, radiation and evaporation). Thus, to predict the one-dimensional steady-state heat transfer rate from one medium to a second medium via a third medium overall heat transfer coefficient (U) is used. This is a very important property of building components: it indicates how great the transmission loss per square meter per degree temperature difference between indoor and outdoor is. In order to reduce the U-value, a layer of insulation can be added as given in Section 1.5.2.4

Further, U-value of a glass can be reduced by the following methods [18]:

- Adding additional layers of glass creating one or more air cavities.
- Apply a low-e coating, which reflects the infrared radiation and reduces the heat transmission through the cavity. This coating is a very thin invisible layer of just a few molecules of metal. By adding this coating the U-value of double-glazing can be reduced from 3 to 1.8 W/m^{2}K.
- The air in the cavity can be replaced by another gas with low thermal conductivity (like argon or krypton).

With these technologies modern insulating glass can have a U-value as low as 1 or even 0.5 W/m²K. This value is very low when compared to single glass but high when compared to well-insulated wall. Thus, reducing the size of the window is an important measure to reduce the transmission losses.

2.8.1 Parallel Slabs [5]

Consider a composite wall in which heat is transferred from hot surface at temperature (T_A) to cold surface at temperature (T_B) through different conducting slabs (Figure 2.2). The heat transfer rate ($\dot{Q}$) through a structure in steady-state condition is the same through each layer and may be expressed as:

$$\dot{Q} = Ah_a(T_A - T_0) = \frac{AK_1(T_0 - T_1)}{L_1} = \frac{AK_2(T_1 - T_2)}{L_2} = \frac{AK_3(T_2 - T_3)}{L_3} = Ah_b(T_3 - T_B) \quad (2.44)$$

where terms like $h\Delta T$ represent heat transfer by either convection or radiation or both depending upon situation, and terms like $K(\Delta T/L)$ represent heat transfer by conduction through various layers. Also, from Equation (2.44) the rate of heat transfer per unit area may be given as:

$$\dot{q} = \frac{\dot{Q}}{A} = \frac{T_A - T_0}{R_a} = \frac{T_0 - T_1}{R_1} = \frac{T_1 - T_2}{R_2} = \frac{T_2 - T_3}{R_3} = \frac{T_3 - T_B}{R_b} \quad (2.45)$$

where R's are thermal resistances, which are inversely proportional to respective heat transfer coefficient at various surfaces and layers and are defined by:

$$R_a = \frac{1}{h_a}, R_1 = \frac{L_1}{K_1}, R_2 = \frac{L_2}{K_2}, R_3 = \frac{L_3}{K_3}, R_b = \frac{1}{h_b} \quad (2.46)$$

Equation (2.45) may also be written in the following forms:

$$T_A - T_0 = \dot{q}R_a \quad (2.46a)$$

$$T_0 - T_1 = \dot{q}R_a \quad (2.46b)$$

$$T_1 - T_2 = \dot{q}R_a \quad (2.46c)$$

$$T_2 - T_3 = \dot{q}R_a \quad (2.46d)$$

$$T_3 - T_B = \dot{q}R_a \quad (2.46e)$$

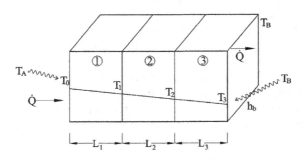

FIGURE 2.2 One-dimensional heat flow through parallel perfect contact slabs [1].

Basic Heat Transfer

Hence, the total thermal resistance (R) in the path of heat flow from temperature (T_A) to temperature (T_B) consists of the sum of several different thermal resistances. From Equations (2.45) and (2.46a–e), we get

$$\dot{q} = \frac{T_A - T_B}{R} = U(T_A - T_B) \tag{2.47}$$

where $R = R_a + R_1 + R_2 + R_3 + R_b$, and U is the overall heat transfer coefficient (W/m²K).

An overall heat transfer coefficient (U) is related to the total thermal resistance (R) of the composite wall by

$$R = \frac{1}{U} = \frac{1}{h_a} + \frac{L_1}{K_1} + \frac{L_2}{K_2} + \frac{L_3}{K_3} + \frac{1}{h_b} \tag{2.48}$$

2.8.2 Parallel Slabs with Air Cavity [5]

Consider a concrete wall/roof structure with an air cavity of air-conductance (C) as shown in Figure 2.3 for the estimation of overall heat transfer coefficient. The thermal air conductance (C) varies non-linearly with thickness of air gap (L) and becomes constant at large air gaps. The graph is given in Figure 2.4.

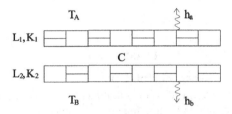

FIGURE 2.3 Configuration of parallel slabs with air cavity [1].

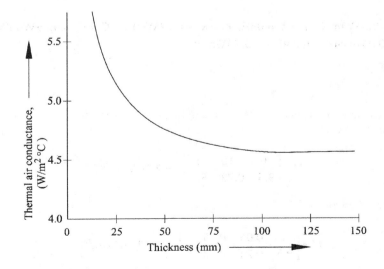

FIGURE 2.4 Variation of thermal air conductance with air gap thickness [1].

The heat is transferred from the hot surface at temperature (T_A) to the cold surface temperature (T_B) as given in Figure 2.3. For steady-state condition, the rate of heat transfer per unit area of walls/roof will be same at each layer boundary and is given by:

$$\dot{Q} = Ah_a\left(T_A - T_0\right) = \frac{AK_1\left(T_0 - T_1\right)}{L_1} = AC\left(T_1 - T_2\right) = \frac{AK_2\left(T_2 - T_3\right)}{L_2} = Ah_b\left(T_3 - T_B\right) \tag{2.49}$$

The equation for U can also be derived for the present case as done in the last case, and its expression is given by:

$$\dot{Q} = UA\left(T_A - T_B\right) \tag{2.49a}$$

$$\text{where } U = \left[\frac{1}{h_a} + \frac{L_1}{K_1} + \frac{1}{C} + \frac{L_2}{K_2} + \frac{1}{h_b}\right]^{-1} = \frac{1}{R} \tag{2.50}$$

In general, the overall heat transfer coefficient (U) for the configuration of parallel slabs with air cavities is given as following:

$$U = \left[\frac{1}{h_a} + \sum_i \frac{L_i}{K_i} + \sum_i \frac{1}{C_i} + \frac{1}{h_b}\right]^{-1} \tag{2.51}$$

EXAMPLE 2.12

Calculate the overall heat transfer coefficient (U) for

(a) Single concrete $(K = 0.72$ W/m °C) slab with thickness (L) of 0.10 m
(b) Two-layered horizontal slab with same material and thickness
(c) Two-layered horizontal slab with air cavity (0.05 m)
(d) Two-layered horizontal slab with two air cavities (each 0.05 m air gap)

for the following parameters:

$$h_a = 9.5 \text{ W/m}^2\text{°C}, \ L_1 = L_2 = 0.05, \ K_1 = K_2 = 0.72 \text{ W/m°C}; \ C_1 = C_2 = 4.75 \text{ W/m}^2\text{°C}$$
for 0.05 m air cavity and $h_b = 5.7$ W/m^2°C.

Solution

An overall heat transfer coefficient in W/m²°C can be calculated as follows:
For single concrete slab,

$$U = \left[\frac{1}{9.5} + \frac{0.10}{0.72} + \frac{1}{5.7}\right]^{-1} = 2.38 \text{ W/m}^2\text{°C}$$

For two-layered horizontal slab,

$$U = \left[\frac{1}{9.5} + \frac{0.05}{0.72} + \frac{0.05}{0.72} + \frac{1}{5.7}\right]^{-1} = 2.38 \text{ W/m}^2\text{°C}$$

Basic Heat Transfer

For two-layered horizontal slab with single air cavity,

$$U = \left[\frac{1}{9.5} + \frac{0.05}{0.72} + \frac{1}{4.75} + \frac{0.05}{0.72} + \frac{1}{5.7}\right]^{-1} = 1.59 \text{ W/m}^2\text{°C}$$

For two-layered horizontal slab with two air cavity gaps separated by metal foil,

$$U = \left[\frac{1}{9.5} + \frac{0.05}{0.72} + \frac{1}{4.75} + \frac{1}{4.75} + \frac{0.05}{0.72} + \frac{1}{5.7}\right]^{-1} = 1.19 \text{ W/m}^2\text{°C}$$

It is clear from above calculation that an increase in the number of air cavities from one to two reduces U from 1.59 W/m²°C to 1.19 W/m²°C, respectively.

2.9 THERMAL CIRCUIT ANALYSIS

The thermal circuit diagram is one of the tools that helps to understand an overall heat transfer coefficient as shown in Figures 2.5 to 2.7 for composite wall/roof.

2.9.1 Composite Wall

A thermal resistance circuit diagram for a composite wall section is given in Figure 2.5(a) is drawn based on the following and given in Figure 2.5(b).

$$\text{Thermal Resistance} \propto \frac{1}{\text{Heat transfer coefficient}}$$

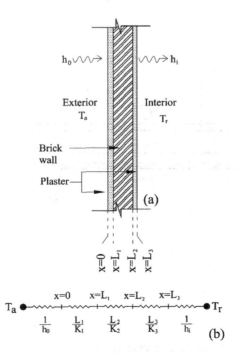

FIGURE 2.5 Thermal resistance circuit diagram of a wall.

Total thermal resistance, $R = \left[\dfrac{1}{h_0} + \dfrac{L_1}{K_1} + \dfrac{L_2}{K_2} + \dfrac{L_3}{K_3} + \dfrac{1}{h_i} \right]$.

Thus, overall heat transfer coefficient, $U = \dfrac{1}{R} = \left[\dfrac{1}{h_0} + \dfrac{L_1}{K_1} + \dfrac{L_2}{K_2} + \dfrac{L_3}{K_3} + \dfrac{1}{h_i} \right]^{-1}$.

2.9.2 Composite Roof

A thermal resistance circuit diagram for a composite roof as given in Figure 2.6(a) with air cavity (C) is drawn based on the following and given in Figure 2.6(b).

Referring to Equations (2.50) and (2.51),

Total thermal resistance, $R = \left[\dfrac{1}{h_0} + \dfrac{L_1}{K_1} + \dfrac{L_2}{K_2} + \dfrac{1}{C} + \dfrac{L_3}{K_3} + \dfrac{1}{h_i} \right]$.

Thus, overall heat transfer coefficient, $U = \dfrac{1}{R} = \left[\dfrac{1}{h_0} + \dfrac{L_1}{K_1} + \dfrac{L_2}{K_2} + \dfrac{1}{C} + \dfrac{L_3}{K_3} + \dfrac{1}{h_i} \right]^{-1}$.

Another example for composite roof with series/ parallel configuration is illustrated in Figure 2.7. In this case, an expression for thermal resistance, and an overall heat transfer coefficient will be as follows:

Total thermal resistance, $R = \left[\dfrac{1}{h_0} + \dfrac{L_A}{K_A} + \dfrac{1}{\dfrac{K_B}{L_B} + \dfrac{K_C}{L_C}} + \dfrac{L_D}{K_D} + \dfrac{1}{h_i} \right]$.

Thus, overall heat transfer coefficient, $U = \dfrac{1}{R} = \left[\dfrac{1}{h_0} + \dfrac{L_A}{K_A} + \dfrac{1}{\dfrac{K_B}{L_B} + \dfrac{K_C}{L_C}} + \dfrac{L_D}{K_D} + \dfrac{1}{h_i} \right]^{-1}$.

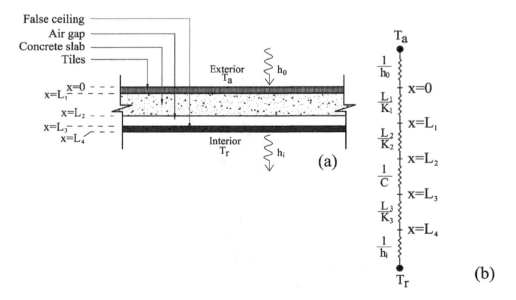

FIGURE 2.6 Thermal resistance circuit diagram of a roof.

Basic Heat Transfer 75

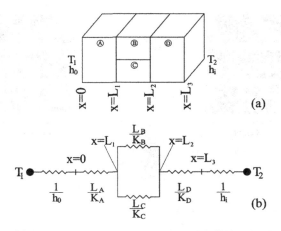

FIGURE 2.7 Thermal resistance circuit diagram of series/parallel configuration of a roof.

2.10 ENERGY BALANCE [19]

Energy balance of any building is comprised of heat gains on one side and heat losses on the other. Let us understand through a few examples.

2.10.1 Energy Balance for Winter's Day

Let us assume that the house is heated at 20°C and the average outside temperature is about 5°C. The temperature difference of about 15°C causes the heat transfer through the building envelope (i.e., roof, wall, windows and the floor). This heat transmission is because of (i) the temperature difference and (ii) thermal properties of the envelope. Energy here is also lost through ventilation, where the outside air is allowed in the house and warm indoor air is exhausted to maintain the indoor air quality. Energy is also lost due to infiltration caused by the cracks and openings in the building envelope. Thus, we can conclude that the heat losses are due to (i) transmission, (ii) ventilation and (iii) infiltration.

Considering the heat gains, we have (i) solar radiation coming inside through the widows/openings, which depends on the orientation and size of the openings and sun shading properties and (ii) heat sources inside the building (like lamps, electrical devices and human beings).

In the winter season, the solar gains and internal heat gains are not sufficient enough to cover the heat losses, thus heating systems (like radiators, heaters, etc.) are used. Typical energy balance in winter season is given in Figure 2.8(a).

Figure 2.8(b) gives the energy balance for a typical winter day with little or no thermal insulation, and it can be noticed that the transmission losses are the highest. And on the gains side, the heating system is the most significant. The heat gains and losses should be equal in any energy balance system. Thus, with addition of an insulation, the transmission losses will be reduced, thus reducing the heating demand by the same amount as given in Figure 2.8(c).

2.10.2 Energy Balance on a Cloudy Day

In continuation of the above for a winter season, on a cloudy day without any sun, the solar gains will reduce to zero, and the heating demand will increase accordingly to compensate for the solar gains as given in Figure 2.8(d), while the losses will remain the same as in Figure 2.8(c).

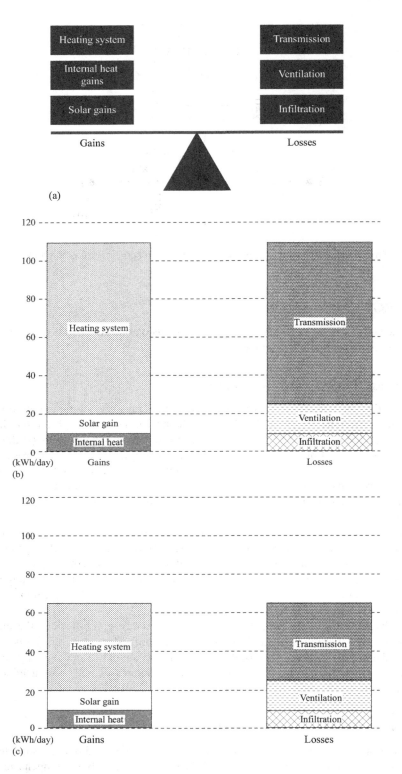

FIGURE 2.8 (a) Energy balance in winter season. (b) Heat gains and losses in winter season without (or little) insulation. (c) Heat gains and losses in winter season with insulation.

(*Continued*)

Basic Heat Transfer

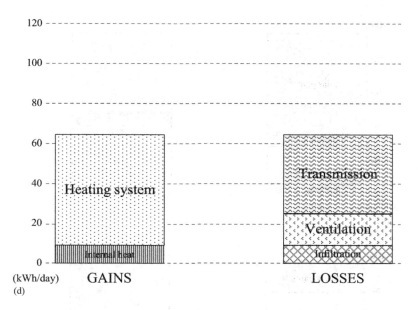

FIGURE 2.8 (*Continued*) (d) Heat gains and losses in winter season with insulation (on a cloudy day).

2.10.3 Energy Balance on a Summer's Day in an Air-Conditioned Building

For a typical day of the summer season, the outside temperature is higher than the inside temperature; thus air conditioning is required to lower the inside temperature to attain comfortable thermal conditions. Here the gains are due to (i) solar gains, (ii) internal heat sources, (iii) the transmission, (iv) ventilation and (v) infiltration. The transmission, ventilation and infiltration in the earlier case (winter season) was at the losses side. They are at the gains side in summers because the temperature of the outside is higher than the inside as already discussed. The heat losses only consist of air conditioning. The energy balance is given in Figure 2.9(a). The heat gains and losses based on Figure 2.9(a) for an air-conditioned residential building are given in

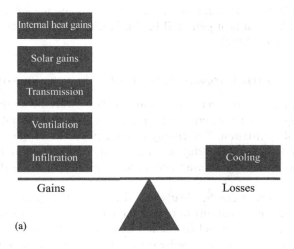

FIGURE 2.9 (a) Energy balance in summer season.

(*Continued*)

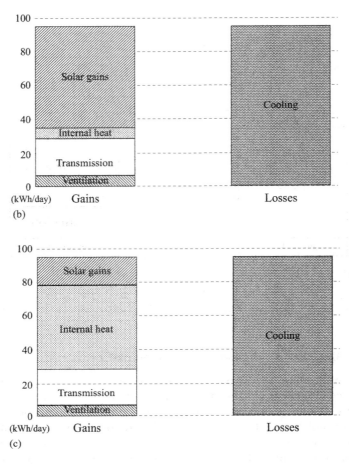

FIGURE 2.9 (*Continued*) (b) Heat gains and losses in summer season for an air-conditioned lecture hall. (c) Heat gains and losses in summer season for an air-conditioned residential building.

Figure 2.9(b). It is evident that internal heat load is very low and the solar gains are significant. For a lecture hall, this internal heat gain will be the largest due to the thermal gains by human beings and is given in Figure 2.9(c).

2.10.4 Energy Balance for Intermediate Season Like Spring and Autumn

The comfortable indoor conditions in these seasons can be achieved without any energy demand for heating and cooling. Indoor temperature is slightly higher than the outside temperature and is controlled by natural ventilation. The energy balance is given in Figure 2.10(a). Energy balance for gains and losses for a typical day in spring or autumn is given in Figure 2.10(b), and it can be seen there is no heating or cooling and mainly ventilation loss if through the windows/openings

From the above examples, it can be clearly stated that different strategies have to be adopted based on the different season conditions to reduce the energy demand. In the winter season, it is mainly about reducing transmission and ventilation losses and optimizing solar heat gain, whereas in summer season, it is about reducing the solar gains. For intermediate seasons, it is about the balancing of solar gains and amount of ventilation.

Basic Heat Transfer

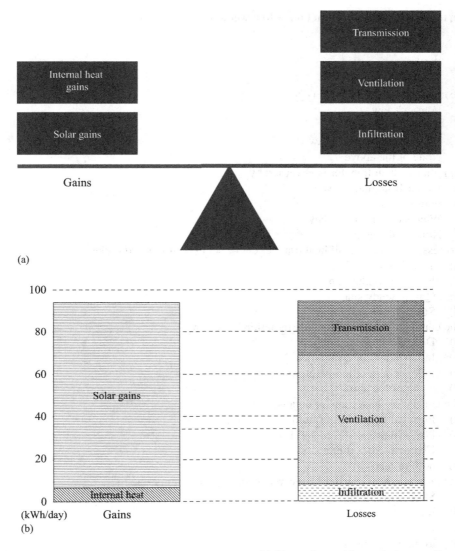

FIGURE 2.10 (a) Energy balance in intermediate season. (b) Heat gains and losses in intermediate season.

OBJECTIVE QUESTIONS

2.1 The thermal conductivity of material depends on:
   (a) Temperature
   (b) Length
   (c) Thickness
   (d) None of the above
2.2 The heat transfer coefficient is inversely proportional to
   (a) Thermal resistance
   (b) Thermal conductivity
   (c) Thickness
   (d) None of the above

2.3 The rate of heat transfer from higher to lower temperature is due to
- (a) Conduction
- (b) Convection
- (c) Radiation
- (d) All of the above

2.4 The conductive heat transfer is governed by
- (a) Fourier's law
- (b) Stefan-Boltzmann law
- (c) Wien's displacement law
- (d) None of the above

2.5 The radiation heat transfer is governed by
- (a) Stefan-Boltzmann law
- (b) Fourier's law
- (c) Wien's displacement law
- (d) None of the above

2.6 Expression for an overall heat transfer coefficient (U) is derived under
- (a) Transient condition
- (b) Periodic conduction
- (c) Quasi-steady state
- (d) Steady-state condition

2.7 The conduction, convection and radiation losses are
- (a) Dependent on each other
- (b) Independent of each other
- (c) Independent of temperature
- (d) None of the above

2.8 The convective heat transfer depends on
- (a) Physical properties of fluid
- (b) Physical properties of solid
- (c) Characteristics dimension
- (d) both (a) and (b)

2.9 The radiation heat transfer between two surfaces is mainly due to
- (a) Short wavelength
- (b) Infrared
- (c) UV
- (d) Long wavelength

2.10 The evaporative heat transfer coefficient (h_{ew}) is
- (a) Proportional to h_{cw}
- (b) Inversely proportional to h_{cw}
- (c) Independent of h_{cw}
- (d) None

2.11 The evaporative heat transfer coefficient (h_{ew}) depends on convective heat transfer coefficient due to
- (a) Lewis relation
- (b) Newton's law
- (c) Fourier's law
- (d) None of the above

2.12 The expression for a radiative heat transfer coefficient (h_r) for a surface having temperatures almost the same but different is
- (a) $4\varepsilon\sigma T^4$
- (b) $4\varepsilon\sigma T^3$
- (c) $1/4\ 4\varepsilon\sigma T^3$
- (d) $0.44\varepsilon\sigma T^3$

Basic Heat Transfer

2.13 For inclined surface, an expression for free convective heat transfer coefficient can be obtained from
- (a) $Nu = C(GrPr)^n$
- (b) $Nu = C(GrPr \sin \theta)^n$
- (c) $Nu = \dfrac{1}{C}(GrPr)^{\frac{1}{n}}$
- (d) $Nu = \dfrac{1}{C}(GrPr\cos\theta)^n$

* Here, $C = 0.54$ and $n = \frac{1}{4}$

2.14 The partial vapor pressure depends on temperature
- (a) Linearly
- (b) Proportionally
- (c) Exponentially
- (d) None

ANSWERS

2.1 (a)
2.2 (a)
2.3 (d)
2.4 (d)
2.5 (a)
2.6 (d)
2.7 (b)
2.8 (d)
2.9 (d)
2.10 (a)
2.11 (a)
2.12 (b)
2.13 (d)
2.14 (c)

PROBLEMS

2.1 Evaluate the convective heat transfer coefficient and the rate of convective heat loss from a horizontal rectangular plate (1.0 m × 0.8 m) at 134°C to a plate at 20°C placed at a distance of 0.10 m above it.
Hint: Use Equation (2.17), Table 2.1 and Appendix E, characteristic length = 0.10 m.

2.2 Evaluate the convective heat transfer coefficient and the rate of convective heat loss from a horizontal rectangular plate (1.0 m × 0.8 m) at 134°C to water at 20°C.
Hint: Use Equation (2.17), Table 2.1 and Appendix E, characteristic length = 0.90 m.

2.3 Evaluate the convective heat transfer coefficient and the rate of convective heat loss from a vertical wall (2.40 m × 1.80 m high) exposed to air at 1 atm pressure and 15°C. The wall is maintained at 49°C.
Hint: Use Equation (2.17), Table 2.1 and Appendix E, characteristic length = 1.8 m.

2.4 Evaluate the convective heat transfer coefficient and the rate of convective heat loss from a vertical wall (2.40 m × 1.80 m high) at 49°C exposed to a plate placed at a distance of $d = 0.10$ m at 15°C.
Hint: Use Equation (2.17), Table 2.1 and Appendix E, characteristic length = 0.10 m.

82

2.5   Calculate the average convective heat transfer coefficient and the rate of convective heat loss from a rectangular (0.91 m × 0.61 m) horizontal plate at 127°C to air flowing with 4.57 m/s at 27°C.

Hint: Use Equation (2.20b), and Appendix E, characteristic length = 0.76 m.

2.6   Calculate the average convective heat transfer coefficient and the rate of convective heat loss from 1 m horizontal flat surface to flowing water (20 m/s). The plate temperature is maintained at 27.8°C above the water temperature.

Hint: Use Equation (2.20b), and Appendix E, characteristic length = 1 m.

2.7   A horizontal plate (1 m × 1 m) is heated to 50°C, and it is exposed to an ambient air at 30°C at atmospheric pressure. Estimate the convective heat transfer coefficient.

Hint: Use $h = \dfrac{K}{L} \times 0.14 \times \left(\mathrm{Gr.Pr}\right)^{1/3}$, evaluate Gr and Pr for thermal properties of air at 30°C (Appendix E) and $L = (L_1 + L_2)/2$.

2.8   Calculate the evaporative heat transfer coefficient for wetted surface at 60°C and the surrounding temperature 35°C and relative humidity of about 60%.

Hint: See Example 2.9.

2.9   Calculate the evaporative heat transfer coefficient for wetted surface at 60°C and the surrounding temperature 35°C and relative humidity of about 60%.

Hint: See Example 2.9.

2.10   Repeat Problem 2.9 to estimate the evaporative heat transfer coefficient for different relative humidity (γ = 20%, 40%, 60% and 80%).

2.11   Calculate the radiative heat transfer coefficient for Problem 2.9 and compare it with an evaporative heat transfer coefficient with justification.

$$h_{rw} = \frac{\varepsilon\sigma\left[\left(T_w + 273\right)^4 - \left(T_{sky} + 273\right)^4\right]}{T_w - T_a}$$

Hint: After linearization of Equation (2.28).

2.12   Compare the radiative, convective and evaporative heat transfer coefficients for wetted surface (Problems 2.9 and 2.11) and find out the total heat transfer coefficient for the same surface.

Hint: Problem 2.9.

REFERENCES

[1] G. N. Tiwari, A. Tiwari and Shyam, *Handbook of Solar Energy*, New Delhi: Springer, 2016.

[2] H. Wong, *Handbook of Essential Formulae and Data on Heat Transfer for Engineers*, New York: Longman London Art, 1977.

[3] W. M. Adams, *Heat Transmission*, 3rd ed., New York: McGraw Hill, 1954.

[4] J. Holman, *Heat Transfer*, UK: McGraw Hill International Ltd., 1992.

[5] G. N. Tiwari, *Solar Energy: Fundamentals, Design, Modelling and Applications*, New Delhi: Narosa, 2002.

[6] F. Y. Bayley, "An analysis of turbulent free convection heat transfer," *Proceedings of the Institution of Mechanical Engineers*, vol. 169, no. 20, p. 361, 1955.

[7] V. T. Morgan, The overall convective heat transfer from smooth circular cylinders, in *Advances in Heat Transfer* (T.F. Irvine and J.P. Hartnett, eds.), vol. 11, New York: Academic Press, Inc., 1975.

[8] T. Fuji and H. Imura, "Natural convection heat transfer from a plate with arbitrary inclination," *International Journal of Heat and Mass Transfer*, vol. 15, p. 755, 1972.

[9] J. V. Clifton and A. J. Chapman, "Natural convection of finite size horizontal plate," *International Journal of Heat and Mass Transfer*, vol. 12, p. 1573, 1969.

[10] E. M. Sparrow and M. A. Ansari, "A refutation of King's rule for multi-dimensional external natural convection," *International Journal of Heat and Mass Transfer*, vol. 26, p. 1357, 1983.

[11] C. K. Brown and W. H. Gauvin, "Combined free and forced convection, I, II," *The Canadian Journal of Chemical Engineering*, vol. 43, no. 6, p. 306, 1965.

[12] J. A. Duffie and W. A. Beckman, *Solar Engineering of Thermal Processes*, New York: John Wiley & Sons, 1991.

[13] J. Watmuff, W. Characters and D. Proctor, "Solar and wind induced external coefficient for solar collectors," *Complex*, vol. 2, 1977.

[14] W. C. Swinbank, "Longwave radiation from clear skies," *Quarterly Journal of the Royal Meteorological Society*, vol. 89, p. 539, 1963.

[15] A. Whillier, "Black painted solar air heaters of conventional design," *Solar Energy*, vol. 8, p. 31, 1964.

[16] J. L. Fernández and N. Chargoy, "Multi-stage, indirectly heated solar still," *Solar Energy*, vol. 44, no. 4, p. 215, 1990.

[17] W. Duangthongsuk and S. Wongwises, "Measurement of temperature-dependent thermal conductivity and viscosity of TiO2-water nanofluids," *Experimental Thermal Fluid and Science*, vol. 33, pp. 706–714, 2009.

[18] E. V. D. Ham, "Thermal insulation," in *Zero-Energy Design: An Approach to Make Your Building Sustainable*, Edx, 2020a.

[19] E. V. D. Ham, "Zero-energy design: An approach to make your building sustainable," *Dezign Ark*, [Online]. Accessed December 2020. Available: https://www.youtube.com/watch?v=HmHzF2paGvE&list=PLuhw3IJ5k9rjwAka_pI5Y-1wJlPcS8yad&index=4.

3 Thermal Comfort

3.1 INTRODUCTION

Understanding thermal comfort is important to architecture, since it not only lays the foundation for building design, but also affects the field of sustainable design. As per the American Society of Heating, Refrigeration, and Air Conditioning Engineers [1] Standard 55 and International Standard ISO 7730 [2], thermal comfort may be defined as "the state of mind that expresses satisfaction with the surrounding environment". Thus, thermal comfort may be related to the comfort level a human being feels in the indoor thermal conditions.

The various factors that influence the thermal comfort are divided into three categories, namely (a) physical, (b) physiological and (c) behavioral aspects.

The physical aspect includes the environmental parameters like air temperature, relative humidity, air movement, mean radiant temperature, air pressure, air ingredients, air electricity, acoustics, daylighting, etc. The aforementioned parameters are known as "human thermal environment". The physiological parameters are influenced by the occupant's nutritional intake, age, ethnic influences, gender differences, constitution, etc. The behavioral aspects are the personal choices of the occupants like clothing, activity levels, adaptation and acclimatization etc. Generally, it is assumed that 80% occupants should be the minimum number of occupants to be satisfied and comfortable in the indoor thermal environment. The thermal chart shown in Figure 3.1 gives the factors governing the thermal comfort.

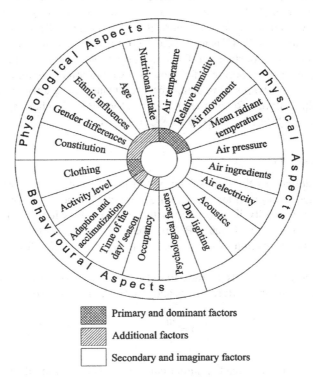

FIGURE 3.1 Thermal comfort chart [3].

The aspects can further be classified as (i) primary and dominant, (ii) additional and (iii) secondary and imaginary factors. For example, the primary and dominant factors in physical aspects are air temperature, mean radiant temperature, relative humidity and air movement whereas in behavioral aspects, clothing and activity level falls into this category. Similarly, all the parameters in the physiological aspects are additional factors, whereas occupancy, adaptation and acclimatization are additional factors in behavioral aspect.

Passive solar concepts should be given the prime concern in any building design in order to reduce the dependency on artificial heating and cooling of buildings in providing comfortable indoor climatic conditions. The conceptual designing of a building is independent of active systems but it is difficult to integrate a passive system to a building once it is designed or constructed. Thus, a necessary precision is required regarding the integration of passive concepts with the building design at every stage of conceptualization and construction. A careful application of the passive solar concepts provides thermal comfort in an economical way.

In this chapter S.I. units are followed. However, the notation of temperature is "T" for which the unit is °C and the convective, radiative, conductive heat transfers from the human body are represented as C, R and K, respectively.

3.2 PHYSICAL ASPECTS

3.2.1 AIR TEMPERATURE

Dry bulb temperature (DBT) and wet bulb temperature (WBT) are the two temperatures used for defining the thermal comfort. DBT is the air temperature measured by an ordinary thermometer (mercury in glass thermometer), which is freely exposed to the air but isolated from any moisture or radiation. It is an indication of the sensible heat content of the air. Sensible heat refers to the temperature load of a given space. WBT is the temperature at which the saturation pressure of air equals the vapor pressure for a parcel of air. It is measured by the mercury in glass thermometer having the bulb enclosed in a wetted cotton sleeve. The WBT is read when the rate of heat transfer and the rate of evaporation is stabilized. WBT is an indication of the enthalpy of moist air, i.e., the enthalpy is approximately the same for constant WBT at constant atmospheric pressure. The DBT is always greater than the WBT due to the cooling offered by the evaporation of the moisture from the wetted cotton, the latent heat for the evaporation being absorbed from the mercury reservoir. Latent heat is the moisture load of the building. Recommended values for DBT for achieving thermal comfort for summers and winters are (25 ± 1)°C and (20 ± 1)°C, respectively.

3.2.2 RELATIVE HUMIDITY

Relative humidity is defined as the ratio of the partial pressure of water vapor to the equilibrium vapor pressure of water for a given air temperature. Relative humidity depends on temperature as well as pressure. Thus, humidity is associated with the moist air which is the mixture of dry air (fixed part) and water vapor (variable part). Relative humidity is always less than one (1). Saturated air is known as when the absorption of water reaches its maximum value for a given volume of dry air.

Specific humidity (ω) is defined as the ratio of mass of water vapor (m_v) to mass of dry air (m_a) in a given volume of the moist air (mixture of dry and water vapor). It is expressed as:

$$\omega = \frac{m_v}{m_a} \tag{3.1}$$

From Equation (3.1), it is clear that the numerical value of specific humidity (ω) is always greater than one (1).

Thermal Comfort

The ratio of actual specific humidity to the specific humidity of saturated air at a given temperature is known as degree of saturation.

Relative humidity (γ) is the ratio of mass of the water vapor (m_v) in a given volume of moist air at a given temperature to the mass of water vapor (m_s) in the same volume of saturated air at same temperature. It is defined as:

$$\gamma = \frac{m_v}{m_s} \tag{3.2}$$

For thermal comfort in both summer and winter months, the recommended value of the relative humidity (γ) is (50 ± 10)%. Its numerical value is always less than 100% [4].

The variation of humidity ratio (specific humidity) with DBT for different relative humidity at atmospheric pressure has been shown in the psychometric chart given in Figure 3.2. The three principle boundaries of a psychometric chart are dry bulb temperature, moisture content (humidity ratio) and saturated air.

3.2.3 Air Movement

The 20°C–25°C is considered to be thermal comfort air temperature for a human being for both winter and summer climatic conditions, respectively. Within the range of 36.9°C to 37.2°C, the body temperature remains constant. Beyond 40°C, the temperature may be fatal. Once we have an idea of thermal comfort, it is easy to eliminate the sources of heating to maintain control for which awareness of air movement is important. It is important to know how air is moving around, how it is getting renewed and refreshed.

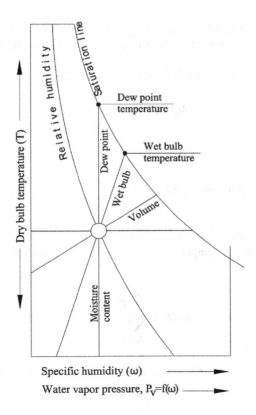

FIGURE 3.2 Psychometric chart [3].

88 Photovoltaic Thermal Passive House System

Air motion may be defined as the rate at which air moves around and touches the skin. With circulation of air close to the human body, heat transfer from the body to the environment takes place in form of respiration which reduces the thickness of the layer of air around the body. With an increase in the air movement near the body, the level of discomfort decreases. This is true only when the ambient air temperature is lower than the body temperature; otherwise, the level of discomfort will increase. The recommended value for air movement is about 0.2 m/s for winters and 0.4 m/s for summers inside the room. This air movement inside the room may be achieved by operating the fan in summer months.

3.2.4 MEAN RADIANT TEMPERATURE

The human body radiates heat to its surroundings by radiation (R), convection (C) and evaporation (E) or respiration. The total heat loss, Q (kWh) may be expressed as:

$$Q = (R + C) + + E \tag{3.3}$$

The above expression may also be written as:

$$Q = Q_s + Q_L \tag{3.4}$$

where

$$(R + C) = Sensible\ heat\ loss = Q_s,\ \text{and}$$
$$E = Latent\ heat\ loss = Q_L.$$

The sensible heat loss is dependent on the temperature difference between the body surface and the ambient (surroundings), whereas the latent heat loss depends on the difference in water vapor pressure.

The heat transfer (kWh) between the human body and its surroundings may be expressed as:

$$M - W = Q + S \tag{3.5}$$

where

M = Metabolic process,
W = Work done by the human body, and
S = Rate of heat storage.

Figure 3.3 gives the rate of heat transfer between the human body and the surrounding temperature.

In Figure 3.3, the surrounding temperature is referred to the mean radiant temperature (MRT) in °C, and may be expressed as:

$$MRT = \frac{\sum_i T_i A_i}{\sum_i A_i} \tag{3.6}$$

where T_i and A_i are the temperature (°C) and area (m²) of the different surfaces of the living enclosure.

Thermal Comfort

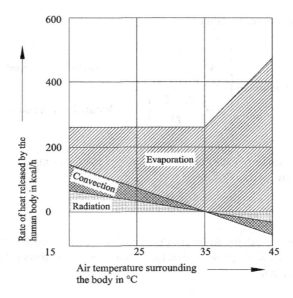

FIGURE 3.3 Effect of surrounding temperature on the rate of heat released by the body [3].

Technically, MRT may be defined as the uniform temperature of a surrounding surface giving off blackbody radiation (emissivity $\varepsilon = 1$), which results in the same radiation energy gain on a human body as the prevailing radiation fluxes, which are usually very varied under open space conditions.

MRT is considered to be the most important parameter influencing human energy balance, especially on hot sunny days. It is also the most governing parameter on thermo-physiological comfort indexes like predicted mean vote (PMV) which are derived from heat exchange models. Human body is highly responsive to the changes in MRT due to the high absorptivity and emissivity (0.97) of the human skin, which is greater than any other known substance.

A globe thermometer, which is a normal dry bulb thermometer encased in a 150 mm diameter matte-black copper sphere is used to measure the radiant temperature. Its absorptivity approaches that of the human skin. MRT may be expressed in terms of globe temperature (T_g) and ambient temperature (T_a) as follows [5]:

$$\bar{T}_r = \sqrt[4]{(T_g + 273)^4 + \frac{h_{cg}}{h_r}(T_g - T_a)} - 273\,°C \tag{3.7}$$

where h_r is the radiant heat transfer coefficient (W/m²°C).

For a globe, convective heat transfer may be given as:

$$h_{cg} = \text{maximum of} \begin{bmatrix} 18(v_{ar})^{0.55} & \text{Forced convection} \\ 3(|T_g - T_a|)^{0.25} & \text{Natural convection} \end{bmatrix} \tag{3.8}$$

For an operative temperature transducer,

$$h_{cg} = maximum \ of \begin{bmatrix} 6.3\dfrac{\left(\upsilon_{ar}\right)^{0.6}}{D^{0.4}} & Forced \ convection \\ 1.4\left(\dfrac{\left|T_g - T_a\right|}{D}\right)^{0.25} & Natural \ convection \end{bmatrix} \tag{3.9}$$

where υ_{ar} is the relative velocity (m/s) between the body and air in m/s, and D is the diameter of matte-black copper sphere.

3.2.5 Air Pressure

Atmospheric pressure is considered to be comfortable for a human body. With increase in the altitude, there is a decrease in the air pressure and the human body feels discomfort due to the increase in the pressure within the body.

3.2.6 Air Ingredients

Every human being produces 0.2 m³/h of carbon dioxide (CO_2) and needs about 0.65 m³/h of oxygen (O_2) under normal conditions for its survival inside and outside of the building. Outside the building, the requirement of oxygen is fulfilled by the plants and vegetation, which release oxygen. On the other hand, the CO_2 released by the human body is consumed by the plants to balance the eco-system. The optimum level of oxygen and carbon-dioxide in the atmosphere are 20.94×10^4 ppm and 320 ppm, respectively. However, the optimum level of oxygen and carbon dioxide should also be maintained inside the building either by passive or active modes. In order to maintain this level inside the building, a roof vent, ventilators, windows or exhaust fans may be proposed to increase the natural ventilation and have a direct contact between the indoors and the environment.

3.2.7 Air Electricity

Sometimes, a shock-like sensation is felt at skin when a thin layer of air passes over it. This may be experienced when the skin is rubbed in the presence of thin layer of air due to the friction between the layer of air and the skin.

3.2.8 Acoustics

A comfortable and effective working environment is considered up to the maximum sound level of $120\,db(0.01\,w/m^2)$. Frequencies within the range of 20–16,000 Hz can be heard by an average human being. Any unwanted sound may be termed as noise pollution and creates a sense of irritation and unpleasantness.

3.2.9 Daylighting

Solar energy in the form of daylight should be the prime concern and heart of any building design. This will not only satisfy the lighting requirement for interiors but also reduce the dependency on the artificial lights. Thus, the cooling demand of the building is also reduced by restricting the use of artificial modes of lighting (like electric bulbs, lamps, tube lights etc.) as a direct consequence of the reduced heat losses. Solar energy can be harvested using photovoltaic (PV) technology, thus electrical energy can also be generated. Daylighting acts as a deciding factor in the quality of space, and great architecture is always associated with natural lighting. The impact of natural light in the interiors has been completely neglected in today's architecture by excessive dependence on artificial means. As reported, artificial lighting constitutes about a 14%, 26% and 31% portion of the total electrical consumption in non-residential buildings in Europe, the United States and Spain respectively [6] and about 19% of the electrical consumption worldwide [7]. The incorporation of daylighting strategies can help achieve about a 30%–77% reduction in energy consumption [8]. The heat gain with use of an electric bulb inside the room is significant since only a fraction of the electrical energy gets converted into light energy. The heating effect of daylighting is 1 W/lm.

Ensuring daylight in the building is a promising solution to the problem of energy consumption. Daylight is one of the cheapest and most efficient ways of utilizing the solar energy in buildings. Architects are also concerned about energy conservation in buildings due to daylight. Daylighting strategies are dependent on architectural characteristics like form, aesthetics and functionality. Moreover, natural light also has positive psychological effects on human behavior and performance. Daylight can be, thus, introduced in the building with the provision of windows, skylights, solar tubes, transparent facades, etc.

3.2.9.1 Windows and Fenestrations

The openings in the buildings may be planned in the walls as windows or in the roof as skylights as shown in Figure 3.4. The purpose of windows/openings is not just to keep the weather outside, but also to allow solar radiation and fresh air to enter the interiors in the form of natural light. The fenestrations, windows and sunshades should be designed in order to make the building self-sustainable. Various shading options are given in Chapter 7. Window to ground ratio should lie in the range of 0.33–0.58 [9]. The amount of natural light entering may be improved by changing the manufacturing process of glass, transmissivity of glass and height of the ceiling. In early architecture, the daylight entrance was limited due to low ceiling heights. With passage of time, ceiling heights increased and the concept of clerestory windows was developed which helped in

FIGURE 3.4 Openings for daylighting.

illuminating the farthest points of the room. A clerestory window (Figure 3.4 [10]) is a horizontal strip of windows with a very high sill level, generally above the lintel level of the door. It allows diffused light to enter the premises minimizing the formation of shadows and direct glare. When the spaces became grander, even these windows were not sufficient, thus the concept of skylights came into existence to introduce the natural light to the interiors that were not reachable by the side windows. Solar radiation, when allowed to enter through glazing, is termed as direct gain and through an opaque surface is referred to as indirect gain. Detailed explanation of direct and indirect gains is given in Chapter 6.

3.2.9.2 Skylights

Skylights or roof lights are the glazed openings created in the roof area to allow natural light inside the building. Earlier skylights were perceived as domes and barrel vaults. Today various shapes and forms are made due to advanced technology. The structure of the skylight may be fixed or openable, made of transparent glass or plastic helping in both top light generation and ventilation (if openable).

3.2.9.3 Solar Tubes

Solar tubes are also known as light tubes or tubular daylight devices (TDD), sun tunnels and in a similar way as skylights but in a different form as given in Figure 3.5 [11, 12]. These are in form of pipes that run through the ceiling. The sunlight is allowed to enter from the top, which is covered with a weatherproof plastic dome. The pipes are generally polished and lined with reflective material from inside, which amplifies the light. The bottom end opens inside the room, and the diffuser spreads the light and fills the room with bright and strong light along with homogenous distribution. They provide more illuminance as compared to traditional skylights with a much larger area. Solar tubes require lesser surface area exposed and thus reduce the heat transfer from the ambient.

3.2.9.4 Semi-Transparent Solar Photovoltaic Lighting System (SSPLS) and Transparent Facades

Integrating the building with semi-transparent photovoltaic (SPV) with roof and facade is a promising approach towards sustainable buildings. It helps in the generation of electrical power, daylighting, and it improves the aesthetics of the elevation of the building. Electrical performance of the SPV

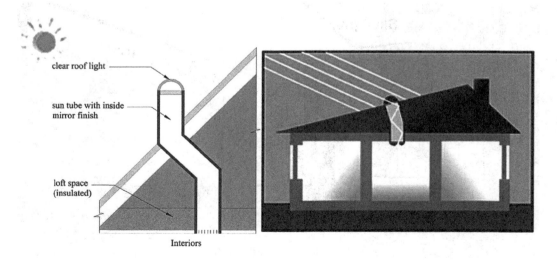

FIGURE 3.5 View of solar pipe/light tube/tubular daylighting device (TDD) [13].

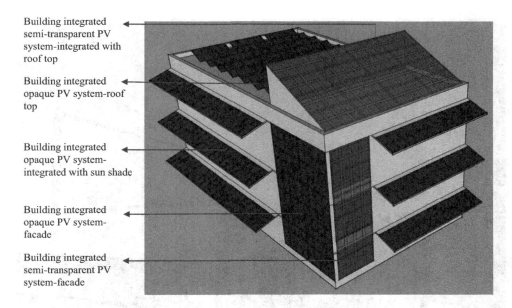

FIGURE 3.6 Configurations of building integrated photovoltaic thermal (BiPVT) system [14].

modules is better than the opaque modules. Various configurations of integrating photovoltaic (PV) modules with the building are given in Figure 3.6.

SPV modules are integrated with the building with a reasonable non-packing factor with inclined roof to receive maximum solar insolation. With decreases in the packing factor, the solar cell electrical efficiency increases. This is due to the decrease in the solar cell temperature. Electrical efficiency increases with increases in the packing factor. Also, with decreases in the packing factor, the daylight savings increases due to increases in the glass area, which allows more solar radiation to enter the building [15].

Integrating SPV modules with the facade of the building gives additional benefit of daylight and availability of large vertical surface area for the installation on architectural elevation. Thin modules are light in weight and thus are considered well for integration with facades. Although they capture less solar radiation (due to vertical orientation) as compared to rooftop and ground installations, therefore, there are lesser diurnal and seasonal variations.

Integrating SPV modules with the rooftop also provides greenhouse effect, which may be used for following applications:

- Daylighting
- Solar crop drying
- Sunbath
- Electrical power

3.2.9.5 Light Shelves

Light shelves are horizontal surfaces placed above eye-level with high reflectance surfaces as seen in Figure 3.7 [16]. They reflect the natural light inside the building. They are generally installed in office complexes and institutional buildings. Since light shelves are overhang structures (projections), not only do they help to illuminate the interiors but they also act as a sunshade over the windows, thus allowing diffused light to enter the interiors. Light shelves make it possible for daylight

FIGURE 3.7 Light shelves [17].

to penetrate the space up to 2.5 times the distance between the floor and the top of the window. These arrangements are suitable for mild climatic conditions only.

3.2.9.6 Sawtooth Roofs

Sawtooth roofs have a series of ridges with dual pitches on either side. The steeper surfaces are glazed, which prevents direct solar exposure and allows diffused light to spread over a larger area. It is a very old concept, generally seen in factories and manufacturing buildings. The major disadvantage of this type of roof is that the exposed area increases unwanted thermal losses during winter months.

3.2.9.7 Heliostats

A heliostat is a device that includes a mirror that rotates with the motion of the sun, reflecting the solar radiation towards a target. This advanced technology reflects the solar radiation to the glazed portion of the building like windows, skylights (Figure 3.4), etc. The light received from these glazed areas is further used to illuminate the building interiors. This is an energy-efficient technique making use of maximum solar radiation all through the day.

3.2.9.8 Smart glass windows

The windows, as discussed in Section 3.2.9.1, when made up of smart glass materials are termed smart glass windows. The smart glass changes its optical properties (transforms from transparent to different levels of tint) by passing a low voltage through intelligent controllers into the panels. Smart glass combined with intelligent controls makes buildings more energy efficient, allows more natural daylight, provides increased comfort for occupants, and eliminates heat and glare with no shades or blinds required.

Thermal Comfort

3.2.9.9 Hybrid Solar Lighting (HSL)

The concept of hybrid solar lighting (HSL) was developed by Oak Ridge National Laboratory (ORNL). The system is made up of optical fiber cables and fluorescent lighting fixtures with transparent rods attached to the optical fiber cables and a light collector system mounted on the rooftop of the building. With this system, during the sunshine hours, complete elimination of artificial lights may be achieved. During the evening, when the solar intensity lessens, the fluorescent fixtures gradually turn on, which maintains nearly constant luminance level in interior space. During off-sunshine hours (nighttime), these fixtures may be electronically operated. The installation cost of this type of system is a concern and major disadvantage.

3.3 PHYSIOLOGICAL ASPECTS

3.3.1 NUTRITIONAL INTAKE

Human comfort is governed by the physiological aspects determined by the rate of heat produced inside body and the rate of heat dissipated to the environment (Figure 3.3). The rate at which the body generates heat is known as metabolic process (Equation 3.5). Heat produced by a healthy person is about 35 W/m^2 ($\approx$60 W) and reach up to 350 W/m^2 ($\approx$600 W) during sleeping and hard work, respectively.

The nutritional intake depends upon the composition of the body and varies from person to person and season to season. During the winters, the body requires more nutritional intake (healthy food) and less during the summer months. The human body is at more thermal comfort during the winter months with a high intake of caloric value food as compared to the intake in summer months.

3.3.2 AGE

With increase in the age of a person, the human body feels colder in winter months and warmer in summer months due to:

* decrease of metabolic process
* change in food habits
* decrease in nutritional intake
* change in activity levels, etc.

3.3.3 ETHNIC INFLUENCES

Thermal comfort is a condition of mind that expresses satisfaction with the thermal environment. Due to its subjectivity, thermal comfort is different for every individual and every region because of a particular way of living like food habits, clothing, etc.

3.3.4 GENDER DIFFERENCES

The author of [4] found that females express more dissatisfaction (especially in cold environments) and are more sensitive than males to a deviation from an optimal temperature. Thus, females should be a base in field studies on thermal comfort.

3.3.5 CONSTITUTION

This refers to the composition of the human body, which varies from person to person. A weak human being feels warmer in summer conditions as compared to a healthy person and vice versa in winter conditions.

3.4 BEHAVIORAL ASPECTS

3.4.1 CLOTHING

The clothing forms a layer between human skin and air movement in the environment, thus it reduces the rate of heat transfers (conduction, convection and evaporation/respiration) from the body to the ambient air. The color and type of clothing is also important with respect to the thermal comfort and climatic conditions as described below:

Desert climate: It is preferred to wear thin, loose-fitted and light color clothes. This type of clothing helps in lower absorption of solar radiation and heat transfer from outside to the skin. The loosely fitted clothes helps improve the air movement across the skin.

Hot and humid climate: It is preferred to wear loose-fitted, porous and light-color clothes. The porosity of the material helps in fast heat transfer from the skin to the ambient air.

Cold climate: It is preferred to wear thick, tight-fitted and dark color clothes. This type of clothing ensures higher absorption of solar radiation and minimum heat transfer from skin to environment air.

CLO value is used to define the insulation levels of clothing which is a numerical representation of a clothing ensemble's thermal resistance.

Clothing insulation are the materials that are used to retain or remove body heat.

$$1\,CLO = 0.155\,m^2\,°C\,/\,W = 0.133\,m^2\,h°C\,/\,kcal = 0.88\,ft^2\,h°C\,/\,Btu$$

Table 3.1 gives CLO values for common articles of clothing. The total insulation value of the final attire may be estimated by adding the CLO value of individual items.

3.4.2 ACTIVITY LEVEL

The rate at which the heat is dissipated from the body depends on the activity level of the individual. The rate of body heat generation because of the oxidation of food is known as metabolism. A unit of metabolism is "met", which corresponds to the metabolism of a relaxed, seated person, i.e.,1 met = 58.2 W/m^2 = 50 kcal/m^{2}h. As mentioned in Section 3.3.1, it varies from 60 W ($\approx$ 35 W/m^2) to 600 W ($\sim \approx$ 350 W/m^2). From this, it can be considered that the human body converts thermal energy into mechanical energy with efficiency of 20%. The value of metabolism for every individual depends on the nutritional intake and the activity level and may be estimated by:

$$M = 2.06 \times 10^4 \, \dot{V} \left(F_{oi} - F_{oe} \right) J \tag{3.10}$$

where

$\dot{V}$ = air breathing rate in per second,

F_{oi} = fraction of oxygen in the inhaled air = 0.209, and

F_{oe} = fraction of oxygen in the exhaled air (varies with the composition of food used in the metabolism); F_{oe} for a diet high in fat is $F_{oe} \approx 0.159$ and for carbohydrates $F_{oe} \approx 0.163$.

The rate of metabolism per unit area of the body surface (A_D) can be estimated by:

$$A_D = 0.202 \, m^{0.425} H^{0.725} \quad m^2 \tag{3.11}$$

TABLE 3.1
CLO Values for Individual Items of Clothing [5, 18, 19]

Clothing Category	Male (M)		Female (F)	
	Clothing Type	CLO Value	Clothing Type	CLO Value
Innerwear	Sleeveless	0.06	Girdle	0.04
	T-shirt	0.09	Bra and panties	0.05
	Briefs	0.05	Half slip	0.13
	Long underwear, upper/ lower	0.10	Full slip	0.19
			Long underwear, upper/lower	0.10
Shirts (M); Blouse/Dress (F)	Short sleeve (light)	0.14	Long sleeve blouse	0.20
	Long sleeve (light)	0.22	Long sleeve blouse (heavy)	0.29
	Short sleeve (heavy)	0.25	Light dress	0.22
	Long sleeve (heavy) (Plus 5% for tie or turtleneck)	0.29	Heavy dress	0.70
Vest (M);	Light	0.15	Light	0.10
Skirt (F)	Heavy	0.29	Heavy	0.22
Trousers (M); Slacks (F)	Light	0.26	Light	0.10
	Medium	0.32	Heavy	0.44
	Heavy	0.44		
Sweater	Light	0.20	Sleeveless (light)	0.17
	Heavy	0.37	Long sleeve (heavy)	0.37
Jacket	Light	0.22	Light	0.17
	Heavy	0.49	Heavy	0.37
Socks (M); Stockings (F)	Ankle length (thin)	0.03	Any length	0.01
	Ankle length (thick)	0.04	Pantyhose	0.01
	Knee length	0.10		
Footwear	Sandals	0.02	Sandals	0.02
	Oxfords	0.04	Pumps	0.04
	Boots	0.08	Boots	0.08
Hat and overcoat	-	2.00	-	2.00
Mask	-	0.03	-	0.03

where,

m = body mass in kg,
H = body height in m, and
A_D = 1.83 m^2 (for an average man of mass 70 kg and height 1.73 m).

In a space of 6 m^3 volume, room air temperature may rise up to 48°C/hr if the body generates energy at the rate of 335 kJ/h. Thus, removal of heat generated by the body via ventilation should be done in order to maintain thermal comfort. Metabolic rates for various activities are given in Table 3.2.

3.4.3 ADAPTATION AND ACCLIMATIZATION

Before actual exposure to harsh climatic conditions, the human body should first become immune with artificially created similar climatic conditions.

TABLE 3.2

Metabolic Rates of Different Typical Activities in Cabins [5]

Activity		Metabolic Rate	
		1.0 met = 50 kcal/m²h	
		Met	W/m²
Rest position	Sleeping	0.7	40.6
	Reclining	0.8	46.4
	Relaxed seated	1.0	58
	Standing at rest	1.2	69.6
Walking	Slow	2.0	116
	Moderate	2.6	150.8
	Fast	3.8	220.4
Equipment handling	Typing/computer	1.1	63.8
	Light machine work	2.0–2.4	116–139.2
	Heavy machine work	4	232
Other activities	Sedentary (Office, dwelling, school, laboratory)	1.2	69.6
	Standing	1.4	81.2
	Cooking	1.6–2.0	92.8–116
	Cleaning	2.0–2.4	116–139.2
	Volleyball, bicycling (15 km/h)	4	232
	Aerobic Dancing, Swimming	6	348

3.4.4 TIME OF THE DAY/SEASON

The recommended air temperature for thermal comfort is 20°C. There is variation in ambient temperature during the day due to the change in the solar insolation levels. Also, ambient temperature is different for all months and varies from season to season. For example, the ambient temperature varies from 5°C–15°C and 30°C–45°C in the winter and summer months, respectively, for northern climatic conditions. Thus, the heating and cooling demand of the building depends on the time and season of the year.

3.4.5 OCCUPANCY

It is clear that the human body generates heat due to energy generation within the body. Therefore, if the occupancy number (i.e., number of persons) for a given volume of enclosed space increases, the total amount of heat generated will also increase resulting in an increase in the room air temperature. Thus, the living and working condition of that space gets effected.

3.4.6 PSYCHOLOGICAL FACTORS

Sometimes, expressing and sharing thoughts with a group of people makes us feel better and psychologically pepped up. For example expressing how cold or warm it is makes the individual feel more comfortable. Thermal comfort is a condition of mind that expresses satisfaction with the thermal environment. Thermal comfort varies from individual to individual. For the simplification of solar energy systems, India has been divided into different climatic conditions, for which the criteria are given in Chapter 1.

Thermal Comfort

3.5 THE COMFORT EQUATION

It is very important to understand thermal comfort in architecture since the final user is human. Thermal comfort thus lays foundation for any building design.

Thermal comfort is maintained when the heat produced by the human body (metabolism) is allowed to dissipate at a rate to maintain thermal balance within the body. Discomfort is felt if there is any heat gain or loss beyond the equilibrium. The human body behaves like a thermodynamic process. For proper functioning of the body, a constant internal temperature of around $37 \pm 0.5°C$ should be maintained. Heat produced in the body should be equal to the heat lost in order to maintain the thermal comfort. It may be assumed, then, that for long exposures to a constant (moderate) thermal environment with a constant metabolic process, a heat balance will exist for the human body, for which the heat production will be equal to the heat dissipation. The heat is released from the skin to the surroundings via convection and radiation (dry heat loss). The skin is the exposed area to the environment and reacts according to the variations in its temperature, sweat and hair coverage. In order to maintain the heat balance, sweat is produced by the skin glands (in case dry heat loss is not sufficient). The sweat gets evaporated, providing an additional cooling effect. The rate of evaporation decreases with increases in the humidity, thus decreasing the cooling effect produced by it.

Comfort equation may be defined as the body's energy balance equation when the person is in thermal equilibrium as expressed in Equation (3.12).

$$M - W = H + \dot{Q}_{se} + \dot{Q}_{res} + E_{res} \tag{3.12}$$

where

M = Metabolic process,
W = Work done by the human body,
H = Heat loss by convection, radiation and conduction,
$\dot{Q}_{se}$ = Evaporative heat exchange at the skin when the subject experiences a sensation of thermal neutrality,
$\dot{Q}_{res}$ = Respiratory convective heat exchange, and
E_{res} = Respiratory evaporative heat exchange.

If the human body is not in thermal equilibrium, the heat will either be stored or released from the body. The rate of heat exchange between the surroundings and the human body is given in Figure 3.3 Heat storage in the body may be expressed as:

$$S = (M - W) - (H + \dot{Q}_{se} + \dot{Q}_{res} + E_{res}) \quad J \tag{3.13}$$

In summer climatic conditions, S being positive, the body temperature rises due to more storage of heat (energy). Thus, the blood flow increases through extremities. Whereas in winter months, S being negative, the body temperature tends to fall, decreasing the blood flow. This results in shivering of the body.

3.5.1 CONDUCTION

The conductive coefficient is often replaced by the reciprocal of the thermal resistance of the clothing. The rate of heat conduction equation through clothing may be expressed as:

$$\dot{Q}_k = h_{cl}(T_s - T_{cl}) \quad W/m^2 \tag{3.14}$$

where,

h_{cl} = heat conductive coefficient of clothing in W/m^{2}K,
T_s = average skin temperature in °C, and
T_{cl} = surface temperature of clothing in °C.

$$h_{cl} = \frac{1}{0.1555 \times I_{cl}} = \frac{6.45}{I_{cl}} \tag{3.15}$$

where I_{cl}, usually given the unit CLO (1 CLO = 0.155 m^{2}K/W)

Typical values of I_{cl} for different clothing ensembles are given in Table 3.1.

The clothing area factor is given by:

$$f_{cl} = \begin{cases} 1.00 + 1.290 I_{cl} \; for \, I_{cl} < 0.078 \; \mathrm{m^2 °C / W} \\ 1.05 + 0.645 I_{cl} \; for \, I_{cl} > 0.078 \; \mathrm{m^2 °C / W} \end{cases} \tag{3.16}$$

The mean skin temperature, T_s, (for metabolic process M between 1 and 4) may be estimated from:

$$T_s = 35.57 - 0.0275(M - W) \, °C \tag{3.17}$$

The surface temperature of clothing is given by:

$$T_{cl} = 35.7 - 0.028(M - W) - I_{cl} \left[3.96 \times 10^{-8} f_{cl} \times \left\{ (T_{cl+273})^4 - (\bar{T}_r + 273)^4 \right\} + f_{cl} \times h_c \times (T_{cl} - T_a) \right] \tag{3.18}$$

3.5.2 CONVECTION

As mentioned in Section 2.3, the heat transfer between a body and the ambient air is primarily by convection. This can be either natural (free) due to buoyancy or forced (mechanical) due to relative movement between the body and the air.

The general rate of heat convection equation may be expressed as:

$$\dot{Q}_c = f_{cl} h_c (T_{cl} - T_a) \; \mathrm{W / m^2} \tag{3.19}$$

where

T_a = air temperature in °C, and
h_c = convective heat transfer coefficient in W/m^{2}K.

Convective heat transfer coefficient under natural and forced mode may be expressed as:

$$h_c \begin{cases} 2.38(T_{cl} - T_a)^{0.25} \; for \, 2.38(T_{cl} - T_a)^{0.25} > 12.1\sqrt{v_{ar}} \; (\text{Natural convection}) \\ 12.1\sqrt{v_{ar}} \quad\quad for \, 2.38(T_{cl} - T_a)^{0.25} < 12.1\sqrt{v_{ar}} \; (\text{Forced convection}) \end{cases} \tag{3.20}$$

where v_{ar} is the relative velocity between the body and air in m/s.

For numbers of air changes ≤ 10, it is considered under natural mode.

Thermal Comfort   **101**

The convective heat transfer from skin or clothing results from an airstream perturbing the insulating boundary layer of air around the surface of the body. In general, the faster the flow of airstream, thinner layer of air will be formed on the surface of the body. This results in lower thermal insulation afforded by the subject. Convective heat transfer from a heated surface like skin or clothing can be further classified into three distinct modes:

(a) Natural convection: When the air movement is driven purely by thermally induced buoyancy and generally confined to low ambient air speeds.
(b) Forced convection: The ambient air speed is generally more than 1.5 m/s
(c) Mixed-mode convection: The air speed ranges between the natural and forced convection.

Heat transfer coefficient values, for both natural and forced convection conditions, defined by a set of laboratory measurements can be referred from [5, 20]. Heat transfer for the whole body (standing); under natural convention is $3.4 \, \text{W/m}^2\text{K}$, and under forced convention is $10.4 \, \text{W/m}^2\text{K}$ (B) and $0.56 \, \text{W/m}^2\text{K}$ (n). The same for in a seated position: (whole body) is $3.3 \, \text{W/m}^2\text{K}$ as natural convention heat transfer, and $10.1 \, \text{W/m}^2\text{K}$ (B) and $0.61 \, \text{W/m}^2\text{K}$ (n) as forced convention heat transfer

3.5.3 RADIATION

The rate of heat transfer between the body (skin and clothing) and the surrounding surfaces via radiation is calculated using the Stefan–Boltzmann equation as follows:

$$\dot{Q}_r = f_{eff} f_{cl} \varepsilon \sigma \left\{ \left(T_{cl} + 273 \right)^4 - \left(\overline{T}_r + 273 \right)^4 \right\} \ \text{W/m}^2 \tag{3.21}$$

where

f_{eff} = factor of effective radiation area, i.e., ratio of the effective radiation area to the total surface area of clothed body,

f_{cl} = factor of clothing area, i.e., ratio of surface area of clothed body to surface area of nude body,

ε = emissivity of clothed body,

σ = Stefan–Boltzmann constant $5.67 \times 10^{-8} \, \text{W/m}^2\text{K}^4$,

T_{cl} = surface temperature of clothing in °C, and

$\overline{T}_r$ = mean radiant temperature, i.e., effective temperature of room surfaces in °C.

The value of f_{eff} is taken as 0.71, which is a mean value of 0.696 (for a seated person) and 0.725 (for a standing position). The value changes in different positions since some of the body parts act as shields to the others.

Thus Equation (3.21) may be reduced to:

$$\dot{Q}_r = 3.96 \times 10^{-8} \, f_{cl} \left\{ \left(T_{cl} + 273 \right)^4 - \left(\overline{T}_r + 273 \right)^4 \right\} \tag{3.22a}$$

The indoor temperature ranges between 10°C–30°C (which is very small variation), and thus may be adequately replaced by the linear equation as:

$$\dot{Q}_r = f_{eff} h_r \left\{ \left(T_{cl} + 273 \right)^4 - \left(\overline{T}_r + 273 \right)^4 \right\} \tag{3.22b}$$

102 Photovoltaic Thermal Passive House System

where the radiant heat transfer coefficient, h_r, may be approximated by

$$h_r = 4.6\{1+0.01\times(\overline{T}_r+273)\} \approx 5.7 \text{ W}/\text{m}^2\text{K} \qquad (3.23)$$

3.5.4 EVAPORATION

As mentioned in Section 2.6, the evaporative heat loss is partially due to the diffusion of water vapor through the skin tissues ($\dot{Q}_e$), and partially due to the evaporation of sweat from the surface of skin ($\dot{Q}_{se}$). The heat is absorbed from the skin in these processes, which prevents the rise in body temperature. The water diffusion can take place even in a cool environment since it is a continuous process, whereas the evaporation of sweat takes place in hot conditions when the body activity is higher than the normal. The heat from the skin is used for the sweat to evaporate.

The amount of water diffusion through the skin, and the corresponding rate of evaporative heat loss ($\dot{Q}_e$) is a function of the difference between the saturated water vapor pressure at skin temperature (p_s) and the water vapor pressure in the ambient air (p_a). Therefore,

$$\dot{Q}_e = 3.05\times10^{-3}\left[p_s - p_a\right] \quad \text{W/m}^2 \qquad (3.24)$$

where p_s and p_a are in Pa (Pascal).

The saturated water vapor pressure at the skin surface temperature (T_s) lies in the range of 27°C–37°C and may be approximated as a linear function, given by:

$$p_s = 256T_s - 3373 \quad \text{Pa} \qquad (3.25)$$

Substituting value of Ps from Equation (3.25) in Equation (3.24) we have,

$$\dot{Q}_e = 3.05\times10^{-3}\left[256T_s - 3373 - p_a\right]\text{W/m}^2 \qquad (3.26)$$

Substituting value of T_s from Equation (3.17) in Equation (3.26), we have

$$\dot{Q}_e = 3.05\times10^{-3}\left[5733 - 6.99(M-W) - p_a\right] \qquad (3.27)$$

Heat loss of about 10 W/m² results from the water diffusion through the skin. For example, for T_s = 33°C, water vapor pressure p_a = 1400 Pa at 23°C ambient temperature, and 50% relative humidity results in 11.2 W/m² of heat loss.

The rate of heat loss by water diffusion through the skin is a continuous process and is not controlled by the thermo-regulatory system. In order to prevent the body temperature rising during high activity level, sweat rate of evaporation from the skin ($\dot{Q}_{se}$) is one of the most effective techniques.

The value of $\dot{Q}_{se}$ may vary from 0 W/m² (rest position) to 400 W/m² (high activity level) in hot and dry climatic conditions depending upon the level of sweat from individual to individual. People adapted to hot climatic conditions or used to high activity level may improve the function of sweat glands and obtain a better control of the body temperature. An acclimatized person is normally not able to sweat more than 1 liter/h, and a total amount of approximately 3.5 liter. In case this sweat gets evaporated, a total heat loss of 375 W/m² and 8505 kJ is generated. This estimation of heat loss due to evaporation of sweat is a complicated process, which is still not fully understood. In hot climatic conditions and high activity levels, drinking water (along with salt intake) is necessary to sweat enough. Sometimes, all the sweat produced is not able to remove the heat from the body via evaporation. This condition occurs when the excessive sweat drips down the body. Only the sweat

Thermal Comfort

that evaporated from the surface of skin is able to remove the body heat. The following is the expression to co-relate sweat evaporation to metabolism by Fanger [21]:

$$\dot{Q}_{se} = 0.42\left(M - W - 58.15\right) \ \text{W/m}^2 \tag{3.28}$$

On combining Equations (3.27) and (3.28), the evaporative heat exchange at the skin may be expressed as:

$$\dot{Q}_{se} = 3.05 \times 10^{-3}\left[5733 - 6.99\left(M - W\right) - p_a\right] + 0.42\left(M - W - 58.15\right) \tag{3.29}$$

3.5.5 RESPIRATION

Air inhaled is both warmed and humidified by its passage through the respiratory system. The sensible and latent losses are proportional to the volume of flow rate of air to the lungs that in turn is proportional to the metabolic process.

The rate of respiratory convective heat exchange may be expressed as:

$$\dot{Q}_{res} = 0.0014M\left(34 - T_a\right) \ \text{W/m}^2 \tag{3.30}$$

where, T_a is the ambient air temperature in °C.

The rate of respiratory evaporative heat exchange may be expressed as:

$$\dot{Q}_{res} = 1.72 \times 10^{-5} M\left(6867 - p_a\right) \ \text{W/m}^2 \tag{3.31}$$

where, p_a is the ambient water vapor pressure in Pa.

Heat loss due to respiration is only significant at high activity levels and under normal sedentary activity. It can be neglected in case the value is less than 6 W/m².

3.6 THERMAL COMFORT INDICES

Thermal comfort indices have been established to measure and predict the thermal comfort. The simplest way of predicting the thermal comfort is via figures and tables from manuals as per ASHRAE 55 and ISO 7730. The other method, which is numerical and produces a rigorous prediction is predicted mean vote (PMV)/predicted percentage dissatisfied (PPD) indexes and two-node models. PMV/PPD gives the quantitative values of the degree of discomfort and the effectiveness of not only environmental factors but also human factors. For mechanically conditioned spaces, PMV/PPD model should be used.

3.6.1 PREDICTED MEAN VOTE (PMV) INDEX

PMV is the most widely used example of thermal comfort performance indicator and was proposed by Fanger [21] by using heat balance equations and empirical studies about skin temperature. PMV index predicts the mean response of a larger group of people according the ASHRAE thermal sense scale and the Bedford's comfort scale. Seven subjective sensations of warmth (thermal comfort) have traditionally been measured using a seven-point scale as given in Table 3.3.

The PMV index has been accepted as international standard since the 1980s [2] and in ASHRAE 55 [1], and consequently a large number of researchers have taken this index as reference for their studies.

TABLE 3.3

Seven-Point Scale for Thermal Comfort [5]

Scale	Bedford's Comfort Scale [22]	ASHRAE Thermal Scale
+3	too much warm	hot
+2	too warm	warm
+1	comfortably warm	slightly warm
0	comfortable	neutral
−1	comfortably cool	slightly cool
−2	too cool	cool
−3	too much cool	cold

PMV index is based on the physical factors (Section 3.2) like air temperature, mean radiant temperature, relative air velocity and partial water vapor pressure. For estimation of PMV, metabolic processes, i.e., activity level and clothing insulation, need to be assessed. The PMV index was developed for steady-state conditions, but according to several researchers it may be applied with good approximation for minor fluctuations of one or more variables, provided that time-weighted averages of the variables are applied. The general expression is given by:

$$PMV = \left(0.303e^{-0.036M} + 0.028\right)\left[\left(M - W\right) - \left(H + E_c + C_{res} + E_{res}\right)\right] \tag{3.32}$$

Substituting the values from Equations (3.19), (3.22a), (3.28), (3.30) and (3.31) in Equation (3.32), we have

$$PMV = \begin{pmatrix} 0.303 \times e^{-0.036M} \\ +0.028 \end{pmatrix} \begin{cases} \left(M - W\right) - 3.05 \times 10^{-3} \times \left\{5733 - 6.99\left(M - W\right) - p_a\right\} \\ -0.42 \times \left\{\left(M - W\right) - 5815\right\} - 1.7 \times 10^{-5} M\left(5867 - p_a\right) \\ -0.0014 \times M\left(34 - T_a\right) - 3.96 \times 10^{-8} \times f_{cl} \\ \times \left\{\left(T_{cl} + 273\right)^4 - \left(\overline{T_r} + 273\right)^4\right\} - f_{cl} \times h_c \times \left(T_{cl} - T_a\right) \end{cases} \tag{3.33}$$

where,

f_{cl} = ratio of man's surface area while clothed to man's surface area while nude
T_a = ambient temperature in °C
T_r = mean radiant temperature in °C
p_a = partial water vapor pressure in Pa
h_c = convective heat transfer coefficient in W/m²K
T_{cl} = surface temperature of clothing in °C

From Equation (3.22b) the PMV can be estimated for various combinations of metabolic processes, clothing, air temperature, mean radiant temperature, air velocity and air humidity. The equations for T_{cl} and h_c may be solved by iteration. The limitation to the range of conditions over which PMV applies is given in Table 3.4.

3.6.2 PREDICTED PERCENTAGE DISSATISFIED (PPD) INDEX

Fanger [21] developed another equation to relate PMV to PPD. Where PMV gives the mean value of the thermal votes of a larger group of people exposed to the same environment, PPD helps to predict

TABLE 3.4
Limitations to the Range of Conditions Over which PMV Applies [5, 19]

Variable	Lower Limit—Upper Limit (met)
Metabolic rate (M)	0.8 met to 4.0 met; 46 W/m² to 232 W/m²
Clothing insulation (I_{cl})	0 CLO to 2 CLO; 0 m²K/W to 0.310 m²K/W
Air temperature (T_a)	10°C to 30°C
Radiant temperature ($\bar{T}_{ar}$)	10°C to 40°C
Relative air velocity (ν_{ar})	0 m/s to 1 m/s
Water vapor pressure (p_a)	0 Pa to 2700 Pa
Predicted Mean Vote (PMV)	−2 to +2

Note: Inside the range for p_a it is furthermore recommended that the relative humidity be kept between 30%–70%.

the number of people dissatisfied with the thermal environment. PPD is required since individual votes are scattered in the mean value used for PMV estimation and it is necessary to understand the number of people uncomfortable with the exposed thermal environment. PPD is thus, found to be more reliable than PMV. PPD index is found as a function of PMV from the given Equation (3.33) or Figure 3.8, may be expressed as:

$$PPD = 100 - 95 \times e^{-\left(0.03353 \times PMV^4 + 0.2179 \times PMV^2\right)} \tag{3.34}$$

The value of PPD does not fall below 5% for any value of PM due to the difference in thermal sensation between individuals; the thermal neutrality for different people is achieved at environmental parameters, which are not identical.

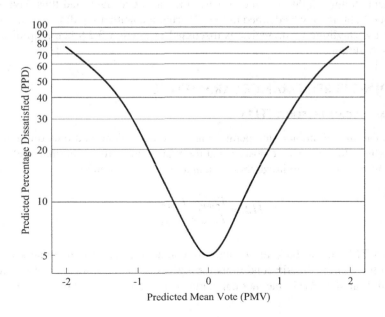

FIGURE 3.8 Predicted percentage dissatisfied (PPD) as a function of predicted mean vote (PMV) [5].

3.6.3 Adaptive Comfort Standard

In order to create a thermally comfortable indoor climate, it is necessary to understand the thermal comfort standards. With changing conditions in the environment, people have a tendency to adapt themselves. This tendency of people is expressed in the adaptive approach to thermal comfort. The adaptive thermal comfort has various approaches as given in the following subsections.

3.6.3.1 Field Studies and Rational Indices

This approach is based on surveys conducted in the field about the thermal environment and response of the people. This approach has a limitation of varying environmental conditions.

3.6.3.2 Rational Approach

The response of people to the thermal conditions is recorded in terms of the physics and physiology of heat transfers in stable conditions in climate chambers. The standards like ISO 7730 [2] and ASHRAE 55 [1] are based on this approach.

The standards should help the architect make decisions in terms of the design of the building. The comfort temperature may be expressed as [23]:

$$T_c = 13.5 + 0.54 T_0 \tag{3.35}$$

where

T_c = Comfort temperature in °C, and
T_0 = Outdoor temperature for free running buildings in °C.

3.6.4 Visual Comfort

Buildings are designed for man, and thus, visual comfort is also a very important parameter. This means people should have enough light for their activities. The light should be of the right quality and balances to ensure good views without strain. Good lighting ensures a pleasant and productive environment. Natural light does much better than artificial light, and thus daylighting is very important. Good lighting may be defined as well distributed, neither too dim nor too strong and uses minimum energy. Daylighting strategies as discussed in Section 3.2.9 can help distribute natural light inside the building.

3.7 BUILDING PERFORMANCE PARAMETERS

3.7.1 Thermal Load Leveling (TLL)

Both solar insolation and ambient temperature are time-dependent, and thus room air temperature varies due to which fluctuations are observed at room air temperature. Thermal load leveling (TLL) is necessary to estimate and minimize these fluctuations and is given by:

$$TLL = \frac{T_{r,\max} - T_{r,\min}}{T_{r,\max} + T_{r,\min}} \tag{3.36}$$

The value of TLL should have more value to a smaller value of denominator (cooling effect) and vice versa for winter climatic conditions for a given value of numerator. When the numerator decreases, there is an increase in the thermal comfort. Thus, TLL should be minimum [24].

3.7.2 Decrement Factor

The decrement factor is defined as reduction ratio of temperature at the inside surface of a room to its outside surface. It is given by [25]

$$f = \frac{\left(T\big|_{x=L}\right)_{max} - \left(T\big|_{x=L}\right)_{min}}{\left(T\big|_{x=L}\right)_{max} + \left(T\big|_{x=L}\right)_{min}} \tag{3.37}$$

It represents the decreasing ratio in the amplitude of heat wave during its propagation from outer surface to the inner surface [26].

3.8 RELATED STANDARDS

The standards related directly or indirectly to the thermal comfort are as follows:

1. Thermal comfort and related thermal environment:

    - ASHRAE Standard 55: Thermal environmental conditions for human occupancy
    - ISO 7730: Moderate thermal environments—Determination of the PMV and PPD indices and specification of the conditions for thermal comfort
    - ISO 7933:1989: Hot environments—Analytical determination and interpretation of thermal stress using calculation of required sweat rate

2. Design of the indoor environment

    - ASHRAE Standard 62.1-2016—Ventilation for acceptable indoor air quality
    - CEN-CR 1752: Ventilation for buildings—Design criteria for the indoor environment

3. Measurement of the indoor thermal environment parameters

    - ASHRAE Standard 55-2020: Thermal environmental conditions for human occupancy
    - ASHRAE Standard 113-2013: Method of testing for room air diffusion
    - ISO 7726:2001: Ergonomics of the thermal environment—Instruments for measuring physical quantities (ISO 7726:1998)

4. Determination of the personal factors

    - ISO 8996:1990: Ergonomics—Determination of metabolic heat production
    - ISO 9920:2007: Estimation of the thermal insulation and water vapour resistance of a clothing ensemble

OBJECTIVE QUESTIONS

3.1 Name the primary and dominant factors for thermal comfort.
   a. Relative humidity
   b. Daylighting
   c. Air pressure
   d. Gender differences

3.2 Choose the physiological aspects for thermal comfort.
    a. Acoustics
    b. Nutritional intake
    c. Air electricity
    d. Occupancy

3.3 Which parameter is not a part of psychometric chart.
    a. Moisture content
    b. Wet bulb temperature
    c. Relative humidity
    d. Specific humidity

3.4 What is the recommended value for dry bulb temperature to achieve the thermal comfort for summer condition.
    a. $(25 \pm 1)°C$
    b. $(50 \pm 1)°C$
    c. $(00 \pm 1)°C$
    d. $(15 \pm 1)°C$

3.5 What is the recommended value of the relative humidity for both summer and winter conditions.
    a. $(25 \pm 10)\%$
    b. $(50 \pm 10)\%, (10 \pm 50)\%$
    c. $(10 \pm 50)\%, (00 \pm 25)\%$
    d. $(00 \pm 25)\%, (25 \pm 10)\%$

3.6 What is the thermal comfort temperature for both summer and winter climatic conditions.
    a. 15°C–25°C
    b. 20°C–25°C
    c. 25°C–30°C
    d. 30°C–35°C

3.7 What is sensible heat loss?
    a. Conduction and radiation
    b. Convection and conduction
    c. Radiation and evaporation
    d. Radiation and convection

3.8 Heat produced by a healthy person in sleeping mode is:
    a. $30 \ W/m^2$, 60 W
    b. $30 \ W/m^2$, 40 W
    c. $35 \ W/m^2$, 60 W
    d. $30 \ W/m^2$, 40 W

3.9 Horizontal surface placed above eye-level with high reflectance surfaces are called
    a. Heliostats
    b. Clerestory windows
    c. Tubular daylight devices
    d. Light shelves

3.10 PMV stands for
    a. Predicted mean vote
    b. Percentage mean vote
    c. Persons mean vote
    d. People mean vote

ANSWERS

3.1   a

3.2   b

Thermal Comfort 109

3.3 d
3.4 a
3.5 b
3.6 b
3.7 d
3.8 c
3.9 d
3.10 a

REFERENCES

[1] ASHRAE-55, *Thermal Environment Conditions for Human Occupancy*, American Society of Heating Ventilating and Air-Conditioning Engineers, USA, 1992.

[2] ISO-7730, *Moderate Thermal Environments—Determination of the PMV and PPD Indices and Specification of the Conditions for Thermal Comfort*, International Organisation for Standardisation, Geneva, 1994.

[3] G. N. Tiwari, *Solar Energy: Fundamentals, Design, Modelling and Applications*, Narosa, 2002.

[4] S. Karjalainen, "Thermal comfort and gender: A literature review," *Indoor Air*, 22, pp. 96–109, 2012.

[5] B. Agrawal and G. N. Tiwari, *Building Integrated Photovoltaic Thermal Systems: For Sustainable Developments*, UK: RSC Publishing, 2010.

[6] M. Zinzi and A. Mangione, "The daylighting contribution in the electric lighting energy uses: EN standard and alternative method comparison," *Energy Procedia*, vol. 78, pp. 2663–2668, 2015.

[7] E. J. Gago, T. Muneer, M. Knez and H. Köster, "Natural light controls and guides in buildings. Energy saving for electrical lighting, reduction of cooling load," *Renewable and Sustainable Energy Reviews*, vol. 41, pp. 1–13, 2015.

[8] P. Ihm, A. Nemri and M. Krarti, "Estimation of lighting energy savings from daylighting," *Building and Environment*, vol. 44, no. 3, pp. 509–514, 2009.

[9] K. S. Y. Wan and F. W. H. Yik, "Building design and energy end-use characteristics of high-rise residential buildings in Hong Kong," *Applied Energy*, vol. 78, no. 1, pp. 19–36, 2004.

[10] Pinterest. [Online]. Available: https://www.pinterest.ch/pin/26177241564818365/. [Accessed 20 March 2020].

[11] Wikipedia. [Online]. Available: https://en.wikipedia.org/wiki/Light_tube#/media/File:Sonnenrohr.svg. [Accessed Month 2020a]

[12] Pinterest. [Online]. Available: https://in.pinterest.com/pin/210402613815504220/. [Accessed 2020].

[13] Wikipedia, 2020. [Online]. Available: https://en.wikipedia.org/wiki/Light_tube.

[14] N. Gupta and G. N. Tiwari, "Energy matrices of building integrated photovoltaic thermal systems: A case study," *Journal of Architectural Engineering*, 05017006-1–05017006-14, 2017b.

[15] N. Gupta and G. N. Tiwari, "A thermal model of hybrid cooling systems for building integrated semi-transparent photovoltaic thermal system," *Solar Energy*, vol. 153, pp. 486–498, 2017a.

[16] Wikipedia. [Online]. Available: https://en.wikipedia.org/wiki/Architectural_light_shelf#/media/File:Bronx_Library_Center_second_floor_interior.jpg. [Accessed 2020].

[17] Wikipedia, "Architectural light shelf," [Online]. Available: https://en.wikipedia.org/wiki/Architectural_light_shelf [Accessed April 2020].

[18] J. Miller, "Wearing the right clothes," *Science News*, vol. 25, no. 11, pp. 396–397, 1980.

[19] M. Jang, C. Koh and I. Moon, "Review of thermal comfort design based on PMV/PPD in cabins of Korean maritime patrol vessels," *Building and Environment*, vol. 42, no. 1, pp. 55–61, 2007.

[20] R. D. Dear, E. Arens, Z. Hui and M. Oguro, "Convective and radiative heat transfer coefficients for individual human body segments," *International Journal of Biometeorology*, vol. 40, pp. 141–156, 1997.

[21] P. Fanger, *Thermal Comfort: Analysis and Applications in Environmental Engineering*, New York: McGraw-Hill, 1970.

[22] T. Bedford, *The Warmth Factor in Comfort at Work*, London, 1936.

[23] J. Nicol and M. Humphreys, "Adaptive thermal comfort and sustainable thermal standards for buildings," *Energy and Buildings*, vol. 34, pp. 563–572, 2002.

[24] N. Gupta and G. N. Tiwari, "Effect of heat capacity on monthly and yearly exergy performance of building integrated semitransparent photovoltaic thermal system," *Journal of Renewable and Sustainable Energy*, vol. 9, 023506, 2017a.

[25] G. N. Tiwari, H. Saini, A. Tiwari, A. Deo, N. Gupta and P. Saini, "Periodic theory of building integrated photovoltaic thermal (BiPVT) system," *Solar Energy*, vol. 125, pp. 373–380, 2016.

[26] H. Asan, "Effects of wall's insulation thickness and position on time lag and decrement factor," *Energy and Buildings*, 28(3) pp. 299–305, 1998.

4 Energy and Exergy Analysis

4.1 INTRODUCTION

The branch of physics that deals with temperature and heat and their relations with work and energy is referred to as "thermodynamics".

Thermodynamic systems are a group of atoms or molecules of a physical system. Their macroscopic properties may be defined as temperature (T), pressure (P), heat capacity (M_w) and mass density (ρ), etc. The condition of the system directs these properties. In principle, the volume of the physical system should be proportional to the number of atoms or molecules. Thus defining the physical system as infinitely large. When the system's volume and the number of atoms or molecules increases so that the density of these atoms/molecules is constant, the thermodynamic limit is attained. In thermodynamic process, the macroscopic properties like thermal energy and the movement of atoms/ molecules of physical system are periodic in nature, which changes with change with time.

Thermodynamics is comprised of three systems: open, closed and isolated. An open system has the ability to exchange both energy and matter with its surroundings. For example, a stove top where heat and water vapor are lost to the surroundings, i.e., air. In a closed system, energy can only be exchanged with its surroundings but not matter. For example, an extremely tight lid over the previous example will convert the system to a closed one. The third system, i.e., the isolated system, is the one which cannot exchange both energy and matter with its surroundings. It is hard to find a true isolated system.

4.1.1 BRIEF HISTORY OF THERMODYNAMICS

French physicist Sadi Carnot, "the father of thermodynamics", initiated the concept of a thermodynamic system and laid the foundation of modern thermodynamics through reflections on the motive power of fire [1]. He typically studied a body of water vapor (i.e., working substance) in steam engines to understand the system's performance with application of heat. Carnot's book addressed the power, heat, energy and engine efficiency and outlined the basic energetic relations between the Carnot engine, the Carnot cycle and motive power. Later, Rudolf Clausius in 1850 [2] referred to the system as a "working body" by including the concept of surroundings to the concept proposed by Sir Carnot. Clausius mentioned that working body can be any fluid or vapor body through which heat (Q) can be introduced to produce work. In 1859, the first textbook on thermodynamics was authored by William Rankine.

4.2 LAWS OF THERMODYNAMICS

The system attains thermal equilibrium when the macroscopic properties do not change with time, i.e., in the absence of any thermodynamic process. This thermodynamic equilibrium is the simplest state of a thermodynamic system. There are few restrictions imposed by the laws of thermodynamics on the possible equations of state and on the characteristic equation. There are four laws of thermodynamics, which define the fundamental physical properties like temperature, energy and entropy that define thermodynamic systems at thermal equilibrium. The behaviors of these quantities under various situations are defined by these laws. Following are the four laws of thermodynamic.

DOI: 10.1201/9780429445903-4

4.2.1 The Zeroth Law of Thermodynamics

The zeroth law states that when two systems, say A and B, are in thermal equilibrium with a third system, say C, then they are said to be in thermal equilibrium with each other as well. This law establishes an equivalence relationship between two systems that is based on a mutual property between these two systems, i.e., temperature. Thus, this law defines the concept of temperature.

This means that the systems are in thermal equilibrium if the small, random exchanges between these systems does not result in a net change in energy.

4.2.2 The First Law of Thermodynamics

The first and second law of thermodynamics together came into existence in the 1850s predominantly out of the works of William Rankine, Rudolf Clausius and William Thomson (Lord Kelvin). The first law of thermodynamics states that "the internal energy of an isolated system is constant." This means that energy can be transformed from one form to another but can neither be created nor destroyed. It is an expression of the principle of conservation of energy. In simpler words, light bulbs transform electrical energy into light energy, trees convert the solar energy into chemical energy that is stored in organic molecules. The law of conservation forms the basis of thermal analysis of solar energy systems. This is a reversible process and can be expressed as:

$$\textit{The available solar energy = the useful energy + the lost energy} \qquad (4.1a)$$

or,

$$\textit{The rate of available solar energy = the rate of useful energy + the rate of lost energy} \qquad (4.1b)$$

The first law of thermodynamics is valid for medium operating temperatures ($<100°C$) of solar thermal systems.

Let us understand the first law of thermodynamics from a simple example as follows:

Suppose there is a macroscopic physical system that reversibly changes its state of equilibrium from P_1 to P_2 when thermal energy, Q, and mechanical work, W, is applied over it. This change can be accomplished in several ways, but following the first law of thermodynamics, the sum of thermal energy and mechanical work, $(Q + W)$, is independent of different paths as long as two equilibrium states are fixed. For an extremely minor deviation, the measure of $(d'Q + d'W)$ depends on the starting and concluding states of the system. Change in internal energy, $(dU = d'Q + d'W)$, of the system as it moves from equilibrium state P_1 to P_2 can be mathematically expressed as:

$$dU = U_{P_2} - U_{P_1} = d'Q + d'W, \text{ closed system} \qquad (4.2)$$

Mathematically, $d'Q$ and $d'W$ individually are path dependent, but their sum $d'Q + d'W$ is path independent.

Referring to sign convention given by Clausius for cyclic processes, Equation (4.2) can be expressed as

$$\Delta U = Q - W \qquad (4.3)$$

where, ΔU is mechanical equivalent of heat.

4.2.3 The Second Law of Thermodynamics

The first law of thermodynamics states that energy cannot be created or destroyed, which means that energy can only be changed from more useful forms into less useful forms (i.e., heat energy).

Energy and Exergy Analysis

A relationship between the supply of heat and work done on application of heat is given by the second law of thermodynamics. The following are the statements coined by eminent scholars in regard to the second law of thermodynamics:

- **Clausius's principle:** Heat cannot spontaneously flow from lower temperature (i.e., colder body) to higher temperature (hotter body) unless some other changes in the system are made.
- **Principle of Caratheodory:** The principle states that in every neighborhood of any state, say, S, of an adiabatically enclosed system there are states that are not accessible from S.
- **Kelvin's principle:** From a hot reservoir, the heat extracted, (Q_H), cannot be entirely put to work (W). The difference between the obtained heat and the work done, i.e., $Q_C = (Q_H - W)$, is equal to the amount of heat that should be rejected to surroundings. This forms the base of a perfect heat engine. The amount of unavailable energy to perform the work, (Q_C), is referred to as entropy as given in Figure 4.1. This is also called the "first-form" of the second law of thermodynamics or the Kelvin-Planck statement of the second law.

In simpler words, every transfer of energy result in the conversion of some part of energy into unusable form. That means with every heat transfer, there is a rise in the entropy of the universe and reduction in the amount of usable energy that is available to perform work. Thus, the second law talks about the direction of the process, i.e., from lower to higher entropy. Entropy can be explained as the measure of the progress of the system. For an isolated system that is not at equilibrium, entropy tends to increase and approach its maximum value at equilibrium. Further, the Carnot efficiency of a heat engine that is functioning between two levels of energy is defined in terms of absolute temperature as:

$$\eta = \left[1 - \frac{T_c}{T_h}\right] \tag{4.4}$$

where

η = thermal efficiency,
T_h and T_c = Hot and cold (surrounding) reservoir temperature in K.

The second law of thermodynamics functions for higher operating temperature range ($\gg 400°C$).

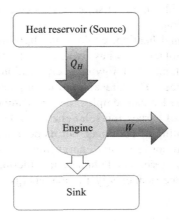

FIGURE 4.1 Schematic diagram of a heat engine system [3].

The concept of exergy also forms its base from the second law of thermodynamics. Exergy may be expressed as the maximum output work as it reaches equilibrium with its surroundings. The numeric value of exergy will be equal to zero for the system in equilibrium with its surroundings. Exergy will be discussed in detail under Section 4.4.

First law confirms that energy cannot be destroyed, and this statement can be true only for reversible processes. On the other hand, exergy of the system is dependent on the irreversibility of the systems, which is directly proportional to the increase in entropy of system in surrounding. Energy is destroyed when the process functions between two temperature ranges and this destroyed energy is referred to as anergy.

4.2.4 THE THIRD LAW OF THERMODYNAMICS

According to the third law of thermodynamics, when the temperature of any system approaches absolute zero, the entropy of that system reaches a minimum constant value. Entropy is related to the degree of disorderedness in the system in statistical thermodynamics. With decrease in the temperature, the thermal motion within the system suppresses, resulting in an ordered state. At a highly-ordered state, entropy reaches its minimum value, which at low temperature is almost at absolute zero. Further, since there is a constant minimum value of entropy near absolute zero, the absolute entropy cannot be defined. Thus, only the change in entropy can be determined for different systems operating at different temperatures. The third law provides an absolute reference point for the determination of entropy.

4.3 ENERGY ANALYSIS

4.3.1 INTRODUCTION

Energy may be defined as the fourth basic need of a human being and plays an important role in the existence of human life for almost all activity types like household, transportation, industrial, medical, etc. The conventional source of energy (fossil fuels) is finite in nature and contribute to climatic changes like global warming and environmental changes since when burned they release harmful pollutants that are responsible for greenhouse effect and raise other concerns like geopolitical and military conflicts, a rise in fuel prices, etc. Thus, with fossil fuels unsustainable in nature, the world has to think about the alternate source of energy that can act as a solution to the previously stated problems with respect to conventional sources of energy. Solar energy emerges as a solution to the growing energy challenges/crisis. Every time, we burn fossil fuel we emit harmful gases to the environment. Thus, technology should be decentralized, eco-friendly, free from contamination, easily obtainable, and solar energy provides the best solution on planet Earth since most parts of world receive high level of solar insolation.

The techniques used for thermal heating, cooling and lighting of buildings in the past were always harmonious with nature. Solar energy has always been part of building design since ancient times. The Gardens of Mesopotamia and Egyptian civilization are still the most spectacular examples, built on series of terraces. The Greeks used town planning and orientation to harness the winter sun. The Roman baths, Roman temples like the Pantheon and the greenhouses of the 18th century are a few examples of harnessing solar energy. Solar energy was harnessed for thermal heating and cooling and lighting of buildings, but its potential was underutilized. Therefore, the technology was developed to convert solar energy into electricity using photovoltaic cells. However, energy analysis of solar thermal and photovoltaic thermal (PVT) systems can determine the impact on the environment. Electrical energy, thermal energy and daylighting can be produced using solar energy.

Energy and Exergy Analysis

4.3.2 Energy Matrices

The performance of buildings is appraised by using essential energy matrices namely energy payback time (EPBT), energy production factor (EPF) and life cycle conversion efficiency (LCCE). These energy matrices are significant for renewable technologies as their use is irrelevant if the energy consumed in the manufacturing process is higher than what will be produced in their life span. Annual energy savings can be calculated as a sum of thermal energy, daylight (in case semi-transparent photovoltaic module is used) and electrical power produced by the photovoltaic modules integrated/mounted with the building.

4.3.3 Embodied Energy Analysis

Thermodynamics gives rise to the concept of embodied energy, initially involving the development of steam engines. Embodied energy is important to study the life cycle and impact of any system on the environment. Embodied energy (EE) is the total energy required to construct and maintain every element and operation of the system over the complete life cycle [4]. It is the amount of consumption of energy to bring the system in operational mode, i.e., starting from the extraction of raw to fabrication and installation of each element. The improvement in design and efficiency of the system can reduce the operational energy consumption, but generally embodied energy in such cases is ignored. Embodied energy is neglected due to lack of data availability, lack of procedures to assess the EE associated with each element and the process and ultimately assuming that the embodied energy of various elements and processes is irrelevant.

Due to the complexity of the supply chain, it becomes more important to evaluate the associated embodied energy with the process. This complexity here implies that the supply chain has to be analyzed for each product and process upstream to the raw materials [5]. In order to calculate the embodied energy (EE), the total energy consumption in the production process of each element is calculated and summed up. For this, the energy densities should be known. They can either be referred from the available literature or provided by the manufacturers of the materials of every element, which are then multiplied to the quantity of the same used in the fabrication process of the component/element. For life cycle analysis, some environmental factors like CO_2 emission during the fabrication of the component of the system etc., in addition to embodied energy are also considered. EE for different materials is explained in Chapter 9.

For the building integrated photovoltaic (BiPV) system, the EE includes indirect and direct energy. Direct energy is the one that is at the site for the operation's purpose, whereas indirect energy may include the energy that is consumed in the manufacturing of photovoltaic modules and structural members of the building. The indirect energy of the solar cells and structure would, in turn, comprise energy embodied directly in the extraction and transportation of silicon, iron ore, etc. The precision and degree of an embodied energy analysis depends on the type of method selection, which includes process analysis, input–output (I-O) analysis and hybrid analysis.

4.3.4 Energy Density Analysis

The energy densities or the intensities are available in the literature, and their values are taken from various national or international texts, particularly in Google search. These values are a necessity to explain the descriptions of technique boundaries or understand whether these values are of primary or delivered energy. To arrive at an exact and authentic energy database for the materials consumed in photovoltaic thermal (PVT) systems is not an easy task. This requires an extensive comparison of materials along with their exact useable quantities. Energy density quantities are a most crucial necessity in the calculations of embodied energy at the stage of design. The following methods may be used to execute the energy analysis.

4.3.4.1 Process Analysis

This is a common method and is used to calculate the needs for direct and indirect energy (via the condition of other services and goods crossing the system periphery and capital equipment along with buildings). In this method, the definition of system periphery is very crucial, thus proper selection of the same can give the significant ranges of results. Basically, process analysis recognizes the inputs and outputs to systems and assigns energy values to the product flows. As discussed, the total embodied energy is the total of direct energy (main manufacturing process) and indirect energy (of the material inputs). This method is associated with major restrictions like incompleteness of the system boundary. The quantification of all the inputs to the system is the most important stage of this process for which a boundary has been drawn around the quantification of inputs that are being evaluated mainly due to complications in procuring necessary data and its understanding. Various inputs are, therefore, ignored, leaving the boundary incomplete. The magnitude of the incompleteness varies with the type of product or process and depth of study but can be 50% or more [5–8].

4.3.4.2 Input-output Analysis

The input-output (I-O) analysis was developed for economic analysis, and this is capable of tracing the economic flows of goods and services. The I-O data, which are the national average statistics that model the financial flows between the sectors of economy can be used to fill the gaps caused by incomplete boundaries of the system [5, 6]. This analysis is based on various assumptions suitable for national modeling; therefore even a perfect I-O model may not give valid results for a particular product. Miller and Blair [9] and Lenzen [8] detailed main limitations and included homogeneity and proportionality assumptions, sector classification and aggregation. The Australian Bureau of Statistics publishes I-O matrices for 109 economic sectors every five years.

4.3.4.3 Hybrid Analysis

The process analysis quantifies and examines the production process including the overall impact on the environment. Following it, the energy densities of services and goods further upstream are assimilated by the second method, input-output analysis. Hybrid analysis is used to minimize the limitations and errors of these traditional methods. This approach is efficient in eliminating the errors that occur in I-O analysis from a large proportion of the results, but the energy intensities are only applicable to materials and products that are manufactured by the specified process audited and is not realistic globally. In calculations of embodied energy, it is important to use realistic conversion factors for the comparison of different energy is given in Appendix F.

4.3.5 An Overall Thermal Energy

Thermal and electrical energy is extracted from solar energy. The former is low-grade energy while the latter is high-grade energy, thus both cannot be summed up directly for first law (of thermodynamics)–based energy analysis. To calculate overall thermal energy from any system, the electrical energy is converted to the equivalent thermal energy. Mathematically, this conversion is expressed as:

$$Q_{overall,\ thermal} = Q_{thermal} + \frac{E_{electrical}}{\eta_{cp}} \tag{4.5}$$

where η_{cp} is the electrical conversion factor.

4.3.5.1 Energy Payback Time (EPBT)

Energy payback time (EPBT) is the total time period needed to recover the entire energy consumed to construct the building (embodied energy, EE). The EPBT of system is defined as the recovery period of EE. The energy sustainability of the building can be measured by EPBT, and it is always

Energy and Exergy Analysis 117

one of the major criteria to compare the feasibility of one technology against the other. Hunt [10] found that EPBT for a PV module is 12 years. Kato et al. [11] have similar results for crystalline silicon (c-Si) PV modules. The EPBT for a c-Si PV module under Indian climatic conditions for an annual peak load duration was found to be 4 years [12]. EPBT should be as minimal as possible to achieve better cost-effectiveness of the system. Mathematically, EPBT is given as [3]:

$$EPBT = \frac{Embodied\ Energy}{Annual\ energy\ output\ from\ the\ system} = \frac{E_{in}}{E_{aout}} \tag{4.6}$$

The embodied energy can be easily calculated for a solar thermal system apart for the PV module since it undergoes several technical processes. The embodied energy of a solar thermal system may be calculated as follows:

- The mass of different materials (m_i) used in the manufacturing process of solar energy technology to be multiplied by its corresponding energy density (e_i), which gives $m_i e_i$.
- Summation of $m_i e_i$ gives the total EE.

In order to reduce the value of EPBT for PV modules, the following shall be considered for other renewable technologies:

(i) Increase of efficiency for increasing annual output (E_{aout})
(ii) Use of cost-effective materials with low-energy densities for longer life to reduce embodied energy (E_{in})
(iii) Minimum annual operating maintenance of system

Several factors are required to evaluate EPBT for BiPVT systems. These factors are solar cell technology, inverter, charge controller, battery, building materials, etc. The performance of BiPVT is determined by irradiation and the performance ratio. However, mathematically, EPBT for the BiPVT system can be expressed as

$$EPBT = \frac{E_{building} + E_{support} + E_{BIPVT} + E_{cc} + E_{inv} + E_{bat} + E_{inst+M\&O} + E_{dec} - E_{rec}}{E_{aout}} \tag{4.7}$$

where $E_{building}$, $E_{support}$, E_{BIPVT}, E_{cc}, E_{inv}, E_{bat}, $E_{inst+M\&O}$, E_{dec}, E_{rec} are the embodied energy of the building, support for the BiPVT system, BiPVT system, charge controller, inverter, battery, installation maintenance and operation, decommissioning and recycling, respectively. E_{aout} is the rate of exergy produced from the system, which is given by

$$E_{aout} = \dot{E}x_{el} + \dot{E}x_{th} \tag{4.8}$$

where $\dot{E}x_{el}$ and $\dot{E}x_{th}$ are the rate of electrical and the thermal exergies equivalent. Substituting its value, we have annual exergy output as

$$E_{aout} = \left[\eta_{ca} \times I(t) \times bL \times n_s \cdot n_p \right] + \left[\dot{Q}_u \times \left(1 - \frac{T_a}{T_{airout}} \right) \right] \tag{4.9}$$

For the sustainable energy character, the EPBT should be less than the entire installation service period.

$$EPBT \leq n_{sys} \tag{4.10}$$

The EPBT as an indicator of energy performance has an appeal because of their similarity with economic payback times. A drawback of EPBT is that it does not account for the energy gain during the rest of the economic lifetime. Indicators that fulfill this requirement are the electricity production factor and the life cycle conversion efficiency.

The EPBT should be less than the entire installation service period for sustainable energy character. EPBT does not include the energy gain during the rest of the economic lifetime. Energy production factor and life cycle conversion efficiency are the indicators that fulfill this requirement.

Few literature works for EPBT studies have been tabulated in Table 4.1.

4.3.6 ENERGY PRODUCTION FACTOR (EPF)

Energy production factor (EPF) can help predict the overall performance of the system and is based on the lifetime of the system. On a lifetime basis, it is defined as the ratio of the total energy output during the service time of the system to the total energy input to the system. EPF should be greater than 1. A higher value of EPF implies better cost-effectiveness of the system.

$$EPF = \chi_{LT} = \frac{Total\ energy\ output\ during\ the\ service\ time\ of\ the\ system}{Total\ energy\ input\ to\ the\ system\ during\ the\ service\ time} = \frac{E_{out}}{E_{tot}} > 1 \qquad (4.11)$$

The total energy output during the service time may be considered equivalent to the product of annual energy produced (E_{aout}) and the lifetime of the system (n_{sys}). Therefore, Equation (4.11) can be rewritten as

$$\chi_{LT} = \frac{\left(E_{aout} \times n_{sys}\right)}{E_{in}} \qquad (4.12)$$

or (using Equation 4.6),

$$\chi_{LT} = \frac{n_{sys}}{EPBT} \qquad (4.13)$$

Thus, the electricity production factor (EPF) is inversely proportional to the EPBT. For the sustainable energy character, the EPF should be greater than unity:

$$\chi_{LT} \geq 1 \qquad (4.14)$$

4.3.7 LIFE CYCLE CONVERSION EFFICIENCY (LCCE)

Life cycle conversion efficiency (LCCE) can be expressed as the ratio of net energy productivity with respect to the solar input (radiation) over the lifetime of the system to the solar input during the same period [13]. LCCE is mostly used by the planners. The numerical value of LCCE is always less than one. However, if the value of LCCE reaches one, the technology is considered to be the best from an energy point of view.

Mathematically,

$$\varphi = \frac{\left(Annual\ saved\ energy \times life\ of\ the\ building\right) - Embodied\ energy}{Total\ annual\ solar\ energy \times life\ of\ the\ building} \qquad (4.15)$$

TABLE 4.1
Literature Review of EPBT for Photovoltaic Systems

Author/Ref	Year	PV Technology	Some Assumptions			EPBT	Carbon Footprint (gCO$_2$/kWh)
			η%	Life-time, years	BOS		
Hunt [10]	1976	Silicon solar cells from the raw material SiO$_2$				12 years for terrestrial cells 24 years for space cells	
Hynes et al. [14]	1991	CIS thin films	10	20		3–48 months	
Palz and Zibetta [15]	1991	mc-Si and a-Si	12 and 6			2.1 and 1.2	
Alsema et al. [16]	2000	mc-Si and a-Si	13.0 and 7.0	30	Y	3–4 and 2.5–4	60 and 50
Bossert et al. [17]	2010	mc-Si	16–13.5			4.1–2.3	
		a-Si a-(Si; Ge:H)	10			1.9–3.0	
		Thin-film Si	14			4.7	
		Cu (in, Ga) (Se, S)2	12			1.8–1.3	
		CdTe	10			0.9–0.5	
		Dye sensitized	8			1	
Knapp and Jester [18]	2001	sc-Si		30		4.1	
		CIS				2.2	
Meijer et al. [19]	2003	InGaP/mc-Si	25			5.3	
		InGaP				6.3	
		mc-Si				3.5	
Jungbluth [20]	2005	c-Si (sc/mc)	14.8–17.5	30	Y	3.0–6.0	39–110
Peharz and Dimroth [21]	2005	Con III–V multi-junction	26		Y	0.67	
Alsema et al. [22]	2006	mc-Si	13.2	30	Y	1.8	32.5
Fthenakis and Alsema [23]	2006	mc-Si	13.2	30	Y	2.2	37
		CdTe	8.0/9.0	30		1.0/1.1	21/25–18
Veltkamp and de Wild-Scholten [24]	2006	Dye sensitized	8			1.3–0.8–0.6	20–120
Raugei et al. [25]	2007	CdTe	9	20	Y	1.5	48
		CIS	11			2.8	95
		mc-Si	14			5.5–2.4–2.5	167–72–57

(Continued)

TABLE 4.1 (Continued)
Literature Review of EPBT for Photovoltaic Systems

Author/ Ref	Year	PV Technology	Some Assumptions			EPBT	Carbon Footprint (gCO$_2$/kWh)
			η%	Life-time, years	BOS		
Pacca et al. [26]	2007	PVL 136 a-Si	6.3		Y	3.2	34.3
		KC 120 mc-Si	12.92			7.5	72.4
Roes et al. [27]	2009	Polymer	5	25		0.93	727
de Wild-Scholten, M. et al. [28]	2010	Rooftop flat-plate PV (multi-junction PV cell)				0.8–1.9 years	8–45 g CO$_{2. eq}$/kWh
Nishimura et al. (2010) [29]	2010	High-concentration PV and multicrystalline Si PV, 100 MW:				The EPBT of the high-concentration PV was found to be 0.27 years longer than that of the multi-crystalline-Si PV system	
Fthenakis and Kim (2013) [30]	2012	Multi-junction GaInP/ GaInAs/Ge cells grown on a germanium substrate		over 30 years		0.9 years;	27
Peharz and Dimroth (2005) [21]	2005	III–V multi-junction				0.7–0.8 years	
Cellura et al. (2011) [31]	–	Crystalline Si				0.7 years	
Sherwani et al. [32]	2011	a-Si				2.5–3.2	15.6–50
		sc-Si				3.2–15.5	44–280
		mc-Si				1.5–5.7	9.4–104
Azzorpadi and Mutale [33]	2010	QD			25	1.51	2.89 g CO$_{2. eq}$/kWh
Burkhardt III et al. (2011) [34]	2011					About 1 year	26 g CO$_{2. eq}$/kWh

Energy and Exergy Analysis

The ratio of annual energy output to the solar energy input is called the efficiency (η) of the system. Therefore, LCCE can be written in the terms of efficiency as:

$$\varphi = \eta\left(1 - \frac{EPBT}{n_{sys}}\right) = \eta\left(1 - \frac{1}{\chi_{LT}}\right) \tag{4.16}$$

Thus, for a higher LCCE of the system, in addition to the improvement in the efficiency, the system should have the lower EPBT and a higher installation service period.

4.3.8 ENERGY MATRICES OF PHOTOVOLTAIC (PV) MODULE

As discussed, the calculations of total embodied energy for a PV module per m² is not an easy task since it requires the energy spent in the manufacturing of each and every component. The energy required in different processes for production of PV modules is given in Table 4.2.

The embodied energy of a PV module for 1 m² is given in Table 4.3. The details of the calculations may be referred from [35–37]. Suppose embodied energy for stand-alone and rooftop per m² is considered to be 500 kWh and 200 kWh, respectively. The total embodied energy of a PV module per m² with installation for stand-alone and rooftop integration to a building will be 1571 and 1271 kWh/m², respectively, for case (i) of Table 4.3.

The data in Table 4.3 excludes the embodied energy of the balance of system (BOS) of a PV system.

The following reasons are stated behind the generation of data as tabulated in Table 4.3.

TABLE 4.2

Energy Requirement (Energy Density) in Different Processes for Production of PV Module [3]

S.No.	Process		Energy Requirement
1	Purification and processing of silicon		
	Type of silicon	Production from	
1a	Metallurgical-grade silicon (MG-Si)	silicon dioxide (quartz, sand)	20 kWh/kg of MG-Si
1b	Electronic-grade silicon (EG-Si)	MG-Si	100 kWh/kg of EG-Si
1c	Czochralski silicon (Cz-Si)	EG-Si	290 kWh/kg of EG-Si
2	Fabrication of solar cell		120 kWh/m² of silicon cell
3	Assembly of PV module		190 kWh/m² of PV module
4	Rooftop integrated PV system		200 kWh/m² of PV module

TABLE 4.3

Embodied Energy of PV Module for 1 m² [3]

Process	Metallurgical-Grade Silicon (MG-Si)	Electronic-Grade Silicon (EG-Si)	Czochralski Silicon (Cz-Si)	Cell Fabrication	Module Assembly	Total
Case (a)	48.0 kWh	230.0 kWh	483.0 kWh	120.0 kWh	190.0 kWh	1071.0 kWh
Case (b)	26.54 kWh	127.3 kWh	267.33 kWh	60.3 kWh	125.4 kWh	607.0 kWh
Case (c)	4.8 kWh	23.0 kWh	48.3 kWh	90.0 kWh	95.0 kWh	261.0 kWh

Reduction in the solar mass: This may be because of development of new materials. Case (iii) has been considered as 10% of case (i).

Cell processing energy: It is reduced by 75% of case (i) to case (iii).

Cell efficiency: It is increased by 4%.

Elimination of wafer trimming and packaging.

Reduction of embodied energy of module assembly by 50%.

For details of the above calculation refer to the book by Tiwari and Ghosal [35] and Tiwari and Dubey [36]. Above results show that an embodied energy of a PV module is reduced significantly from 1071 to 261 kWh/m^2 due to various reason described above.

The PV module itself is called the system. Other components are called balance of system. BOS includes wiring, electronic components, foundation, support structure, battery, installation, etc. For open field installation, the concrete, cement and steel are the main components used for foundation and frame, which requires maximum energy. The energy requirement (energy density) for open field installation is 500 kWh/m^2 of panel. For rooftop integrated PV systems, the energy requirement is reduced to 200 kWh/m^2 of panel due to the absence of foundation and structure for frame. Further the present embodied energy can have lower value (say about 75% of its present value) in the future due to development of new materials used for BOS.

$$Annual\ output\ of\ PV\ module = \eta_m \times \overline{I} \times A_m \times N \times n_0 \tag{4.17}$$

where

A_m and η_m are area and electrical efficiency of PV module,
N and n_0 are number of sunshine hours and clear days in a year.

The $\overline{I}$ is the annual average solar intensity, which is different for different places, e.g., $\overline{I}$ for Port Hedland (NW Australia, India and Sydney are 2494, 1800 and 1926 kWh/m^2/year, respectively. Thus, the annual electrical output will be maximum for Port Hedland (NW Australia).

Annual output for PV module for $\eta_m = 0.12$, $A_m = 1$ m^2 for different locations is given below:

$$\textbf{\textit{Annual output of PV module}} (Port\ Hedland\ (NW\ Australia) = 0.12 \times 2494 \times 1 = 299.28\ kWh$$

$$\textbf{\textit{Annual output of PV module}}\ for\ 1m^2 (Sydney) = 0.12 \times 1926 \times 1 = 231.12\ kWh$$

$$\textbf{\textit{Annual output of PV module}}\ for\ 1m^2 (India) = 0.12 \times 1800 \times 1 = 216\ kWh$$

The matrices for a PV module can be obtained from Table 4.4 for lifespan of PV module as $T = 30$ years.

From Table 4.4, it can be concluded the PV module is best suited for Port Hedland (NW Australia) because of lowest value of EPBT and highest value of EPF and LCCE.

4.4   EXERGY ANALYSIS

All macroscopic processes are irreversible as per the second law of thermodynamics and this law gave the concept of exergy. A non-recoverable loss of exergy is involved in every such process. It is expressed as the product of the ambient temperature, T_a, and the entropy generated (the sum of

TABLE 4.4
Matrices of PV Module for Different Places [3]

Port Hedland

Energy Matrices	Case (i)	Case (ii)	Case (iii)
EPBT	3.6	4.64	4.96
EPF	8.3	6.47	6.05
LCCE	0.105	−0.101	0.100

Sydney

Energy Matrices	Case (i)	Case (ii)	Case (iii)
EPBT	2	2.62	2.81
EPF	15	11.45	10.68
LCCE	0.112	0.109	0.109

India

Energy Matrices	Case (i)	Case (ii)	Case (iii)
EPBT	0.87	1.13	1.21
EPF	34.5	26.55	24.79
LCCE	0.116	0.110	0.115

the values of the entropy increases for all the bodies that form a part of the process). Some of the components of entropy generation can be negative, but the overall sum is always positive. Let us understand this taking a simple example of a waterfall. With fall of water from a height, potential energy is converted into kinetic energy. Finally, this kinetic energy is converted into thermal energy. The energy is conserved during this process since it is converted from one form to another. At the same time, no work is produced during the process, thus energy is destroyed or lost. This lost energy is referred as exergy.

As explained, exergy is the measure of energy quality. Let us consider a space of 20 m^3 at 20°C. The energy stored in the movement of air particles would probably be more than the energy stored in a standard 12 V car battery (3 numbers). In spite of more energy contained in the space, this energy cannot be put into useful work (apart from thermal heat) because of its random distribution; it cannot be easily put into use, and it is not easily accessible. However, on the other hand, despite having lower amounts of stored energy, these batteries can be used for performing various works. The reason behind this is the electrical energy in the batteries is concentrated, controllable and available for performance of work. Thus, even though the quantity of energy in batteries may be lesser or even equal to the one stored in the room space, the quality or usefulness is much higher. This difference is taken into account by exergy.

The quality of energy is more significant than the quantity. Saloux et al. [38] have explained exergy as a qualitative aspect of energy that is the available share of energy. Exergy analysis is a powerful tool to assess the system that involves several sources of energy. Both the first and second law of thermodynamics are used to calculate the exergy of a system, and this helps to recognize the main source of exergy losses [39–41]. The exergy analysis helps us to identify the sources of irreversibility and inefficiencies so that the losses can be reduced, ultimately maximum resources, efficiency and capital savings may be achieved. Thus, to achieve the same, a careful selection of technology, design optimization shall be done.

> Exergy is the property of a system, which gives the maximum amount of useful work obtained from the system when it comes in equilibrium with a reference to the environment.

This definition concludes that some part of total thermal energy is converted into useful work and the rest is lost (which cannot be recovered) to the environment. Therefore,

$$Heat\ supplied\ (energy) = Available\ energy\ (exergy) + unavailable\ energy\ (anergy) \tag{4.18}$$

4.4.1 LOW-GRADE AND HIGH-GRADE ENERGY

As discussed in Section 4.3.5, the energy output from solar technologies can be broadly classified into low-grade and high-grade energy.

- **Low-grade energy:** The thermal energy that is a low-grade energy is grounded on the first law of thermodynamics. However, the concentrating collector, which is operating at high temperature, should be examined using second law of thermodynamics. An example of low-grade energy is the thermal energy obtained from the rear side of the PV module or the thermal energy available to the room from the floor.
- **High-grade energy:** The electrical energy output of a PV system is an example of high-grade energy. This is based on the second law of thermodynamics, which is referred as exergy analysis.

Application of both low-grade (thermal energy) and high-grade (electrical energy) can be seen in PVT systems.

To evaluate the overall performance of the system, either electrical energy is converted to low-grade energy [41] or thermal energy is converted into high-grade energy by using the concept of Carnot's efficiency as in Equation (4.5).

4.4.1.1 Exergy as a Process

The maximum work available (W_{max}) from the heat source at T_1 (in K) and sink at (ambient) temperature T_2 (in K), is expressed as:

$$W_{max} = exergy = \left(1 - \frac{T_0}{T_1}\right) \times Q_1 \tag{4.19a}$$

where, Q_1 is the available thermal energy at T_1.

For a given ambient temperature T_a, the unavailable energy is proportional to entropy change (Δs) during the process and is given by:

$$unavailable\ energy = T_0 \Delta s \tag{4.19b}$$

or,

$$\Delta s = \frac{\Delta Q}{T} \tag{4.19c}$$

where, Δs is the change in the entropy of the system during change in process.

Energy and Exergy Analysis

EXAMPLE 4.1

Calculate W_{max} (maximum work available) from the heat source at T_1 = 30 °C, 50 °C, 70 °C and ambient temperature =25 °C when Q_1 = 150 kWh.

Solution

Using Equation (4.19a) for 40 °C, we have

$$W_{max} = \left(1 - \frac{20 + 273}{40 + 273}\right) \times 150 = 2.47\,\text{kWh}$$

Similarly, for 50 °C and 70 °C, W_{max} = 11.60 kWh and 19.67 kWh, respectively.
   It can be concluded that higher-source temperature with constant sink temperature delivers maximum work.

4.4.2 EXERGY EFFICIENCY

Performance of the thermodynamic system can be evaluated using exergy efficiency. As discussed, the thermal efficiency of the system is based on the first law of thermodynamics, which is comprised of energy balance of the system to account energy input, desired energy output, and energy losses. The exergy efficiency of the system is based on the second law of thermodynamics and accounts for total exergy inflow, exergy outflow and exergy destruction for the process. Exergy efficiency is defined as the ratio of energy output to the exergy input.

The general exergy balance for the system is given by:

$$\sum \dot{Ex}_{in} - \sum \dot{Ex}_{out} = \sum \dot{Ex}_{dest} \tag{4.20a}$$

or,

$$\left(\sum \dot{Ex}_{heat} + \sum \dot{Ex}_{mass,\,in}\right) - \left(\sum \dot{Ex}_{work} + \sum \dot{Ex}_{mass,out}\right) = \sum \dot{Ex}_{dest} \tag{4.20b}$$

or,

$$\left(\sum\left(1 - \frac{T_0}{T_k}\right) \times \dot{Q}_k + \sum \dot{m}_{in}\psi_{in}\right) - \left(\sum \dot{W} + \sum \dot{m}_{out}\psi_{out}\right) = \sum \dot{Ex}_{dest} \tag{4.20c}$$

where $\dot{Q}_k$ is the rate of thermal energy transfer at the boundary of interest at location, k, and temperature T_k (in K).

The ψ is exergy flow per unit mass as defined as follows:

$$\psi = (h - h_0) - T_0(s - s_0) \tag{4.21}$$

where,

   h and s are the specific enthalpy and entropy, respectively, and
   h_0 and s_0 are the specific enthalpy and entropy at dead state temperature T_0.

Now, the exergy destruction $\left(\dot{Ex}_{dest}\right)$ or the irreversibility $\left(\dot{I}\right)$ can be expressed as:

$$\dot{Ex}_{dest} = \dot{I} = T_0 \dot{S}_{gen} \tag{4.22}$$

where rate of entropy generation $\dot{S}_{gen} = \sum \dot{m}_{out} s_{out} - \sum \dot{m}_{in} s_{in} - \sum \dfrac{\dot{Q}_k}{T_k}$

With minimization of the term $\left[\sum \dot{Ex}_{in} - \sum \dot{Ex}_{out}\right]$, the exergy efficiency can be improved since the inefficiency and irreversibility will be minimized.

The concept of "improved potential (IP)" because of the improvement in exergy efficiency is directly related to the term $\sum \dot{Ex}_{in} - \sum \dot{Ex}_{out}$. It is defined as follows:

$$IP = \left(1 - \varepsilon\right)\left(\sum \dot{Ex}_{in} - \sum \dot{Ex}_{out}\right) \tag{4.23}$$

where ε is the exergy efficiency or second law efficiency also defined as the ratio of actual performance of the system to the ideal performance of the system. It may also be defined as the ratio of exergy output (product exergy) to exergy input and expressed as [42]

$$\varepsilon = \frac{Rate\ of\ useful\ product\ energy}{rate\ of\ exergy\ input} = \frac{\dot{Ex}_{out}}{\dot{Ex}_{in}} = 1 - \frac{\dot{Ex}_{dest}}{\dot{Ex}_{in}} \tag{4.24}$$

where $\dot{Ex}_{dest}$ is the rate of exergy destruction given in Equation (4.22).

4.4.3 SOLAR RADIATION EXERGY

Exergy of any system is property of matter (substance) or a field (radiation, magnetic, acoustic, gravitational field, etc.). The matter can either be a substance with rest mass more than zero or a field matter for which the rest mass is equal to zero. Exergy depends on the source temperature in case sink temperature is constant in the entire process. Therefore, it should be "change in exergy of the heat source" instead of "exergy of heat" [43]. Solar radiation exergy is defined as the input exergy from the sun, which is available to any solar system or device.

The solar radiation received by Earth is used in almost all processes by humans, plants, animals, etc. The solar intensity (insolation) is converted into work, heat and other necessary processes. The energy efficiency and exergy efficiency can be evaluated from the work or heat equivalent of thermal radiation using the relations given in Table 4.5.

TABLE 4.5
Conversion Efficiency of Thermal Radiation [43, 44]

S.No.	Efficiency	Conversion from Radiation to Work	Conversion from Radiation to Heat
1.	Energetic efficiency (η_e)	$\eta_e = W/e$ $\eta_{e\ max} = W_{max}/e$	$\eta_e = 1 - (T_d/T)^4$
2.	Exergetic efficiency (η_{ex})	$\eta_{ex} = W/W_{max}$	$\eta_{ex} = b_d/W_{max}$

Note: W is the work performed due to utilization of the radiation, and W_{max} is the exergy of radiation.

Energy and Exergy Analysis

TABLE 4.6

The Input, Output and Unified Efficiency Expression (U_{ee}) of Utilization of Thermal Radiation [3]

S.No.	Input	Output	U_{ee}	Reference
1.	Radiation Energy	Absolute work	$1-\dfrac{4}{3}\times\left(\dfrac{T_0}{T_s}\right)+\dfrac{1}{3}\times\left(\dfrac{T_0}{T_s}\right)^4$	[43]
		Useful work Radiation Exergy	$1-\dfrac{4}{3}\times\left(\dfrac{T_0}{T_s}\right)$	[45]
2.	Heat	Network of a heat engine	$1-\left(\dfrac{T_0}{T_s}\right)$	[46]

Note: T_s and T_0 are the surface temperature of the sun and the environment temperature in Kelvin, respectively.

The unified efficiency expression (U_{ee}) gives the ability to use thermal radiation. U_{ee} for different input and output given by different researchers is tabulated in Table 4.6. The following relation can be used to assess the solar radiation exergy.

$$\dot{Ex}_{sun} = b = e \times U_{ee} \tag{4.25}$$

If solar radiation, $I(t)$, is falling on a unit surface area, A, the energy contained in solar radiation will be expressed as $\{I(t) \times A\}$. The equivalent exergy from the incident solar radiation is then expressed as [47]:

$$\dot{Ex}_{sun} = \left\{A \times I\left(t\right)\right\} \times U_{ee} = \left\{A \times I\left(t\right)\right\} \times \left[1-\frac{4}{3}\times\left(\frac{T_0}{T_s}\right)+\frac{1}{3}\times\left(\frac{T_0}{T_s}\right)^4\right] \tag{4.26}$$

T_0 = Surrounding or ambient temperature (K) = T_a, T_s = Sun surface temperature = T_{sun} = 6000 K

EXAMPLE 4.2

Using the expression of the Petela model [43] and radiation exergy when surrounding temperature = 25°C, A = 2 m², $I(t)$ = 750 W/m². Calculate the unified efficiency (U_{ee}).

Solution

From Table 4.5 and Equation (4.26), we have

$$\dot{Ex}_{sun} = \left\{2\times750\right\} \times \left[1-\frac{4}{3}\times\left(\frac{20+273}{6000}\right)+\frac{1}{3}\times\left(\frac{20+273}{6000}\right)^4\right] = 1.4 \, kW$$

4.4.3.1 Exergy Analysis Methods

Thermal energy, which is low-grade energy, can be converted into high-grade energy (exergy) using the following methods:

- **Method I**

  For systems operating at a high temperature range, say 400°C–500°C for concentrating systems to generate power following expression is used:

$$\dot{Ex} = \dot{Q}_u \left[1 - \frac{T_{fi} + 273}{T_{fo} + 273} \right] \tag{4.27}$$

where $\dot{Q}_u = \dot{m}_f C_f \left(T_{fo} - T_{fi} \right)$ is rate of thermal energy (i.e., low grade energy) at outlet temperature T_{fo} (°C), and T_{fi} (°C) is inlet temperature at surrounding air temperature.

- **Method II**

  For systems operating at medium temperature range of about 100°C–150°C like flat plate collectors, method II is considered. This method incorporates the entropy losses in the system. The exergy $\left(\dot{Ex} \right)$ for a solar thermal collector system is expressed as following:

$$\dot{Ex}_c = \dot{m}_f C_f \left(T_{fo} - T_{fi} \right) - \dot{m}_f C_f \left(T_a + 273 \right) \ln \frac{T_{fo} + 273}{T_{fi} + 273} \tag{4.28}$$

where,

T_{fo} and T_{fi} are outlet and inlet temperature, respectively, in K
$\dot{m}_f$ is the mass flow rate of fluid in kg/s.

The exergy efficiency is the ratio of exergy available from the solar thermal system to the exergy of sun and can be written as:

$$\varepsilon_c = \frac{\dot{Ex}_c}{\dot{Ex}_{sun}} \tag{4.29}$$

where $\dot{Ex}_{sun}$ is given by Equation (4.26).

The exergy efficiency can be improved by reducing the inefficiency and reversibility of each critical component of the system.

EXAMPLE 4.3

Calculate maximum work available (exergy) from the collector system operating at 50°C and 650°C and ambient air temperature at 25°C for producing thermal energy at a rate of 150 kW.

Solution

By using Equation (4.27), referring to Method I, one can get

$$\dot{Ex} = 150 \left[1 - \frac{25 + 273}{50 + 273} \right] = 11.60 \, \text{kW}$$

$$and \ \dot{Ex} = 150 \left[1 - \frac{25 + 273}{650 + 273} \right] = 101.57 \, \text{kW}$$

It can be concluded that an exergy (maximum work available) will be reduced significantly, which is high-grade energy. It can also be seen that higher exergy is obtained at higher temperature, which is as per expectation.

Energy and Exergy Analysis

129

EXAMPLE 4.4

Calculate exergy of the system operating temperature = 60 °C and T_a = 25 °C for mass flow rate of 0.05 kg/s, and C = 4190 J/kgK.

Solution

Method I
Using Equation (4.27), we get

$$\dot{Ex} = 0.05 \times 4190 \times 35 \times \left[1 - \frac{25 + 273}{60 + 273}\right] = 770.6\,\text{W} = 0.770\,\text{kW}$$

Method II
Using Equation (4.28), we get

$$\dot{Ex}_c = \left[0.05 \times 4190 \times 35 - 0.05 \times 4190 \times (298)\ln\frac{60 + 273}{25 + 273}\right] = 4321\,\text{W} = 4.32\,\text{kW}$$

It can be seen from the example that exergy obtained from Method II is significantly higher than the one obtained from Method I at medium temperature range.

From the above examples, the following conclusions can be made:

1. Equation (4.27) should be used for high operating temperatures to convert low-grade energy (thermal energy) into exergy.
2. Equations (4.28–4.29) should be used for low operating temperatures to convert low-grade energy (thermal energy) into exergy.

4.4.4 Exergy Analysis of Photovoltaic Thermal (PVT) Systems

As discussed, exergy analysis is used to evaluate the performance of a thermodynamic system. The energy efficiency of a thermal system may be defined as the ratio energy recovered from the product to the original energy input. On the other hand, exergy efficiency can be defined as the ratio of the product of exergy to exergy inflow. Based on the second law of thermodynamics, exergy analysis includes the total exergy inflow, exergy outflow and exergy that is destructed from the system.

$$\sum \dot{Ex}_{in} - \sum \left(\dot{Ex}_{th} + \dot{Ex}_{ee}\right) = \sum \dot{Ex}_{dest} \tag{4.30}$$

where
Exergy of radiation +

$$\dot{Ex}_{in} = \left\{A_c \times N_c \times I(t)\right\} \times \left[1 - \frac{4}{3} \times \left(\frac{T_0}{T_s}\right) + \frac{1}{3} \times \left(\frac{T_0}{T_s}\right)^4\right] \tag{4.31}$$

$$Thermal\ exergy = \dot{Ex}_{th} = \dot{Q}_u\left[1 - \frac{T_a + 273}{T_{fo} + 273}\right] \tag{4.32}$$

$$Electrical\ exergy = \dot{Ex}_{ee} = \eta_c \times A_c \times N_c \times I(t) \tag{4.33}$$

$$\dot{E}x_{overall} = \dot{E}x_{th} + \dot{E}x_{ee} \tag{4.34}$$

where, A_c is area of collector.

4.5 CASE STUDY WITH ROOF-MOUNTED BiPVT SYSTEM

The following assumptions have been considered:

- The lifespan of the system to work is equal to the life of the building. Nominal maintenance is required by the building during its lifespan.
- The lifespan for the group of elements/components is similar. This means when the life of a group is completed, it is replaced by an identical group of new components.
- The replaced components are produced using similar technology as the previous ones. This means that the embodied energy of the system remains same with time.
- About 50% of the average electrical power generated is stored in the batteries. The excess energy that is generated in the daytime is not used, and the excess demand during the night-time is satisfied from the grid.
- To determine the size of the invertor, peak load demand is considered to be 5 kW.
- Charge controller capacity is as per the peak electrical energy produced from the PV array.

4.5.1 DESCRIPTION

The life of the building is 60 years. The quantity and the total mass of materials used to construct the building are tabulated in Table 4.7. Referring to Table 4.7, the total embodied energy of the building is 367,366 MJ.

TABLE 4.7
Embodied Energy of the Material Used for Structure and Frame [37, 48]

Building Construction Material	Purpose	Quantity Used	Specific Heat Content		Embodied Energy, MJ
			Range, MJ/kg	Assigned Value	
Clay bricks (230 × 110 × 77 mm³)	Foundation	9353 kg (5300 no.)	2–7	4.25 MJ/kg	39,750
Clay bricks (230 × 110 × 77 mm³)	Walls	23,830 kg (13,500 no.)	2–7	4.25 MJ/kg	101,248
Cement	Mortar	2250 kg (45 bags)	4–8	4.2 MJ/kg	9450
Sand	Mortar	13,500 kg	<0.5	0.1 MJ/kg	1350
Concrete	Foundation	10,750 kg	0.8–1.5	1 MJ/kg	10,750
Lime	Whitewash	1.5 kg	3–5	4 MJ/kg	6
Mild steel	Columns	4800 kg	30–60	36 MJ/kg	172,800
Glass (for windows)	Windows	395 kg	12–25	15.9 MJ/kg	6280
Paint	Doors and windows	0.5 kg	80–150	93.3 MJ/kg	47
Plywood	Doors	18 kg	8–12	10.4 MJ/kg	187
Stabilized mud block masonry	Flooring	61 m²	-	418 MJ/m²	25,498
Total					367,366

Energy and Exergy Analysis

TABLE 4.8
Module Efficiency, Service Period and Specific Energy Content Assigned for Silicon and Non-Silicon PV Modules [37, 48]

Solar Cell Technology	Assigned Value for Module Efficiency, η_{PV}	Expected Life, n_{pv}	Assigned Value for Specific Energy Content, ε_{PV}		
	%	Years	MJ/m²	MJ/kWp	Range kWh/m²
c-Si	16	30	4294	26,775	1120–1260
p-Si	14	30	3276	23,400	840–980
rc-Si	12	25	2214	18,450	570–650
a-Si	6	20	1360.8	22,680	308–448
CdTe	8	15	957.6	11,970	196–336
CIGS	10	5	87.48	874.8	20–29

TABLE 4.9
Energy Content Coefficients and Service Period Assigned for the BiPVT System Components [37, 48]

BIPVT System Components	Lifespan, years	Assigned Value for Specific Heat Content	
		Average Value	Range
Charge controller (n_{cc}, ζ_{cc})	10	756 MJ/kW	130–290 kWh/kW
Inverter, safety factor 0.3 (n_{inv}, ζ_{inv})	10	756 MJ/kW	140–280 kWh/kW
PbA battery, depth of discharge 80%	5		0.24–0.33 kWh/Wh
(n_{bat}, k_{bat})		1.03 MJ/Wh	
PV frame (n_{frame}, ε_{frame})	30	403.2 MJ/m²	84–140 kWh/m²
Installation of junction box, cabling,	30		23.6–35 kWh/m²
etc.		80.8 MJ/m²	
Maintenance and operation	-	5.4 MJ/m²-year	1–2 kWh/m²/year
Foundation, array support, etc.	30		168–252 kWh/m²
(n_{sup}, ε_{sup})		756 MJ/m²	
Diesel incl. installation	5	3600 MJ/kW	400–800 kWh/kW

The PVT system consists of PV arrays to produce the electrical energy. The duct is constructed from plywood to store the thermal energy and cool the solar panels. The module efficiency, specific energy of the PV module with different technologies, is tabulated in Table 4.8.

The rest of the system is comprised of a charge controller, lead acid battery (PbA) with 80% depth of discharge, an inverter, frame to mount PV panels, junction box, cabling etc. The designated specific heat content of these components is tabulated in Table 4.9.

4.5.2 Overall Embodied Energy, EPBT, EPF

For the propose BiPVT system, the overall embodied energy for different PV technologies is summarized in Table 4.10 for a lifespan of 60 years. It can be seen that maximum embodied energy (equal to 1,215,225 MJ) has been found for mono-crystalline silicon.. The overall exergy calculations for the climatic conditions of New Delhi shows that for c-Si, p-Si, r-Si, a-Si, CdTe and CIGS BiPVT systems for a roof area of 45 m² generates 16,224 kW, 14,352 kW, 12,512 kW, 7790 kW,

TABLE 4.10

Embodied Energy in MJ for Lifespan of 60 Years [37, 48]

PV Technology	n_{sys}	E_{PV}	E_{cc}	E_{inv}	E_{bat}	$E_{ins+M\&O}$	E_{frame}	$E_{building}+E_{support}+E_{duct}$	EE_{total}	EPBT (Years)	EPF (Years)	LCCE
Amorphous silicon (a-Si)	20	183,708	3402	29,484	128,383	25,227	36,288	410,481	816,973	29.13	2.06	0.132
Cadmium telluride (CdTe)	15	172,368	4536	29,484	168,426	26,352	36,288	410,481	847,935	24.67	2.43	0.185
Copper indium gallium selenide (CIGS)	5	47,239	5670	29,484	202,712	35,352	36,288	410,481	767,226	19.31	3.10	0.247
Mono-crystalline silicon (c-si)	30	385,560	9072	29,484	320,238	24,102	36,288	410,481	1,215,225	20.81	2.8	0.349
Multi-crystalline silicon (p-Si)	30	294,840	7938	29,484	278,121	24,102	36,288	410,481	1,081,254	20.93	2.87	0.308
Ribbon silicon (r-Si)	25	239,112	6804	29,484	236,597	24,552	36,288	410,481	983,318	21.83	2.75	0.262

Energy and Exergy Analysis

9547 kW and 11,037 kW of overall exergy output. Minimum EPBT has been estimated for CIGS with 19.31 years. The higher value of EPBT for a-Si (29.13 years) is due to (i) using exergy as the divisor and (ii) including the embodied energy of the building itself, which has been excluded by the researchers previously. Maximum value of EPF corresponds to CIGS with 3.10 years. Thus, it can be concluded that from the energy point of view, the CIGS technology is the most suitable for use in a BIPVT system.

OBJECTIVE QUESTIONS

4.1 The first law of thermodynamics process is
   (a) Irreversible
   (b) Reversible
   (c) Both
   (d) None of the above

4.2 The system that allows exchange of both energy and matter with its surroundings
   (a) Open
   (b) Closed
   (c) Isolated
   (d) None of the above

4.3 The first law of thermodynamics is related to
   (a) Temperature
   (b) Principle of conservation of energy
   (c) Entropy
   (d) Exergy

4.4 The concept of temperature was defined by this law of thermodynamics:
   (a) Zeroth
   (b) First
   (c) Second
   (d) Third

4.5 The second law of thermodynamics states that heat transfer takes place
   (a) From a hotter body to a colder body
   (b) From a colder body to a hotter body
   (c) From same temperature of both bodies
   (d) None of the above

4.6 The second law of thermodynamics process is
   (a) Irreversible
   (b) Reversible
   (c) Both
   (d) None of the above

4.7 Embodied energy of most of renewable energy technologies is always
   (a) Equal to energy density
   (b) Higher than energy density
   (c) Less than energy density
   (d) None of the above

4.8 Energy payback time (EPBT) should be preferably
   (a) $= 0$
   (b) < 1
   (c) > 1
   (d) $= 1$

4.9 The numerical value of LCCE is always
   (a) = 0
   (b) < 1
   (c) > 1
   (d) = 1
4.10 Energy density of solar thermal system is
   (a) Higher than PV module
   (b) Lower than V module
   (c) Equal to PV module
   (d) None of the above
4.11 Electrical energy is
   (a) High-grade energy
   (b) Low-grade energy
   (c) Both low- and high-grade energy
   (d) All of the above
4.12 Energy payback time (EPBT) of a solar thermal system is
   (a) Equal to the PV module
   (b) Higher than the PV module
   (c) Lower than the PV module
   (d) None of the above
4.13 Energy payback time (EPBT) should be calculated under
   (a) Real weather conditions
   (b) Standard test condition (STC)
   (c) Both condition
   (d) All of the above
4.14 Fundamental energy matrices are
   (a) Energy payback time
   (b) Energy production factor
   (c) Life cycle conversion efficiency
   (d) All of the above

ANSWERS

4.1 (b)
4.2 (a)
4.3 (b)
4.4 (a)
4.5 (a)
4.6 (a)
4.7 (c)
4.8 (a)
4.9 (b)
4.10 (b)
4.11 (a)
4.12 (c)
4.13 (a)
4.14 (d)

Energy and Exergy Analysis

PROBLEMS

4.1 Calculate maximum work available W_{max} from the heat source at $T_1 = 40°C$, $60°C$, $90°C$ and ambient temperature $= 20°C$ when $Q_1 = 125$ kWh.
Hint: Example 4.1

4.2 Calculate the unified efficiency (U_{ee}) using the expression of the Petela model and radiation exergy when surrounding temperature $= 25°C$, $A = 1$ m^2, $I(t) = 450$ W/m^2.
Hint: Example 4.2

4.3 Calculate exergy (maximum work available) from the collector system operating at $110°C$ and $450°C$ and ambient air temperature at $25°C$ for producing thermal energy at rate of 150 kW.
Hint: Example 4.3

REFERENCES

[1] S. Carnot, *Reflections on the Motive Power of Fire and on Machines Fitted to Develop That Power*, London: Dover Publications, 1824.

[2] R. Clausius, *The Mechanical Theory of Heat: With Its Applications to the Steam-Engine and to the Physical Properties of Bodies*, London: Taylor & Francis, 1867.

[3] G. N. Tiwari, A. Tiwari and Shyam, *Handbook of Solar Radiation: Theory, Analysis and Applications*, Singapore: Springer, 2016.

[4] N. Gupta and G. N. Tiwari, "Energy matrices of building integrated semitransparent photovoltaic thermal systems: A case study," *Journal of Architectural Engineering*, 2017.

[5] R. H. Crawford, G. J. Treloar, R. J. Fuller and M. Bazilian, "Life-cycle energy analysis of building integrated photovoltaic systems (BiPVs) with heat recovery unit," *Renewable and Sustainable Energy Reviews*, vol. 10, no. 6, pp. 559–575, 2006.

[6] G. J. Treloar, "Comprehensive embodied energy analysis framework—Thesis," Deakin University, 1998.

[7] G. J. Treloar, "Extracting embodied energy paths from input-output tables: Towards an input- output-based hybrid energy analysis method," *Economic Systems Research*, vol. 9, no. 4, pp. 375–391, 1997.

[8] M. Lenzen, "Errors in conventional and input-output-based life-cycle inventories," *Journal of Industrial Ecology*, vol. 4, no. 4, pp. 127–148, 2000.

[9] R. E. Miller and P. D. Blair, *Input-Output-Analysis: Foundations and Extensions,* 1st edition, New Jersey: Prentice Hall, 1985.

[10] L. P. Hunt, "Total energy use in the production of silicon solar cells from raw materials to finished product," in *12th IEEE Photovoltaic Specialists Conference*, Baton Rouge, LA, 1976.

[11] K. Kato, A. Murata and K. Sakuta, "Energy pay-back time and life cycle CO_2 emission of residential PV power system with silicon PV module," *Progress in Photovoltaics*, vol. 6, no. 2, pp. 105–115, 1998.

[12] H. A. Aulich, F. Schulz and B. Strake, *"Energy pay-back time for crystalline silicon photovoltaic modules using new technologies,"* in *18th IEEE PV Specialists Conference*, Las Vegas, NV, 1985.

[13] M. Sudan and G. N. Tiwari, "Daylighting and energy performance of a building for composite climate: An experimental study," *Alexandria Engineering Journal*, vol. 55, pp. 3091–3100, 2016.

[14] K. M. Hynes, N. M. Pearsall and R. Hill, "The sensitivity of energy requirements to process parameters for $CuInSe_2$ module production," in *12th IEEE PV Specialists Conference, v1*, Las Vegas, NV, 1991.

[15] W. Palz and H. Zibetta, "Energy pay-back time of photovoltaic modules," *International Journal of Solar Energy*, vol. 10, pp. 211–216, 1991.

[16] E. A. Alsema, "Energy payback time and CO_2 emissions of PV systems," *Progress in Photovoltaics: Research and Applications*, vol. 8, no. 1, pp. 17–25, 2000. doi: 10.1002/(SICI)1099-159X(200001/02)8: 13.0.CO;2-C

[17] R. H. Bossert, C. J. J. Tool, J. A. M. van Roosmalen, C. H. M. Wentink and M. J. M. de Vaan, *Thin-Film Solar Cells. Technology Evaluation and Perspectives*, BP, London, UK: The Netherlands Agency for Energy and the Environment (Novem), Berenschot and the Netherlands Energy Research Foundation ECN, 2010.

[18] K. Knapp and T. Jester, "Empirical investigation of the energy pay back time for photovoltaic modules," *Solar Energy*, vol. 71, no. 3, pp. 165–172, 2001.

[19] Arjen Meijer, Mark A. J. Huijbregts, J. J. Schermer and L. Reijnders, "Life-cycle assessment of photovoltaic modules: Comparison of mc-Si , InGaP and InGaP/mc-Si solar modules," *Progress in Photovoltaics: Research and Applications*, vol. 11, no. 4, pp. 275–287, 2003.

[20] N. Jungbluth, "Life cycle assessment of crystalline photovoltaics in the Swiss ecoinvent database," *Progress in Photovoltaics: Research and Applications.*, pp. 429–446, 2005.

[21] G. Peharz and F. Dimroth, "Energy payback time of the high-concentration PV system FLATCON1," *Progress in Photovoltaics: Research and Applications*, vol. 13, pp. 627–634, 2005.

[22] E. A. Alsema, M. J. de Wild-Scholten and V. M. Fthenakis, "Environmental impacts of PV electricity generation–a critical comparison of energy supply options," in *The 21st European Photovoltaic Solar Energy Conference*, Dresden, Germany, 2006.

[23] V. E. Fthenakis and E. Alsema, "Photovoltaics energy payback times, greenhouse gas emissions and external costs: 2004–early 2005 status," *Progress in Photovoltaics: Research and Applications*, vol. 14, pp. 275–280, 2006.

[24] A. C. Veltkamp and M. J. de Wild-Scholten, "Dye sensitised solar cells for large-scale photovoltaics: Determination of durability and environmental performances," in *Renewable Energy*, 2006.

[25] M. Raugei, S. Bargigli and S. Ulgiati, "Life cycle assessment and energy pay-back time of advanced photovoltaic modules: CdTe and CIS compared to poly-Si," *Energy*, vol. 32, pp. 1310–1318, 2007.

[26] S. Pacca, D. Sivaraman and G. A. Keoleian, "Parameters affecting the life cycle performance of PV technologies and systems," *Energy Policy*, vol. 35, no. 6, pp. 3316–3326, 2007.

[27] A. Roes, E. A. Alsema, K. Blok and M. K. Patel, "Ex-ante environmental and economic evaluation of polymer photovoltaics," *Progress in Photovoltaics: Research and Applications*, vol. 17, no. 6, pp. 372–393, 2009.

[28] M. de Wild-Scholten, M. Sturm, M. A. Butturi, M. Noack, K. Heasman and G. Timò, "Environmental sustainability of concentrator PV systems: Preliminary LCA results of the Apollon project," in *25th European Photovoltaic Solar Energy Conference and Exhibition, 5th World Conference on Photovoltaic Energy*, Valencia, Spain, 2010.

[29] A. Nishimura, Y. Hayashi, K. Tanaka, M. Hirota, S. Kato, M. Ito, K. Araki and E. J. Hu, "Life cycle assessment and evaluation of energy payback time on high-concentration photovoltaic power generation system," *Applied Energy*, vol. 87, no. 9, pp. 2797–2807, 2010.

[30] V. M. Fthenakis and Hyung Chul Kim, "Life cycle assessment of high-concentration photovoltaic systems," *Progress in Photovoltaics: Research and Applications*, vol. 21, no. 3, pp. 379–388, 2012.

[31] M. Cellura, V. Grippaldi, V. L. Brano, S. Longo and M. Mistretta, "Life cycle assessment of a solar PV/T concentrator system," in *Life Cycle Management Conference LCM 2011*, Berlin, 2011.

[32] A. F. Sherwani, J. A. Usmani, and Varun, "Life cycle assessment of solar PV based electricity generation systems: A review," *Renewable and Sustainable Energy Reviews*, vol. 14, no. 1, pp. 540–544, 2010.

[33] B. Azzorpadi, J. Mutale, "Life cycle analysis for future photovoltaic systems using hybrid solar cells," *Renewable and Sustainable Energy Reviews*, vol. 14, pp. 1130–1134, 2010.

[34] J. J. Burkhardt III, G. A. Heath, and C. S. Turchi, "Life cycle assessment of a parabolic trough concentrating solar power plant and the impacts of key design alternatives," *Environmental Science & Technology*, vol. 45, pp. 2457–2464, 2011.

[35] G. N. Tiwari and M. Ghosal, *Renewable Energy Resources: Basic Principles and Applications*, New Delhi: Narosa Publishing House, 2005.

[36] G. N. Tiwari and S. Dubey, *Fundamentals of Photovoltaic Modules and Their Applications*, U.K: RSC Publications, 2010.

[37] B. Agrawal and G. N. Tiwari, *Building integrated Photovoltaic Thermal Systems*, U.K: RSC Publications, 2010.

[38] E. Saloux, A. Teyssedou and M. Sorin, "Analysis of photovoltaic (PV) and photovoltaic/thermal (PV/T) systems using the exergy method," *Energy and Buildings*, vol. 67, pp. 272–285, 2013.

[39] A. S. Joshi, I. Dincer and B. Reddy, "Thermodynamic assessment of photovoltaic systems," *Solar Energy*, vol. 83, pp. 1139–1149, 2009.

[40] A. D. Sahin, I. Dincer and M. Rosen, "Thermodynamic analysis of solar photovoltaic cell systems," *Solar Energy Materials and Solar Cells*, vol. 91, pp. 153–159, 2007.

[41] N. Gupta, A. Tiwari and G. N. Tiwari, "Exergy analysis of building integrated semitransparent photovoltaic thermal (BiSPVT) system," *Engineering Science and Technology, an International Journal*, vol. 20, pp. 41–50, 2016.

[42] E . K. Akpinar, A. Midilli and Y. Bicer, "The first and second law analyses of thermodynamic of pumpkin drying process," *Journal of Food Engineering*, vol. 72, no. 4, pp. 320–331, 2006.

[43] R. Petela, "Exergy of undiluted thermal radiation," *Solar Energy*, vol. 74, pp. 469–488, 2003.

[44] G. N. Tiwari and R. K. Mishra, *Advanced Renewable Energy Sources*, London: Royal Society of Chemistry, 2011.

[45] R. Battisti and A. Corrado, "Evaluation of technical improvements of photovoltaic systems through life cycle assessment methodology," *Energy*, vol. 30, no. 7, pp. 952–967, 2005.

[46] G. Lewis and G. Keoleian, *National Pollution Prevention Centre*, School of Natural Resources and Environment, University of Michigan, 1996.

[47] J. T. Szargut, "Anthropogenic and natural exergy losses (exergy balance of the Earth's surface and atmosphere)," *Energy*, vol. 28, no. 11, pp. 1047–1054, September 2003.

[48] G. N. Tiwari, A. Deo, V. Singh and A. Tiwari, "Energy efficient passive building: A case study of Sodha Bers Complex," *Renewable Energy*, vol. 1, no. 3, pp. 109–183, 2016.

5 Solar Cell Materials, PV Modules and Arrays

5.1 INTRODUCTION

A photovoltaic cell (PV cell) converts the incident solar radiation on it into the electricity due to the photovoltaic effect. The solar radiation incident on the solar cell transmitted through n-type material into depletion region to separate the positive and negative charge carriers due to high energy content in photon. The electric fields created due to separation of electrons and holes present at the junctions (depletion region) in material provide the required EMF for flow of electric current in the external circuit opposite to the current produced in without illumination and hence the electrical power is generated. Basically, photovoltaic cells are made of the silicon semiconductor junction devices. When PV cells are connected in series then they are referred to as a PV module. A PV generator is the system consisting of PV modules connected in different combinations (series connection, parallel connection or connected in both configurations) depending upon the requirement.

Depending upon the nature of electricity conduction in solids, it can be divided into three categories: conductors, semiconductors and insulators. These can be differentiated on the basis of energy band gaps between valence bands and conduction bands. The energy band gap is very large in the case of insulators ($E_g > h\upsilon$, h is the Planck's constant, and υ is the frequency). Thus, electrons in a valence band cannot jump to the conduction band, no conduction current. For semiconductor ($E_g < h\upsilon$), this gap is lesser than that of an insulator, and the valence electron can jump to the conduction band through this gap on acquiring thermal or light energy (Figure 5.1). For conductors ($E_g \approx 0$), no forbidden gap exists; hence the electron can easily move to the conduction band.

The variation of band gap with temperature can be written as:

$$E_g(T) = E_g(0) - \frac{aT^2}{T+b} \tag{5.1}$$

where the numerical value of constants, a and b for different materials are given in Table 5.1. At $T = 0$, $E_g(T) = E_g(0)$ that means the material behaves as an insulator.

Effect of temperature on band gap is shown in the Figure 5.2 for germanium, silicon and gallium arsenide.

5.2 BASICS OF SEMICONDUCTORS AND SOLAR CELLS

In this section, the basics of semiconductors and solar cells will be discussed. The conductivity of the intrinsic (pure) semiconductor can be increased by adding (doping) the controlled amount of specific impurity ion known as extrinsic semiconductors. Extrinsic semiconductors are of two types, namely (a) p-type and (b) n-type, respectively. In the case of p-type semiconductors, the doping ions have valency less than the semiconductor's valency, which produces an electron acceptor, and they have an energy level near to the valence band in the forbidden energy band. These traps (vacancies) produce positively charged states called holes. In this case, holes are in majority carriers and electrons are in minority carriers. In case of n-type material, the impurity (doping) ions have valency greater than that of the semiconductor's valency, which produces an extra electron. In this case, electrons are the majority charge carriers and holes are the minority charge carriers.

DOI: 10.1201/9780429445903-5

139

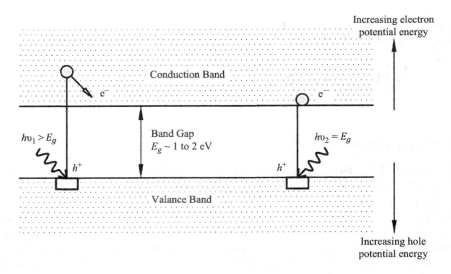

FIGURE 5.1 Semiconductor band structure of intrinsic material: (i) Photon absorption $h\upsilon < E_g$, no photoelectric absorption; (ii) $(h\nu_1 - E_g)$ excess energy dissipated as heat and (iii) $h\nu_2 = E_g$, photon energy equals band gap [1].

TABLE 5.1
The Values of Constants a and b [2]

	Gallium Arsenide (GaAs)	Germanium (Ge)	Silicon (Si)
$E_g(0)$	1.519 eV	0.7437 eV	1.166 eV
a	5.8×10^{-4} eVK^{-1}	4.77×10^{-4} eVK^{-1}	7×10^{-4} eVK^{-1}
b	204 K	235 K	636 K

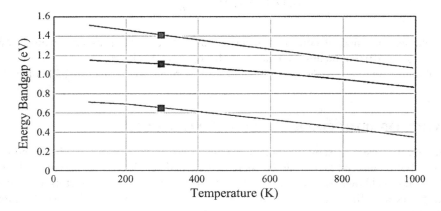

FIGURE 5.2 Temperature dependence of the energy band gap of germanium (bottom curve), silicon (middle curve) and GaAs (top curve) [2].

The semiconductor can be broadly categorized into intrinsic and non-intrinsic semiconductors as follows.

Solar Cell Materials, PV Modules and Arrays

141

EXAMPLE 5.1

Find out the energy band gap in silicon at 50°C.

Solution

With the substitution of an appropriate value placed in Equation (5.1), one gets

$$E_g(T) = 1.16 - \frac{7 \times 10^{-4} \times (323)^2}{323 + 1100} = 1.11\,\text{eV}$$

5.2.1 Intrinsic Semiconductor

In this case, Fermi-level is in the middle of the conduction and valence band, i.e., the probability of a state being occupied is 0.5. At thermal equilibrium, concentrations of electrons (n_e) and holes (n_h) are the same and equal to the intrinsic carrier concentration, i.e.,

$$n_e = n_h = n_i \tag{5.2}$$

Intrinsic semiconductors are also known as pure semiconductors. The doping in pure semiconductors affects the carrier concentration, as well as electrical properties. Following Maxwell-Boltzmann statistics, the electrons (n_e) and holes (n_h) concentrations are given by the following equations:

$$n_e = N_c(T)\exp\left[\frac{(E_F - E_C)}{kT}\right] \tag{5.3}$$

$$\text{and } n_h = N_V(T)\exp\left[\frac{(E_V - E_F)}{kT}\right] \tag{5.4}$$

where E_F is energy of Fermi level; E_C is energy at the bottom of the conduction band; E_V is energy at the top of the valance band as shown in Figure 5.1. From Equations (5.1–5.4), one can easily get the following relation

$$n_i^2 = n_e \times n_h = N_V(T) \times N_C(T)\exp\left[\frac{-(E_C - E_V)}{kT}\right] \tag{5.5}$$

where k is Boltzmann's constant, can be defined as the ratio of gas constant $R(8.314)$ and the Avogadro constant, N_A (6.022×10^{22}) and it is expressed as $k = R/N_A$. The values of k in different units are given in Table 5.2. Equation (5.5) shows that intrinsic carrier concentration (n_i) depends on E_c and E_v, independent of doping concentration.

Expressions for concentration factors $N_C(T)$ and $N_V(T)$ are given by

$$N_C(T) = N_C = \left[\frac{2\pi m_e kT}{h^2}\right]^{3/2} \tag{5.6a}$$

and

$$N_V(T) = N_V = \left[\frac{2\pi m_h kT}{h^2}\right]^{3/2} \tag{5.6b}$$

where $m_e = m_h = (9.11 \times 10^{-31}$ kg) are the effective masses of electrons and holes at constant temperature. The h is the Plank's constant ($6.62607015 \times 10^{-33}$ Js).

From Equations (5.2) and (5.5), one can have

$$n_e = n_h = \sqrt{\left[N_V(T) \times N_C(T) \right]} \times \exp\left[-\frac{\Delta E}{2kT} \right] \qquad (5.7)$$

which is proportional to $\exp(-\Delta E/2kT)$ and $\Delta E = E_C - E_V$.

5.2.2 Non-Intrinsic Semiconductor

For a non-intrinsic semiconductor, there is low doping. If n_0, p_0 and n_i are the electron, hole and intrinsic carrier concentrations respectively then at thermal equilibrium:

$$n_0 p_0 = n_i^2 \qquad (5.8)$$

where intrinsic carrier concentration n_i depends on material characteristics and the operating temperature. For Si, $n_i \approx 1.08 \times 10^{10}$ cm^{-3} at 300 K [3].

5.2.3 Fermi Level in Semiconductor

The Fermi level is the energy level between valance, E_V, and conduction band, E_c, of an extrinsic semiconductor. At a given temperature T, the probability for the majority charge carriers to be excited for conduction of current varies as $\exp[-e\varphi/kT]$, where e is the electronic charge, and φ is the electric potential difference between the Fermi level and the valence or conduction band, k the Boltzmann constant (Table 5.2).

For n-type semiconductor, an expression for Fermi level, E_F, is given by

$$E_F = E_C + kT \ln \frac{N_D}{N_C} \qquad (5.9)$$

where N_D is the donor concentration and N_C is the effective density of states in conduction band, Equation (5.6a) and is constant at fixed temperature T.

For p type semiconductor, an expression for Fermi level, E_F, is given by

$$E_F = E_V - kT \ln \frac{N_A}{N_V} \qquad (5.10)$$

where N_A is the acceptor ion concentration and N_V is the effective density of states in the valence band, Equation (5.6b).

TABLE 5.2

Value of Boltzmann Constant [2]

Boltzmann Constant

1.38×10^{-23} JK^{-1}

8.62×10^{-5} eVK^{-1}

1.38×10^{-16} ergK^{-1}

Solar Cell Materials, PV Modules and Arrays

> **EXAMPLE 5.2**
>
> Evaluate the shift in Fermi energy level, E_F, in a silicon crystal, C-Si, doped with a Vth group impurity of concentration 10^{15} cm^3.
> Given: The effective density (N_C) of states in the conduction band = 2.82×10^{19} cm^{-3}; band gap E_C = 1.1 eV; room temperature (T) = 27°C = 300 K.
>
> **Solution**
>
> From Equation (5.9), one gets
>
> $$E_F = E_C + kT \ln(N_D / N_C)$$
>
> Here E_C = 1.1 eV. Substitution of these values in the above equation gives
>
> $$E_F = 1.1 + kT \ln(N_D / N_C)$$
>
> $$E_F = 1.1 + \left[8.62 \times 10^{-5} \times 300 \times \ln\left(10^{15} / (2.82 \times 10^{19})\right)\right] = 1.1 - 0.265 = 0.835$$
>
> The shift is 0.835 − 0.55 = 0.285

5.2.4 p-n Junction

At the junction between *p*-type and *n*-type semiconductors, the majority charge carriers flow in the opposite direction to create a positive charge carrier in n-region and a negative charge carrier in p-region. During the flow of charge carriers, the recombination process occurs in a region having no mobile charges carriers known as depletion region. The *p-n* junction is connected either in forward bias or in reverse bias as per the required application.

5.2.5 Photovoltaic Effect

Incident solar radiation transmitted into depletion region creates the electron–hole pairs in the depletion region. An internal electric field produces a photo current (I_L) in depletion region. The direction of the photocurrent (I_L) is in a direction opposite to the forward dark current as shown in Figure 5.3. This photocurrent flows continuously even in the absence of external applied voltage and known as short circuit current (I_{SC}). Absorption of more light produces more electron-holes pair hence this current depends linearly on the light intensity. This effect is referred as photovoltaic (PV) effect.

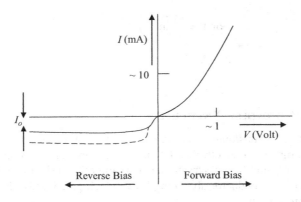

FIGURE 5.3 Dark characteristics for *p-n* junction [1].

The *p-n* junction with effect of solar radiation is referred as solar cell due to its working behavior as s conventional cell.

The overall solar cell current, I, is the difference of light (solar radiation) induced current, I_L and diode dark current, I_D and can be expressed as follows:

$$I = I_D - I_L \tag{5.11}$$

$$\text{Then}, I = I_0 \left[\exp\left(\frac{eV}{kT} \right) - 1 \right] - I_L \tag{5.12}$$

EXAMPLE 5.3

Determine the value of the overall cell current (*I*) in the limiting case $V \longrightarrow 0$.

Solution

From Equation (5.12), one gets as $V \longrightarrow 0$, $\exp(eV/kT) \longrightarrow 1$ and hence, $I \longrightarrow -I_L$.

EXAMPLE 5.4

Evaluate voltage for zero overall solar cell current.

Solution

For $I = 0$ in Equation (5.12), we have

$$I_0 \left[\exp\left(\frac{eV}{kT} \right) - 1 \right] - I_L = 0$$

or

$$\exp\left(\frac{eV}{kT} \right) = \frac{I_L}{I_0} + 1$$

or $V = \dfrac{kT}{e} \ln\left[\dfrac{I_L}{I_0} + 1 \right]$

5.2.6 SOLAR CELL (PHOTOVOLTAIC) MATERIALS

The solar cells consist of various/many materials with different atomic structure to minimize the initial cost and achieve maximum electrical efficiency. There are various types of solar cell materials namely (i) the single crystal, (ii) polycrystalline, (iii) amorphous silicon and (iv) compound thin film material, and other semiconductor absorbing layer which give highly electrical efficient solar cells for specialized applications.

Solar Cell Materials, PV Modules and Arrays **145**

Thin film solar cells can be manufactured by using a variety of compound semiconductors. These compound materials are

(a) Copper-Indium Selenide ($CuInSe_2$),
(b) Cadmium Sulfide (CdS),
(c) Cadmium Telluride (CdTe),
(d) Copper Sulfide (Cu_2S),
(e) Indium Phosphate (InP).

The copper-indium selenide solar cell stability appears to be excellent. The combinations of different band gap material in the tandem configurations lead to photovoltaic generators of higher efficiencies.

Here we will only briefly discuss silicon-based solar cell.

5.2.6.1 Silicon (Si)

Crystalline **silicon** (c-Si) is the most extensively used bulk material for manufacturing of solar cells. Bulk silicon can be processed to obtain (i) monocrystalline silicon, (ii) polycrystalline silicon and (iii) ribbon silicon using advanced processing technologies.

(i) **Monocrystalline silicon (c-Si):** Monocrystalline silicon is cut from the cylindrical ingots made from the Czochralski process. The solar cells are cut in pseudo-square shapes to minimize the wastage of processed monocrystalline silicon. Therefore, in solar panels manufactured from monocrystalline silicon, some portion of the module area is uncovered from the cell. This uncovered area is known as non-packing area of a PV module.

(ii) **Poly or multicrystalline silicon (poly-Si or mc-Si):** In poly- or multicrystalline silicon the crystal structure is not the same throughout. The solar cells made from polycrystalline silica have grain boundaries. Polycrystalline silica is made from square ingots. The ingots are made by cooling and solidifying the molten silicon in a controlled environment. The wafers (of thickness ~ 180–350 micrometer) are cut from the square ingots, and they are used for manufacturing polycrystalline solar cells. Polycrystalline solar cells are less expensive compared to monocrystalline solar cells, but these solar cells have lower efficiency due to grain boundaries present in solar cells. In this case, the packing factor of PV module is one.

(iii) **Ribbon silicon:** Ribbon silicon is thin film made from molten silicon. These are polycrystalline in nature. In processing of ribbon silicon, there is no waste of processed silicon as well as no sawing required, therefore solar cell manufactured from ribbon silicon are further less expensive than polycrystalline solar cells. The reflection losses are reduced by using antireflection coating, which allows higher absorbance of light into solar cells. Thin film solar cells with transparent top and bottom covers are potentially used in BiPV systems. These thin film solar cells can be used at glazed portions of the building, such as windows or facades. In spite of all the current technologies (generations of solar cells), first-generation solar cells abundantly cover the photovoltaic market, hence efforts are being made to achieve the lowest cost per watt solar cell.

c-Si cells are expensive but most popular due to easy availability. The amorphous silicon thin-film solar cells are less expensive. But cost of c-Si is decreasing very fast due to mass production and demand. The efficiency of a-Si modules lies between 6–8%.

5.2.6.2 Single-Crystal Solar Cell

Single-crystalline solar cells made from high-purity material (solar grade) show excellent efficiencies and long-term stability, but they are expensive.

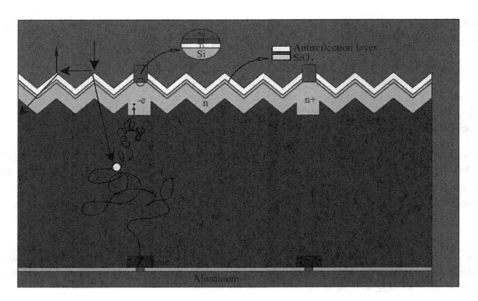

FIGURE 5.4 The structure of silicon solar cell and working mechanism [2].

Figure 5.4 shows the diagram of a silicon solar cell structure and mechanism. The electric current generated in the semiconductor is extracted by contact to the front and rear of the cell. The cell is covered with a thin layer of dielectric material, the anti-reflecting coating (ARC). ARC minimizes the reflection from the top surface; hence electrical efficiency increases.

The total series resistance of the single crystal solar cell can be expressed as:

$$R_s = R_{cp} + R_{bp} + R_{cn} + R_{bn} \tag{5.13}$$

where R_{cp} is the metal contact to p-type semiconductor resistance; R_{bp}, the bulk p-type resistance (bulk of p-type region is where most electron/hole pairs are generated by the absorption of solar light and where minority carriers (electrons) are transported by diffusion and partially lost by recombination); R_{cn}, the contact to n-type semiconductor resistance; and R_{bn}, the bulk n-type resistance.

The idealized junction current is given by

$$I = I_0 \left[\exp\left(\frac{e(V - IR_s)}{kT} \right) - 1 \right] \tag{5.14}$$

In addition, a shunt path may exist for current flow across the junction due to surface effect or a poor junction region. This alternate path for current constitutes a shunt resistance, R_p, across the junction. Then,

$$I = I_L - I_0 \left[\exp\left(\frac{e(V - IR_s)}{AkT} \right) - 1 \right] - \left(\frac{V - IR_s}{R_p} \right) \tag{5.15}$$

where A is an empirical non-idealist factor and is usually 1.

Solar Cell Materials, PV Modules and Arrays

EXAMPLE 5.5

What is the condition for zero idealized junction current ($I = 0$)?

Solution

Substitute $I = 0$ in Equation (5.14), one gets,

$$\exp\left(\frac{eV}{kT}\right) = 1 \Rightarrow V = 0$$

5.2.7 BASIC PARAMETERS OF SOLAR CELLS

In this section, some basic and important parameters essential for understanding characteristics curves of solar cells will be discussed. These parameters are as follows:

Overall current (I)

Overall current (I) flowing through solar cell is given by

$$I = I_D - I_L \tag{5.16}$$

or

$$I = I_0 \left[\exp\left(\frac{eV}{kT}\right) - 1 \right] - I_L \tag{5.17}$$

where I_D is diode dark current, I_L is light-induced current and I_0 is leakage current.

Short-circuit current (I_{SC})

If both the terminals (positive and negative) of solar cells are connected together (zero load) then the light-induced current is a short-circuit current (I_{SC})

Open-circuit voltage (V_{OC})

Open-circuit voltage is the voltage across the solar cell when there is no current flowing in the circuit (infinite resistance between the terminals of solar cells). With the help of Equation (5.17), one can have the following equation:

$$V_\infty = \frac{kT}{e} \ln\left(\frac{I_L}{I_0} + 1\right) \tag{5.18}$$

I–V characteristics

The I–V characteristics curve with illumination and without illumination is shown in Figure 5.5. The I–V curve for both cases has been plotted using $I = I_0 \left[\exp\left(\frac{e(V - IR_s)}{kT}\right) - 1 \right]$ [4]. For ideal solar cell

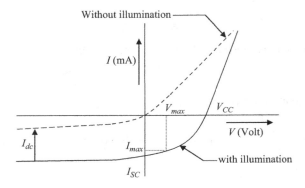

FIGURE 5.5 *I–V* characteristics of a solar cell with and without illumination [1].

series, resistance must be zero, and shunt resistance must be infinite. For better performance of solar cells, the series resistance must be kept as minimal as possible, and the shunt resistance must be as large as possible. In commercial solar cells, the shunt resistance is very large and is neglected in comparison to the forward resistance of diode.

The optimum load resistance $R_L(P_{max}) = R_{pmax}$ corresponding to the maximum power delivered from the solar cell is obtained as following:

$$P_{max} = V_{Pmax} I_{Pmax} \tag{5.19}$$

$$R_{Pmax} = \frac{V_{Pmax}}{I_{pmax}} \tag{5.20}$$

The electrical efficiency of solar cell is defined as,

$$\eta = \frac{P}{I \times A_c} \tag{5.21}$$

where

$P = V \times I$, is the power delivered by a PV generator
$I \times A_c$, is the solar radiation falling on the PV generator
I is the solar intensity, and A_c is the solar cell surface area

Fill factor (FF)

The fill factor (FF) gives an idea about the maximum power output withdrawn from the solar cell for a given V_{oc} and I_{sc}. It can also be expressed as the sharpness of an *I–V* curve. The value of FF in ideal condition is unity. Deviation from the ideal value is due to defects and contact resistance. The lower the value of the *FF*, the less sharp will be the *I–V* curve (Figure 5.6). For Si solar cells the maximum value of *FF* is 0.88. Mathematically, it is given by Equation (5.22).

$$FF = \frac{P_{max}}{V_{oc} \times I_{sc}} = \frac{I_{max} \times V_{max}}{V_{oc} \times I_{sc}} \tag{5.22}$$

Solar Cell Materials, PV Modules and Arrays

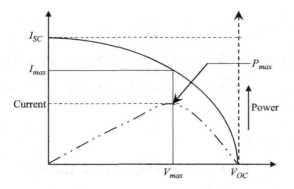

FIGURE 5.6 Characteristic and power curve for determining the fill factor (FF) [2].

Maximum power (P_{max})

The output power from a solar cell is obtained by multiplying the output voltage and output current as:

$$P_{out} = V_{out} \times I_{out} \qquad (5.23)$$

Similarly, the maximum power is obtained by multiplying V_{max} and I_{max}, and it corresponds to the maximum value of product I and V on a characteristics curve. Hence,

$$P_{max} = V_{max} \times I_{max} \qquad (5.24)$$

From Equation (5.21), one can obtain:

$$= V_{oc} \times I_{sc} \times FF \qquad (5.24a)$$

Solar cell electrical efficiency (η_{ec})

The solar cell electrical efficiency (η_{ec}) can be given as

$$\eta_{ec} = \frac{P_{max}}{P_{in}} = \frac{V_{max} \times I_{max}}{Incident\ solar\ radiation \times Area\ of\ solar\ cell} = \frac{V_{oc} \times I_{sc} \times FF}{I(t) \times A_c} \qquad (5.25)$$

where, I_{max} and V_{max} are corresponding to given ($I(t)$).

EXAMPLE 5.6

Calculate fill factor (FF) for a solar cell for a given following parameters:
V_{OC} = 0.2 V, I_{SC} = − 5.5 mA, V_{max} = 0.125 V, I_{max} = − 3 mA

Solution

From Equation (5.22), one gets,

$$Fill\ factor\ (FF) = \frac{V_{max} \times I_{max}}{V_{oc} \times I_{sc}} = \frac{0.125 \times 3}{0.2 \times 5.5} = 0.34$$

EXAMPLE 5.7

Evaluate the maximum power, P_{max}, and solar cell electrical efficiency (η_{ec}) at an intensity of 200 W/m².

Given parameters: $V_{OC} = 0.24\ V$, $I_{SC} = -9\ mA$, $V_{max} = 0.14\ V$, $I_{max} = -6\ mA$ and $A_C = 4\ cm^2$

Solution

From Equation (5.24), we have

$$P_{max} = V_{max} \times I_{max} = 0.14 \times (-6) = -0.84\,mW$$

and from Equation (5.25), we have,

Solar cell electrical efficiency(η_{ec}) = output/input = $(0.14 \times 6 \times 10^{-3})/(200 \times 4 \times 10^{-4})$
= 0.0105 = 1.05%

EXAMPLE 5.8

Calculate the power output from a solar cell at standard test condition (I) = 1000 W/m² and $T_c = 25°C$, when (η_{ec}) = 16%, FF = 0.782, aperture area = 4.02×10^{-4} m².

Solution

Power output = $0.16 \times 1000 \times 4.02 \times 10^{-4} \times 0.782 = 0.05$ W.

Effect of solar cell temperature on solar cell electrical efficiency

The solar cell electrical efficiency (η_{ec}), as a function of temperature, is given by [5]

$$\eta_{ec} = \eta_0 \left[1 - \beta_0 \left[T_c - 298\right]\right] \tag{5.26}$$

where η_0 is an electrical efficiency of a solar cell at standard test condition (STC) [solar flux of 1000 W/m² and surrounding temperature of 25°C]; β_0 is the silicon efficiency temperature coefficient (0.0045 K^{-1} or 0.0064 K^{-1}) and T_c is solar cell temperature (K).

Different solar cell materials with their module electrical efficiency and temperature coefficients are given in Table 5.3 [6].

TABLE 5.3
Values of Module Electrical Efficiency and Temperature Coefficients [6]

Sr. No.	Module Type	η_0	β_0 (°C⁻¹)	Reference
1	Mono-silicon (Mono-Si)	0.15	0.0041	[7]
		0.13	0.004	[8]
		0.15	0.0045	[9]
2	Photovoltaic thermal (PVT) system	0.127	0.0063	[10]
3	Polycrystalline silicon (p-Si)	0.11	0.004	[8]
4	Amorphous silicon (a-Si)	0.05	0.0011	[8]

Solar Cell Materials, PV Modules and Arrays

5.3 PHOTOVOLTAIC (PV) MODULES AND PV ARRAYS

PV module: Single solar cells cannot generate enough power. Hence, many solar cells are connected in series, and they are sandwiched between top transparent and bottom opaque/transparent covers to protect them from adverse weather conditions. Such series-connected solar cells are known as photovoltaic modules (PV module). The PV modules are available in different sizes and shapes depending on the required electrical output. Each PV module has a junction box with positive and negative terminals for proper connection to other PV modules.

Photovoltaic (PV) array: This is a collection of series/parallel/both series and parallel-connected photovoltaic (PV) modules. The size of the PV array depends on the requirement of electrical power. The DC power produced from the PV array is converted into the AC power by using a charge controller and an inverter, and it is further connected to the different electrical loads. The PV modules are connected in series to achieve the desired voltage; then such series-connected strings are connected in parallel to enhance the current and hence power output from the array. The size of the PV array decides the capacity of such array; it may be in watts, kilowatts or megawatts.

5.3.1 SINGLE-CRYSTAL SOLAR CELLS PV MODULE

The series-connected crystalline solar cell is sandwiched between a top glass cover with high transmittivity and a low iron glass, encapsulate (ethylene vinyl acetate (EVA)) and back cover usually a foil of Tedlar® as shown in Figure 5.7. EVA is about 100% transparent and insulating.

The crystalline PV modules are divided into two categories depending on the material of the back cover of the module. If the back cover of the module is made of opaque Tedlar, it is referred as glass to Tedlar or opaque PV module. If the back cover is made of glass, it is referred to as a glass-to-glass or semi-transparent PV module. The amount of light transmitted from the semi-transparent PV module depends on the packing factor of the semi-transparent PV module. The lower the packing factor, the lower is the area covered by the solar cell. Therefore, for electrical/thermal/daylighting applications, the packing factor of the semi-transparent PV module can be tailored as per the desired requirement of thermal/electrical/daylighting from the system.

5.3.2 THIN-FILM PV MODULES

Thin-film PV modules are made of thin-film solar cells. The thin-film solar cells are manufactured at lower temperature compared to crystalline solar cells, hence these technologies are less energy-intensive. Also, the production cost of thin-film solar cells is lower than that of crystalline solar cells.

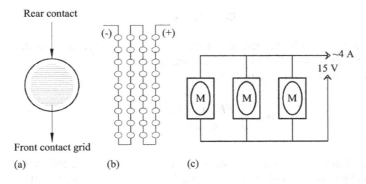

FIGURE 5.7 Typical arrangements of commercial Si solar cells: (a) cell, (b) module of 36 cells, (c) array of PV module [1].

Further, the thin films can be easily deposited on different substrates such as glass, metal or even plastic, and this flexibility leads to the greater interest in manufacturing of thin-film PV modules. Another advantage of thin-film PV modules is that they can bend; therefore, they can be used at different structures of the building facade and other glazing areas. The major drawback of this technology is low-energy conversion efficiency and degradation on exposure to adverse weather conditions (Staebler-Wronski effect). The major challenge for this technology is improvement of conversion efficiency of commercially-made thin-film PV modules [11].

Initially, amorphous silicon thin-film solar cell were used in thin-film PV modules as production costs are low, and production processes are simpler than that of polycrystalline silicon [12]. However, the energy-conversion efficiency of amorphous silicon thin-film PV modules is only 6%–7% [13]. The other materials used for thin-film solar cell technology for manufacturing thin-film PV modules are copper–indium–diselenide (CIS), copper–gallium–diselenide (CGS), copper–indium– gallium– diselenide (CIGS) and cadmium telluride (CdTe). The energy conversion efficiency achieved by these thin-film technologies are up to about 20% (19.9% for CIGS (NREL, U.S.), 16.5% for CdTe (NREL, U.S.), 13% for a CIGS (Wurth Solar, Germany). The thin-film solar cell technology is getting more interest, and large-capacity manufacturing plants are already in operation around the globe [14].

5.3.3 PACKING FACTOR (β_c) OF PV MODULE

This is defined as the ratio of total solar cell area to the total PV module area, and it can be expressed as follows:

$$\beta_c = \frac{area\ of\ solar\ cells}{area\ of\ PV\ module} \tag{5.27}$$

It is clear that β_c is less than unity (pseudo-solar cells), and it has maximum value of one when all area is covered by solar cells (rectangular solar cells). For thin-film solar cell PV modules, it is always one.

5.3.4 EFFICIENCY OF PV MODULES

The electrical efficiency of PV module, η_{em}, in fraction can be expressed as follows:

$$\eta_{em} = \tau_g \times \beta_c \times \eta_{ec} \tag{5.28}$$

For $\beta_c = 1$, Equation (5.28) reduces to

$$\eta_{em} = \tau_g \times \eta_{ec} \tag{5.29}$$

This shows that the electrical efficiency of PV module (η_{em}) is less than electrical efficiency of a solar cell due to the presence of glass over solar cell (η_{ec}).

It is also expressed in percentage as follows:

$$\eta_{em} = \left(\frac{FF \times I_{sc} \times V_{oc}}{A_m \times I_p} \right) \times 100 \tag{5.30}$$

where A_m = area of the PV module; I_p = incident solar intensity on the PV module and FF, I_{sc} and V_{oc} are fill factor (FF), short-circuit current, I_{SC}, and open-circuit voltage, V_{OC} of the PV module. The maximum value of fill factor (FF) of the c-Si PV module is 0.88.

The temperature-dependent electrical efficiency of the PV module η_{em} can be expressed as follows:

Solar Cell Materials, PV Modules and Arrays

$$\eta_{em} = \eta_{mo}\left[1 - \beta_0\left[T_c - 298\right]\right] \tag{5.31}$$

where $\eta_{mo} = \tau \times \eta_0$ is electrical efficiency of the PV module under standard test condition (STC).

The electrical load efficiency, η_{load} n percentage may be expressed as:

$$\eta_{load} = \left(\frac{I_L \times V_L}{A_m \times I_p}\right) \times 100 \tag{5.32}$$

5.3.5 ENERGY BALANCE EQUATIONS FOR PV MODULES

In this section, we will consider crystalline solar cells–based PV module. As we mentioned earlier, there are two types of PV modules, namely (a) opaque PV modules and (b) semi-transparent PV modules.

The energy-balance equations for PV modules have been written with following assumptions:

(i) One-dimensional heat conduction across thickness
(ii) The PV module is in quasi-steady state condition
(iii) The ohmic losses between two solar cells connected in series of PV modules are negligible

5.3.5.1 For Opaque (Glass to Tedlar) PV Modules [15]

In this case, solar radiation, $I(t)$, is transmitted by the top glass of the PV module as $\tau_g I(t)$, and further, it is absorbed by the solar cell of the PV module having area A_m and packing factor, β_c as $\tau_g \alpha_c \beta_c I(t) A_m$ and remaining solar radiation $\tau_g(1 - \beta_c)I(t)$ is absorbed by Tedlar (α_T) on the non-packing portion of the PV module as $\tau_g \alpha_T(1 - \beta_c)I(t)A_m$. The temperature of the solar cell increases; hence there will be (i) the rate of overall upward heat loss $[U_{t,ca}(T_c - T_a)A_m]$ and (ii) the rate of overall back heat loss $[U_{b,ca}(T_c - T_a)A_m]$, in addition to electrical power generation as $\tau_g \eta_c \beta_c I(t)A_m$. This can be mathematically summarized as follows with the thermal circuit diagram shown in Figure 5.8(a).

$$\tau_g\left[\alpha_c\beta_c I(t) + (1 - \beta_c)\alpha_T I(t)\right] = \left[U_{tc,a}(T_c - T_a) + U_{bc,a}(T_c - T_a)\right] + \tau_g\eta_c\beta_c I(t) \tag{5.33}$$

From Equation (5.33), one can get

$$T_c - T_{ref} = (T_a - T_{ref}) + \frac{\left[\tau_g\left\{\alpha_c\beta_c + (1 - \beta_c)\alpha_T - \eta_c\beta_c\right\}\right]I(t)}{U_{Lm}} \tag{5.34}$$

The temperature-dependent electrical efficiency of the solar cell, Equation (5.34), is given as,

$$\eta_c = \eta_0\left[1 - \beta_0(T_c - T_0)\right] \tag{5.35}$$

where, η_0 is the solar electrical efficiency at the reference temperature, T_0, and at solar radiation of 1000 W/m², β_0 is the temperature coefficient. The values of η_0 and β_0 are given in Table 5.4 [16].

Using Equation (5.34) and Equation (5.35), one can obtain,

$$\eta_c = \frac{\eta_0\left[1 - \beta_0\left\{(T_a - T_0) + \frac{\left[\tau_g\left\{\alpha_c\beta_0 + (1 - \beta_0)\alpha_T\right\}\right]I(t)}{U_{Lm}}\right\}\right]}{\left[1 - \frac{\eta_0\beta_0\tau_g\beta_c}{U_{Lm}}I(t)\right]} \tag{5.36}$$

TABLE 5.4
Values of Module Electrical Efficiencies and Temperature Coefficients [2]

Module Type	T_{ref} (°C)	η_{Tref}	β_{ref}	References
Mono-Silicon (Mono-Si)	25	0.15	0.0041	[17]
		0.11	0.003	[18]
		0.13	0.004	[8]
		0.12	0.0045	[19]
Photovoltaic thermal (PVT) system	25	0.13	0.0041	[20]
			0.005	[21]
		0.10	0.0041	[22]
		0.178	0.00375	[23]
		0.097	0.0045	[24]
		0.09	0.0045	[25]
		0.12	0.0045	[25]
		0.12	0.0045	[26]
		0.127	0.0063	[10]
	20	0.10	0.004	[27]
		0.125	0.004	[28]
PVT system (unglazed)	25	0.127	0.006	[10]
PVT system (glazed)	25	0.117	0.0054	[29]
Amorphous silicon (a-Si)	25	0.125	0.0026	[30]
		0.05	0.0011	[8]
Polycrystalline silicon (p-Si)	25	0.11	0.004	[8]
Average of Sandia and other commercial cells	28	0.117 (average)	0.0038 (average)	[31]

After obtaining electrical efficiency of the solar cell by Equation (5.36), one can get an electrical efficiency of the PV module as

$$\eta_m = \tau_g \eta_c \beta_c \tag{5.37}$$

The threshold intensity $I(t)_{th}$ can be obtained by putting the denominator of Equation (5.37) equal to zero and is given as following:

$$I(t)_{th} = \frac{U_{Lm}}{\eta_0 \beta_0 \tau_g \beta_c} \tag{5.38}$$

5.3.5.2 For Semi-Transparent (Glass-to-Glass) PV Modules

In this case, some fraction of solar radiation is transmitted through the bottom glass cover from the non-packing area as $\tau_g^2(1 - \beta_c)I(t)A_m$. This can be mathematically summarized as follows with the thermal circuit diagram shown in Figure 5.8(b).

$$\alpha_c \tau_g \beta_c I(t) = \left[U_{tc,a} \left(T_c - T_a \right) + U_{bc,a} \left(T_c - T_a \right) \right] + \tau_g \eta_c \beta_c I(t) \tag{5.39}$$

Solar Cell Materials, PV Modules and Arrays

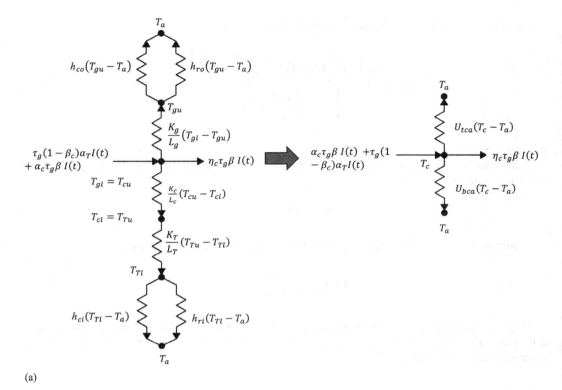

(a)

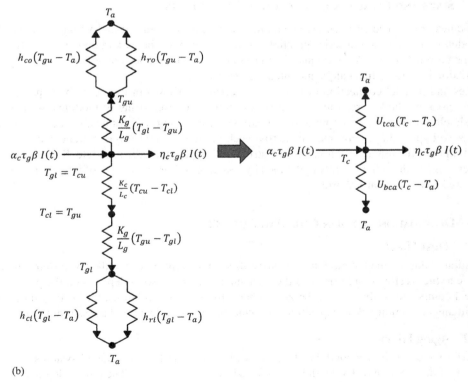

(b)

FIGURE 5.8 (a) Thermal circuit diagram of opaque PV module. (b) Thermal circuit diagram of semi-transparent PV module [2].

From Equation 5.39 and 5.35, one gets the following equation

$$\eta_c = \frac{\eta_0 \left[1 - \beta_0 \left\{ (T_a - T_0) + \frac{\alpha_c \tau_g \beta_c}{U_{Lm}} I(t) \right\} \right]}{\left[1 - \frac{\eta_0 \beta_0 \tau_g \beta_c}{U_{Lm}} I(t) \right]} \tag{5.40}$$

where $U_{Lm} = (U_{tc,a} + U_{bc,a})$.

After obtaining the electrical efficiency of the solar cell by Equation (5.40), one can get electrical efficiency of the PV module as

$$\eta_m = \eta_c \tau_g \beta_c \tag{5.41}$$

Similarly, in this case also, the threshold intensity $I(t)_{th}$ is given as following:

$$I(t)_{th} = \frac{U_{Lm}}{\eta_0 \beta_0 \tau_g \beta_c} \tag{5.42}$$

Hourly variation of solar cell temperature and solar cell efficiency (Figure 5.9a) reveals that solar cell electrical efficiency decreases with increasing temperature, which is in accordance with the studies concluded by Evans [17].

5.3.6 Series and Parallel Combination of PV Modules

PV modules are connected in series or parallel to increase the current and voltage ratings. When PV modules are connected in series/parallel, it is desired to have the maximum power production at the same current/voltage. A solar panel is a group of several modules connected in series–parallel combination in a frame that can be mounted on a structure.

Series and parallel connection of modules in a panel is shown in Figure 5.9b. In parallel connection, blocking diodes are connected in series with each series string of modules, so that if any string should fail, the power output of the remaining series string will not be absorbed by the failed string. Also bypass diodes are installed across each module, so that if one module should fail, the power output of the remaining modules in a string will bypass the failed module. Some modern PV modules come with such internally embedded bypass diodes. Large numbers of interconnected solar panels are known as solar PV array.

5.3.7 Degradation of Solar Cell Materials [32]

5.3.7.1 Dust Effect

Deposition of dust on the PV modules seriously affects the performance of the PV systems. It is very sensitive to the weather conditions and the local environment. Also, it depends on the physical and chemical composite of the dust in the atmosphere. In outdoor conditions, the surface finish, optimum tilt angle, humidity, dew deposition and wind speed also affect the dust accumulation.

5.3.7.2 Aging Effect

The performance of the PV modules degrades with increasing life of installed systems. This phenomenon of degradation of PV modules is referred as the aging effect. The rate of degradation over time period depends on the materials and technology used for the fabrication of the PV module. The aging effect is due to (i) the weather condition, (ii) fluctuation in the ambient temperature, (iii) rain,

Solar Cell Materials, PV Modules and Arrays

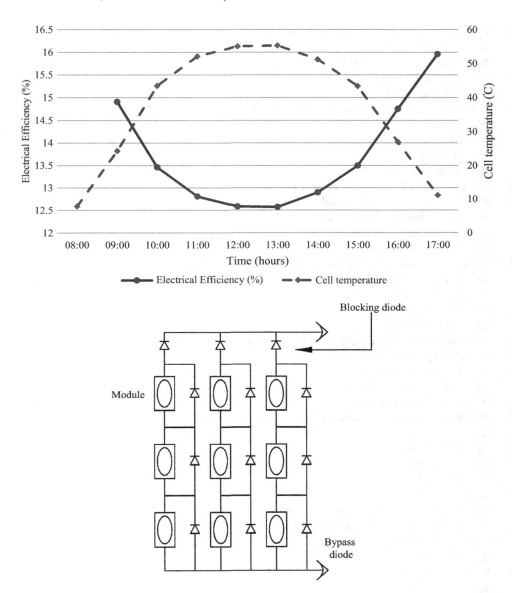

FIGURE 5.9 (a) Hourly variation of cell temperature and cell efficiency for a typical day of summer [2]. (b) Series and parallel connection of modules in a panel [16].

(iv) hailstorm and (v) dust deposition, etc. It has been seen that there is 1% degradation of PV module per year for the life performance of 30 years.

OBJECTIVE QUESTIONS

5.1 What is the most common material used for solar cell manufacturing?
   (a) Silver
   (b) Iron
   (c) Aluminum
   (d) Silicon

158 Photovoltaic Thermal Passive House System

5.2 The electrical output of a solar cell depends on
  (a) Solar radiation
  (b) Thermal part of solar radiation
  (c) Ultraviolet radiation
  (d) Infrared radiation
5.3 Photovoltaic cells convert solar energy directly into
  (a) Mechanical energy
  (b) Electricity
  (c) Heat energy
  (d) Transportation
5.4 What does SPVT stand with respect to solar energy?
  (a) Solar photovoltaic thermal
  (b) Solar plate voltaic thermal
  (c) Solar plate voids thermal
  (d) None of the above
5.5 Which of the following appliances are used by a PV module?
  (a) Solar lantern
  (b) Biogas plant
  (c) Solar water heater
  (d) Solar air heater
5.6 Which material has the highest solar cell electrical efficiency?
  (a) Amorphous silicon
  (b) Thin-film silicon
  (c) Polycrystalline silicon
  (d) Single-crystal silicon
5.7 Maximum electrical efficiency of a commercial solar cell is?
  (a) 3%
  (b) 12%–30%
  (c) 50%–65%
  (d) 65%–70%
5.8 A PV system is
  (a) A non-renewable source of energy
  (b) A renewable source of energy
  (c) A finite source of energy
  (d) All of the above
5.9 A PV system provides
  (a) Clean power
  (b) Lack of dependence on fossil fuels
  (c) Sustainable climate
  (d) All of the above
5.10 Electrical efficiency of a semi-transparent PV module is
  (a) More than in an opaque PV module
  (b) Equal to an opaque PV module
  (c) Less than an opaque PV module
  (d) None of the above

ANSWERS

5.1 (d)
5.2 (a)
5.3 (b)

Solar Cell Materials, PV Modules and Arrays

5.4 (a)
5.5 (a)
5.6 (d)
5.7 (b)
5.8 (b)
5.9 (d)
5.10 (a)

REFERENCES

[1] G. N. Tiwari, *Solar Energy: Fundamentals, Design, Modelling and Applications*, New Delhi: Narosa, 2002.

[2] G. N. Tiwari, A. Tiwari and Shyam, *Handbook of Solar Energy*, Springer, 2016.

[3] J. D. Mondol, Y. G. Yohanis and B. Norton, "Optimal sizing of array and inverter for grid-connected photovoltaic systems," *Solar Energy*, vol. 80, pp. 1517–1539, 2006.

[4] P. Tsalides and A. Thanailakis, "Direct computation of the array optimum tilt angle in constant-tilt photovoltaic systems," *Solar Cells*, vol. 14, pp. 83–94, 1985.

[5] J. Kern and I. Harris, "On the optimum tilt of a solar collector," *Solar Energy*, vol. 17, no. 2, pp. 97–102, 1975.

[6] N. Gupta, G. N. Tiwari, A. Tiwari and V. S. Gupta, "New model for building-integrated semitransparent photovoltaic thermal system," *Journal of Renewable and Sustainable Energy*, vol. 9, pp. 043504, 2017.

[7] D. Evans and L. Florschuetz, "Cost studies on terrestrial photovoltaic power systems with sunlight concentration," *Solar Energy*, vol. 19, pp. 255–262, 1977.

[8] RETScreen© International, "Photovoltaic project analysis," vol. PV.22, 2001.

[9] N. Gupta, A. Tiwari and G. N. Tiwari, "Exergy analysis of building integrated semitransparent photovoltaic thermal (BiSPVT) system," *Engineering Science and Technology, an International Journal*, vol. 20, pp. 41–50, 2016.

[10] J. K. Tonui and Y. Tripanagnostopoulos, "Improved PV/T solar collectors with heat extraction by forced or natural air circulation," *Renewable Energy*, vol. 32, pp. 623–637, 2007.

[11] C. M. Fortman, T. Zhou, C. Malone, M. Gunes and C. R. Wronski, "Deposition conditions, hydrogen content, and the Staebler-Wronski effect in amorphous silicon," *IEEE Conference on Photovoltaic Specialists*, 1990, pp. 1648–1652 vol. 2, doi: 10.1109/PVSC.1990.111888.

[12] M. A. Green, "Solar cells operating principles technology and system application," Englewood Cliffs, NJ: Prentice-Hall, Inc., 1982.

[13] B. von Roedern and H. S. Ullal, "The role of polycrystalline thin-film PV technologies in competitive PV module markets", *2008 33rd IEEE Photovoltaic Specialists Conference*, 2008, pp. 1–4, doi: 10.1109/PVSC.2008.4922493.

[14] H. Schock, "Chalcopyrite (CIGS)-based solar cells and production in Europe." in: *Technical Digest 17th International Photovoltaic Science and Engineering Conference (PVSEC-17)* 2007, 40–43.

[15] A. Tiwari and M. Sodha, "Performance evaluation of hybrid PV/thermal water/air heating system: A parametric study," *Renewable energy*, vol. 31, no. 15, pp. 2460–2474, 2006.

[16] G. N. Tiwari and R. K. Mishra, *Advanced Renewable Energy Sources*, UK: RSC publishing, 2012.

[17] D. L. Evans, "Simplified method for predicting photovoltaic array output," *Solar Energy*, vol. 27, no. 6, pp. 555–560, 1981.

[18] N. T. Truncellito and A. J. Sattolo. "An analytical method to simulate solar energy collection and storage utilizing a flat plate photovoltaic panel." *General Electric Advanced Energy Department*, 1979.

[19] T. T. Chow, "Performance analysis of photovoltaic-thermal collector by explicit dynamic model," *Solar Energy*, vol. 75, no. 2, pp. 143–152, 2003.

[20] R. Mertens "Hybrid thermal-photovoltaic systems," *Proc. UK-ISES Conference on C21 Photovoltaic Solar Energy Conversion*, September, 1979, p. 65.

[21] D. Coiante and L. Barra, "Can photovoltaics become an effective energy option?," *Solar Energy*, vol. 27, no. 1, pp. 79–89, 1992.

[22] H. P. Garg and R. K. Agrawal, "Some aspects of a PV/T collector/forced circulation flat plate solar water heater with solar cells," *Energy Conversion and Management*, vol. 36, no. 2, pp. 87–99, 1995.

[23] K. Nagano, T. Mochida, K. Shimakura, K. Murashita and S. Takeda, "Development of thermal-photovoltaic hybrid exterior wallboards incorporating PV cells in and their winter performances," *Solar Energy Materials and Solar Cells*, vol. 77, no. 3, pp. 265–282, 2003.

[24] H. A. Zondag, D. W. de Vries, W. G. J. van Helden, R. J. C. van Zolingen and A. A. van Steenhoven, "The yield of different combined PV-thermal collector designs," *Solar Energy*, vol. 74, no. 3, pp. 253–269, 2003.

[25] A. Tiwari and M. Sodha, "Performance evaluation of solar PV/T system: An experimental validation," *Solar Energy*, vol. 80, no. 7, pp. 751–759, 2006a.

[26] Y. B. Assoa, C. Menezo, G. Fraisse, R. Yezou and J. Brau, "Study of a new concept of photovoltaic–thermal hybrid collector," *Solar Energy*, vol. 81, no. 9, pp. 1132–1143, 2007.

[27] J. Prakash, "Transient analysis of a photovoltaic-thermal solar collector for co-generation of electricity and hot air/water," *Energy Conversion and Management*, vol. 35, no. 11, pp. 967–972, 1994.

[28] A. A. Hegazy, "Comparative study of the performances of four photovoltaic/thermal solar air collectors," *Energy Conversion and Management*, vol. 41, pp. 861–881, 2000.

[29] M. Y. Othman, B. Yatim, K. Sopian and M. N. A. Bakar, "Performance studies on a finned double-pass photovoltaic-thermal (PV/T) solar collector," *Desalination*, vol. 209, no. 1–3, pp. 43–49, 2007.

[30] T. Yamawaki, S. Mizukami, T. Masui and H. Takahashi, "Experimental investigation on generated power of amorphous PV module for roof azimuth," *Solar Energy Materials and Solar Cells*, vol. 67, pp. 369–377, 2001.

[31] Office of Technology Assessment, *Application of Solar Technology to Today's Energy Needs, Energy Conversion with Photovoltaics*, Princeton, pp. 406, 1978.

[32] G. K. Singh, "Analysis of environmental impacts on the performance of PV modules," Ph.D. Thesis, I.I.T Delhi, 2013.

6 Static Design Concept for a Light-Structured Building for Cold Climatic Conditions

6.1 INTRODUCTION

Cold and cloudy climatic conditions may be defined as when the mean monthly temperature is less than 25°C, relative humidity is more than 55%, and precipitation (either rain or snowfall) is more than 5 mm with less than 20 clear days. Such climatic conditions may be found in cities like Srinagar, Shimla in India. Cold and sunny climatic conditions may be defined as when the mean monthly temperature, relative humidity and precipitation (either rain or snowfall is less than 25°C, 55% and 5 mm, respectively, with more than 20 clear days. Such climatic conditions may be found in cities like Leh in India as given in Table 1.5. In such climatic conditions, maximum heating is required, in case there is a need for cooling, and either natural ventilation or forced convection may be used. Electricity for forced convection can be drawn by integrating photovoltaic modules with the buildings. For thermal heating, light-structured buildings are preferred because of minimum U-value. Generally, wooden or insulating material constructions are preferred. Preferably, south-facing double-glazed windows are used to transmit the radiation and reduce the U-value due to the availability of maximum solar radiation. Design criteria for these climatic conditions were briefly discussed in Chapter 1.

6.2 SOL-AIR TEMPERATURE

Sol-air temperature (T_{sa}) combines the effect of solar radiation, ambient air temperature and long-wave radiant heat exchange with the environment. Physically sol-air temperature is the temperature of the surroundings, which produces a similar heating effect as the incident radiation in conjunction with the actual external air temperature [1].

A horizontal surface can be treated in three ways for various applications of heating and cooling of a building. For all the three cases, sol-air temperature must be defined.

6.2.1 BARE SURFACE

Let us assume a horizontal surface exposed to solar radiation and ambient air temperature (T_a) with wind velocity as V in m/s (Figure 6.1a)

Sol-air temperature for horizontal bare surface can be written as:

$$T_{sa} = \frac{\alpha}{h_o} I(t) + T_a - \frac{\varepsilon \Delta R}{h_o} \tag{6.1a}$$

where

α is the absorptivity of the surface (dimensionless)
h_0 is the outside heat transfer coefficient in W/m²°C
$I(t)$ is the solar radiation in W/m²

DOI: 10.1201/9780429445903-6

161

ε is the emissivity of the surface; A blackbody has an emissivity of 1, while a perfect reflector or white body has an emissivity of 0

ΔR is the difference between the long-wavelength radiation exchange between the horizontal surface at temperature T_a to the sky temperature at T_{sky} as explained in Chapter 2

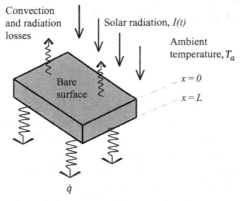

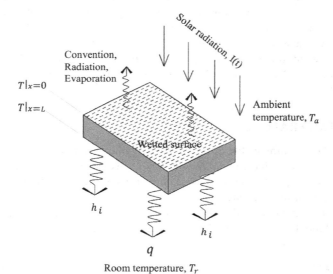

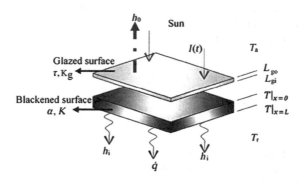

FIGURE 6.1 (a) A bare surface exposed to solar radiation. (b) A horizontal wetted surface exposed to solar radiation. (c) Schematic view of blackened and glazed surface [2].

Equation (6.1a) can be discussed for following cases:

Case (i): For $\dfrac{\alpha}{h_o} I(t) - \dfrac{\varepsilon \Delta R}{h_o} \geq 0$ or $\alpha I(t) - \varepsilon \Delta R \geq 0$, the sol-air temperature (T_{sa}) is greater than ambient air temperature (T_a). In this case, heating is possible by solar energy from a horizontal roof surface.

Case (ii): For $\dfrac{\alpha}{h_o} I(t) - \dfrac{\varepsilon \Delta R}{h_o} \leq 0$ or $\alpha I(t) - \varepsilon \Delta R \leq 0$, then sol-air temperature (T_{sa}) is less than ambient air temperature (T_a), i.e., $T_{sa} < T_a$. In this case, cooling is possible by solar energy from the horizontal roof surface.

Case (iii): For $\dfrac{\alpha}{h_o} I(t) - \dfrac{\varepsilon \Delta R}{h_o} = 0$ or $\alpha I(t) - \varepsilon \Delta R = 0$, then sol-air temperature (T_{sa}) is equal to ambient air temperature (T_a). In this case, neither heating nor cooling is possible by solar energy from a horizontal surface.

For a vertical surface, $\varepsilon \Delta R = 0$, expression for sol-air temperature is given by

$$T_{sa} = \frac{\alpha}{h_o} I(t) + T_a \tag{6.1b}$$

In this case only heating of a room is possible through a vertical wall for $I(t) > 0$.

6.2.2 Wetted Surface

Referring to Figure 6.1(b), there is a water film on horizontal surface, thus in addition to convective and radiative heat transfer, there is evaporative heat transfer from exposed surface to ambient air.

Sol-air temperature can be defined as:

$$T_{sa} = \frac{\alpha}{h_1} I(t) + T_a - \frac{\varepsilon \Delta R}{h_1} \tag{6.2}$$

where, h_1 is defined in Equation (2.43).

The value of T_{sa} in a case of wetted surface will be a lower value due to a higher value of h_1 in comparison to h_o of bare surface.

Equation (6.2) can be discussed for following cases:

Case (i): For $\dfrac{\alpha}{h_1} I(t) - \dfrac{\varepsilon \Delta R}{h_1} \geq 0$ or $\alpha I(t) - \varepsilon \Delta R \geq 0$, the sol-air temperature (T_{sa}) is greater than ambient air temperature (T_a). However, the values of T_{sa} in a wetted surface case will be lower compared to bare surface. Hence, in this case, heating level will be reduced.

Case (ii): For $\dfrac{\alpha}{h_1} I(t) - \dfrac{\varepsilon \Delta R}{h_1} \leq 0$ or $\alpha I(t) - \varepsilon \Delta R \leq 0$, sol-air temperature (T_{sa}) is less than ambient air temperature (T_a), i.e., $T_{sa} < T_a$. In this case, cooling level will be increased in the wetted surface case.

Case (iii): For $\dfrac{\alpha}{h_1} I(t) - \dfrac{\varepsilon \Delta R}{h_1} = 0$ or $\alpha I(t) - \varepsilon \Delta R = 0$, then sol-air temperature (T_{sa}) is equal to ambient air temperature (T_a). In this case, heating level will be reduced while cooling level is increased by solar energy from horizontal wetted surfaces.

6.2.3 Blackened and Glazed Surface

Figure 6.1(c) shows a schematic diagram where the top of a bare surface is painted black and the glass cover is placed on it at proper distance. In this case, the solar radiation after transmission

$[\tau I(t)]$ is absorbed $[\alpha \tau I(t)]$ by blackened surface. After absorption, there will be a rate of heat transfer from blackened surface to ambient through the glass cover and the rest is transferred by conduction through bottom.

Sol-air temperature can be defined as:

$$T_{sa} = \frac{\alpha \tau}{U_t} I(t) + T_a \tag{6.3}$$

It may be noted that in Equation (6.3), the $\varepsilon \Delta R$ is missing. This is because the blackened and glazed surface is not directly exposed to sky, thus $\varepsilon \Delta R = 0$ where, U_t is the overall heat loss coefficient from the blackened surface to an ambient through the glass cover.

$$U_t = \left[\frac{1}{h_1} + \frac{1}{h_2} \right]^{-1} \tag{6.3a}$$

6.3 THERMAL GAIN

In order to understand the thermal gain, the following concepts should be analyzed and studied.

6.3.1 DIRECT GAIN

Direct gain is the most common passive heating concept in which the solar radiation is directly transmitted through transparent material (glass or plastic etc.) through either roof or wall into the interiors of the building for thermal heating as shown in Figure 6.2(a). During the sunshine hours (daytime), the structure of the building collects, absorbs, stores and distributes the heat received. This stored heat is released during the off-sunshine hours to reach an equilibrium as shown in Figure 6.2(b). To avoid overheating of the space during daytime and cooling during nighttime, the walls and floors should be constructed of materials that can act as an energy reservoir medium for use during the nighttime. The most commonly used materials for heat storage are masonry and water.

To reduce the loss of trapped thermal energy, especially at the places that experience severe winters, double and sometimes triple-glazing are preferred. Moreover, the windows are covered with movable insulation during the off-sunshine hours to reduce the U-value. Building-integrated semi-transparent photovoltaic system with movable insulation during off-sunshine hours has been explained in Section 8.9.5. If movable insulation is not accepted for a particular application infrared heat mirrors may be used. Solar-shading devices like overhangs, awnings, louvers, etc., may be provided to avoid overheating during the summer months. Openable windows and vents may be installed for natural and cross-ventilation to maintain comfortable temperature. Thus, a direct gain system has the following components:

1. Glazing – to transmit and trap the incoming solar radiation
2. Thermal mass – to store heat for night-time use
3. Insulation – to reduce losses at night
4. Ventilation – for summertime cooling
5. Shading – to reduce overheating in summer

Static Design Concept for a Light-Structured Building for Cold Climatic Conditions

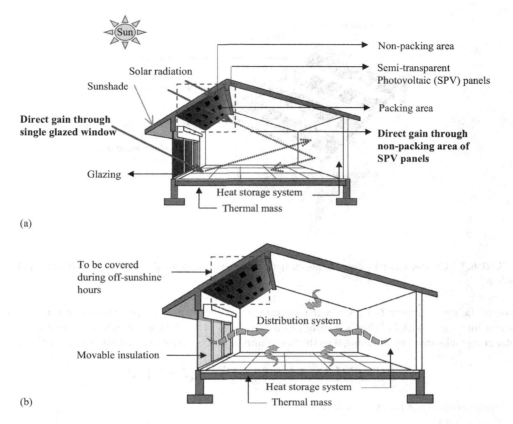

FIGURE 6.2 (a) Thermal heating during sunshine hours through direct gain (single-glazed window and non-packing area of semi-transparent photovoltaic modules). (b) Thermal heating during off-sunshine hours through direct gain.

Advantages and disadvantages of direct gain systems are listed below:
Advantages:

- They provide natural daylighting and ventilation (if openable).
- They permit an outdoor view.
- They are simplest in design and inexpensive.

Disadvantages:

- There is a possibility of glare problem.
- The color of the furnishing/objects exposed to direct sunlight may fade.
- The fiber materials may deteriorate in physical strength by extended exposure to ultraviolet radiation.

6.3.1.1 Direct Gain through Semi-Transparent Photovoltaic (SPV) System

In case of semi-transparent photovoltaic (SPV) modules integrated with the roof or façade of the building, the solar radiation transmitted through the non-packing area, $(1 - \beta)A_m$, of the module is termed as direct gain as shown in Figure 6.2(a). The solar gain (i.e., direct gain) is absorbed by the

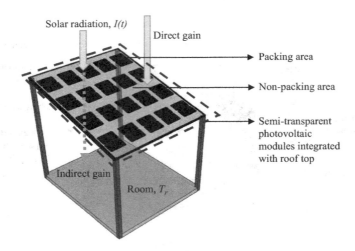

FIGURE 6.3 Cross-sectional view of semi-transparent photovoltaic thermal system integrated with building's rooftop.

floor of the room. Figure 6.3 gives a cross-sectional view of semi-transparent photovoltaic thermal system integrated with building's rooftop. Equation (6.4) gives the energy-balance equation for the solar energy absorbed by the floor, thus the floor temperature is calculated and given in Equation (6.5).

$$\alpha_f \tau_g^2 (1-\beta) I(t) A_m = h_c A_f (T_f - T_r) + U_b (T_f - T_0) A_f \quad (6.4)$$

Equation (6.4) can be explained as:

$$\begin{bmatrix} \text{Rate of solar energy} \\ \text{absorbed by the floor through} \\ \text{through non-packing area} \end{bmatrix} = \begin{bmatrix} \text{Rate of transfer} \\ \text{of heat from} \\ \text{floor to} \\ \text{room} \end{bmatrix} + \begin{bmatrix} \text{Steady-state heat} \\ \text{loss from floor to} \\ \text{bottom slab} \\ \text{of room} \end{bmatrix}$$

where, $(1-\beta)A_m$ is the non-packing area of an SPV module that is also responsible for daylighting and direct gain.

In Equation (6.4), if the value of packing factor (β) equals zero, then the entire roof will be responsible for direct gain. This holds true for wall assembly also, where, U_b = overall heat transfer coefficient from floor to room in W/m²K.

Overall heat transfer coefficient (refer to Section 2.8) is applicable where the heat transfer is involved through a medium composed of several different parallel layers each having a thermal conductivity or involving two or more of the heat transfer modes, namely conduction, convection and radiation.

Therefore, U_b in Equation (6.4a) can be expressed as:

$$U_b = \left[\frac{1}{h_c} + \frac{L_f}{K_f} + \frac{1}{h_i} \right]^{-1}, h_i = 2.8 + 3V, \ V = 0 \text{ m/s} \quad (6.4a)$$

Equation (6.4) can be rearranged and written as:

$$T_f = \frac{\alpha_f \tau_g^2 (1-\beta) I(t) A_m + U_b A_f T_0 + h_c A_f T_r}{(h_c + U_b) A_f} \quad (6.4b)$$

Static Design Concept for a Light-Structured Building for Cold Climatic Conditions **167**

where,

A_f and A_m are the floor and module area in m²
h_c and h_i are the convective heat transfer and inside heat transfer coefficient, respectively, in W/m²K
L_f and K_f are the thickness and thermal conductivity of the floor in m and W/mK, respectively
T_a, T_c, T_f, T_r, and T_0 are the ambient, solar cell, floor, room and below the floor room temperatures, respectively in °C
V is the wind velocity in m/s
α_f is the absorptivity of floor (dimensionless)
τ_g is the transmissivity of the glass (dimensionless)

6.3.1.2 Direct Gain through Glazed Windows

As shown in Figure 6.2(a), the solar gain within the building through windows (single- or double-glazed) or openings preferably south-facing (for northern hemisphere) is termed as direct gain, which is also responsible for daylight and thermal heating of the interiors.

Sol-air temperature of single-glazed systems can be calculated using Equation (6.5a)

$$T_{sa} = \frac{\alpha \tau_g I(t)}{U_L} + T_a \text{ and } U_L = \left[\frac{1}{h_0} + \frac{L_g}{k_g} + \frac{1}{h_i}\right]^{-1} \tag{6.5a}$$

where,

U_L is the overall heat transfer coefficient in W/m²K
$I(t)$ is the solar radiation in W/m²
h_0 is the inside heat transfer coefficient respectively in W/m²K

The rate of heat flux into the room in W/m²°C can be expressed as follows:

$$\dot{q} = U_L\left(T_{sa} - T_r\right) \tag{6.5b}$$

For thermal heating, double-glazed system is recommended for south-facing glazing as shown in Figure 6.4 and the sol-air temperature becomes,

$$T_{sa} = \frac{\alpha \tau^2_g I(t)}{U_L} + T_a \text{ and } U_L = \left[\frac{1}{h_0} + \frac{L_g}{K_g} + \frac{1}{Cc} + \frac{L_g}{K_g} + \frac{1}{h_i}\right]^{-1} \tag{6.5c}$$

where,

Cc is the air conductance in W/m²K
L_g and K_g are the thickness and thermal conductivity of the floor in m and W/mK, respectively

The air gap (i.e., air conductance) helps reduce the heat transfer by conduction since air is a poor conductor of heat. Thus, a reduction of about 9% in heat gain and 28% in losses may be achieved using a double-glazed system when compared to a single-glazed system [1]. The air gap acts as an insulator and reduces the U-value.

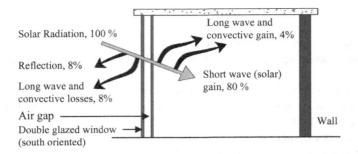

FIGURE 6.4 Double-glazed system [2].

6.3.1.3 Net Thermal Energy Gains

The rate of net useful thermal energy gain into the room is given by:

$$\dot{q} = U_t \left[T_{sa} - T_b \right] \tag{6.6}$$

where the expression for U_t is given in Equation (6.3a), and T_b is the base or reference temperature in the building.

Equation (6.6) can also be used to find out the total net energy gain during sunshine hours (t_T is the total sunshine hour). This can be obtained as follows:

$$q_{ud} = \int_0^{t_T} \dot{q}\, dt = \left[\tau \int_0^{t_T} I(t)\, dt - U_t \left(T_b t_T - \int_0^{t_T} T_a\, dt \right) \right] \tag{6.7}$$

Now if average solar intensity $\left(\bar{I}(t) \right)$ and ambient air temperature $\left(\bar{T}_a \right)$ are considered for $0 - t_T$ time interval then

$$\bar{I}(t) = \frac{1}{t_T} \int_0^{t_T} I(t)\, dt \text{ and } \bar{T}_a = \frac{1}{t_T} \int_0^{t_T} T_a\, dt \tag{6.8}$$

The accuracy of assumption depends on the size time interval, say, a minimum of about one hour. Equation (6.7) can be written as:

$$q_{ud} = \left[\tau \bar{I}(t) - U_t \left(T_b - \bar{T}_a \right) \right] t_T \times 3600 \tag{6.9}$$

During off-sunshine (or low sunshine) hours, the glazed windows or walls are covered with movable insulation, which is generally thick curtains to reduce the losses from the enclosed space to the outside ambient air. In this case, the net daily loss can be calculated as follows:

$$q_{un} = U_t \left(T_b - \bar{T}_a \right) \left(24 - t_T \right) \times 3600 \tag{6.10}$$

The daily net heat/thermal energy gain is given by

$$q_T = q_{ud} + q_{un} \tag{6.11}$$

Static Design Concept for a Light-Structured Building for Cold Climatic Conditions 169

EXAMPLE 6.1

Calculate the overall heat transfer coefficient for a double-glazed window with air cavity of 0.05 m, $C = 4.75$ w/m²°C. Given $K_g = 0.78$ W/m²°C and $L_g = 0.003$ m, $v = 1$ m/s, $h_0 = 9.5$ W/m²°C

Solution

From Equation (6.4a) and (6.5c), one obtains

$$U_L = \left[\frac{1}{9.5} + \frac{0.003}{0.78} + \frac{1}{4.75} + \frac{0.003}{0.78} + \frac{1}{5.8}\right]^{-1}$$
$$= \left[0.1053 + 0.0038 + 0.2105 + 0.0038 + 0.1724\right]^{-1} = 2.01 \text{ W/m}^2°C$$

EXAMPLE 6.2

Calculate the sol-air temperature for a double-glazed system for ambient temperature 20°C using the input values from Example 6.1 for the following parameters:

$$\alpha = 0.9, \tau_g = 0.9, I(t) = 250 \text{ W/m}^2$$

Also, prove that the double-glazed system is better for thermal heating.

Solution

For double glazed system, using Equation (6.5c) $T_{sa} = 109.11°C$
  For single-glazed system, using Equation (6.5a), overall heat transfer is calculated to be:

$$U_L = 3.59 \text{ W/m}^2°C \text{ and } T_{sa} = 76.3°C$$

From Example 6.1 and 6.2, it can be seen that overall heat transfer in a single-glazed system is higher than in a double-glazed system. This is due to the air gap (i.e., air conductance) that helps reduce the heat transfer by conduction and reduces the U-value. As a result, sol-air temperature is higher in double-glazed system.

EXAMPLE 6.3

Calculate the rate of heat flow through a south-facing concrete wall with mean incident solar radiation of 250 W/m², $T_{sa} = 13.0°C$, wall thickness = 30 cm, wall conductivity = 0.72 W/m²°C, mean room temperature = 20°C, $h_c = 8.7$ W/m²°C, $h_0 = 12.5$ W/m²°C, $h_i = 8.0$ W/m²°C, $\alpha = 0.6$. The south wall is covered with a 4-cm thick movable night insulation (I) with $K_{in} = 0.025$ W/m²°C. Calculate the heat flux into the room.

Solution

$$U = \left[\frac{1}{h_0} + \frac{L_{in}}{K_{in}} + \frac{L_g}{K_g} + \frac{1}{h_i}\right]^{-1} = \left[\frac{1}{12.5} + \frac{0.04}{0.025} + \frac{0.30}{0.72} + \frac{1}{8}\right]^{-1} = 0.45 \text{ W/m}^2°C$$

The rate of heat loss from the room to ambient during the night can be calculated using Equation (6.5b).

$$\dot{q} = U(T_{sa} - T_r) = -3.15 \, W/m^2 °C$$

EXAMPLE 6.4

Evaluate the total thermal energy gain q_{ud} in terms of kWh for a direct gain system at 25°C base temperature and exposed to solar radiation of 800 W/m² for 5 hours. Given parameters are ambient temperature = 15°C, transmissivity = 0.9 and overall heat loss coefficient is = 5 W/m²°C.

Solution

Using Equation (6.9), q_{ud} can be calculated as follows:

$$q_{ud} = [0.9 \times 800 - 5 \times (25-15)] \times 5 \times 3600 = 12.06 \, J = 3.35 \, kWh \, (1J = 2.778 \times 10^{-7} \, kWh)$$

6.3.2 Indirect Gains

Solar radiation is indirectly transmitted into the interiors of the building. It may be absorbed by a thermal mass located between the sun and the living space and then transferred into the room, creating a time lag or transmitted inside the room by the rear side of the PV modules (either integrated with the wall or roof). This thermal storage mass is placed in between the transparent material (i.e., glazing), preferably south-oriented, and the living space. This stored heat, absorbed by the mass is converted into thermal energy and is then transferred to the room via conduction and convection as shown in Figure 6.5. The flow of heat inside the room is controlled by the thickness, surface area, material and thermal properties of the thermal mass.

Advantages and disadvantages of indirect gain are listed below:
Advantages:

- The system helps in maintaining more uniform room-air temperature than the direct thermal gain system.
- Due to time lag, the storage material emits the heat to the living space at night when it is needed most.

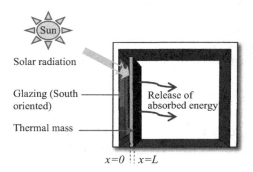

FIGURE 6.5 Indirect gain [2].

Static Design Concept for a Light-Structured Building for Cold Climatic Conditions 171

- Problems of glare and ultraviolet degradation of materials are eliminated.
- Reduces the thermal load leveling (TLL).

Disadvantages:

- Requires more collector area than a direct gain system.
- The system is more expensive owing to the cost of materials and the structural modifications required.
- Thermal storage material blocks the view and the daylight entering from the south.

6.3.2.1 Thermal Storage Wall/Roofs

Both walls and roofs may be used for thermal storage but south-facing walls are preferred over the roof. This is because of the abundance of solar radiation available on the south-oriented façade in the northern hemisphere. Also, in case of roof, the structure needs to be heavy, which is a bit problematic. A thermal storage wall absorbs sun's radiation on its outer surface, and then transfers this heat into the building through conduction. Thermal storage wall may be either masonry or water. The thermal energy is allowed to enter the inside of the building through glazing/openings and is stored in the thermal mass located between the sun and the living space. This stored heat, absorbed by the mass, is converted into thermal energy and then transferred to the roof. The concept of thermal mass has been used in ancient times as an effective passive cooling technique that reduced the temperature fluctuations inside, thus decreasing the thermal load leveling (TLL). Since solar radiation and ambient air temperature are periodic in nature, the room air temperature will fluctuate. TLL is used as a measure to indicate this fluctuation and is given in Equation (3.36). The concept of TLL has also been discussed in Section 3.7.

The thermal mass should be dark in color for maximum absorption of solar radiation, thus resulting in maximum efficiency of the system. In case the numeric value of the packing factor for the proposed system is zero, the system is then referred to as a glazed system.

6.3.2.2 Trombe Walls

A Trombe wall is a thick sun-facing wall popularized in 1964 by French engineer Félix Trombe and architect Jacques Michel. The first example of a Trombe wall system was used in the Kelbaugh House in Princeton, New Jersey. It is basically a thermal storage wall made of materials having high heat storage capacity like concrete, adobe, stone or composites of brick blocks and sand as shown in Figure 6.6(a). In order to increase the absorption, the outer surface is painted black and the outer surface are double glazed with an air gap in between. The solar radiation falling on the outer surface of the Trombe wall is absorbed as sensible heat and transferred to the interior of the storage mass by conduction or convection. During off-sunshine hours, heat stored in the thermal mass wall is then radiated and convected into the living space to be heated. Insulation may be provided between the glass and the Trombe wall to reduce potentially large heat losses. If properly designed, the system can provide adequate heat to the living space throughout the night.

For each 100 mm of concrete, there is a lag of about 2–2.5 hours between peak solar absorption and heat delivery inside. With wall thickness below about 300 mm the temperature swing of the interior would be excessive. Increasing the thickness above about 40 cm would result in higher cost while having only a small effect on the indoor swing. Considering this aspect and also the time of peak heating, the optimal thickness for concrete in residential buildings is approximately 30–40 cm [3].

A modified concept of a Trombe wall with storage was developed by [4] to reduce the heat lost to the outside as well as to provide additional storage space. An air cavity is provided within the exposed wall (Trombe wall) via a cupboard to reduce the U-value as shown in Figure 6.6(b).

Various parameters to improve the efficiency of Trombe wall are discussed in Chapter 7.

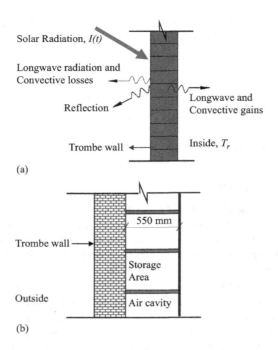

FIGURE 6.6 (a) Trombe wall [2]. (b) Trombe wall with storage and air cavity [2].

EXAMPLE 6.5

Calculate the rate of heat flow through a south-facing concrete wall for the following parameters (i) mean incident solar radiation = 250 W/m², (ii) ambient air temperature =13°C, (iii) wall thickness = 30 cm, (iv) wall thermal conductivity = 0.72 W/m°C, (iv) mean room temperature = 20°C, h_c = 8.7 W/m²°C, h_r = 3.8 W/m²°C, α = 0.6, V = 1 m/s and h_i = 8 W/m²°C.

Solution

Here, considering $h_o = h_c + h_r$ = (8.7 + 3.8) = 12.5 W/m²°C. The U value should be calculated as

$$\frac{1}{U} = \frac{1}{8.7+3.8} + \frac{0.30}{0.72} + \frac{1}{8} = 0.622$$

or

$$U = 1.609 \text{ W/m}^2°C$$

From Equation (6.1a), we have,

$$T_{sa} = \frac{0.6 \times 250}{12.5} + 13 = 25°C$$

Here, $\varepsilon\Delta R$ = 0 for the south wall.
The rate of heat flow can be obtained from

$$\dot{q} = U\left[T_{sa} - T_b\right]$$

Static Design Concept for a Light-Structured Building for Cold Climatic Conditions 173

$$\dot{q} = 1.609 \times (25-20) = 8.045 \text{ W/m}^2$$

The positive sign indicates that the heat is gained from the environment to the living space. Heat has to be removed at the above rate from the living space to maintain the space at 20°C.

The same problem, if done with an ambient air temperature of 5°C gives

$$T_{sa} = \frac{0.6 \times 250}{12.5} + 5 = 17°C$$

and $\dot{q} = 1.609 \times (17-20) = -4.827 \text{ W/m}^2$

This indicates that the heat has to be added at this rate to the living space to maintain the space at 20°C.

EXAMPLE 6.6

Calculate the mean heat flux into a room through Trombe wall of 450 mm thick made up of concrete ($K = 0.62$ W/m°C) with $\alpha = 0.8$ and $\tau = 0.71$ for room air temperature = 20°C, the external wall heat transfer coefficient from glazed surface = 5 W/m²°C, $h_i = 8$ W/m²°C, average ambient temperature = 14°C and mean solar radiation on the south face = 250 W/m².

Solution

Using $\dot{q} = U(T_{sa} - T_b)$, we have

$$\dot{q} = \left[\frac{1}{5.0} + \frac{0.450}{0.62} + \frac{1}{8} \right]^{-1} \times \left[\frac{0.8 \times 0.71 \times 250}{5} + 14 - 20 \right] = 0.951 \times (28.4 + 14 - 18) = 21.31 \text{ W/m}^2$$

EXAMPLE 6.7

For Example 6.6 and average ambient temperature of 8°C, calculate the night loss for the above system if the inside temperature (T_b) is maintained at 18°C and also calculate the net heat flux into the room for the same duration of day/night.

Also calculate net energy in terms of kWh when the Trombe wall is exposed to solar radiation for 8 and 4 hours, respectively.

Solution

$$\text{The rate of night losses} = 0.951 \times (8-18) = -9.5 \text{ W/m}^2$$

Since the duration of the day and night are the same, i.e. ,12 hours each, then

$$\text{net heat flux} = \text{the rate of heat flux gain} - \text{the rate of night losses} = 21.31 - 9.5 = 11.81 \text{ W/m}^2$$

$$\text{The net energy gain in terms of joules} = 11.81 \times 12 \times 3600 \text{ J} = 5.10 \times 10^5 \text{ J} = 0.141 \text{ kWh}$$

Now, when the Trombe wall is exposed to solar radiation for 8 hours,

$$\text{net energy gain} = (21.31 \times 8 - 9.5 \times 16) \times 3600 \, \text{J} = 6.65 \times 10^4 = 0.018 \, \text{kWh}$$

Now, when Trombe wall is exposed to solar radiation for 8 hours,

$$\text{net energy gain} = (21.31 \times 4 - 9.5 \times 20) \times 3600 \, \text{J} = -3.7 \times 10^5 = 0.104 \, \text{kWh}$$

This means that some external thermal energy is to be fed for 4-hour exposure to maintain the living space temperature.

6.3.2.3 Waterwalls

The principle of a waterwall is similar to that of the Trombe wall. The water tanks are stacked one above the other and placed behind a glazing that acts as a collector of solar energy as shown in Figure 6.7. Waterwalls have the advantage of large heat capacity unlike Trombe walls. It is used as a storage medium to release the heat into living space during night.

The solar heating factor is a function of thickness of the storage wall, and any increase in thermal conductivity results in a proportional increase in the efficiency of the solar wall based on the experiments conducted using different materials, like water, concrete, bricks and hollow bricks. Exposed surface of the waterwall is painted black to increase absorption of solar radiation. Sometimes a thin concrete wall or insulating layer is also placed adjacent to the inner surface of wall. Owing to higher specific heat of water than the concrete, the waterwall stores more heat than the Trombe wall. As waterwall is a convective thermal storage, heat is transmitted to the living space rapidly as compared to the Trombe wall. Translucent waterwalls may provide diffuse lighting. For further details see [1].

6.3.2.4 Trans Walls

A trans wall is a transparent thermal storage wall partially admitting solar energy. A layer of concrete wall is placed behind the water mass to not only reduce the average absorption of solar energy but also the temperature fluctuations (Figure 6.8). Thus, trans walls absorb marginally less energy as compared to the pure Trombe wall. Trans walls are more useful for heat transfer when daytime heating load is significant. For energy-balance equations refer to [1].

6.3.2.5 Solariums

A solarium is a combination of direct gain and thermal storage concepts. It consists of sun space (with thick thermal mass on the south side), linking living and sun space for Northern hemisphere

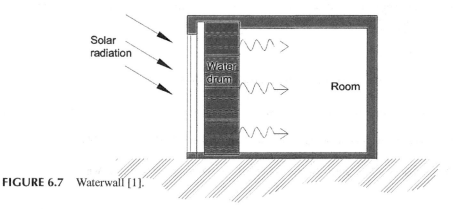

FIGURE 6.7 Waterwall [1].

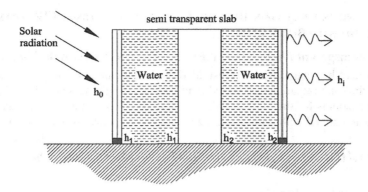

FIGURE 6.8 Trans wall [1].

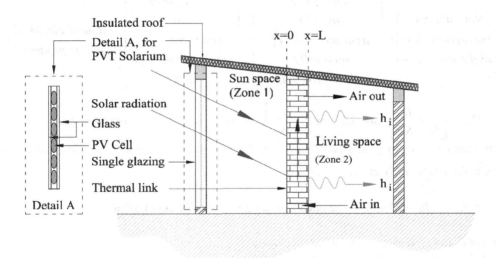

FIGURE 6.9 Schematic diagram for photovoltaic solarium [2].

(Figure 6.9). The thermal-link (like air collector, trans wall, water wall, metallic sheet) between the collector space and the storage mass, helps in heat retention and distribution and also enhances the efficiency of the system. The sunspace collects the energy through the glazing, absorbs it and pre-warm air for the living space. The sunspace works on the direct gain principle, in which the heat is used to maintain the temperature suitable for its transfer to the living space. The overheating of sunspace during the summer can be avoided by the use of shading. The installation of movable insulation/shutters may also be done to minimize the heat losses. For energy balance equations, see [1].

6.3.3 Isolated Gain

Greater flexibility in the design and operation can be achieved by isolating the collection unit of solar energy and the storage from the building. The most common method of this is the natural convective loop with thermo syphoning water heater. A rock bed storage system integrated with the building where the hot air from solar collector enters the building, and the cooled air is returned to this solar collector unit through basement and rock bed. When this hot air is not allowed to enter the building, it enters the rock bed and gets stored to be used later, preferably during nighttime. Details can be referred from [1].

6.3.4 Direct and Indirect Gain through Photovoltaic Thermal (PVT) Systems Integrated with Building

6.3.4.1 Semi-Transparent Photovoltaic (SPV) Roof Integrated with Building's Rooftop

As explained earlier, the solar radiation transmitted through the non-packing area is responsible for direct gain, and the solar radiation transferred to the room through the rear-side of the SPV module inside the room accounts for the indirect gain. Equation (6.6) gives the energy balance of indirect gain for building integrated with semi-transparent photovoltaic systems with rooftop (Figure 6.3).

$$\alpha_c \tau_g I(t) \beta A_m = \left[U_{tca}(T_c - T_a) + U_{bcr}(T_c - T_r) \right] A_m + \tau_g I(t) \beta A_m \eta_c \tag{6.6}$$

Equation (6.6) can be explained as:

$$
\begin{bmatrix} \text{Rate of solar} \\ \text{energy absorbed} \\ \text{by the solar cells} \end{bmatrix} = \begin{bmatrix} \text{Rate of heat} \\ \text{transfer} \\ \text{from top surface of} \\ \text{solar cell to} \\ \text{ambient through glass} \end{bmatrix} + \begin{bmatrix} \text{Rate of heat} \\ \text{transfer} \\ \text{from solar cell} \\ \text{to room air through} \\ \text{glass} \end{bmatrix} + \begin{bmatrix} \text{Rate of electrical} \\ \text{energy produced} \end{bmatrix}
$$

where $U_{tca} = \left[\dfrac{1}{h_0} + \dfrac{L_g}{K_g} \right]^{-1}$ and $U_{bcr} = \left[\dfrac{1}{h_i} + \dfrac{L_g}{K_g} \right]^{-1}$ are the overall heat transfer coefficients from solar cell to ambient through glass cover and overall heat transfer coefficient from solar cell to room through glass cover in $W/m^2 K$, respectively,

$$h_0 = 5.7 + 3V \text{ is the outside heat transfer coefficient in } W/m^2$$

α_c is the absorptivity of solar cell (dimensionless)
T_c is the temperature of solar cell in °C
$\eta_c = \eta_0[1 - \beta_0(T_c - T_{ref})]$ is the solar cell electrical efficiency in percent
β_0 is the temperature coefficient in $°C^{-1}$
η_0 is the solar cell electrical efficiency at standard test conditions (dimensionless)

EXAMPLE 6.8

Calculate the overall heat transfer coefficient from solar cell to ambient for building integrated semi-transparent photovoltaic system from solar cell to ambient for the glass thickness of 3 mm and thermal conductivity of 0.9 W/m²K. Consider wind velocity = 0 m/s.

Solution

$$U_{tca} = \left[\frac{1}{h_0} + \frac{L_g}{K_g} \right]^{-1} = \left[\frac{1}{5.7} + \frac{0.003}{0.9} \right]^{-1} = [0.1754 + 0.0033] - 1 = 5.6 \text{ W/m}^2\text{K}$$

6.3.4.2 Photovoltaic Thermal (PVT) Trombe Walls

As shown in Figure 6.10, a PVT-Trombe wall is integrated with the semi-transparent photovoltaic (SPV) modules for thermal heating of the interiors during winter season. The Trombe wall receives and stores heat from the non-packing area of the SPV modules (direct gain) and from the rear side of the SPV module (via conduction, indirect gain). This stored heat is released during the off-sunshine hours via convection and radiation, which leads to an increase in the indoor temperature.

Referring to Figure 6.10, energy balance equation for $x = 0$ may be written as follows:

$$\tau^2{}_g \alpha_b (1-\beta) I(t) + U\left(T_c - T\big|_{x=0}\right) = -K \frac{\partial T}{\partial x}\bigg|_{x=0} \tag{6.7a}$$

The energy balance equation for $x = L$ may be written as follows:

$$-K \frac{\partial T}{\partial x}\bigg|_{x=0} = h_i \left(T\big|_{x=L} - T_r\right) \tag{6.7b}$$

Equation (6.7b) can be defined as

$$\left[\text{Heat leaving by conduction}\right] = \left[\text{Heat entering by convection}\right]$$

A photovoltaic Trombe wall can be used for thermal heating and cooling. For thermal cooling, it is installed with vents, and for thermal heating, it should be installed without vents (Figure 7.10).

Variation of thermal load–leveling and decrement factor with change in thickness of Trombe wall is given in Figure 6.11 [5]. It can be seen with increase in the thickness of the wall, the fluctuations in the room air temperature reduces. This result indicates that with greater thickness of PVT wall, the thermal stability of the room is improved, and the attenuation in room air temperature is also reduced. The optimum wall thickness as suggested by Taffesse et al. [5] was 400 mm.

6.3.4.3 Integration of Roof (with Vent) with Semi-Transparent Photovoltaic Modules

The photovoltaic module gets heated when solar radiation falls on it. Inlet vent is provided at the bottom through which the cold air enters. This heat gets heated due to the thermal energy produced by the rear of the PV module and escapes through the outlet vents. Since there is a difference in

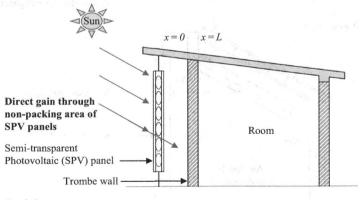

FIGURE 6.10 PVT-Trombe wall is integrated with the semi-transparent photovoltaic modules.

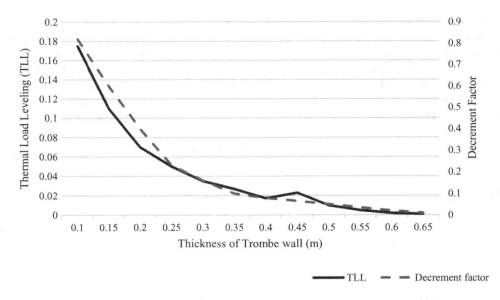

FIGURE 6.11 Variation of thermal load leveling with thickness of Trombe wall.

density between air (at inlet) and hot air (at outlet), the hot air (at outlet) enters the interiors and replaces the inside cold air. A fan may also be provided for better circulation of air. Thus, the thermal energy produced by the PV modules may be utilized for space heating. Further, this also reduces the temperature of the PV module as a result of which the efficiency of the system increases. Thus, the room air temperature is lower that the similar case without an air duct. This is due to the indirect heating because of the presence of insulated roof between the PV module and room air.

Figure 6.12 shows a schematic diagram for a roof with vents and integrated with semi-transparent photovoltaic modules. The energy balance of SPVT roof with air duct [6].

$$\alpha_c \tau_g I(t) \beta b dx = \left[U_{tca}(T_c - T_a) + U_{bcf}(T_c - T_f) \right] b dx + \tau_g I(t) \beta \eta_c \alpha_c b dx \tag{6.8}$$

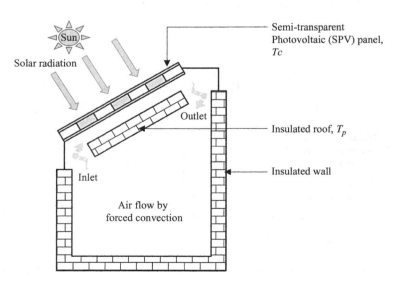

FIGURE 6.12 Semi-transparent photovoltaic modules integrated with roof with air duct [6].

Equation (6.8) can be explained as:

$$
\begin{bmatrix} Rate\ of\ solar \\ radiation\ available \\ on\ the\ solar\ cells \end{bmatrix} = \begin{bmatrix} Rate\ of\ transfer \\ of\ heat\ from \\ top\ surface\ of \\ solar\ cell\ to \\ ambient\ through\ glass \end{bmatrix} + \begin{bmatrix} Rate\ of\ transfer \\ of\ heat\ from \\ solar\ cell\ to \\ flowing\ air\ through \\ glass\ cover \end{bmatrix} + \begin{bmatrix} Rate\ of \\ electrical\ energy \\ produced \end{bmatrix}
$$

The energy balance equation for air flowing through air duct may be written as:

$$
\dot{m}_f C_f \frac{dT_f}{dx} dx = \left[h_{pf}\left(T_p - T_f\right) + U_{bcf}\left(T_c - T_f\right) \right] bdx \tag{6.9a}
$$

$$
\dot{q}_u = \dot{m}_f C_f \left(T_{fo} - T_r\right) \tag{6.9b}
$$

where,

$\dot{m}_f$ is the mass flow rate of fluid (air) in kg/s

C_f is the specific heat of the fluid (air) in J/kgK

T_p is the temperature of insulated roof in °C

$U_{tcf} = \left[\dfrac{1}{h_i} + \dfrac{L_g}{K_g} \right]^{-1}$ is the overall heat transfer coefficient from solar cell to flowing air through glass cover in W/m²K.

6.3.4.4 Integration of Roof with Opaque Photovoltaic Modules

The energy balance equation of an opaque photovoltaic thermal (OPVT) system may be written as [6]:

$$
\left[\alpha_c \tau_g \beta + (1-\beta) \right]\left[\tau_g \alpha_{td} \right] I(t) bdx = \left[U_{tca}\left(T_c - T_a\right) + h_{ctd}\left(T_c - T_{td}\right) \right] bdx + \tau_g I(t) \beta \eta_c \alpha_c bd \tag{6.10}
$$

Equation (6.10) can be explained as:

$$
\begin{bmatrix} Rate\ of\ solar \\ radiation\ available \\ on\ the\ solar\ cells \end{bmatrix} = \begin{bmatrix} Rate\ of\ transfer \\ of\ heat\ from \\ top\ surface\ of \\ solar\ cell\ to \\ ambient\ through\ glass \end{bmatrix} + \begin{bmatrix} Rate\ of\ transfer \\ of\ heat\ from \\ solar\ cell\ to \\ tedlar \end{bmatrix} + \begin{bmatrix} Rate\ of \\ electrical\ energy \\ produced \end{bmatrix}
$$

The energy balance equation for air flowing through the air duct may be written as:

$$
\dot{m}_f C_f \frac{dT_f}{dx} dx + U_{pr}\left(T_f - T_r\right) bdx = \left[h_{ctd}\left(T_{td} - T_f\right) \right] bdx \tag{6.11}
$$

where

h_{ctd} is the heat transfer coefficient from solar cell to Tedlar in W/m²K

T_{td} is the temperature of Tedlar in °C

α_{td} is the absorptivity of Tedlar (dimensionless)

U_{pr} is the overall heat transfer coefficient from brick wall and insulation to room air in W/m²K

The room air temperature is higher in SPVT systems than OPVT systems. The reason behind this is that a double layer of glass in SPVT systems reduces the heat losses from the room to outside air. In addition to this, the transmission of solar energy (direct gain) through the non-packing area of SPV module further increases the heat gains. Since Tedlar has low conductivity, the rise in temperature is lower in OPVT system.

Another advantage of BiSPVT over BiOPVT is the natural light let into the building, thus reducing the building's demand on artificial lighting. Solar radiation is allowed to penetrate inside the room through the non-packing area of the SPV module. This leads to a reduction in the dependence on artificial lighting. Hence, daylight savings are attained. It is high impact because not only daylight savings are achieved but also the cooling demand of the building is reduced by minimizing the use of artificial lighting as a direct consequence of the reduced heat losses. Since the dependence on the artificial lighting reduces, the dependence on conventional fuels also reduces, thus making the system more sustainable. Moreover, daylight also has positive psychological effects on human behavior and performance. These daylight savings depend on the packing factor of the photovoltaic module, and packing factor should be minimum.

6.3.4.5 PVT Solariums

As discussed in Section 6.3.2.5, solariums consist of a sun space and a living space with a linking wall. In PVT solariums, the south-oriented façade of the sun space is integrated with the photovoltaic modules, and the roof is made up of glass as shown in Figure 6.9 (detail A). The south orientation of the PV modules allows maximum solar radiation inside which leads to an increase in the temperature of the sun space, and thermal heat is thus stored in the linking wall i.e., Trombe wall ion this case. Trombe walls absorb solar radiation from sides, roofs, and non-packing areas of the SPV modules.

OBJECTIVE QUESTIONS

6.1   A non-packing area of a semi-transparent module is responsible for
    (a)   Electricity production
    (b)   Cooling of building
    (c)   Direct gain
    (d)   Efficiency of solar cell

6.2   Which of these concepts is not an indirect gain?
    (a)   Double-glazed window
    (b)   Waterwall
    (c)   Thermal mass
    (d)   Trans wall

6.3   A Trombe wall is used to
    (a)   Increase the TLL
    (b)   Reduce the U-value
    (c)   Allow direct gain
    (d)   Allow day lighting

6.4   Semi-transparent photovoltaic systems shall be used over opaque photovoltaic systems due to
    (a)   Daylighting
    (b)   Lower room air temperature
    (c)   Reduction in direct gain
    (d)   None of the above

6.5   Following is a direct gain system
    (a)   Trans wall
    (b)   Waterwall
    (c)   Rock bed storage
    (d)   Windows

Static Design Concept for a Light-Structured Building for Cold Climatic Conditions 181

6.6 Which system is best for thermal heating amongst the following?
(a) Single-glazed
(b) Double-glazed
(c) Single-glazed with PV integration
(d) Double-glazed with PV integration

6.7 For the northern hemisphere, PV modules shall be oriented towards
(a) North
(b) South
(c) East
(d) West

6.8 When the value of packing factor in semi-transparent photovoltaic system integrated with a building equals 1, the system will be responsible for:
(a) Direct gain
(b) Indirect gain
(c) Daylighting
(d) Glazed system

6.9 PVT-Trombe walls can be used for thermal heating
(a) With vent
(b) Roof as duct
(c) Without vent
(d) None of the above

6.10 For thermal heating, light-structured buildings are preferred because of
(a) Minimum U-value
(b) Maximum U-value
(c) U-value = 0
(d) None of the above

ANSWERS

6.1 (c)
6.2 (a)
6. 3 (b)
6. 4 (a)
6. 5 (d)
6. 6 (d)
6. 7 (d)
6. 8 (b)
6. 9 (c)
6. 10 (a)

REFERENCES

[1] G. N. Tiwari: *Solar Energy: Fundamentals, Design, Modelling and Applications*, Delhi: Narosa, 2002.

[2] G. N. Tiwari and N. Gupta, "Review of passive heating/cooling systems of buildings," *Energy Science & Engineering*, vol. 4, no. 5, pp. 305–333, 2016.

[3] B. Agrawal and G. N. Tiwari, *Building Integrated Photovoltaic Thermal Systems: For Sustainable Development*, U.K: RSC Publications, 2010.

[4] N. Gupta and G. N. Tiwari, "Energy matrices of building integrated semitransparent photovoltaic thermal systems: A case study," *Journal of Architectural Engineering*, vol. 23, no. 4, pp. 05017006-1–05017006-14 2017.

[5] F. Taffesse, A. Verma, S. Singh and G. N. Tiwari, "Periodic modeling of semi-transparent photovoltaic thermal-trombe wall (SPVT-TW)," *Solar Energy*, vol. 135, pp. 265–273, 2016.

[6] K. Vats and G. N. Tiwari, "Performance evaluation of a building integrated semitransparent photovoltaic thermal system for roof and facade," *Energy and Buildings*, vol. 45, pp. 211–218, 2012.

7 Dynamic Design Concepts for Hot Climatic Conditions

7.1 INTRODUCTION

Reducing the peak cooling demand of a building can be referred to as thermal cooling of a building. A number of natural heat rejection mechanisms, including ventilation, evaporation, infrared radiations to the sky, and earth contact cooling, etc. are used to reduce the peak cooling demand. The overall efficiency of such systems depends on the specific climatic conditions in the area, the cooling needs and patterns in the building, as well as on the efficiency of the technology used.

The first step in the direction of passive cooling is the reduction of unnecessary thermal loads:

(a) **Exterior load:** The exterior loads due to climate on the buildings. The climate-dependent loads involve conduction of heat through the building skin, infiltration of outside air and penetration of short-wavelength radiations directly. This is achieved by having an optimized Trombe wall without ventilation during day.
(b) **Interior loads:** Thermal emission due to people activities/occupancy on the building.

The cooling concepts may be divided into two categories:

(a) **Direct cooling:** The direct cooling concepts include phase change materials, ventilation/infiltration/courtyards, thermotropic and thermochromic coatings, courtyards, air cavities, radiative cooling, wind towers, air vents, earth coupling, Trombe walls and direct skin façade.
(b) **Indirect cooling:** The indirect cooling concept includes shading, green/cool roof, evaporative cooling, movable insulation, rock bed regenerative cooling, earth air heat exchanger, roof pond, thermosiphon and solar photovoltaic cooling.

These concepts are described in the subsequent sections.

7.2 PHASE CHANGE MATERIALS (PCMs)

We know that the solar energy available is periodic in nature, thus it is important to ensure solar energy storage both in the long and short term. Heat is produced because of photothermal conversion happening within the components of the building's envelope, which is then stored depending on the thermal capacity of the storage material. The capacity of heat storage of the material is defined by its specific heat capacity (expressed in kJ/kgK). This value describes the amount of heat stored by 1 kg of a material at a temperature difference (ΔT) of 1 K. Except for specific heat capacity, storage capacity depends upon density and volume of the storage medium/material. For example, in places at higher latitude, massive, large and heavy density external walls (often internal) are used to certify high heat capacity. These massive structures keep the heat inside the building for a longer duration of time (to reduce the heat losses).

A building's heat capacity can have an effect increase without increasing the mass and volume of the construction materials but even by reducing these parameters. This can be achieved by phase change materials (PCMs). These PCMs can store large amounts of energy at relatively constant temperature, at which they remain in a liquid state, thus having high heat capacity. Therefore, with

DOI: 10.1201/9780429445903-7

183

decreases in the temperature, a solidification process takes place, and latent heat is released out of the PCM. These materials are characterized by their ability to absorb heat in the liquid state at room temperature. During off-sunshine hours (nighttime), due to the solidification process, PCM starts releasing heat stored in it as discussed and the room is warmed. During the sunshine hours (daytime), the temperature is increased due to solar gains, and the PCM melts and transforms from solid to liquid state, thus absorbing heat from the surroundings depending upon the thermal capacity of the PCM, resulting in a drop in room air temperature [1].

Using PCMs decreases the consumption of conventional energy carriers and also reduces the peak energy demand, thus improving the energy balance of a building by allowing the storage of solar energy in building components like walls, floors, etc., during the day and releasing the stored energy at night with a time lag.

Table 7.1 gives the density and specific heat capacity of major building construction materials and an example of the PCM applied in buildings. In case of PCM, the effective specific heat capacity refers to the phase change heat. It is also clear that the phase change material has considerably higher heat capacity than other building materials.

Phase change materials have high phase change enthalpy for phase changes at room temperature. They can absorb huge amounts of heat flux at nearly constant temperature. By adding a determined dose of the PCM to the construction materials (concrete, plaster, etc.), PCM can be used in construction.

PCMs may also be used within BiPV systems. Flat containers filled with PCM are attached to the rear side of the PV module, where they act as a reservoir of solar energy, which is later supplied to the building's interior as heat with a time lag (in off-sunshine hours). Alongside, the PCM container behaves as a cooler for PV panels, thus stabilizing the temperature of the module and preventing the loss of efficiency [1]. This can be explained as the efficiency of PV module drops with a rise in temperature as explained in Chapter 8.

7.3 INFILTRATION/NATURAL VENTILATION

Energy conservation and natural ventilation should be the prime concern of any building design. As buildings are often planned to be sealed and well-insulated, with low heat gain or loss, often this means extreme use of HVAC systems to improve air quality and to dilute the VOCs emitted by the building materials and furniture. In buildings, there are two types of airflows induced due to pressure difference leading to natural air flow [2].

TABLE 7.1
Density and Heat Capacity of Different Construction Materials and PCM [1]

Construction Material	Density (kg/m³)	Specific Heat Capacity		Remarks
		(kJ/kgK)	(MJ/m³K)	
Brick	1600	0.84	1.18	-
Concrete	2300	1.0	2.3	-
PCM	870–1000	18.0	0.96	PCM here refers to a mixture of fatty acids with melting point 22°C and effective heat capacity in range of 18°C–28°C
Steel	7800	0.47	3.67	-
Wood	600	1.6	0.96	-

Dynamic Design Concepts for Hot Climatic Conditions

Infiltration means to admit outside air into the living space of a room through door and/or window openings and cracks and interstices around the doors and windows. It can also be said that there is exchange of air between room air and outside air through infiltration/ventilation.

The infiltration may be due to

- The pressure difference generated by the difference in temperatures and humidity of inside and outside air of a building
- Wind pressure
- Entry and exit of occupants

Infiltration can be minimized with draft sealing, air locks, and airtight and quality construction of doors and windows (Section 1.5.2.5).

Wang et al. [3] pointed out that airtightness alone can significantly impact the heating and cooling performance of the building. In the study, infiltration of hot and humid air led to an increase of 9.4% and 56% in total cooling load and latent load, respectively. The heating load was reduced by 1.4% due to higher outside temperature.

The other type of air flow is ventilation. Natural ventilation is an important and simple technique that when appropriately used may improve thermal comfort conditions in indoor spaces, decrease the energy consumption of air conditioned buildings, and contribute to fight problems of indoor air quality by decreasing the concentration of indoor pollutants [4]. Natural ventilation is not an alternative to air conditioning but is a much more effective instrument to improve the indoor air quality, provide better comfort, health and decrease unnecessary energy consumption for artificial cooling of the space.

Windows play a vital role for air circulation within the building premises with a recommended value of air movement being 0.2–0.4 m/s during winters and summers, respectively.

The rate of transfer from roof bottom to room air with ventilation losses can be expressed as:

$$M_a C_a \frac{dT_r}{dT} = h_r \left(T_{|x=L} - T_r\right) A_r + A_{win} U_t \left(T_{|sa,win} - T_r\right) - 0.33 NV \left(T_r - T_a\right) \tag{7.1a}$$

where

M_a is the mass of air in kg C_a is the specific heat of the air in J/kgK
h_r is the radiative heat transfer coefficient in W/m²K
$T_{|x=L}$ is the roof bottom temperature
T_r and T_a are the room air temperature and ambient temperature, respectively, in °C
A_{win} is the area of the window in m²
U_t is the overall heat transfer coefficient in W/m²K
$T_{|sa,win}$ is the sol-air temperature for glazed window
N is the number of air changes per hour in h⁻¹
V is the volume of room air in m³

Convective heat losses due to ventilation are attributable to the air exchange rate, temperature difference between inside and outside the building and heat capacity of air. Infiltration and ventilation are also explained in Chapter 1.

Equation (7.1a) gives the relation to determine the room air temperature and then the ventilation losses as:

$$\dot{Q} = 0.33 NV \left(T_r - T_a\right) \tag{7.1b}$$

Local topography and surface texture affect the wind conditions considerably. Spacing of buildings, at six times their height in a grid iron pattern, results in proper wind movement due to uniform flow and removal of stagnant zones. Windows play a dominant role in inducing ventilation. The ventilation rate is affected by the parameters namely climate, wind direction, size of inlet and outlet, volume of the room, shading devices and the internal partitions. These are also explained in Chapters 1 and 3.

Three basic functions of ventilation, viz. (i) inflow of fresh air to replace the used internal air and removal of the products of combustion of fuel-less cooking and gas cooking, (ii) to cool the body by increasing evaporation of moisture from the skin and (iii) heating or cooling the interior of the building, may be achieved in three different ways: the stack effect (temperature difference), wind pressure and mechanical means, respectively.

7.3.1 SMART WINDOWS

A smart window controls and regulates the solar energy influx through double-skin facades to the interior of the building. Glass coated with thin films that can change their optical properties reversibly from transparent to opaque when heated and cooled or when subject to an applied electrical current are of great interest. Again, however, they do not utilize the incident solar energy for electricity production [5].

7.3.2 LITERATURE STUDY: INFILTRATION/NATURAL VENTILATION

Wan and Yik [6] suggested the window-to-ground ratio should lie within the range of 0.33–0.58 (average 0.44) for residential buildings with mean window-to-wall ratio for the dining/living room and bedroom and are 0.34 and 0.27, respectively. Wang and Malkawi [7] developed a natural ventilation evaluation index for early-stage design of office buildings, which included both indoor air quality and thermal environment aspects

Ventilated roof was studied by Ciampi et al., and 30% of energy savings were achieved when compared to a non-ventilated structure [8]. A major focus of the study was the case of a small-sized thickness duct with laminar air flow (microventilation). Another study for ventilated roof was done by Ibrahim et al. [9] with 50% and 100% openings in the roof and a corresponding reduction in room air temperature up to 3°C–6°C and 2°C–10°C, respectively. Thermal performance of a ventilated roof was studied by Li et al. [10] based on different parameters like air-gap thickness, roof slope, exhaust outlet size and absorption coefficient of external roof surface. The roof slope 0.33–0.4 has been recommended. Ventilated roof made up of phase change materials was proposed by Li et al. [11] to reduce the heat gains and thus the building load. It was found that indoor peak temperature can be reduced by 16.9%–18.8%.

Tiwari [12] suggested the value of air movement to be 0.2 m/s for winters and 0.4 m/s for summer months. For temperate climatic conditions, Parys et al. [13] suggested to cool office buildings by opening the windows. According to the study, when a combination of diurnal window operation and night ventilation is used, the heat gains can be limited to about 1500 kJ/m² per working day. Various design strategies were implemented and an increase in air velocity up to 0.5 m/s was achieved by cross-ventilation while a drop of 2.0°C–2.5°C in the air temperature was found using night ventilation [14].

7.3.3 SHADING

Different methods can be used to shade the building to provide passive cooling and also help in energy conservation in buildings. Shading reduces the incident solar radiation and thus, cools the building effectively, affecting the building's energy performance.

Criteria for shading for various climatic zones is given in Table 7.2.

Dynamic Design Concepts for Hot Climatic Conditions

TABLE 7.2

Criteria of Shading for Various Climatic Zones [15]

Climate	Requirements
Hot and dry (HD)	Complete shading all year-round
Warm and humid (WH)	Complete shading year-round, but design should be such that ventilation is not affected
Temperate	Complete shading year-round but only during major sunshine hours
Cold and cloudy (CC)	No shading
Cold and sunny (CS)	Shading during summer months only
Composite (CO)	Shading during summer months only

7.3.4 WINDOWS

An important building element that greatly influences the thermal environment is the window. To prevent passive solar heating, when it is not wanted, a window must always be shaded from the direct solar component and often so from the diffuse and reflected components. Windows/openings are the most important element of any building in terms of comfort and energy use per unit area [16] and are the major contributor for the solar heat gain of the interiors, thus shading devices are the common strategies to prevent the same. The openings are consciously designed and installed to provide fresh air, oxygen and extract the pollutants along with odor to maintain good air quality and achieve maximum benefits with minimal capital cost and environmental loss. Location and orientation of these fenestrations (openings/windows) have a significant influence on the natural ventilation.

Window openings have to be designed in relation to sunlight, ventilation and air-movement. Window design with relation to the sunlight can be done in two ways:

- Design of shading devices to prevent radiation from entering
- Design of openings to permit the adequate natural lighting of the interior

Although shading of the whole building is beneficial, shading of the window is crucial. The shading systems contribute toward lowered transmittance values of the window. There are a number of options available for windows that we explore in the following subsections.

7.3.4.1 Self-Inflating Curtains

The curtain consists of a number of layers of thin flexible material of high reflectivity and low emissivity. Solar radiation warms the air between the layers, increasing the pressure and decreasing the density in the upper part of the system. The pressure pushes the layers apart and causes fresh air intake of air from the bottom. Thus, the system of reflecting layers separated by air gaps provides good insulation, and when the insulation is not required the air is evacuated from the sides.

The rate of thermal energy transfer ($\dot{q}$) through a self-inflating curtain is written as follows:

$$\dot{q} = \left(1 - f\right)\tau I\left(t\right) - U\left(T_r - T_a\right) \tag{7.2}$$

where f is the shading factor, and the value of f is 1 for complete shading and less than 1 for partial shading. T_r and T_a are the room and ambient temperatures, respectively.

7.3.4.2 Window Quilt Shade

This type of shade is a sandwich of fine layers, assembled with an ultrasonic fiber welder and often enclosed in decorative polyester fiber.

7.3.4.3 Venetian Blind between the Glasses

Blinds are frequently used within façades to control the intensity of the incident direct solar radiation component by blocking the radiation from entering the building and thus reducing the cooling loads.

This is an effective system to reduce the heat loss through a double-glazed window. In this case, the characteristic dimension of the unit is small; hence the convective heat transfer is stopped. In this case, an expression for $\dot{q}$ is given below with a lower value of transmissivity (τ) and overall heat transfer from room to ambient through glass (U_t)

$$\dot{q} = U_t \left[\frac{\tau}{U_t} I(t) + T_a - T_b \right] = U_t \left[T_{sa} - T_b \right] \tag{7.3}$$

where $I(t)$ is the solar radiation and T_a, T_{sa} and T_b are the ambient, sol-air and base temperatures, respectively.

7.3.4.4 Transparent Heat Mirrors

In order to reduce the heat loss from the glazed surface, the glass is coated by a film which to a large extent reflects the infrared radiation from the surface. However, this coating also reduces the transmissivity of the glass window for solar radiation. The coating may consist of single or multiple layers of different substances, deposited by vacuum evaporation or spray technique. These heat mirrors give much less heat loss and a higher transmission when compared with multi-pane systems.

7.3.4.5 Solar Shading Devices

These devices can be classified as vertical (vertical louvers, projecting fins), horizontal devices (canopy, awnings, horizontal louvers and overhangs), egg crate devices (concrete grille blocks, metal grills) and screenings (venetian blinds, double-glass windows, window quilt shades, movable insulation curtains, natural vegetation, etc.). Horizontal shading devices are best suited for south-oriented openings, whereas vertical shading devices for east- and west-facing facades to protect from intense sun at low angles. Details are given in Section 1.5.2.3.

7.3.4.6 Roofs

Shading of roofs is a necessity to reduce the heat gains. They can be shaded by a roof cover of concrete, sheet, plants, canvas or earthen pots (Figure 1.9). Shading by roof cover (made of concrete or galvanized iron sheets) does provide protection against direct solar gains, but it does not permit heat to escape to the sky during off-sunshine hours. Another inexpensive and effective device is a removable canvas cover mounted close to the roof. During daytime it prevents entry of heat, and its removal at night offers radiative cooling. Painting the canvas white minimizes the radiative and conductive heat gain [17]. Shading by a cover of deciduous plants and creepers is a better alternative.

7.3.5 WALLS

7.3.5.1 Heat Trap

A reasonable thickness of transparent insulating material with good transmissivity shall be used to reduce the heat transfer.

7.3.5.2 Optical Shutter

An optical shutter is made up of three layers of transparent sheets and one layer of cloud gel. It is opaque at high temperatures. It can be used for reducing air conditioning loads and preventing overheating in greenhouses and solar collector systems.

Dynamic Design Concepts for Hot Climatic Conditions

7.3.5.3 Shading by Textured Surface

Surface shading, like highly textured walls, have a portion of their surface in shade. The increased surface area of such a wall results in an increased outer surface coefficient, which permits the sunlit surface to stay cooler as well as to cool down faster at night [17].

7.3.5.4 Trees and Vegetation [17]

Landscaping (vegetation and trees) are very effective in providing shading to reduce the heat gains. Trees can be used with advantage to shade roofs, walls and windows. Shading and evapotranspiration from trees can reduce the surrounding temperature as much as 5°C. The following shall be considered for shading during summer months:

- Deciduous trees and shrubs provide summer shading and also allow winter access. They should be planted on the south and southwest side of the building. During winter months, these drop off their leaves, allowing sunlight to enter and provide thermal heating to the interiors.
- Trees with heavy foliage are very effective in obstructing the sun's rays and casting a dense shadow. Dense shade is cooler than filtered sunlight. High branching canopy trees can be used to shade the roof, walls and windows
- Evergreen trees should be grown on the south and west sides of the building as they provide best shading from the setting summer sun and cold winter winds [18].
- Shading and insulation for walls can be provided by plants that adhere to the wall, such as English ivy, or by plants supported by the wall, such as jasmine.
- Horizontal shading should be done for south-facing windows, e.g., deciduous vines (which lose foliage in the winter), such as ornamental grape or wisteria, can be grown over a pergola for summer shading.
- Vertical shading is suited for east and west facades and windows in summer, to protect from intense sun at low angles, e.g., screening by dense shrubs, trees, deciduous vines supported on a frame, shrubs used in combination with trees.

7.4 LITERATURE STUDY: SHADING

Solar shading techniques serve as both functional and aesthetic element if designed properly. They help reduce the thermal gains and provide comfortable indoor temperature.

- **Solar shading devices and techniques:**

Various passive cooling techniques were studied by Kumar et al. [19] that solar shading alone was capable of reduction of about 2.5°C–4.5°C of inside room temperature. A further drop of 4.4°C–6.8°C was observed with insulation and controlled air exchange rates in the study. The impact of four different shading types (namely overhangs, blind system, light shelf and experimental shading device) was analyzed by Kim et al. [20] and maximum cooling energy saving was found to be 11% for experimental shading with 76° of solar altitude. Grynning et al. [21] performed simulations for north- and south-oriented office cubicles with different floor areas, openings and shading schemes. The simulations revealed that the cooling load increases with increase in the window size from 41% to 61%; therefore, heating demand decreased. The results also show that by using a proper and appropriate shading technique, energy demands can be reduced for south-facing facades by 9%, whereas an improper technique may lead to an increase of 10%. An experimental investigation of the thermal performance of a building solar shading device was conducted by Evangelisti et al. [22] and was found that the shielding system achieves a reduction in terms of incoming thermal energy equal to 38.7% during summertime. Pearlmutter and Rosenfeld [23] studied the passive cooling of roof in hot, arid climatic conditions. The authors studied the performance of a watered soil with two types of shading for roof cooling.

- **Trees and vegetation:**

The ambient temperature near the outer wall may be substantially reduced by 2°C–2.5°C without excess use of supplementary energy [24]. Shading and evapotranspiration from trees can reduce the surrounding temperature by 5°C [18]. Ca et al. [25] found that the air temperature can be reduced by 2°C with presence of a park. Papadakis et al. [26] experimented with tree buffering as solar control in a south-east oriented building. The solar irradiation peak in the non- shaded area and the shaded area at the same time was observed to be 600 W/m^2 and 100 W/m^2, respectively. Solar gains increased on the shaded wall on the mid-day and reached 180 W/m^2 as the sun was at high horizon. Various parameters like air and wall surface temperatures, wind speeds, humidity and heat exchange between the wall surface and surrounding environment were measured in the shaded (by deciduous trees) and unshaded areas for a hot summer period.

7.5 THERMOTROPIC AND THERMOCHROMIC COATINGS

Transparent components provide solar control and adequate daylight due to their low transmissivity to the solar radiation and high transmissivity to the visible spectrum, thus making them appropriate for warm climates. Solar control coatings with good g-value and U-value include thermotropic and thermochromic coatings. G-value is the sum total of solar transmittance and a factor for heat gains resulting from sunlight absorbed in the glazing unit [4].

These coatings have a variable transparency, which acts as a function of temperature. Thermotropic layers are made of a mixture of a polymer and water or two polymers. Since both the components have different refractive indices, they segregate and become opaque when the temperature reaches above the threshold value. Wilson [27] expressed that a polymer blend in combination with a low-e coating can have a visible transmittance range of 73% at 30°C to 31% at 50°C. Thermochromic glazing can reduce the energy consumption of a building by allowing natural daylighting, reducing unwanted solar gains during the cooling season and allowing the useful solar gains in heating season [28].

7.6 COURTYARDS

A courtyard or a court is an enclosed area within a building and open to sky. The concept of a courtyard can enhance the thermal comfort in hot, dry climatic conditions, where the temperature drops considerable after the sun sets from re-radiation to the night sky. Courtyard planning is amongst the oldest passive cooling strategy and the most effective to achieve visual aesthetics, functional fulfillment and thermal comfort. It is a design element in most of the vernacular buildings, which originally started in the Mediterranean, Middle Eastern and Tropical Regions. In vernacular architecture, the courtyards were integrated with vegetation and water bodies to enhance the humidity, evaporative cooling, and provision of shade. Localized heating within buildings may be reduced to a large extent by exploiting the thermal interaction due to the difference in temperature of courtyard and the building core depending upon the aspect ratio of the court, wind speed and direction [29]. The design and the average size of the courtyard depends on the size of the plot. They are narrow enough to maintain a shaded area during the daytime in summer months but wide enough to allow the penetration of solar radiation during winter months. It generally acts as the nucleus of the house and serves as a common circulation space and meeting areas.

The heat transfer between the courtyards and adjoining areas is shown in Figure 7.1. During the off-sunshine hours (evening time), the warm air of the courtyard, which was heated directly by solar radiation (direct gain) and indirect gain by the warm buildings, rises and is gradually replaced by the

Dynamic Design Concepts for Hot Climatic Conditions

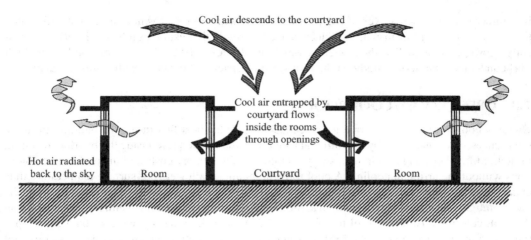

FIGURE 7.1 Heat transfer in courtyard planning.

cool night air above. The cool air accumulates in the courtyard and flows inside the rooms through the openings towards the courtyard. During the daytime, the air in the courtyard, which is shaded by four walls around it and the surrounding rooms, heats slowly and remains cool until late in the day when the sun shines directly into the courtyard.

7.7 AIR CAVITIES

Air cavities in walls act as an insulation, which reduces the heat transfer. The insulation effect created by an air cavity is due to its low thermal conductivity, therefore as an application of thermal insulation, the concept of air cavity has been introduced in building configurations to reduce the thermal conductivity of the exterior envelope of the building and improve the thermal resistance. Cavity walls are lighter and have higher thermal resistance than solid masonry walls. Inside an air cavity, convection takes place. The heat transfer by convection does increase with the thickness of the cavity and will limit the thermal resistance values of such cavities in roofs and walls. The temperature difference inside the air cavity and their thickness determines the relevance of one mechanism over the other. Air cavities reduce the incoming heat flux and also reduce the thermal load leveling.

Figure 6.6(b) shows a modified Trombe wall with an almirah (as an air cavity of 550 mm) to reduce the U-value.

7.7.1 Literature Study: Air Cavity

Ciampi et al. [30] evaluated thermal performance of six different designs of an opened and ventilated double-layer masonry wall and found that with an increase in the width of the air cavity, the energy savings capacities of the different ventilated masonry walls increases. But, when the width exceeds 150 mm, the thermal performance of the wall starts to decline. In summer months, the energy demand for cooling was reduced by 40% with a ventilated double-layer masonry wall design. Mahlia and Iqbal [31] found that optimal thickness of different insulation materials and having air gaps of 2 cm, 4 cm and 6 cm, energy consumption and emissions can be reduced by 65%–77%, in comparison to a wall without insulation or air gaps. Thermal performance of a proposed design of static sunshade and brick cavity wall has been done by Charde and Gupta [32] for composite climatic conditions. It was found that the proposed brick cavity wall with brick projections is more useful to lower the indoor air temperature in summer and increase it in winter evenings and nights. The difference in indoor air temperature due to proposed brick cavity wall with brick projections is more in summer with the open air vents rather than in winter with closed air vents. Hong et al. [33] found that cavity walls and loft insulation reduce space heating fuel consumption by 45%–49% theoretically. However, in actuality, the reduction was limited to only 10%–17%. Zhang and Wachenfeldt [34] conducted a numerical study on the heat storing capacity of concrete walls with air cavities.

7.8 GREEN ROOFS/COOL ROOFS

Green or cool roofs can be referred to the concept where the heat flux inside the building from the top reduces. The solar gains by a traditional roof raises the temperature inside the building resulting in high artificial cooling requirements of the building, high energy costs and uncomfortable conditions without the artificial cooling. A cool roof can maintain a lower roof surface temperature thus enhancing the comfortable thermal conditions inside the building. The surface of a cool roof reflects more sunlight and absorbs less heat than a traditional roof. This may be done either by vegetation on the roof, coloring/coating the roof to reflect the radiation back to the sky and cool the roof, evaporative cooling or solar ponds, etc. [35]. In some cases, roof ventilation has also been used to cool the roof. The green roofs are the passive cooling techniques, which do not have any ill effect on the environment or on the health of human beings.

- Roof cooling by vegetation is an effective way to provide thermal comfort. The evaporation of water from the surface of leaves helps reduce the temperature of the roof to a level lower than that of the daytime air temperature. At night, it is even lower than the sky temperature.
- Another method of roof cooling can be by covering the entire roof with closely-packed earthen pots increases the surface area for radiative emissions. An insulating cover over the roof obstructs heat flow inside the building. But this method leaves the roof unusable and difficult to maintain [17].
- Cool roof coatings/paints may also be applied to improve the solar reflectance and thermal remittance, which significantly reduces the temperature of the roof. These can be divided in two categories: (a) cementitious coatings, which contain concrete or ceramic particles, and (b) elastomeric coatings, which contain added polymers that make them less brittle and more adhesive to building surfaces. The former coating should be applied over roofing materials that have already been thoroughly waterproofed while the latter acts act as a watertight membrane. White tiles with high reflectivity and low thermal conductivity may also be laid on the roof surface to reduce the U-value of the roof.
- Roof cooling by false ceiling and air cavity: For thermal cooling, the placement of insulation either below or above is important, as discussed in walls, configuration. However, the insulation in the case of a roof should be rigid and waterproof if placed outside unlike a wall configuration. It is always preferred in an air-conditioned room to reduce the load on the air conditioner.
- A provision of ventilated roof by creating air cavity between roof can be also considered since it provides insulation along with the protection against the solar gains. Thus, it is more advantageous during the summer season.
- Wetted roof: In the case of wetted roof, the availability of water in the arid region is an important issue due to cooling of the roof in the summer season. If water is available, then the flowing water or roof pond system over roof can be adopted. Wetted roof surface provides the evaporation (mass transfer) from the roof surface to air due to unsaturated ambient air. This is more effective in arid (dry) regions due to the higher difference in dry and wet bulb temperatures. It prevents the heat transfer from exposed surface to inside room and simultaneously provides the evaporative cooling. Therefore, the wetted roof temperatures are lower than the ambient air temperature.

Dynamic Design Concepts for Hot Climatic Conditions

The sol-air temperature is given in Equation (7.3a). The roof shading techniques of (a) earthen pots, (b) solid cover and (c) plant cover have been given in Chapter 1 (Figure 1.8).

7.8.1 Literature Study: Cool Roof

Site experiments were conducted by Al-Hemiddi and Mohammed [36] based on roof with moist soil where the roof was shaded by 10 cm of pebbles for hot dry climate of South Arabia. A reduction of 5°C in indoor temperature was observed. Thirty-seven roof designs were studied by Dabaieh et al. [37] for the hot, dry climate of Cairo. Roughly a 10%–40% reduction in air conditioning energy was observed. For a flat roof, heat gain and average indoor temperature were 414 kWh and 32.5°C, respectively, were recorded. For a domed roof, heat gain and average indoor temperature were 310 kWh and 32.2°C, respectively, was achieved. For a vaulted roof, a drop 1.5°C in the average indoor temperature along with a fall of 53% and 826 kWh savings during summers in discomfort hours with a rim angle of 70° with high albedo coating as compared to the reference case of the conventional non-insulated roof was observed.

7.8.2 Evaporative Cooling

The interior spaces are cooled by passing the hot ambient air over damped surface to cool an air stream (by evaporating the water) before its introduction to the interior spaces. It is one of the oldest, effective and efficient techniques with the potential to reduce indoor temperature using air as the heat sink. Evaporative cooling may be (a) direct or (b) indirect.

In a direct evaporative cooling (passive) system, the room air should be in direct contact with the water surface. The air temperature is thus reduced by about 70%–80% of the difference between the dry bulb temperature (DBT) and the wet bulb temperature (WBT). Therefore, direct evaporative cooling is effective when there is a large difference between DBT and WBT [4]. The relative humidity inside the room increases due to the evaporation of water. In such cases, it is possible to cool a small building by placing wetted pads in the windows or porches, facing the wind direction.

Either water film or intermittent spray of water or flowing water over the roof is referred to as indirect evaporative cooling. As the water draws heat from the roof surface, it leaves a cooler ceiling surface below, which acts as a radiative cooling panel for the space. The indoor temperature is lowered without elevating the humidity level as in the case of direct evaporative cooling. Energy savings of up to 60% compared to mechanical A/C may be achieved in hot dry regions [38]. The indirect evaporative cooling by a roof pond is more effective if the roof is covered with plants and movable insulation.

Qingyuan and Yu [39] concluded that the potential of evaporative cooling is subjective to the difference between humidity ratio of outdoor air and wet bulb temperature at saturation.

Heat transfers for evaporative cooling are shown in Figure 6.1(b), where a wetted surface is exposed to solar radiation. Evaporative cooling applies to all processes in which the sensible heat in an air stream is exchanged for the latent heat of water droplets or wetted surfaces [4]. The rate of thermal energy per unit area can be expressed as:

$$\dot{q} = U_L \left(T_{sa} - T_r \right) \tag{7.4}$$

where

$$T_{sa} \text{ is the sol-air temperature, } T_{sa} = \frac{\propto I(t)}{h_1} + T_a - \frac{\varepsilon \Delta R}{h_1} \tag{7.5}$$

194   Photovoltaic Thermal Passive House System

h_1 is the is total heat transfer coefficient from wetted surface to ambient air, which includes the evaporative heat transfer coefficient.

$$h_1 = h_{ra} + h_{ca} + h_{ea} \text{ in W/m}^2 \text{°C}$$

$$U_L = \left(\frac{1}{h_1} + \frac{L}{K} + \frac{1}{h_i} \right)^{-1}$$

where $h_{ra} + h_{ca} + h_{ea}$ are radiative, convective and evaporative heat transfer coefficients, respectively, in W/m²°C. L and K are the length and thermal conductivity of the surface, respectively. In the case of heating, the rate of thermal energy per unit area is enhanced while it is reduced for cooling.

In the Middle East, evaporative cooling is used with combination of wind towers to channelize the cool wind over water cisterns into the building. The evaporative cooling is extensively used in the form of dessert coolers in areas with arid climatic conditions in daytime when the ambient temperature is between 37°C and 42°C. This leads to an increase in the indoor humidity. This concept is thus limited to areas with low outdoor humidity with availability of 0.40.5 m³ water per dwelling [40].

EXAMPLE 7.1

Calculate the total heat transfer coefficient (h_1) and the rate of heat loss due to radiation, convection and evaporation for a wetted surface.

    Given: Wetted surface temperature = 20°C, T_a = 12°C, relative humidity = 0.6, wind velocity = 3 m/s, emissivity = 0.9, h_{ra} = 4.93 W/m²°C

Solution

The total heat transfer coefficient, h_1, can be calculated using $h_1 = h_{ra} + h_{ca} + h_{ea}$
    From Equation (2.22a)

$$h_{ca} = 2.8 + 3V = 2.8 + 3 \times 3 = 11.8 \text{ W/m}^2 \text{°C}$$

Using Equation (2.38)

$$h_{ew} = \frac{16.276 \times 10^{-3} \times 2.8 \times (2346.5 - 0.6 \times 1433.5)}{(20 - 12)} = 8.47 \text{ W/m}^2 \text{°C}$$

Therefore, $h_1 = 4.93 + 11.8 + 8.47 = 25.2 \text{ W/m}^2 \text{°C}$.

The rate of heat loss due to radiation, convection and evaporation is given by:

$$\dot{q}_{ra} = 4.93 \times (20 - 12) = 39.44 \text{ W/m}^2$$

$$\dot{q}_{ca} = 11.80 \times (20 - 12) = 94.40 \text{ W/m}^2$$

$$\dot{q}_{ea} = 8.47 \times (20 - 12) = 67.76 \text{ W/m}^2$$

7.8.3 Literature Study: Evaporative Cooling

With implementation of passive evaporative cooling for hot and humid climatic conditions, reduction of 2°C–6.2°C in room air temperature was found in comparison with ambient temperature and was found effective for metal ceilings [41]. Better results can be obtained with higher ambient temperature and solar insolation. Raman et al. [42] analyzed the configuration of solar air heater and evaporative cooling for composite climatic conditions and did not found this proposal as an effective solution for passive cooling. The performance of the system was improved by adding a roof duct and wetted south wall collector.

The effects of various porous and non-porous roofing materials have been analyzed by different researchers [43–45]. Wanphen and Nagano [43] observed a reduction of 6.8°C–8.6°C in roof temperature due to high evaporation rate (0.3 kg/m²/h) and adsorption rate (0.07 kg/m²/h) of siliceous shale. Siliceous shale yields more latent heat, whereas silica sand, volcanic ash, and pebbles release sensible heat of about 0.62, 0.18 and 0.18 kg/m²/h, respectively. Wetted roofs (wetted gunny bags) may reduce the roof temperature by 15°C in relation to the ambient temperature [45].

Qiu and Riffat [46] investigated the impact of fountains for hot and arid regions with outside temperatures of about 45°C. The study concluded that about 20°C indoor temperature was achieved, which is within the comfortable range with reduction in cooling load by 9% and the annual energy consumption by 23.6%.

The concept of passive downdraft evaporative cooling (PDEC) tower has been studied by various researchers. According to Chiesa and Grosso [47], cooling load can be reduced by 25%–85%. The concept is valid for about 70% of the European buildings [47–49]. Badran [50] suggested tower height be minimum 9 m to maintain a proper airflow in desert areas. PDEC is an energy-efficient and cost-effective concept, however it is difficult to maintain the constant performance.

Integration of evaporative cooling with wind towers for hot, arid regions has been studied by [51], and indoor temperature was reduced by 12°C with use of wet columns of 10 m height. However, relative humidity may increase by 22%. In another study by Bouchahm et al. [52], the temperature was reduced by 17.6°C with proposed wetted surface design for residential bioclimatic housing.

Indirect evaporative cooling resulted in reduction of 1°C in mean daily temperature and 2.5°C in room air temperature was found for Brazil [53]. In the proposed setup, an insulated high-reflectance sheet was installed over the roof slab with fan assembly, to enhance the evaporation It was also suggested that thermal discomfort due to excess solar gain can be reduced for almost 95%–100% of the year. For hot and dry climatic conditions, energy demand by air conditioning can be reduced by 20% in the next 20 years [54].

7.9 RADIATIVE COOLING

The sky temperature is always lower than ambient air temperature by 12°C with clear night sky. Effective radiation from the exposed horizontal surface to ambient air via convective and radiative heat transfer is referred to as radiative cooling. The northern sky in the northern hemisphere is often cool enough even during the day. It acts as a cooler heat sink with respect to ambient air the day/night. A horizontal surface is the most effective radiative configuration. Exposed horizontal surface losses heat ambient air by convection and radiation till its temperature equals that of dry bulb air temperature. Also, there are heat losses from ambient air/surface to the sky by long wavelength radiation exchange. If the net heat exchange between surface and sky reduces the roof surface temperature to the wet bulb temperature of surrounding air, condensation of moisture of air starts taking place on the roof. The condensation will further cool the roof due to fast heat loss from enclosed room to surface of the roof.

If the surface is inclined as shown in Figure 7.2, then the cooled air will trickle down towards an internal courtyard from the inclined surface due to its high density. Then the trickled cooled air

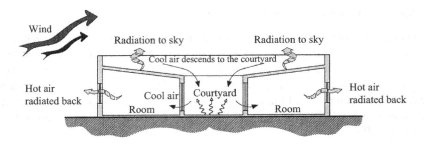

FIGURE 7.2 Cooling by radiation through open loop [12].

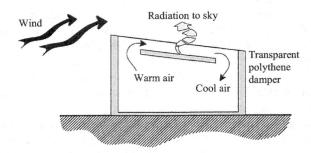

FIGURE 7.3 Cooling by radiation through closed roof loop [12].

enters the room through the openings at lower level as shown in the figure. To avoid this, the roof should be covered with a transparent polyethylene sheet. The polyethylene sheet also has limited life. This problem can be solved by covering the roof with corrugated metallic sheet with the provision of openings at the lower and upper end from inside the roof as shown in Figure 7.3.

The radiant heat exchange between sky and a body/surface can be expressed in Equation (2.28).

The rate of long-wavelength radiation exchange between ambient air and sky can be expressed in Equation (2.29c).

Note that the emissivity of the roof surface (ε) should have maximum value near 0.9 for fast heat transfer.

Radiative cooling can be classified as (i) direct or passive and (ii) hybrid radiative cooling. In passive radiative cooling, the building envelope radiates towards the sky and gets cooler, producing a heat loss from the interior of the building. However, in hybrid radiative cooling, the radiator is not the building envelope but usually a metal plate. The operation of such a radiator is the opposite of an air flat-plate solar collector. Air is cooled by circulating under the metal plate before being injected into the building. Other systems are combinations of these two configurations [4].

7.9.1 Literature Study: Radiative Cooling

Hanif et al. [55] found out that 25% of the power consumption can be saved with radiative cooling independent of all locations. Also, it was concluded that the cooling power decreases with decreases in the difference between the ambient and sky temperature. The specific cooling power was measured by Cavelius et al. [56], which ranged between 2080 W/m² with movable insulations, air-based systems and open or closed water-based systems. The water-based radiative cooling plate concept was investigated by Juchau [57] and Erell and Etzion [58] and the net cooling power of 81 W/m² was achieved with removal of cover during operations of 7 hours. The study was based upon conventional cooling systems with flat plate collectors attached with a storage tank. Another study was conducted by Beck and Büttner [59] for an open water-based system and the specific cooling power

Dynamic Design Concepts for Hot Climatic Conditions

of the plant came out to be 120 W/m^2. Cooling outputs were observed to be higher for the closed system due to absence of the thermal resistance between the water and the ambient. Zhang et al. [60] concluded that cooling load decreases with increases in the elevation, but the potential of the radiative cooling is large. The long-wave terrestrial radiation shows no correlation with the elevation while the short-wave incoming radiation shows a proportionate decrease at the normal lapse rate. Thus, leading to an increase in the value of radiative cooling. Liu et al. [61] evaluated sub-ambient radiative cooling in low-latitude seaside and found that it has the potential to meet the solar peak cooling demand of 65% areas in China alone.

7.10 MOVABLE INSULATION

Thermal insulation is another solar and heat protection technique. Movable insulation in nighttime shall be used to reduce the heat losses from room air to the ambient.

7.11 DYNAMIC INSULATION WALLS

There are many well-known studies in the area of dynamic insulation integration with wall. For example, an external wall should be covered with insulation during daytime to reduce indirect solar flux reaching inside rooms by conduction, but it should be removed during nighttime for removal of heat from inside rooms to outside. However, it was observed that it is not a practical solution, though it has been practiced in other areas of heating and cooling, such as greenhouse heating and cooling. So provision of insulation integration should be permanent, which depends on climatic conditions. For harsh, cold climatic conditions, it should be integrated from either outside or inside to reduce U-value. The optimization of thickness of insulation inside and outside walls has been carried out by Sodha et al. [62].

7.11.1 EXTERIOR INSULATION

An external insulation integration to exposed wall should be generally recommended for harsh, warm conditions to minimize the heat flux entering the room for cooling purposes. This can also be done by using thicker wall (heavy structure).

7.11.2 INTERIOR INSULATION

An interior insulation integration with wall is recommended for harsh, cold climatic conditions to minimize the heat loss from inside to outside. For harsh, cold climatic conditions, external heat is supplied to room for thermal comfort.

7.12 WIND TOWERS

Wind towers are designed to harness cool air through the wind and circulate cool air inside the building. A wind tower operates according to the time of the day and the presence or absence of wind for both thermal heating and cooling. During the day, hot ambient air enters the tower through the openings in the sides and is cooled as it comes in contact with the cool tower. The cool air, being denser than the warm air, sinks down through the tower creating a downdraft. The draft is faster in the presence of wind. At night, the tower operates like a chimney. The thermal energy that was stored in the wind tower during sunshine hours, warms the cool night air in the tower, the pressure at the top of the tower being reduced, an updraft is created. The concept of wind towers work well in individual units and not in multistory apartments unless it is designed accordingly. A major limitation to the cooling efficiency of a wind tower is the climate and location.

7.12.1 LITERATURE STUDY: WIND TOWERS

A drop of about 12°C in inside temperature and an increase by 22% in the relative humidity was found with use of wet columns of 10 m height [63]. Various cooling techniques integrated with wind towers were studied by Hughes et al. [64]. To understand the feasibility of the proposals for their respective use, the key parameters of the study included ventilation rate and temperature. Drop of temperature within range of 12°C–15°C was found by integrating the concept of evaporative cooling with wind towers. This drop in temperature was greater than the solitary wind tower arrangement. Another arrangement, a two-sided wind tower, was examined by Montazeri et al. [65]. The maximum performance was achieved during an experiment at a 90° angle. Chaudhry et al. [66] studied a novel closed-loop thermal cycle embedded inside a circular wind tower with internal cross-sectional area of 1 m^2 with 1 m height installed at the rooftop to achieve internal thermal comfort. The louvers were angled at 45° located at the openings of the wind towers. The exit temperature using traditional cooling was increased up to 4°C without any impact of the height in case of the proposed heat pipe design with water and ethanol as working fluids. Bahadori [67] proposed few improvements in the design of wind towers for hot, arid climatic areas with variable wind direction. The wind towers were integrated with the concept of evaporative cooling with tower heads able to trap the wind from any direction. For given parameters, 306 kg of mass storage material per cubic meter of tower was used, which was capable of storing 36 m^2 of heat. The energy storage system was made up of baked non-glazed clay in form of long conduits to increase the heat transfer area. The proposed design was capable of releasing the air at a higher flow rate. Other novel designs introduced were wetted columns with cloth curtains and wetted surface with evaporative cooling pads for hot arid regions of Middle East [51, 68–69]. For high and low wind conditions, wetted column and wetted surface designs, respectively, were recommended. The proposed design released air at much lower temperature to the interior space as compared to the conventional design. Benhammou [51] suggested that with wet columns of 10 m height, indoor temperature and relative humidity can be reduced by 12°C and 22%, respectively. It may also be concluded that with integration of cooling devices with the conventional wind towers is beneficial with air exiting the towers at significantly lower temperature than the outside temperature. Bouchahm et al. [52] proposed design with clay conduits to improve the mass and heat transfer and water pool to increase the humidification for hot dry regions of Ouargla having a maximum temperature of 47°C–52°C. Wind towers at the rooftop for passive ventilation in windy areas were studied by Dehghani-Sanij et al. [70] and the proposed tower is able to rotate and align itself in the direction of the predominant wind to compensate for low wind speeds. A study of wind towers with different funnels attached to increase natural ventilation in an underground building has been conducted by Varela-Boydo et al. [71]. Another study by Varela-Boydo and Moya [72] was conducted to study the inlet extensions for wind towers to improve natural ventilation in buildings.

7.13 AIR VENTS [12]

Air vents are used in the areas where dusty winds make the working of wind tower impossible. These are suited for single units and work well in both hot and dry and warm and humid climates. A typical air vent is shown in Figure 7.4. A typical air vent is a hole in the apex of domed or cylindrical roof with a protective cap over it. Openings in the protective cap over the vent direct the wind across it. When air passes through the openings of protective caps, it creates low pressure at the apex of the curved roof due to increase in the velocity, thereby inducing the hot air under the roof to flow out through the vent. Thus, the air is kept circulating through the room. Air vents are usually placed over living rooms to cool air moving through the room and are suitable for hot and dry, as well as for warm and humid climatic conditions. Air vents are preferred for single units.

Dynamic Design Concepts for Hot Climatic Conditions   199

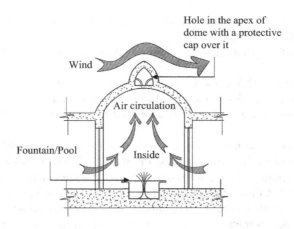

FIGURE 7.4 Air vent under operation [12].

7.14 ROCK BED REGENERATIVE COOLER

A regenerative cooler basically consists of two rock beds set side by side. It also acts as a heat exchanger and is separated by an air space. A damper has been used between two rock beds to divert the incoming air from the house towards cooling rock bed by water spray. The rock beds are cooled alternatively and also absorb water during cooling. The rock beds are cooled alternatively. It also absorbs water during cooling. The air passing through a dry rock bed (already cooled in the first cycle) is cooled by transferring its heat to the rock bed and the cooled air is allowed to pass into the room. The humid air is produced during its evaporation cycle from rock bed, and it is vented to the outside. After getting the rock bed warm, the damper is reversed for further cooling. In the meantime, other cooled rock bed is used for cooling similarly.

Schematic diagram explaining the working principle can be referred from [12].

7.15 EARTH COUPLING

Earth coupling is an effective passive cooling technique when earth temperatures is lower than ambient air temperature mostly in hot climatic conditions. It uses the moderate and consistent temperature of the soil to act as a heat sink to cool a building through conduction. Earth coupling can be classified in two categories:

(a) Direct coupling: It can also be referred to as earth sheltering. This is when a building uses earth as a buffer for the walls. The earth acts as a heat sink and is used to mitigate the extreme temperature effectively. It reduces both heat losses and heat gains by limiting infiltration and thus improves the performance of the building envelope.
(b) Indirect coupling: This strategy is when the building is coupled with the earth by means of earth ducts. This can also be referred to as an earth-air heat exchanger (EAHE).

7.15.1 EARTH-AIR HEAT EXCHANGER (EAHE)

An earth-air heat exchanger (EAHE) exploits the constant ground temperature a few meters below Earth's surface. The ground temperature remains constant throughout the year at appropriate depth, which can be harnessed for both cooling and heating purposes during winter and summer seasons, respectively. Parameters like surface area of pipe, length and depth of the tunnel below ground, dampness of the earth, humidity of inlet air and its velocity affect the exchange of heat between air and the surrounding soil.

Ground temperatures for various surface conditions at a depth of ≥4 m [12] can be given as

- Ground temperature of 27.5°C for dry sunlit surface conditions
- Ground temperature of 21.5°C for wet sunlit surface conditions
- Ground temperature of 21.0°C for wet shaded surface conditions

Figure 7.5 shows the cross-sectional view of an earth tunnel below the ground at 4 m depth, and air available from the atmosphere is allowed to pass through it. The shape of the tunnel is cylindrical having a radius of r and length L. As the air passes through the tunnel there is a heat transfer from the inner surface of the tunnel to flowing air by forced convection. Depending upon the air temperature, the air is either heated or cooled. If the temperature of air is below the temperature of the inner surface of the tunnel, then heat is transferred from surface to air for heating. This occurs in the winter season. For summer it is vice versa. Figure 7.6 gives the flow directions through the heat exchanger.

Referring to Figure 7.6, the energy balance for an elemental length dx in terms of W can be written as

$$\dot{m}_a C_a \frac{dT(x)}{dx} dx = 2\pi r h_c \left(T_0 - T(x)\right) dx \tag{7.6}$$

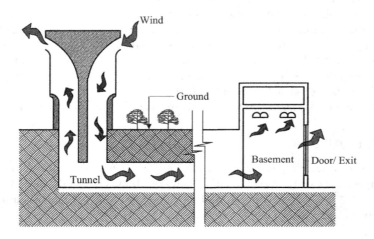

FIGURE 7.5 Earth air heat exchanger integrated with wind tower [2].

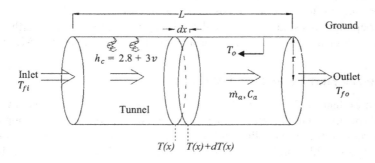

FIGURE 7.6 Flow directions through EAHE [12].

Dynamic Design Concepts for Hot Climatic Conditions

where

$\dot{m}_a$ is the mass flow rate of air; $\dot{m}_a = \pi r^2 \rho v$ in kg/s
r is the radius of tunnel in m
ρ is the density of air in kg/m^3
v is the speed of air flowing through the tunnel in m/s
C_a is the specific heat of air in kJ/kg
h_c is the convective heat transfer coefficient; $h_c = 2.8 + 3\ V$ in W/m^2°C
T_0 is the inside ground temperature at 3–4 m depth outside pipe in °C
$T(x)$ is the temperature of air as a function of x in m through pipe

The solution of above equation with initial condition $T(x = 0) = T_{fi}$ can be written as

$$T(x) = T_0 \left[1 - \exp\left(-\frac{2\pi r h_c}{\dot{m}_a C_a} x \right) \right] + T_{fi} \exp\left(-\frac{2\pi r h_c}{\dot{m}_a C_a} x \right) \tag{7.7}$$

Now the outlet air temperature at exit of EAHE can be obtained as follows:

$$T_{fo} = T(x = L) = T_0 \left[1 - \exp\left(-\frac{2\pi r h_c}{\dot{m}_a C_a} L \right) \right] + T_{fi} \exp\left(-\frac{2\pi r h_c}{\dot{m}_a C_a} L \right) \tag{7.8}$$

where L is the length of EAHE in m.

The rate of thermal energy in W carried away by the flowing air is given by

$$\dot{Q}_u = \dot{m}_a C_a (T_{fo} - T_{fi}) = \dot{m}_a C_a \left[1 - \exp\left(-\frac{2\pi r h_c}{\dot{m}_a C_a} L \right) \right] (T_0 - T_{fi}) \tag{7.9}$$

The energy available in 1 hour $= \dot{Q}_u \times 3600\,\text{J}$
The volume of hot air $= \pi r^2 v \times 3600\ \text{m}^3$
If V_0 is the volume of a room to be heated then, the number of air change (N) per hour can be written as
$N = (\pi r^2 v \times 3600)/V_0$ Number per hour
The following special cases should be discussed:

• Case (a)

For $L \Rightarrow \infty$ and $\dot{m}_a$ is very small
This case refers when the tunnel length is very large and a very low speed of air,

$$\text{then } T_{fo} = T_0 \text{ and } \dot{Q}_u = \dot{m}_a C_a (T_0 - T_{fi})$$

This indicates the withdrawal of maximum thermal energy to heat living space.

• Case (b)

For $L \Rightarrow 0$ and $\dot{m}_a \Rightarrow \infty$
This case refers when the tunnel length is very small and a very large speed of air,

$$\text{then } T_{fo} = T_{fi} \text{ and } \dot{Q}_u = 0.$$

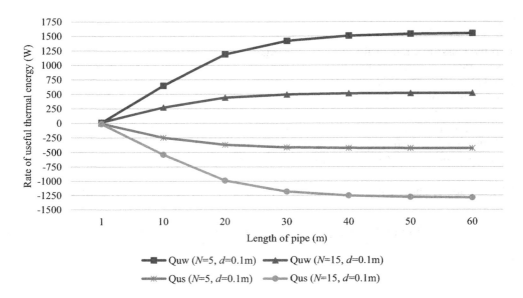

FIGURE 7.7 The variation of $\dot{Q}_u$ with length of heat exchanger [73].

This indicates no withdrawal of heat from tunnel, hence there is a need to optimize the length, radius and velocity of air for thermal heating/cooling of a living space. The variation of $\dot{Q}_u$ with L for winter and summer conditions for a typical set of parameters has been shown in Figure 7.7.

A combination of wind-tower and earth-air tunnel is effective in increasing the draft of air in the earth-air tunnel for the cooling rate of building air (Figure 7.5).

7.15.1.1 Literature Study: EAHE

A drop of about 5°C–6°C in the outlet temperature during summer months was observed by Tiwari et al. [74]. In this study for given parameters with five air changes and 100 mm diameter and 210 mm length of pipe was used. An analytical model to calculate the cooling potential was modelled by F Al-Ajmi et al. [75] for hot and arid climatic conditions and found that 30% of the cooling demand can be met by EAHE. Also, a reduction of 1700 W in the peak cooling load and 2.8°C in the indoor temperature was found during summer months. Benhammou and Draoui [76] buried a pipe with 5-mm thickness in soil with 22.27°C temperature. Various parameters were investigated, and it was found that with an increase in the length of pipe, there is a decrease in the outlet temperature with improvement in the daily mean efficiency and drop in the coefficient of performance (COP). With an increase in the pipe length from 10 m to 30 m, for an inlet temperature of 29°C, there was a reduction of 2°C in the outlet temperature. The reduction rate was not found to be constant. COP dropped by 10.5% and a rise in daily mean efficiency was found to be 142%. With an increase in the cross-section and air velocity (from 1 m/s to 3 m/s) daily mean efficiency decreased by 31.65% with a drop in outlet temperature. EAHE integrated with wind tower was studied by [51] for hot and arid regions and the daily cooling potential reached a maximum of 30.7 kWh and found to be proportionate with the diameter of the pipe. Energy conservation of EAHE was studied by Kumar et al. [77] for Indian climatic conditions. Cooling potential of 19 kW was recorded to maintain an average room temperature 27.65°C for the proposed design with 80 m of pipe length, 0.53 m^2 of cross-sectional area and 4.9 m/s air flow velocity. Another study by [78] for Indian climatic conditions established 512 kWh and 269 kWh cooling and heating capacity for an 80 m long tunnel with 0.528 m^2 cross-sectional area [78]. Rodrigues et al. [79] found that thermal performance of EAHE can be improved by 73% and 11% for cooling and heating purposes, respectively, by increasing the number of ducts, while keeping the area occupied by the ducts and the mass flow rate of air fixed.

Wei et al. [80] also performed a field experiment on the cooling capacity of EAHE in hot and humid climate. For an EAHE of 0.075 m diameter and 3 m depth, the maximum reductions in the air temperature and moisture content were found to be approximately 22.13°C and 7.41 g/kg, respectively.

7.16 ROOF POND

Convectional approaches to reduce the heat gains via roof includes false ceilings, roof shading, roof coatings etc. as already discussed. Roof pond is another concept, which is used for both thermal heating and thermal cooling of the building for composite climate found in 1920s and was first investigated at the University at Texas [81].

A schematic diagram of a roof pond integrated with water circulating columns working under natural circulation mode is given in Figure 7.8. It is covered with a thin layer exchange membrane with a louvered shade above the membrane.

For thermal cooling, this louvered shade is opened to increase the heat loss during the nighttime. The evaporation of water allows the water to cool. It is closed during daytime to minimize solar radiation to incident on the roof pond. The cooled water is allowed to pass through a downcomer. The thermal energy is gained from room air during passes through the downcomer. Thus, the room is cooled.

For thermal heating, this louvered shade is closed to reduce the heat loss during nighttime and is opened to allow solar radiation to fall on the roof pond during daytime to heat the water of roof pond. The hot water is allowed to pass through the downcomer. During passes through the downcomer, the thermal energy is released to water outside and cooled down towards lower ends due to high density.

7.16.1 LITERATURE STUDY: ROOF POND-PASSIVE COOLING

Tiwari et al. [82] found that with use of an open roof pond, the roof surface temperature can be reduced to 42.2°C and 39.4°C with 0.05 m and 0.15 m of pond depth, respectively. Without any treatment, the roof surface temperature can reach up to 65.6°C [83]. A drop of 20°C in the room temperature was observed by installation of roof ponds in arid areas [84]. A room of $4 \times 7 \times 2.75$

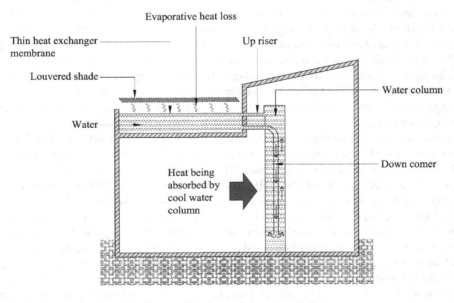

FIGURE 7.8 Schematic view of roof pond with thermosiphon flow [12].

m was analyzed to study the effectiveness of the roof pond with mechanical ventilation for thermal cooling in the dry summer season of Baghdad [85]. It was found that space temperature was significantly reduced during the peak time outside temperatures reaching 6.0°C between the room without the pond and with a ventilated one and 6.5°C at 18:00 during peak inside temperatures. An experimental roof pond building with room size of $3 \times 2.5 \times 2$ m and pool depth of 60 mm was studied by Ahmad [86], wherein the vapors were discharged from the pond in order to improve the building's summer performance, whereas the same will reduce the efficiency of the system in winters. Cooling load (500 kJ/m²) is at its maximum at noon because the most significant component of the cooling load is the direct heat gained through the south glazing. A significant contribution of sensible heat around 11:00 a.m. was observed, which was responsible for cooling. Also, at noon, the cooling is due to latent heat of evaporation. However, a need to design the vapor leakages from the pond is required. Site experiments were conducted by Al-Hemiddi and Mohammed [36] based on walkable roof ponds along with night water circulation for the hot, dry climate of south Arabia. In the former case, reduction of 5°C in indoor temperature was observed where it was shaded by 10 cm of pebbles. A reduction of 6°C was seen in the indoor temperature compared to the outdoor temperature. Tang and Etzion [81] studied the roof pond with gunny bags floating on the surface of water and found this better than the roof pond with wetted gunny bags in terms of cooling performance of room temperature and heat flux through the roof into the pond because of thermal satisfaction inside the pond during daytime irrespective of the building type. This technique was also found better than movable insulation. The optimum water depth for the roof pond with floating gunny bags was found to be 200 mm and 50 mm for concrete and metal-decked roofs, respectively. Spanaki et al. [87] studied the ventilated roof pond (depth 1.10–1.12 m) with a reflecting aluminum layer at 1.15 m height from the free water surface. A reduction of 30% in the maximum inside temperature was observed when compared to corresponding temperature without any application of the roof cooling concept.

7.16.2 Trombe Walls

As discussed in Chapter 6, a Trombe wall is a sun-facing wall popularized in 1964 by French engineer Félix Trombe and architect Jacques Michel. It is regarded as part of the green building envelope (passive concept), which is sustainable for cooling and heating to address the environmental and energy crises throughout the world. A thick thermal mass used as an exterior façade reduces the decrement factor and leads to a time lag. A heavy structure is preferred for thermal cooling as the mass acts as an insulator and a heat storage medium. The unglazed Trombe wall may be constructed with different materials like stone, brick, reinforced cement concrete, mud, etc. For thermal cooling of the building, the exposed solid thick wall should be bare (absorptivity, $\alpha \leq 0.4$) and for thermal heating of the building, it should be a blackened surface (absorptivity, $\alpha \geq 0.9$)

The expression for sol-air temperature for bare surface Trombe wall (for thermal cooling) is given in Equation (7.5).

In history, this concept was used frequently all over the world, particularly in India. Usually, the ventilation system used to be arranged above the lintel level of the windows and the inside temperature is found to be comfortable all year round [2]. All exposed as well as partition solid walls of buildings at Banaras Hindu University (BHU), Varanasi (India), have thickness of more than 600 mm, and these building provide better thermal comfort in summer condition with less fluctuation in room air temperature (~25°C–28°C) against outside temperature of (~45°C–48°C). The concepts of thick solid walls (Trombe wall) (>0.60 m) were very common in buildings till 1950 in India. Thick solid mud wall was used to have minimum U-value, but it requires annual maintenance. Currently, most buildings in India use brick wall of about 0.115 m thickness (higher U-value) including partition walls for minimum annual maintenance on the cost of using expensive grid power for good thermal comfort in summer as well as winter. This phenomenon is common due to cost-effectiveness

Dynamic Design Concepts for Hot Climatic Conditions

of brick materials in construction of housing. However, exposed wall to solar radiation can either have a cupboard or thin layer of insulation to have minimum load on air conditioners to sustain environment and climate.

Saadatian et al. [88] reviewed the characteristics of Trombe walls, including various Trombe wall configurations and technologies. Efficiency analysis of various parameters like vent, fan, size, thickness, color, insulation, materials of Trombe walls have also been done. For hot climatic conditions, Ozel [89] determined optimum insulation thickness according to cooling requirements of buildings in a hot climate by using a cost analysis. The optimum thickness of insulation was found to be 3.1 cm for cooling seasons with north orientation to be most economic. Stazi et al. [90] studied the thermal behavior of Trombe walls during summer months of Mediterranean climate for a low or highly insulated building. A drop of 1.4°C in the internal surface temperature of the wall and 0.5 MJ/m^2 in daily heat gains was observed with use of screening (rolling shutters). The heat gains of the screened Trombe walls were about 18 times lower than the ones that were unscreened. A reduction of 72.9% and 65% in cooling energy needs was observed with the combination of overhangs, rolling shutters and cross-ventilation, respectively, in comparison with unvented Trombe walls without solar shadings.

7.17 DIFFERENT COMPOSITIONS OF TROMBE WALL

7.17.1 VENTED TROMBE WALL

A part of the heat generated in the air space between the glazing and the Trombe wall is lost back to the atmosphere through the glazing. The higher the temperature of the air trapped in the air space, the greater is the heat loss. This heat loss can be reduced by venting the storage wall at the top and bottom as shown in Figure 7.9(a). This trapped air between the glazing and the Trombe wall gets heated up and enters the living space through the vents located at the top side. Cool room air takes its place through the lower vents, thus establishing a natural convective loop (thermo-circulation). During the off-sunshine hours (nighttime), the vents can be closed to keep the cold air out and the interior space is then heated by the storage mass, which gives up its heat by radiation. Such systems are preferrable for buildings having daytime use like offices or shops.

The vented Trombe wall can provide induced ventilation for summer cooling of the space as shown in Figure 7.9(b). In this case, the heated air in the collector space flows out through exhaust vents at the top of the outer glazing, and air from outside enters the space through openings on the cooler side to replace the hot air. This continuous air movement cools the living space. A perspective view of the arrangement with double-glazing is given in Figure 7.9(c).

The overall efficiency of the vented Trombe wall is about 10% higher than the unvented Trombe wall. This efficiency can be reduced in case the vents are not closed properly during the off-sunshine hours due to the reverse air flow. There are few problems associated with vented wall like accumulation of dust on the inner surface of the glazing and on the dark surface of the absorbing portion of the wall. Since the cleaning and dust removal from these areas are near to impossible, this may reduce the performance of the system eventually along with aesthetic appearance.

7.17.2 PHASE CHANGE MATERIAL (PCM) TROMBE WALL

As discussed, phase change materials (PCMs) are those who possess the thermal energy transfer by change in its state, from solid to liquid or liquid to solid. PCMs, having melting temperature between 20°C and 32°C, are useful for thermal storage in conjunction with both passive storage and active solar storage for heating and cooling in buildings. A thick massive wall is heavy and increases the building's dead load, and thus is considered a problem for structural engineers. This problem is thus addressed by PCM, which can store more energy and are smaller in volume and are lighter in weight than normal building materials.

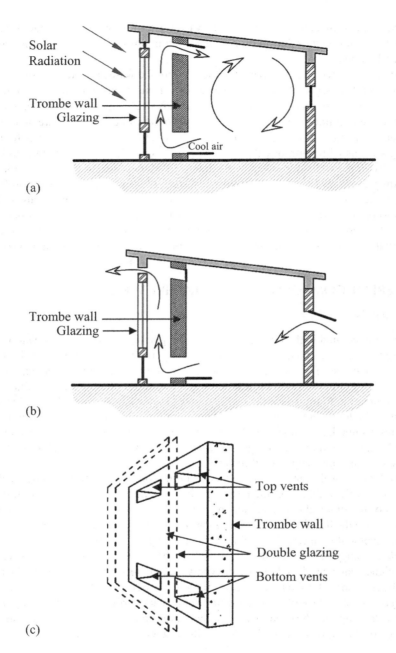

FIGURE 7.9 Vented Trombe wall. (a) Heating space during sunshine hours during winter, (b) Ventilation during summer months and (c) Perspective view with double-glazing [91].

We have already learned about Trombe walls in Section 6.3.2.2. A wall filled with PCM (such as phase eutectic salts or salt hydrates) to enhance efficiency is constructed in the south side of the house similar to a Trombe wall. PCM-Trombe wall can reduce the energy consumption of buildings.

This wall may be vented or non-vented depending upon the passive heat required. During the day, the wall is exposed to the solar insolation, thus the temperature of the wall rises, which melts the PCM filled inside the wall. During the off-sunshine hours (nighttime), the absorbed heat is

Dynamic Design Concepts for Hot Climatic Conditions

withdrawn and released to rise the indoor temperature of the room. For a given amount of heat storage, the phase change units require less space than water walls or mass Trombe walls and are much lighter in weight. These are, therefore, more convenient to use in retrofit applications of buildings.

A large number of PCMs are known to melt with a heat of fusion in the required range. Few of them used in the construction of Trombe wall are salt hydrates and hydrocarbons. A calcium chloride hexahydrate (melting point 29°C) wall with 8.1 cm thickness has thermal performance slightly better than a masonry wall with 40 cm thickness. Stearic acid has better transmittance than the glass for the same thickness and can be used in windows/walls as a transparent insulating material. However, for their employment as latent heat storage materials they must exhibit certain desirable thermodynamic, kinetic and chemical properties. Moreover, economic consideration and easy availability of these materials has to be kept in mind.

According to a study conducted by Bourdeau [92], a 150 mm concrete wall can be replaced by 350 mm wall made of PCM with similar performance. Different PCMs effect the performance of a Trombe wall in a different manner. The Trombe wall was constructed from a polyethylene container on a wooden shelf and double-glazed and the study established that a Trombe wall with latent heat storage is more efficient than a concrete wall because latent heat becomes saturated during the operation. A computerized dynamic simulation taken up by Khalifa and Abbas [93] for Baghdad, Iraq, with different PCMs had different reactions for a south-facing Trombe wall. In the study, the room temperature was recorded to be 15°C–25°C for concrete, 18°C–22°C for hydrated salt and 15°C–25°C for paraffin wax. The variation of temperature in the mass wall during night hours was recorded to be 1°C–3°C for concrete, 4°C–6°C for hydrated salt and 3°C–7°C for paraffin wax. The variation of temperature in the mass wall during day hours was recorded to be 7°C–15°C for concrete, 10°C–18°C for hydrated salt and 10°C–22°C for paraffin wax. The thickness of storage walls were 200 mm, 80 mm and 50 mm for concrete, hydrated salt and paraffin wax, respectively. Zalewski et al. [94] replaced a concrete wall with a wall that contained hydrated salt and found that the latter was able to release the solar gains with a time lag of 2 hours and 40 minutes.

7.17.3 Photovoltaic Integrated Phase Change Materials (PV-PCM) Wall

A photovoltaic (PV) Trombe wall is where PV panels are integrated with glazing to convert the solar energy into electrical energy and is considered to be an aesthetic approach to Trombe walls (refer to Section 6.3.4.2).

As already discussed, PCMs have high potential of heat storage per unit mass and with the use of a PV panel, they can produce sufficient electrical energy for household purposes. Figure 7.10 (a) and (b) show the PV-PCM wall without vent and with vent, respectively. The photovoltaic (PV) array can be both semi-transparent (glass to glass) or opaque (glass to Tedlar) type.

7.17.4 Heat Transfer in Trombe Walls

7.17.4.1 *U*-Value

(a) *U*-value for single-layered Trombe wall

$$U = \left[\frac{1}{h_0} + \frac{L}{K} + \frac{1}{h_i} \right]^{-1} \tag{7.10a}$$

where h_o is convective, and the radiative heat transfer coefficient (W/m²°C) from the exterior wall surface to the ambient air-facing wind speed (V), and h_i is a convective heat transfer coefficient from the interior wall surface to room air, with L as thickness (m) and K as thermal conductivity (W/mK) of the material of wall.

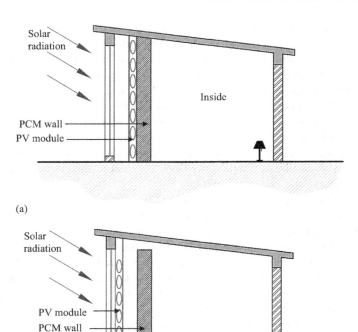

FIGURE 7.10 Photovoltaic integrated phase change material (PV-PCM) wall. (a) Unvented type and (b) vented type [91].

For brick wall thickness $(L) = 0.60$ m and 0.115 m with thermal conductivity of $K = 0.69$ W/mK, and zero wind speed, the U-value from Equation (7.10a) can be obtained as

$$U = [1.40]^{-1} = 0.71\,\text{W}/\text{m}^2{}^\circ\text{C for } L = 0.60\text{ m} = 1.43\,\text{W}/\text{m}^2{}^\circ\text{C for } L = 0.115\text{m}$$

It can be seen that there is increase in U-value by 101% by reducing the thickness of solid wall from 0.60 to 0.115 m. Hence, one can conclude that for thermal cooling of a building, solid wall thickness plays an important role in reducing the incoming flux into room.

(b) U-value for multilayered Trombe wall

$$U = \left[\frac{1}{h_0} + \sum_{j=1}^{N}\frac{L_j}{K_j} + \frac{1}{h_i}\right]^{-1} \quad (7.10\text{b})$$

If brick solid wall of thickness $L = 0.115$ m is covered with waterproof insulated material of thermal conductivity $K = 0.04$ W/m°C and thickness $L = 0.075$ m from either exterior or interior wall, then Equation (7.10b) can be used to evaluate the U-value for a waterproof insulated wall as follows:

$$U = [0.175 + 0.167 + 1.875 + 0.357]^{-1} = [2.7405]^{-1} = 0.38\,\text{W}/\text{m}^2{}^\circ\text{C}$$

Dynamic Design Concepts for Hot Climatic Conditions

One can further note from this calculation that the U-value of waterproof insulated wall of brick thickness of 0.115 m is even half the value in comparison with brick wall of thickness 0.60 m.

7.17.4.2 Rate of Heat Transfer

Now, the rate of heat transfer from exposed wall to room air can be expressed as follows:

$$\dot{Q}_u = UA_w \left(T_{sa} - T_r \right) \tag{7.11a}$$

where $\dot{Q}_u$ is the heat transfer in W, U is the thermal transmittance in W/m²°C, T_r is the room air temperature in °C on one side of the structure, A_w is the exposed (exterior) wall area in m², and T_{sa} is the sol-air temperature in °C on the exposed (solar radiation) side of wall.

An expression for sol-air temperature (T_{sa}) is given by

$$T_{sa} = \frac{\alpha}{h_o} I(t) + T_a - \frac{\varepsilon \Delta R}{h_o} \tag{7.11b}$$

where α and ε are the absorptivity and emissivity of exposed surface of wall, $\Delta R = 60$ W/m² is long-wavelength radiation exchange between ambient and sky, and it is zero for vertical wall. $I(t)$ is incident solar radiation on wall in W/m².

In absence of solar radiation during off-sunshine hours (night), Equation (7.11a) is reduced to

$$\dot{Q}_u = UA \left(T_r - T_a \right) \tag{7.11c}$$

7.17.5 Efficiency Analysis of Trombe Wall [88]

Following are various parameters that affect the efficiency of a Trombe wall.

7.17.5.1 Vent

As already discussed, Trombe wall can be of two types: vented and unvented. For vented, here are two thermocirculation vents at top and bottom of the vented Trombe wall for heat circulation and to control the heat loss in the air space between glazing and the wall through convection, conduction or radiation back to the atmosphere. The heat loss is directly proportional to the temperature in the air space, which means the higher the temperature, the higher will be the heat loss. The hot air in the air space being lighter enters the room through the top vent, and cool air replaces it through bottom vent [91].

The heat transfer coefficients between the air in the gap and wall and glazing is changed by closing/opening of the vents. Heat transfer coefficients shall be measured for optimization when the vents are closed. While the vents are open, a mathematical equation is used to calculate the heat transfer coefficient. Balcomb and McFarland [95] concluded that vents produce a reverse flow during off-sunshine hours (night) and reduces the efficiency under certain circumstances. Vent size is an important parameter and depends on solar saving fraction (SSF).

Also, external vents installed in the exterior part of the Trombe wall can facilitate the air circulation and enhances ventilation, thus cooling the air space between the glazing and main wall during summer months [96].

7.17.5.2 Size

The ratio of Trombe wall to the total wall area is an important parameter and directly impacts the thermal efficiency. If this ratio is 20%, annual savings of up to 22.3% of heating auxiliary energy can be achieved for Mediterranean climatic conditions and given parameters in the research [97]. It was also found that if the ratio increases to 37%, the annual savings of heating auxiliary energy reached up to 32.1%, and for ratios more than 37%, the amount of savings were observed to be negligible. Therefore, the optimal Trombe wall area ratio was found to be 37% [97].

The optimum thickness of Trombe wall depends on latitude, climate and heat loss. And is an important parameter to determine the effectiveness of a Trombe wall. There is a lag of 120 minutes to 150 minutes for heat transfer from outside to inside for each 100 mm of concrete wall [91]. Insufficient thickness of the Trombe wall results in inside temperature fluctuations. For Indian climatic conditions, optimal thickness has been proposed to be 300 mm–400 mm by Agrawal and Tiwari [91].

7.17.5.3 Fan

Research is available to understand the efficiency of using fans to assist the circulation of heat through vents depends on computer modeling. Sebald et al. [98] concluded that efficiency of 37 m^2 of room was improved by using a fan by 22% in Albuquerque, 20% in Santa Barbara and 7% in Madison. Another study by Sebald [99] concluded that the performance of a Trombe wall was improved by 8% when the room required heating. Efficiency of PV–Trombe walls with and without fans was studied by Jie et al. [100] and found that fans can reduce the temperature of PV cells by 1.28°C between 7:00 and 17:00 hours, thus improving the efficiency.

7.17.5.4 Material and Color

The material used for construction of Trombe walls contributes to the efficiency of heat storage, convection and conduction. Knowles [101] replaced paraffin-metal mixture wall with a concrete wall and found that storage mass was reduced by concrete wall by 90%, thus improving the efficiency by more than 20%. Experimental study was carried by Hassanain et al. [102] and results established that various adobe materials have better efficiency in Trombe walls. High-absorption coating materials improve the storage capacity and enhance the efficiency of Trombe wall [103, 104].

Effect of three color-types (dark, natural and light) were evaluated by Özbalta and Kartal [105]. The authors used brick wall, concrete wall and autoclaved aerated concrete wall as the main material of the solar mass. It was found that annual heat gain of concrete wall was 26.9% for dark, 20.2% for natural and 9.7% for light color. The annual heat gain of brick wall was 20.5% for dark, 16.4% for natural and 7.1% for light color. The annual heat gain of autoclaved aerated concrete wall was 13.0% for dark, 7.9% for natural and 4.3% for light color.

The thermal mass should be dark in color for maximum absorption of solar radiation, resulting in maximum efficiency of the system for winter months.

7.17.5.5 Insulation

A Trombe wall has low thermal resistance and loses a large amount of heat at nighttime. In hot climatic conditions and specifically for well-insulated buildings, they might act as a source of undesired heat gains and result in over heating because of reverse hear transfer. In order to prevent this reverse heat transfer, the Trombe walls should be well insulated. Proper insulation is important to maximize the ventilation rate during summer months. Additionally, interior of Trombe walls should also be insulated. The integration of insulation and composite Trombe wall (made up of glass, thermal mass and an insulated wall) with vents between the mass and insulated wall can perform better than a conventional Trombe wall in cold and cloudy climatic conditions. This combination can also reduce the thickness of the Trombe wall [106].

Thermal insulation can improve the efficiency of PV–Trombe wall for both heating and cooling season. Numerical study for climatic conditions of Xining, China, for composite Trombe walls with internal insulation and air cavities showed that efficiency was improved by 56% when compared to normal Trombe wall [100]. Stazi et al. [107] conducted an experimental and numerical study and analyzed the impact of insulation and compared the results of normal Trombe wall and super-insulated Trombe wall. For cooling, the amount of energy required by normal Trombe wall was approximately 9.19 kWh/m^2 and for super-insulated wall, it was approximately 23.32 kWh/m^2. However,

Dynamic Design Concepts for Hot Climatic Conditions

the results were opposite for heating-energy demand for normal Trombe wall, which came out to be approximately 58.33 kWh/m² and for super-insulated wall 16.21 kWh/m².

7.18 SOLAR COOLING

In hot climates, thermal comfort cooling is important for space conditioning of buildings. Feasibility of solar insolation and cooling load exists nearly in a similar phase. Solar air conditioning can be attained by (a) absorption cycles, (b) desiccant cycles and (c) solar-mechanical processes. A continuous or intermittent cycle, hot- or cold-side energy storage, diverse control strategies, various temperature ranges of operation, different collectors, etc. can be possible under above-mentioned categories. Theoretically, numerous thermal efficiencies of solar cooling processes can be achieved. The temperature constraints in mechanism of solar collectors restrict their thermal efficiencies. Compared to provision of additional cooling, minimization of cooling loads through cautious building design and insulation are much cheaper, which makes them preferable. Proficient design and construction of a building can minimize the load on any air conditioning or heating system but we are concerned for cooling loads that should really be considered in a decent building design.

Solar cooling has the following advantages:

- It saves conventional primary energy source based on electricity.
- It leads to a reduction of peak electricity demand for cost saving.
- It is environmentally sound without ozone depletion to sustain global warning.
- The demand and supply matches, i.e. hottest periods have sufficient maximum solar energy.

7.18.1 SOLAR PHOTOVOLTAIC COOLING

Solar photovoltaic has been used to provide electrical power to absorption/adsorption-based vapor compressor. In absorption/adsorption-based vapor compressor process, energy is transferred through phase-change process. For small residential/commercial cooling (≤ 5 MWh/year), solar photovoltaic-power cooling is the most suitable/successful solar cooling technology. The reason for this may be the incentive provided by government agencies. The cost of solar photovoltaic-power cooling effectiveness mainly depends on the cooling system and use of electrical processes. Partially replacing the grid supply for air conditioning, with PV systems can be significantly cost-effective. This limits carbon emission in the atmosphere too.

OBJECTIVE QUESTIONS

7.1 Which of the following is not a direct cooling technique?
(a) Ventilation
(b) Courtyard
(c) Evaporative cooling
(d) Wind tower

7.2 The value of shading factor for complete shading should be
(a) >1
(b) 0
(c) 1
(d) <1

7.3 Ventilation losses can be defined by:
(a) $0.33NV\left(T_r - T_a\right)$
(b) $3.3NV\left(T_r - T_a\right)$
(c) $0.33NV\left(T_a - T_r\right)$
(d) $3.3NV\left(T_a - T_r\right)$

212   Photovoltaic Thermal Passive House System

7.4   Deciduous trees for summer shading should be planted towards this direction of the building
      (a)   North and northwest
      (b)   South and southwest
      (c)   North and northeast
      (d)   South and southeast
7.5   The sky temperature is always lower than ambient air temperature with clear night sky by
      (a)   10°C
      (b)   11°C
      (c)   12°C
      (d)   13°C
7.6   The following concept can be used for both passive heating and cooling
      (a)   Roof pond
      (b)   Wind tower
      (c)   Air cavity
      (d)   Green roof
7.7   Effective radiation from the exposed horizontal surface to ambient air via convective and radiative heat transfer is referred to as which passive concept?
      (a)   Evaporative cooling
      (b)   Cool roof
      (c)   Radiative cooling
      (d)   Courtyard
7.8   For thermal cooling, roof pond is operated under
      (a)   Opened louvered shade to increase the heat loss during the night
      (b)   Closed louvered shade to increase the heat loss during the night
      (c)   Opened louvered shade to decrease the heat loss during the night
      (d)   Closed louvered shade to decrease the heat loss during the night
7.9   Cool roof coating should have
      (a)   Low reflectivity and low thermal conductivity
      (b)   High reflectivity and high thermal conductivity
      (c)   High reflectivity and low thermal conductivity
      (d)   Low reflectivity and high thermal conductivity
7.10 Following is an example of indirect coupling
      (a)   Wind tower
      (b)   Earth-air heat exchanger
      (c)   Earth shelter
      (d)   Roof pond

ANSWERS

7.1   (c)
7.2   (c)
7.3   (a)
7.4   (b)
7.5   (c)
7.6   (a)
7.7   (c)
7.8   (a)
7.9   (c)
7.10   (b)

PROBLEMS

7.1 Calculate the U-value for brick wall thickness $(L) = 230$ mm and 460 mm with thermal conductivity of $K = 0.69$ W/mK, and 1 m/s wind speed
Hint: Use Equation (7.10a)

7.2 Calculate U-value for a solid 230 mm brick wall with external plaster of 12 mm and waterproof insulation layer of 75 mm with thermal conductivity 0.18 W/m°C and 0.040 W/m°C. respectively.
Hint: Use Equation (7.10b)

7.3 Calculate the total heat transfer coefficient (h_1) and the rate of heat loss due to radiation, convection and evaporation for wetted surface.

Given: Wetted surface temperature = 25°C, $T_a = 10$°C, relative humidity = 50%, wind velocity = 2 m/s, emissivity = 0.9, $h_{ra} = 4.93$ W/m²C
Hint: Example 7.1

REFERENCES

[1] D. Chwieduk, *Solar Energy in Buildings. Thermal Balance for Efficient Heating and Cooling*, Amsterdam, The Netherlands: Elsevier, 2014.

[2] N. Gupta and G. N. Tiwari, "Review of passive heating/cooling systems of buildings," *Energy Science & Engineering*, vol. 4, no. 5, pp. 305–333, 2016.

[3] Z. Wang, Y. Ding, G. Geng and N. Zhu, "Analysis of energy efficiency retrofit schemes for heating, ventilating and air-conditioning existing office buildings based on the modified bin method," *Energy Conversion and Management*, vol. 77, pp. 233–242, 2014.

[4] M. Santamouris, "Passive cooling of buildings," *Klimatizacija Grejanje Hlađenje*, vol. 34, issue 4, 2005.

[5] M. Sabry, P. C. Eames, H. Singh and Y. Wu, "Smart windows: Thermal modelling and evaluation," *Solar Energy*, pp. 200–209, 2014.

[6] K. S. Y. Wan and F. W. H. Yik, "Building design and energy end-use characteristics of high-rise in Hong Kong," *Applied Energy*, vol. 78, pp. 19–36, 2004.

[7] B. Wang and A. Malkawi, "Design-based natural ventilation evaluation in early stage for high performance buildings," *Sustainable Cities and Society*, vol. 45, pp. 25–37, 2019.

[8] M. Ciampi, F. Leccese and G. Tuoni, "Energy analysis of ventilated and microventilated roofs," *Solar Energy*, vol. 79, no. 2, pp. 183–192, 2005.

[9] S. H. Ibrahim, N. A. Azhari, M. N. M. Nawi, A. Baharun and R. Affandi, "Study on the effect of the roof opening on the temperature underneath," *International Journal of Applied Engineering Research*, vol. 9, no. 23, pp. 20099–20110, 2014.

[10] D. Li, Y. Zheng, C. Liu, H. Qi and X. Liu, "Numerical analysis on thermal performance of naturally ventilated roofs with different influencing parameters," *Sustainable Cities and Society*, vol. 22, pp. 86–93, 2016.

[11] H. Li, J. Li, C. Xi, W. Chen and X. Kong, "Experimental and numerical study on the thermal performance of ventilated roof composed with multiple phase change material (VR-MPCM)," *Energy Conversion and Management*, vol. 213, p. 112836, 2020.

[12] G. N. Tiwari, *Solar Energy—Fundamentals, Design, Modelling and Applications*, Delhi: Narosa Publishing House Pvt. Ltd., 2002.

[13] W. Parys, H. Breesch, H. Hens and D. Saelens, "Feasibility assessment of passive cooling for office buildings in a temperate climate through uncertainty analysis," *Building and Environment*, vol. 56, pp. 95–107, 2012.

[14] D. Gupta and V. R. Khare, "Natural ventilation design: Predicted and measured performance of a hostel building in composite climate of India," *Energy and Built Environment*, vol. 2, issue 1, pp. 82–93, 2021.

[15] N. K. Bansal, M. S. Sodha, P. K. Bansal, and A. A. Kumar. *Solar Passive: Building Design Science*. New York: Oxford: Pergamon Press, 1988.

[16] M. Bessoudoa, A. Tzempelikos, A. K. Athienitis and R. Zmeureanua, "Indoor thermal environmental conditions near glazed facades with shading devices – Part I: Experiments and building thermal model," *Building and Environment*, vol. 45, no. 11, pp. 2506–2516, 2010.

[17] M. A. Kamal, "Shading: A simple technique for passive cooling and energy conservation in buildings," *Architecture - Time, Space & People*, pp. 18–23, January 2011.

[18] M. A. Kamal, "Energy conservation with passive solar landscaping," in *National Convection on Planning for Sustainable Built Environment*, Bhopal, 2003, pp. 92–99.

[19] R. Kumar, S. N. Garg and S. C. Kaushik, "Performance evaluation of multi-passive solar applications of a non air-conditioned building," *International Journal of Environmental Technology and Management*, vol. 5, pp. 60–75, 2005.

[20] G. Kim, H. S. Lim, T. S. Lim, L. Schaefer and J. T. Kim, "Comparative advantage of an exterior shading device in thermal performance for residential buildings," *Energy and Buildings*, vol. 46, pp. 105–111, 2012.

[21] S. Grynning, B. Time and B. Matusiak, "Solar shading control strategies in cold climates - Heating, cooling demand and daylight availability in office spaces," *Solar Energy*, vol. 107, pp. 182–194, 2014.

[22] L. Evangelisti, C. Guattari, F. Asdrubali and R. d. L. Vollaro, "An experimental investigation of the thermal performance of a building solar shading device," *Journal of Building Engineering*, vol. 28, p. 101089, 2020.

[23] D. Pearlmutter and S. Rosenfeld, "Performance analysis of a simple roof cooling system with irrigated soil and two shading alternatives," *Energy and Buildings*, vol. 40, pp. 855–864, 2008.

[24] G. Hauser G. Minke and N. K. Bansal, *Passive Building Design - A Handbook of Natural Climate Control*, Amsterdam: Elsevier Science, 1994.

[25] V. T. Ca, T. Asaeda and E. M. Abu, "Reductions in air conditioning energy caused by a nearby park," *Energy and Buildings*, vol. 29, pp. 83–92, 1998.

[26] G. Papadakis, P. Tsamis and S. Kyritsis, "An experimental investigation of the effect of shading with plants for solar control of buildings," *Energy and Buildings*, vol. 33, pp. 831–836, 2001.

[27] H. Wilson, "*Solar Control Coatings for Windows*," in *Proceedings of the Conference EUROMAT 99*, Munich, Germany, 1999.

[28] M. Saeli, C. Piccirillo, I. P. Parkin, R. Binions and I. Ridley, "Energy modelling studies of thermochromic glazing," *Energy and Buildings*, vol. 42, issue 10, pp. 1666–1673, 2010.

[29] P. Littlefair, M. Santamouris, S. Alvarez, A. Dupagne, D. Hall, J. Teller, J.-F. Coronel and N. Papanikolaou, *Environmental Site Layout Planning: Solar Access, Microclimate and Passive Cooling in Urban Areas*, London: Construction Research Communications Ltd. BRE Publications, 2000.

[30] M. Ciampi, F. Leccese and G. Tuoni, "Ventilated facades energy performance in summer cooling of buildings," *Solar Energy*, vol. 75, pp. 491–502, 2003.

[31] T. M. I. Mahlia and A. Iqbal, "Cost benefits analysis and emission reductions of optimum thickness and air gaps for selected insulation materials for building walls in Maldives," *Energy*, vol. 35, no. 5, pp. 2242–2250, 2010.

[32] M. Charde and R. Gupta, "Design development and thermal performance evaluation of static sunshade and brick cavity wall: An experimental study," *Energy and Buildings*, vol. 60, pp. 210–216, 2013.

[33] S. Hong, T. Oreszczyn and I. Ridley, "The warm front study group, the impact of energy efficient refurbishment on the space heating fuel consumption in English dwellings," *Energy and Buildings*, vol. 38, pp. 1171–1181, 2006.

[34] Z. L. Zhang and B. J. Wachenfeldt, "Numerical study on the heat storing capacity of concrete walls with air cavities," *Energy and Buildings*, vol. 41, pp. 769–773, 2009.

[35] M. Zinzi and S. Agnoli, "Cool and green roofs. An energy and comfort comparison between passive cooling and mitigation urban heat island techniques for residential buildings in the Mediterranean region," *Energy and Buildings*, vol. 55, no. 66, 2012.

[36] Al-Hemiddi and N. A. Mohammed, "Passive cooling systems applicable for buildings in the hot-dry climate of Saudi Arabia," Los Angeles, CA: Graduate School of Architecture and Urban Planning, UCLA, 1995, ProQuest Dissertations Publishing, 9604212.

[37] M. Dabaieh, O. Wanas, Mohd A. Hegazy and E. Johansson, "Reducing cooling demands in a hot dry climate: A simulation study for non-insulated passive cool roof thermal performance in residential buildings," *Energy and Buildings*, vol. 89, pp. 142–152, 2015.

Dynamic Design Concepts for Hot Climatic Conditions

[38] M. Santamouris and D. Asimakopoulos (Eds), *Passive Cooling of Buildings*, London, UK: James and James Science Publishers, 1996.

[39] Z. Qingyuan and L. Yu, "Potentials of passive cooling for passive design of residential buildings in China," *Energy Procedia*, vol. 57, pp. 1726–1732, 2014.

[40] P. Agrawal, "A review of passive systems for natural heating and cooling of buildings," *Solar and Wind Technology*, pp. 557–567, 1989.

[41] S. Chungloo and B. Limmeechokchai, "Application of passive cooling systems in the hot and humid climate: The case study of solar chimney and wetted roof in Thailand," *Building and Environment*, vol. 42, no. 9, pp. 3341–3351, 2007.

[42] P. Raman, S. Mande and V. V. N. Kishore, "A passive solar system for thermal comfort conditioning of buildings in composite climates," *Solar Energy*, vol. 70, no. 4, pp. 319–329, February 2001.

[43] S. Wanphen and K. Nagano, "Experimental study of the performance of porous materials to moderate the roof surface temperature by its evaporative cooling effect," *Building and Environment*, vol. 44, no. 2, pp. 338–351, 2009.

[44] W. Chen, S. Liu and J. Lin, "Analysis on the passive evaporative cooling wall constructed of porous ceramic pipes with water sucking ability," *Energy and Buildings*, vol. 86, pp. 541–549, 2015.

[45] M. A. Kamal, "An overview of passive cooling techniques in buildings: Design concepts and architectural interventions," *Acta Technica Napocensis: Civil Engineering & Architecture*, vol. 55, no. 1, pp. 84–97, 2012.

[46] G. Q. Qiu and S. B. Riffat, "Novel design and modelling of an evaporative cooling system for buildings," *International Journal of Energy Research*, vol. 30, pp. 985–999, May 2006.

[47] G. Chiesa and M. Grosso, "Direct evaporative passive cooling of building. A comparison amid simplified simulation models based on experimental data," *Building and Environment*, vol. 94, pp. 263–272, 2015.

[48] J. M. Salmeron, F. J. Sánchez, J. Sánchez, S. Álvarez, J. L. Molina and R. Salmeron, "Climatic applicability of downdraught cooling in Europe," *Architectural Science Review*, vol. 55, no. 4, pp. 259–272, 2012.

[49] H. Xuan and B. Ford, "Climatic applicability of downdraught cooling in China," *Architectural Science Review*, vol. 55, no. 4, pp. 273–286, 2012.

[50] A. Badran, "Performance of cool towers under various climates in Jordan," *Energy and Buildings*, vol. 35, no. 10, pp. 1031–1035, 2003.

[51] M. Benhammou, B. Draoui, M. Zerrouki and Y. Marif, "Performance analysis of an earth-to-air heat exchanger assisted by a wind tower for passive cooling of buildings in arid and hot climate," *Energy Conversion and Management*, vol. 91, pp. 1–11, February 2015.

[52] Y. Bouchahm, F. Bourbia and A. Belhamri, "Performance analysis and improvement of the use of wind tower in hot dry climate," *Renewable Energy*, vol. 36, no. 3, pp. 898–906, March 2011.

[53] E. G. Cruz and E. Krüger, "Evaluating the potential of an indirect evaporative passive cooling system for Brazilian dwellings," *Building and Environment*, vol. 87, pp. 265–273, 2015.

[54] Z. Duan, C. Zhan, X. Zhang, M. Mustafa, X. Zhao, B. Alimohammadisagvand and A. Hasan, "Indirect evaporative cooling: Past, present and future potentials," *Renewable and Sustainable Energy Reviews*, vol. 16, no. 9, p. 6823–6850, 2012.

[55] H. Marzuki, T. Mahlia, A. Zare, T. J. Saksahdan and H. S. C. Metselaar, "Potential energy savings by radiative cooling system for a building in tropical climate," *Renewable and Sustainable Energy Reviews*, vol. 32, pp. 642–650, 2014.

[56] R. Cavelius, C. Isaksson, E. Perednis and G. E. F. Read, *Passive Cooling Technologies*, Vienna: Austrian Energy Agency, pp. 125, 2005.

[57] B. Juchau, "Nocturnal and conventional space cooling via radiant floors," in *International Passive and Hybrid Cooling Conference*, Miami Beach, 1981.

[58] E. Erell and Y. Etzion, "A radiative cooling system using water as a heat exchange medium," *Architectural Science Review*, vol. 35, pp. 39–49, 1992.

[59] A. Beck and D. Büttner, "Radiative cooling for low energy cold production," in *Proceedings of the Annual Building Physics Symposium*, Hochschule für Technik Stuttgart, 2006.

[60] Q. Zhang, K. Asano and T. Hayashi, "Regional characteristics of heating loads for apartment houses in China," *Journal of Architecture, Planning and Environmental Engineering*, vol. 555, pp. 69–75, 2002.

[61] J. Liu, Z. Zhou, D. Zhang, S. Jiao, Y. Zhang, L. Luo, Z. Zhang and F. Gao, "Field investigation and performance evaluation of sub-ambient radiative cooling in low latitude seaside," *Renewable Energy*, pp. 90–99, 2020.

[62] M. S. Sodha, S. C. Kaushik, G. N. Tiwari, I. C. Goyal, M. A. S. Malik and A. K. Khatry, "Optimum distribution of insulation inside and outside the roof," *Building and Environment*, vol. 14, issue 1, pp. 47–52, 1979.

[63] H. Saffari and S. Hosseinnia, "Two-phase Euler-Lagrange CFD simulation of evaporative cooling in a wind tower," *Energy and Buildings*, vol. 41, pp. 991–1000, 2009.

[64] B. R. Hughes, J. K. Calautit and S. A. Ghani, "The development of commercial wind towers for natural ventilation: A review," *Applied Energy*, vol. 92, pp. 606–627, 2012.

[65] H. Montazeri, F. Montazeri, R. Azizian and S. Mostafavi, "Two-sided wind catcher performance evaluation using experimental, numerical and analytical modeling," *Renewable Energy*, vol. 35, pp. 1424–1435, 2010.

[66] H. N. Chaudhry, J. K. Calautit and B. R. Hughes, "Computational analysis of a wind tower assisted passive cooling technology for the built environment," *Journal of Building Engineering*, vol. 1, pp. 63–71, 2015.

[67] M. N. Bahadori, "An improved design of wind towers for natural ventilation and passive cooling," *Solar Energy*, vol. 35, pp. 119–129, 1985.

[68] M. N. Bahadori, "Viability of wind towers in achieving summer comfort in the hot arid regions of the Middle East," *Renewable Energy*, vol. 5, pp. 879–892, 1994.

[69] M. N. Bahadori, M. Mazidi and A. R. Dehghani, "Experimental investigation of new designs of wind towers," *Renewable Energy*, vol. 33, pp. 2273–2281, 2008.

[70] A. R. Dehghani-Sanij, M. Soltani and K. Raahemifar, "A new design of wind tower for passive ventilation in buildings to reduce energy consumption in windy areas," *Renewable and Sustainable Energy Reviews*, vol. 42, pp. 182–195, 2015.

[71] C. A. Varela-Boydo, S. L. Moya and R. Watkins, "Study of wind towers with different funnels attached to increase natural ventilation in an underground building," *Frontiers of Architectural Research*, vol. 9, issue 4, pp. 925–939, 2020.

[72] C. A. Varela-Boydo and S. L. Moya, "Inlet extensions for wind towers to improve natural ventilation in buildings," *Sustainable Cities and Society*, vol. 53, p. 101933, 2020.

[73] G. N. Tiwari, A. Tiwari and Shyam, *Handbook of Solar Energy: Theory, Analysis and Applications*, New Delhi: Springer, 2016.

[74] G. N. Tiwari, V. Singh, P. Joshi, A. Deo, Prabhakant and A. Gupta, "Design of an Earth Air Heat Exchanger (EAHE) for climatic condition of Chennai, India," *Open Environmental Sciences*, vol. 8, pp. 24–34, 2014.

[75] F. Al-Ajmi, D. L. Loveday and V. I. Hanby, "The cooling potential of earth-air heat exchangers for domestic buildings in a desert climate," *Building and Environment*, vol. 41, issue 3, pp. 235–244, 2006.

[76] M. Benhammou and B. Draoui, "Parametric study on thermal performance of earth-to-air heat exchanger used for cooling of buildings," *Renewable and Sustainable Energy Reviews*, vol. 44, pp. 348–355, 2015.

[77] R. Kumar, S. Ramesh and S. C. Kaushik, "Performance evaluation and energy conservation potential of earth–air–tunnel system coupled with non-air-conditioned building," *Building and Environment*, vol. 38, issue 6, pp. 807–813, 2003.

[78] M. S. Sodha, A. Sharma, S. Singh, N. Bansal and A. Kumar, "Evaluation of an earth-air for cooling/heating of a hospital complex," *Building and Environment*, vol. 20, pp. 115–122, 1985.

[79] M. K. Rodrigues, R. da. S. Brum, J. Vaz, L. A. O. Rocha, E. D. d. Santos and L. A. Isoldi, "Numerical investigation about the improvement of the thermal potential of an Earth-Air Heat Exchanger (EAHE) employing the Constructal Design method," *Renewable Energy*, vol. 80, pp. 538–551, 2015.

[80] H. Wei, D. Yang, J. Wang and J. Du, "Field experiments on the cooling capability of earth-to-air heat exchangers in hot and humid climate," *Applied Energy*, vol. 276, p. 115493, 2020.

[81] R. Tang and Y. Etzion, "On thermal performance of an improved roof pond for cooling buildings," *Building and Environment*, vol. 39, pp. 201–209, 2004.

[82] G. N. Tiwari, A. Kumar and M. S. Sodha, "A review-cooling by water evaporation over roof," *Energy Conversion and Management*, vol. 22, pp. 143–153, 1982.

[83] G. Sutton, "American social heating ventilation guide," 1950.

[84] D. Jain, "Modeling of solar passive techniques for roof cooling in arid regions," *Building and Environment*, vol. 44, pp. 277–287, 2006.

[85] S. N. Kharrufa and Y. Adil, "Roof pond cooling in buildings in hot arid climates," *Building and Environment*, pp. 82–89, 2008.

[86] I. Ahmad, "Improving the thermal performance of a roof pond system," *Energy Conversion and Management*, vol. 25, pp. 207–209, 1985.

[87] A. Spanaki, D. Kolokotsa, T. Tsoutsos and I. Zacharopoulos, "Assessing the passive cooling effect of the ventilated pond protected with a reflecting layer," *Applied Energy*, vol. 123, pp. 273–280, 2014.

[88] O. Saadatian, K. Sopian, C. H. Lim, N. Asim and M. Y. Sulaiman, "Trombe walls: A review of opportunities and challenges in research and development," *Renewable and Sustainable Energy Reviews*, vol. 16, pp. 6340–6351, 2012.

[89] M. Ozel, "Determination of optimum insulation thickness based on cooling transmission load for building walls in a hot climate," *Energy Conversion and Management*, vol. 66, pp. 106–114, 2013.

[90] F. Stazi, A. Mastrucci and C. di Perna, "Trombe wall management in summer conditions: An experimental study," *Solar Energy*, vol. 86, pp. 2839–2851, 2012.

[91] B. Agrawal and G. N. Tiwari, *Building Integrated Photovoltaic Thermal Systems: For Sustainable Developments*, United Kingdom: RSC Publishing, 2010.

[92] L. E. Bourdeau, "Study of two passive solar systems containing phase change materials for thermal storage," in *Fifth Natl Passive Solar Conference*, Amherst, Mass: Smithsonian Astrophysical Observatory, 1980.

[93] A. J. N. Khalifa and E. F. Abbas, "A comparative performance study of some thermal storage materials used for solar space heating," *Energy and Buildings*, vol. 41, pp. 407–415, 2009.

[94] L. Zalewski, A. Joulin, S. Lassue, Y. Dutil and D. Rousse, "Experimental study of small-scale solar wall integrating phase change material," *Solar Energy*, vol. 86, pp. 208–219, 2012.

[95] J. D. Balcomb and R. D. McFarland, "Simple empirical method for estimating the performance of a passive solar heated building of the thermal storage wall type," in *2nd National Passive Solar Conference*, Philadelphia PA, 1978.

[96] R. Stepler, "Trombe Wall-Retrofit," New Delhi: Popular Science Bonnier Corporation, p. 140, 1980.

[97] S. Jaber and S. Ajib, "Optimum design of Trombe wall system in Mediterranean region," *empirical Solar Energy*, vol. 85, pp. 1891–1898, 2011.

[98] A. V. Sebald, J. R. Clinton and F. Langenbacher, "Performance effects of Trombe wall control strategies," *Solar Energy*, vol. 23, pp. 479–487, 1979.

[99] A. V. Sebald, "Efficient simulation of large, controlled passive solar systems: Forward differencing in thermal networks," *Solar Energy*, vol. 34, pp. 221–230, 1985.

[100] J. Jie, Y. Hua, P. Gang, J. Bin and H. Wei, "Study of PV-Trombe wall assisted with DC fan," *Building and Environment*, vol. 42, pp. 3529–3539, 2007.

[101] T. R. Knowles, "Proportioning composites for efficient thermal storage walls," *Solar Energy*, vol. 31, pp. 319–326, 1983.

[102] A. A. Hassanain, E. M. Hokam and T. K. Mallick, "Effect of solar storage wall on the passive solar heating constructions," *Energy and Buildings*, vol. 43, pp. 737–747, 2010.

[103] N. P. Nwachukwu and W. I. Okonkwo, "Effect of an absorptive coating on solar energy storage in a Trombe wall system," *Energy and Buildings*, vol. 40, pp. 371–374, 2008.

[104] N. P. Nwosu, "Trombe wall redesign for a poultry chick brooding application in the equatorial region-analysis of the performance of the system using the Galerkin finite elements," *International Journal of Sustainable Energy*, vol. 29, pp. 37–47, 2010.

[105] T. G. Özbalta and S. Kartal, "Heat gain through Trombe wall using solar energy in a cold region of Turkey," *Scientific Research and Essays*, vol. 5, pp. 2768–2778, 2010.

[106] Z. Zrikem and E. Bilgen, "Theoretical study of a composite Trombe-Michel wall solar collector system," *Solar Energy*, vol. 39, pp. 409–419, 1987.

[107] F. Stazi, A. Mastrucci and C. di Perna, "The behaviour of solar walls in residential buildings with different insulation levels: An experimental and numerical study," *Energy and Buildings*, vol. 47, pp. 217–229, 2011.

8 Building Integrated Photovoltaic Thermal Systems (BiPVT)

8.1 INTRODUCTION

Photovoltaic (PV) modules, as discussed in Chapter 5, convert a part of solar energy into electrical energy with an efficiency of 10%–15%. When this absorbed heat is put into use, the system is termed a photovoltaic thermal (PVT) system. When PV modules are incorporated into the buildings as part of the building envelope (such as roof, façade, skylights) by replacing conventional building materials like tiles, etc., for generation of electrical energy, the system is referred to as a building integrated photovoltaic (BiPV) system. Integration of the PVT system with buildings is therefore termed as building integrated photovoltaic thermal (BiPVT) system. The efficiency of the system depends on the type of PV module used. If opaque PV modules (OPV) are integrated with the building, the system is termed as building integrated opaque photovoltaic thermal (BiOPVT) system and if it is a semi-transparent photovoltaic (SPV) type, the system is called a building integrated semi-transparent photovoltaic thermal (BiSPVT) system. The major advantage of BiPV system is that it exploits available area on the building envelope for production of energy and serves as a function of shading, daylighting, thermal insulation, primary weather impact protection.

In cases where PV modules do not form an integral part of the building, they are referred to as building attached PVs (BaPV) systems. In BaPV systems, PV modules are added on the building and do not have any direct impact on the structure's performance. Standalone PV modules, whether opaque or semi-transparent, can be used to produce electrical energy when the main grid is not present or not available at a reasonable price. The 3-D model was previously shown in Figure 3.6 [1].

Net zero energy buildings (as discussed in Chapter 1) are a combination of high level energy performance and a balance of its energy demands by in-situ or nearby renewable energy sources, and among the various solutions available, solar BIPV systems is the most widely considered [2]. The BiPV system has experienced an extraordinary growth and has proved to be a practical and promising solution towards energy efficiency and energy generation. To maximize the energy savings, the most practical approach is to exploit the solar source by integrating buildings with PV panels and respecting the sun, glazing and the lighting preferably in harsh cold climatic conditions. With this integration, the dependency of building on the conventional grid will reduce, thus encouraging the dependence on non conventional sources of energy, and reducing the harmful impact of conventional source of energy on environment. This is a step towards energy conservation, self sustainable building. Passive solar designs should be implemented as an economic and resource efficient techniques for achieving natural harmony between climate, buildings and anthropos. The building structure should be self-sustainable, i.e., able to generate the energy for its own consumption. Various works demonstrate that there is reduction in building's energy consumption with BiPVT systems. BiPVT also leads to reduction in CO_2 emissions and can act as a path to move towards net-zero energy buildings. The details are discussed in Chapter 9.

Generally, the efficiency of the PV module is low at high temperatures, thus the efficiency is increased by utilizing the heat from the rear side of the PV module. The studies were conducted with working fluid that can be in the form of air or water to create a forced convection. The opening alternative inputs between cold and hot areas in photovoltaics provide hot air and air gap to set the

DOI: 10.1201/9780429445903-8

219

optimal level. This helps to increase the annual power output and life of the module associated with the system. Further, to increase the production and efficiency of the system in building applications, ambient temperature, shadowing effect, building direction and slope of the module also has a significant impact. Yang et al. [3] examined the BiPV system's performance with an air gap between the PV modules and wall to allow cooling and increase the efficiency. The total annual energy output of the proposed system was estimated to be 6878 kWh.

8.2 LITERATURE REVIEW OF BiPV/BiPVT SYSTEMS

Various studies have been conducted to analyze the integration of PV modules with the buildings to understand the performance of BiPV and BiPVT system in terms of system design and building performance. BiPVT systems were formed in early '90s. and because of its potential towards designing a net-zero building, the interest in this sector has risen since 2000 [4]. Jelle and Breivik [5] investigated the future opportunities for BiPV systems and their development. Less impact on the environment, and low manufacturing costs with high efficiency were the influencing factors for future BiPV systems.

Architectural aspects of BiPV system in terms of designing, lifetime and choice between BaPV or BiPV system were discussed by Peng et al. [6]. The study concluded that function, cost, technology and aesthetics play a vital role.

The BiPV system may or may not have a working fluid. PVT air systems can be used to produce electricity, thermal energy for space heating and night purging (cooling). The BiPVT system uses ventilated fluid as the working fluid, which collects the heat from the panel and releases it for thermal heating. Pantic et al. [7] examined a BiPVT system with three configurations: (i) unglazed BiPV with air flow (base case), (ii) 1500 mm vertical glazed solar air collector as an addition to the base case and (iii) a glazing over the PV module. It was concluded that the pre-heated air in case (i) was suitable for HVAC system and pre-heating of water. However, in case (ii) and (iii), significant improvement in thermal efficiency was attained. In the (iii) configuration, the generation of electrical power reduced due to extremely high temperature of PV modules. Agrawal and Tiwari [8] integrated glazed PVT/air collectors into a building for space heating or into a dryer for crop drying.

Li and Wang [9] assumed BiPV systems as building-integrated energy storage systems and divided them into three categories: (i) with solar battery, (ii) connected with grid and (iii) PV-Trombe wall. Corbin and Zhai [10] experimented with BiPVT systems with absorbers, thermal storage tanks and a pump to circulate the water to the absorbers. The study concluded that BiPVT systems with liquid-cooled tube fin absorbers located in the cavity attained 5.3% more electrical efficiency when compared to a naturally ventilated system. A BiPV Trombe wall with an air gap of 500 mm between the PV modules and the wall was assessed by Koyunbaba et al. [11]. The proposed system was comprised of two 0.8 m^2 vents located on the wall for winter heating. The electrical and thermal efficiency for the proposed system was found to be 4.5% and 20.3%, respectively.

8.3 TYPES OF PV INTEGRATIONS WITH BUILDINGS

The type of building integration (façade, roof, etc.) is an important parameter since it is associated with the efficiency of a solar system.

8.3.1 ROOFTOP

Integration of PV modules with the rooftop depends on several conditions. In tropical locations, they should be oriented south to have maximum exposed hours to solar radiations for the northern hemisphere. A north-oriented roof surface will not collect enough solar radiations for the system to operate, for it depends on the latitude as well. Thus, the location and orientation of PV modules is one of the major criteria to ensure higher productivity. Also, the irradiance value at the locations,

Building Integrated Photovoltaic Thermal Systems (BiPVT)

prevalent weather conditions, shading by nearby structures, trees etc. also determine the performance of the system. The design of the roof (flat, pitch, semi-pitch) will influence the performance of the rooftop BiPVT system. The roof pitch should be such that it does not have shade. The slope or the tilt angle of the roof is one of the major parameters that evaluates the performance of the system. The tilt angle should be equal to the latitude of the place for annual energy generation. The optimum tilt angle is calculated by adding 15° to your latitude during winter, and subtracting 15° from your latitude during summer.

8.3.2 FAÇADE

Integration of photovoltaics with façade is an effective approach for a highly urbanized city/state, where there is a limitation of rooftops but there is large surface area of façade of skyscrapers. Integration of SPVT systems with façade of the building has various advantages like the availability of a large vertical area for installation along with the combination of energy production, heat insulation and illumination. Thermal analysis of double-skin façade with BiPV panels was studied by Agathokleous and Kalogirou [12]. The outside air is allowed to enter the system from the bottom and escapes from the top when the BiPV system is integrated with facade. While moving up, the air absorbs the heat of the PV module, thus reducing its temperature. As a result, the solar cell efficiency improves due to reduction in solar cell temperature. In a few applications, to further improve the efficiency of the system, a fan or an air duct is employed. This may be done for both roof- and façade-integrating systems.

Integrating SPV modules with multifunctional PV façades has following advantages [13]:

- Generation of electricity at site and avoiding the losses coming from transportation of energy.
- Thermal envelope benefits can be attained with energy savings of about 15%–35%.
- Daylight savings.
- Acoustic benefits of multifunctional façades are comparable to conventional glass facades.
- Aesthetically pleasing.
- A simple structure and mounting system allows easy installations.
- A ventilated façade or a double-skin solution enables it to act as a protective layer against action of harmful atmospheric conditions.

However, this cannot be incorporated in all buildings since it depends on various requirements as listed below:

- Aesthetic requirements depend on the urban setting of the area.
- Dimensional requirements are focused on customization capacity, distance between multifunctional facade and existing solid wall, the electrical/storage equipment area and minimum area needed for the integration (minimum installed power: 1 kW).
- Functional requirements majorly focus on the shadow limitation, appropriate orientation and compatibility with existing insulating material or the state of electricity grid.

8.3.3 OTHER APPLICATIONS

PV systems can be integrated in many more innovative ways. A few examples are listed below:

- A PV skylight not only ensures electricity production and daylight but also ensures bioclimatic properties of thermal comfort inside the building. This is because the PV module material, which is a silicon base, absorbs most of the UV and IR radiations, and thus, these acts as a protective screen. Semi-transparent photovoltaic glass integrated as skylight for

Somerset Development's Bell Works Complex in Holmdel, New Jersey is one of the largest photovoltaic skylight in the world approximately 60,000 sq.ft. of area [14].
- Instead of using conventional glass in curtain walls, an SPV curtain wall can be installed, thus optimizing the performance of the building's envelope and in situ electricity generation. It should be kept in mind that the transparency of the photovoltaic glass should be sufficient enough to allow the entrance of natural light. Figure 8.1(a) illustrates an example of PV curtain wall in a future business centre at Cambridge comprised of 57 m² of area [15].

FIGURE 8.1 (a) PV Curtain wall, Cambridge [15], (b) PV panels used as external shading device [16], (c) PV parking shed [17], (d) PV streetlight [18], (e) Bench integrated with PV modules at D-21 Corporate Office, DMRC and (f) PV on the roof of pedestrian over-bridges [19]. Source: Saurenergy.com.

Building Integrated Photovoltaic Thermal Systems (BiPVT)

- Solar-shading devices, canopies or pergolas can be replaced by integrating them with PV modules. This solution will not only generate electrical power but also will provide protection against the adverse weather conditions. The additional energy produced by the system can be transferred to the grid, thus financial benefits can also be achieved. The parameters like orientation, minimum slope, dimensions, wind velocity, etc., should be kept in mind while designing such systems. Figure 8.1(b) gives an example of PV panels used as external shading devices in zero-energy building, Singapore. Another example is the PV pergola at Tanjong Pagar, Singapore. The pergola is over 2600 m^2 located at the entrance of the building with power capacity of 125 kWp made up of amorphous silicon photovoltaic glass modules [20].
- Parking sheds can also be integrated by PV modules and guarantee in situ generation of electricity, which can be supplied to batteries of electric cars (Figure 8.1c).
- Diverse PV urban mobility is another trend to integrate PV systems or infrastructure like benches, tables, street lights, etc. Incorporating PV in street furniture can help turn traditional outdoor furniture into charging points for electronic devices [21]. A USB charger in the furniture gets its power from the photovoltaic kit attached [20]. Figure 8.1(d) shows an image of a PV streetlight installed at a bus stop at Llandysiliogogo. The city of Las Vegas, Nevada, was the first city in the world that tested new EnGoPlanet Solar Street Lights, which are coupled with kinetic tiles that produce electricity when people walk over them. Figure 8.1(e) shows a bench integrated with PV modules along with a provision of a charging point.
- PV modules can also be integrated as a shading roof on pedestrian walkways, footpaths over bridges, etc. An example of such system can be seen at the North Sydney Coal Loader where monocrystalline glass has been installed to provide a shaded walkway. Another example can be seen in Figure 8.1(f), where solar panels have been installed over about 250 m over bridge [19]
- A very creative innovation is a PV walkable floor, which was developed by ONYX and is a patented floor. Photovoltaic tiles, triple laminated glazing units based on a-Si solar cells were integrated with floor. The PV tiles were anti-slippery and supported 400 kg in point test load [13]

8.4 BUILDING INTEGRATED OPAQUE PHOTOVOLTAIC SYSTEMS (BiOPV)

Opaque PV modules may be integrated with the building as rooftop, i.e., either grid (for annual electricity production) or off-grid (for seasonal electricity production), façade and PVT systems (rooftop and façade). The grid connection is only practical/feasible if grid power is available during the sunshine hour.

8.4.1 OPAQUE PHOTOVOLTAIC SYSTEM INTEGRATED WITH ROOFTOP

Various research has been carried out to understand the performance of building integration with opaque photovoltaic modules. Integration of opaque PV modules with rooftop is given in Figure 8.2.

Mainzer et al. [22] evaluated the potential of PV systems integrated with rooftop of a residential building for each municipality in Germany. It was found to be 148 TWh/a with an installation capacity of 208 GW_p. About 30% of the municipalities had the potential to become independent from the electrical grid, if a storage system equivalent to their 57% of daily electrical demand is installed. The performance of rooftop BiPV systems has been examined by Madessa [23] in Osloon. The study was based on the module-type, row spacing, tilt angle, electrical energy production, and environmental and economic impacts. Optimum tilt angle was found to be in the range of 30°–40°. Sadineni et al. [24] explored the influence of PV installation on 200 residences due to their orientation on the peak load and found that the 40° southwest orientation to be the most economically viable. About 38% of

FIGURE 8.2 Integration opaque PV modules with rooftop [25].

annual energy demand was found to be reduced. Othman and Rushdi [26] conducted a survey on the energy production with BiPV systems. The average daily electricity generation was dependent upon the number of modules, pitch angle, location and orientation, and it ranged from 11.18–29.18 kWh. Ordenes et al. [27] established in their study that 45% of the energy consumed can be produced by the installation of a PV module on the rooftop in Brazil. It has also been projected that the PV modules are capable of generating more energy than the building's demand for about 30% of the year.

8.4.2 Opaque Photovoltaic System Integrated with Façade

Multiple studies have been carried out to study the thermal behavior of BiOPVT with façade. Figure 8.3 is an example of a BiOPVT system with façade at the CIS Tower, Manchester, England. It started feeding electrical power to the National Grid in the year 2005 [28]. OPV panels were integrated with the solar-shading devices (sun shades) of the building in Korea by Yoo et al. [29]. The annual average efficiency of these solar shades were found to be 9.2% with a maximum of 20.2% in the winter months and a minimum of 3.6% in summer months. Ordenes et al. [27] have evaluated the effect of BiPV on the energy demand of multifamily dwellings in Brazil. The BiPV was integrated with each of the opaque surfaces of the building. The study concluded that the integration of PV modules with façade is appropriate for various latitudes, which is contrary to the common belief of its suitability only for high latitudes. A case study on an existing residential building was conducted by Memari et al. [30] to analyze the performance of BiPV system, and it was found to be about 12%–15% efficient when used as wall panels. Saranti et al. [31] integrated the PV system with shading devices for residential buildings concentrating on the optimization of the PV technology and the architectural design. The purpose was to fulfill the energy demands with regard to artificial lighting and other uses. Integration of opaque PV modules with the façade of a house in Taiwan was studied by Huang et al. [32]. The PV modules were attached to the ceramic tile and dry suspended on the wall. The configuration of the ceramic tiles and the dry, suspended method offered good insulation and the inside temperature was about 10°C higher than the temperature of the exterior façade. The temperature was found to be almost constant to the entire length of the façade with a 2°C difference between the top and bottom. Quesada et al. [33] reviewed the research related to opaque solar facades by dividing them into two subgroups, namely active and passive facades. The study concluded that both BiPV and BiPVT systems are favorable and substantial energy can be produced with higher efficiency due to the cooling effect offered by the air flowing behind the PV panels.

FIGURE 8.3 Building integrated opaque photovoltaic systems with façade at CIS Tower, Manchester [28].

8.5 BUILDING INTEGRATED SEMI-TRANSPARENT PHOTOVOLTAIC (BiSPVT) SYSTEM

Semi-transparent PV modules may be integrated with the building envelope to generate electricity via photovoltaic modules and allowing daylight to enter in the interior spaces. It may be integrated as rooftop or façade. Use of semi-transparent modules and their impact on human comfort was examined by [34].

Fung and Yang [35] found that the packing area of the SPV modules have a significant impact on the total heat gain. About 70% of total heat gain may be reduced by adjusting the solar cell area ratio to 0.8. Song et al. [36] characterized the power output of a transparent thin-film solar cell integrated with building. Optimum annual power performance of the system was found at a slope of 30° and south orientation. About 2.5 times higher power output was produced when compared with the module having vertical slope. Park et al. [37] studied the electrical and thermal performance of BiSPVT system and found that the property of glass have a significant impact on the module temperature and thus affecting its electrical performance. The study confirmed that PV module with a bronze glass for a winter day had a higher temperature when compared with a module having a clear glass.

8.5.1 Semi-transparent Photovoltaic System Integrated with Rooftop

Rooftop BiSPVT systems prove to be a promising technology for transition towards a low-carbon power and buildings sector due to their higher electrical efficiency and ability to provide additional daylighting [38]. Figure 8.4 is the image of the renovated head office of Alliander in Arnhem, Netherlands, where SPV modules are integrated with rooftop of the building. The PV cells seamlessly blend into the fabric of the glazing system and run centrally through the main building allowing natural light to enter.

FIGURE 8.4 Integration of SPV modules with the rooftop of a building [39].

Many studies including exergy and energy performance of BiSPVT systems have been carried out to understand the impact of semi-transparent PV modules integrated into the building with respect to the solar panel type, packing factor [40–42]. Li et al. [43] found that 1203 MWh of electricity can be saved annually with use of BiSPVT. Tiwari et al. [44] established a periodic model for BiSPVT systems for residential buildings. The study incorporated the effect of the number of air changes on the inside room air temperature, decrement factor and thermal load leveling for thermal comfort of the occupants. It has been found that the optimum roof thickness should lie within the range of 300–400 mm in order to minimize the decrement factor. The first floor was proposed to be used for the purpose of crop drying since the maximum inside temperature was 47°C during cold climatic conditions [42].

8.5.2 Facade-Building Integrated Semi-Transparent Photovoltaic (BiSPVT) System

Several studies have been carried out to understand the performance of semi-transparent PV modules integrated with a building's façade. Kamthania et al. [45] executed a case study on a double-pass semi-transparent PV system in India for winter months and found that the indoor temperature increased by 5°C–6°C. It was also seen that the semi-transparent PV module gave better electrical performance when compared to the opaque PV module. Thermal and electrical energy produced by the system was 480.81 kWh and 469.87 kWh, respectively. Memari et al. [30] performed a case study to analyze the integration of the BiSPVT system on an existing residential building and found that 5%–8% efficiency is achieved when used as a thin film contained within the windows. Quesada et al. [46] reviewed semi-transparent BiPV and BiPVT systems as active façade systems. Mercaldo et al. [47] analyzed architectural issues along with technology development of thin-film silicon photovoltaic. Chow et al. [48] evaluated the energy performance of "see-through" photovoltaic glazing

Building Integrated Photovoltaic Thermal Systems (BiPVT) 227

for an open-plan office building situated in Hong Kong. The proposed setup for a single-glazing PV system was able to achieve electricity savings of about 23% per annum. The single-glazed SPV system saved 23% of the electricity that was being consumed in space cooling per annum, whereas 28% was saved by the naturally ventilated double-glazing SPV system. Li et al. [43] studied the thermal, visual, energy and financial performance of BiSPVT systems for façade. Annual savings of 1203 MWh were achieved with reduction in peak cooling load of 450 kW.

8.6 BiOPVT AND BiSPVT SYSTEM ON ROOFTOP AND FAÇADE

The thermal behavior of BiPVT with roof and façade has been studied by various researchers [49–51]. Baljit et al. [52] compared BiPV and BiPVT technologies and discussed their installation on roof and façade. It was found that BiPVT systems are more efficient in terms of electricity production and thermal space heating. Nagano et al. [53] replaced a wall-mounted BiPVT system with a roof-mounted BiPVT system for cost-effectiveness and elimination of roof leakage due to snowfall. Buonomano et al. [54] analyzed the performance of BiPVT systems with facades/rooftops for residential buildings in European climates. Implementation of the BiPVT system produced a reduction of 67%–89% in the primary energy demand. The BiPVT system integrated with roof turned out to be more feasible economically when compared with its integration with the façade. The payback period for South European countries was estimated to be 11 years and the same for North European countries, which were calculated to be 20 years.

Vats and Tiwari [40] evaluated the performance of BiSPVT and BiOPVT systems for roof and façade with and without an air duct for cold climatic conditions of Srinagar, India. It was found that in case of the SPVT roof without an air duct, a maximum rise of 18°C in room temperature was recorded. In case of OPVT with an air duct, the same accounted for 2.3°C. On the other hand, the maximum room temperature was observed to be 22°C with SPVT integrated with roof without an air duct (ambient temperature = 4.4°C). The study also concluded that the rise in room temperature for both SPVT and OPVT with an air duct is greater than without an air duct due to the presence of insulation between the PV module and the air. It was also found that with an increase in air mass flow rate from 0.85–10 kg/s via duct, the room air temperature increases from 9.4°C to 15.2°C in the case of SPVT. The air temperature measured for SPVT and OPVT façade differed by 1.46°C, while it differed by 1.13°C in case of SPVT and OPVT roofs, The temperature difference between room air for SPVT and OPVT façade without an air duct was found to be 9.8°C. The same for SPVT and OPVT roofs was found to be 9.55°C.

8.7 USE OF PV MODULES IN AN URBAN SETTINGS

Energy distribution systems reduce the losses and enhance the efficiency of obtained energy. Energy is partially lost in the grid throughout electronic components, transformers, and long distribution lines. Thus, unnecessary integration of grid and backup resources on the grid with a BiPV system should be avoided.

Ondeck et al. [55] proposed integration of combined heat and power for a residential district in a hot climatic zone in order to reduce the carbon dioxide emissions and improve the efficiency of the power generation sector. The study found that the electricity produced from PV can be used in the morning when it is uneconomical to use gas turbine. Dávi et al. [56] evaluated the energy performance of a residential building integrated with a PV system (connected with the grid) in Brazil and found that the electrical demand varied from 29%–51%. Franco and Fantozzi [57] analyzed the combination of PV and a ground source heat pump for a residential building. Various studies have been done in order to study and examine the economic impact of PV system [58–60]. Darghouth et al. [59] estimated the savings in a bill if the PV system is implemented for 226 households in California under two different types of metering. Lang et al. [61] have reviewed the economic performance of rooftop PV modules for residential buildings globally. Performance of on-grid BiPV system was evaluated by Wittkopf et al. [62]. It was grouped into 22 arrays with different orientation and tilt angles. An array consisted of two parallel strings with 18 modules in a series. The average monthly energy generated was estimated to be 12.1 MWh.

Further, examples for installation of BiOPVT and BiSPVT systems around the globe are discussed in detail in Chapter 11.

8.8 ENERGY AND EXERGY ANALYSIS OF BiSPVT SYSTEM

Figure 8.5 is a schematic 3-D representation of semi-transparent PV modules integrated with the roof of a building. Thermal modeling of the BiSPVT system has been done considering four cases: (a) BiSPVT system, (b) BiSPVT system with water flow, (c) BiSPVT system with heat capacity and (d) BiSPVT system with heat capacity and water flow for different climatic conditions. Various literatures are available on energy and exergy performances of BiPVT and BiSPVT systems, but none have considered the impact of daylighting, natural ventilation, water mass and water-cooled SPV modules on one system integrated with a semi-transparent photovoltaic thermal system along with their individual effect. The solar insolation and ambient temperature for different places is given in Appendix C. Thermal modeling, comparative analysis and performance evaluation for all the cases has been done for similar input parameters to understand the impact of each additional element in different cases and discussed the subsequent sections.

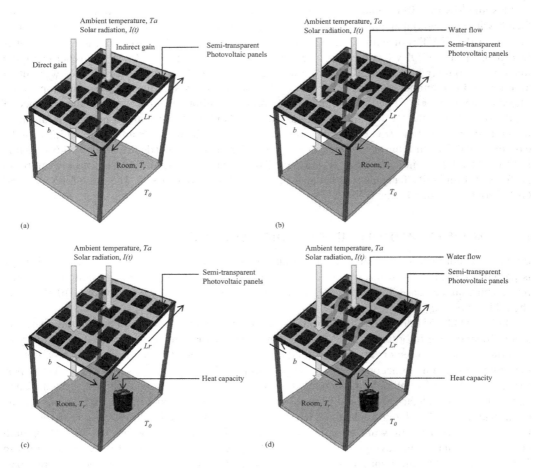

FIGURE 8.5 (a) BiSPVT system, (b) BiSPVT system with water flow, (c) BiSPVT system with heat capacity, (d) BiSPVT system with water flow along with heat capacity [63].

Building Integrated Photovoltaic Thermal Systems (BiPVT)

The proposed cases are valid for all the discussed climatic and weather conditions. The proposed BiSPVT system can be used for both thermal heating (system without the water flow over the PV modules) for cold climatic conditions and thermal cooling (with water flow over the PV modules) for hot climatic conditions. PV modules have been integrated at the rooftop.

8.8.1 Working Principle

The solar radiation enters the room under study after reflection from the top surface through the non-packing area of the semi-transparent photovoltaic modules, $(1 - \beta)A_m$, in all the four cases (Figure 8.5). This gain is referred to as direct gain (as discussed in Chapter 6) and is also responsible for natural light to penetrate inside the room. Most of the transmitted radiations are absorbed by the floor of the room in all cases, and the rest are absorbed by the water tank placed (for cases c and d). In addition to this, thermal energy from the back side of the PV module is transferred to the floor of the room through convective heat transfer. The floor temperature rises, which leads to an increase in the room air temperature. BiSPVT systems are best suited for cold climatic conditions due to high temperatures, thus in order to lower the temperature range, cases (b), (c) and (d) are equipped with passive cooling techniques, i.e., water flow over SPV modules and water mass (or heat capacity).

In case (b), the water flowing over the SPV roof helps cool down the SPV panels and the rooms beneath roof integrated with SPV panels. While flowing over the hot roof, the heat from SPV panels is transferred to water via convection, thus the temperature of water rises and it evaporates. Taking advantage of evaporative cooling effect helps to improve the solar cell electrical efficiency and reduce the room temperature (Figure 8.5(b)).

In case (c), a water tank has been placed inside the room which acts as a heat storage system (Figure 8.5(c)). This water mass as discussed absorbs some part of the solar radiation entering the room through the non-packing area and from the room (both direct and indirect gain as discussed in case (a), the base case) during the sunshine hours and releases the heat in off-sunshine hours. Since solar radiations are periodic in nature, fluctuations in the room temperature are noticed; this thermal mass helps to reduce these fluctuations and reduce the thermal load leveling (TLL) for the proposed system. Further, the thermal heat produced by the system is used to heat the water inside the tank, and this pre-heated water can be then re-heated by conventional means, thus saving energy. Case (d) as shown in Figure 8.5(d) is a combination of cases (b) and (c), and thus incorporates all the added advantages of both the cases, resulting in an enhanced system with improved room temperature and increased solar cell electrical efficiency.

8.8.2 Thermal Modeling

Thermal modeling for the proposed cases has been carried out. A thermal model is basically a mathematical model which dynamically predicts the temperature of various components of the building or space under study. The accuracy of this model depends upon the algorithm and the precision of the input parameters. Thus, to minimize the energy consumption of the space, it becomes necessary to understand the dynamics of energy production and losses. For a BiPVT system, sources of heat gains include the solar radiation, heat-generating equipment installed inside the building.

The thermal modeling of the BiSPVT system has been based on following assumptions:

- Thermal side losses have been ignored.
- Steady-state heat conduction has been assumed for the floor of the room.
- Glass thickness of the solar cell (both top and bottom surface) is very small. Thus, solar cell temperature has been assumed to be equal to glass temperature.
- No temperature gradient has been assumed along the thickness of the SPV module and the room air column.
- Air changes are considered to be constant for other given parameters.
- Room air temperature does not vary with mass flow rate of water.

230　　　　　　　　　　　　　　　　　　Photovoltaic Thermal Passive House System

8.8.3　Basic Energy Balance Equations

I. The energy balance equation for the roof-integrated with SPV modules of the system can be expressed as:

(a) For the system which has a bare SPV roof integrated, i.e., there is no water flow (valid for cases (a) and (c) [64]

$$\alpha_c \tau_g I(t) \beta A_m = \left[U_{tca}(T_c - T_a) + U_{bcr}(T_c - T_r) \right] A_m + \tau_g I(t) \beta A_m \eta_c \qquad (8.1)$$

where,

α_c is the absorptivity of solar cell (dimensionless)
τ_g is the transmissivity of the glass (dimensionless)
$I(t)$ is the solar intensity in W/m^2
β is the packing factor (dimensionless)
A_m is the area of the module in m^2
U_{tca} is the overall heat transfer coefficient from solar cell to ambient through glass

cover in W/m^2K, $U_{tca} = \left[\dfrac{1}{h_0} + \dfrac{L_g}{K_g} \right]^{-1}$

U_{bcr} is the overall heat transfer coefficient from solar cell to room through glass

cover in W/m^2K, $U_{bcr} = \left[\dfrac{1}{h_i} + \dfrac{L_g}{K_g} \right]^{-1}$

h_0 and h_i are the outside and inside heat transfer coefficient, respectively. in W/m^{2}K
L_g and K_g are the thickness and thermal conductivity of the glass in m and W/mK, respectively
T_c, T_a, T_r are the solar cell, ambient and room air temperatures, respectively, in °C
η_c is the solar cell electrical efficiency (dimensionless)

$$\begin{bmatrix} Rate\ of\ solar \\ energy\ available \\ on\ solar\ cells \end{bmatrix} = \begin{bmatrix} Rate\ of\ heat \\ loss\ from\ top \\ surface\ of \\ solar\ cells\ to \\ ambient\ through \\ glass\ cover \end{bmatrix} + \begin{bmatrix} Rate\ of\ heat \\ transfer\ from \\ solar\ cell \\ to\ room\ air \end{bmatrix} + \begin{bmatrix} Rate\ of \\ electrical \\ energy \\ produced \end{bmatrix}$$

Equation (8.1) can be explained as:

(b) For the system, which has an additional water flow over the SPV-integrated roof to lower the solar cell temperature (valid for cases b and d) [65, 66]

$$\alpha_c \tau_g I(t) \beta b dx = \left[U_{tcw}(T_c - T_{w1}) + U_{bcr}(T_c - T_r) \right] b dx + \tau_g I(t) \beta \eta_c b dx \qquad (8.2)$$

where,

b is the width of the roof in m
dx is the elemental length in m
U_{tcw} is the overall heat transfer coefficient from solar cell to water flowing on the roof in W/m^{2}K
T_{w1} is the temperature of the water flowing on the roof in °C

Building Integrated Photovoltaic Thermal Systems (BiPVT)

$$\begin{bmatrix} \text{Rate of solar} \\ \text{energy available} \\ \text{on solar cells} \end{bmatrix} = \begin{bmatrix} \text{Rate of heat} \\ \text{transfer from} \\ \text{solar cells} \\ \text{to water flowing} \end{bmatrix} + \begin{bmatrix} \text{Rate of heat} \\ \text{transfer from} \\ \text{solar cell} \\ \text{to room air} \end{bmatrix} + \begin{bmatrix} \text{Rate of} \\ \text{electrical} \\ \text{energy} \\ \text{produced} \end{bmatrix}$$

Equation (8.2) can be explained as:

In cases when there is no water flow (Equation 8.1), the heat is transferred from the solar cell to the ambient, whereas with additional flow of water (Equation 8.2), the heat is transferred from the solar cell to the flowing water. In the latter case, due to the flow of water, the solar cell temperature reduces, further improving the solar cell electrical efficiency.

(c) For cases (b) and (d), energy balance equation for an elemental area bdx with water flow over the roof integrated with SPV modules (Figure 8.6) is expressed as:

$$\dot{m}_{w1} C_w \frac{dT_{w1}}{dx} dx = U_{tcw}(T_c - T_{w1}) bdx - (h_{ew1} + h_{cr})(T_{w1} - T_a) bdx \tag{8.3}$$

where,

$\dot{m}_{w1}$ is the mass flow rate in kg
C_w is the specific heat of water in J/kgK
T_{w1} is the temperature of water flowing on the roof in °C
h_{ew1} is the evaporative heat transfer coefficient in W/m²K
h_{cr} is the convective heat transfer coefficient to room in W/m²K

$$\begin{bmatrix} \text{Rate of} \\ \text{heat} \\ \text{transfer} \\ \text{to water} \end{bmatrix} = \begin{bmatrix} \text{Rate of heat} \\ \text{transfer} \\ \text{from solar cell} \\ \text{to water} \end{bmatrix} - \begin{bmatrix} \text{Rate of heat} \\ \text{transfer} \\ \text{from water} \\ \text{to ambient air} \end{bmatrix}$$

Equation (8.3) can be explained as:

Arranging Equation (8.3) as a first-order differential equation to evaluate temperature of the water flowing over SPV modules integrated with the building. Consider Lr as length of the roof in meters.

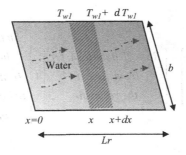

FIGURE 8.6 Cross-section of elemental area bdx [63].

$$\frac{dT_{w1}}{dx} + aT_{w1} = f(x) \tag{8.4}$$

In Equation (8.4), water temperature, T_{w1}, directly depends on x. Solar cell temperature directly depends on the water temperature. Other output parameters like room temperature, floor temperature and solar cell electrical efficiency are dependent on the solar cell temperature. Thus, it is important to consider x as it has a direct or indirect effect on all the output parameters.

To determine the water temperature of the system, the solution for Equation (8.4) becomes

$$\bar{T}_{w1} = \frac{1}{Lr} \int_0^L T_{w1} dx = \frac{f(t)_1}{a_1} \left[1 - \frac{1 - e^{-a_1 Lr}}{a_1 Lr} \right] + \left[\frac{1 - e^{-a_1 Lr}}{a_1 Lr} \right] T_{w1,0} \tag{8.5}$$

Further details can be referred from [65].

If the numeric value of mass flow rate of water, $\dot{m}_{w1}$, is substituted as 0 (zero), then the value for $a_1 = \infty$. This means there is no water flowing; thus, the value of $T_{w1} = T_c$.

Therefore, solar cell temperature can be derived from Equation (8.5) for the system without water flow (cases a and c) as given in Equation (8.6)

$$T_c = \frac{\tau_g I(t) \beta (\alpha_c - \eta_c) + U_{tca} T_a + U_{bcr} T_r}{\left(U_{tca} + U_{bcr} \right)} \tag{8.6}$$

II. Energy balance equation for floor of the room will remain the same for all cases and is given as:

$$\alpha_R \tau_g^2 (1 - \beta) I(t) A_m = h_c A_f (T_f - T_r) + U_b (T_f - T_o) A_f \tag{8.7}$$

$$U_b = \left[\frac{1}{h_c} + \frac{L_R}{k_R} + \frac{1}{h_i} \right]^{-1}$$

where,

α_R is the absorptivity of the roof (dimensionless)
h_c is the convective heat transfer coefficient in W/m²K
A_f is the area of the floor in m²
T_f is the floor temperature in °C
T_o is the air temperature below the room under study °C
L_R and K_R are the thickness and thermal conductivity of the roof in m and W/mK, respectively

$$\begin{bmatrix} Rate\ of\ solar \\ energy\ absorbed \\ by\ roof \end{bmatrix} = \begin{bmatrix} Rate\ of\ heat\ transfer \\ from\ floor\ to \\ the\ room\ air \end{bmatrix} + \begin{bmatrix} Steady\text{-}state \\ heat\ loss \\ from\ floor\ to \\ to\ bottom\ slab \\ of\ room \end{bmatrix}$$

Equation (8.7) can be explained as:

Using Equation (8.7), floor temperature can be derived as:

$$T_f = \frac{\alpha_R \tau_g^2 (1-\beta) I(t) A_m + U_b A_f T_0 + h_c A_f T_r}{(h_c + U_b) A_f}$$ (8.8)

III. For a BiSPVT system with heat capacity (Figure 8.5(c) and (d)), the energy balance equation for tank surface temperature, (T_t), can be given by [67]:

$$A_f h_c (T_f - T_r) + U_{bcr} (T_c - T_r) A_m = h_0 A_t (T_r - T_t)$$ (8.9)

where,

A_t is the area of the water drum in m^2
T_t is the temperature of the water drum in °C

$$\begin{bmatrix} Rate\ of\ heat \\ transfer \\ from\ floor\ to \\ the\ room\ air \end{bmatrix} + \begin{bmatrix} Rate\ of \\ heat\ transfer \\ from\ solar \\ cell\ to \\ room\ air \end{bmatrix} = \begin{bmatrix} Rate\ of\ heat \\ transfer\ from \\ room\ to \\ water\ tank \end{bmatrix}$$

Equation (8.9) can be explained as:

The energy-balance equation for transfer of heat from the surface of water tank to the water inside is given by:

$$F\alpha_b (1-\beta) \tau^2 I(t) + h_0 A_t (T_r - T_t) = h_w A_t (T_t - T_w)$$ (8.10)

where,

α_b is the absorptivity of the blackened surface (dimensionless)

$F = \dfrac{Cross\text{-}sectional\ area\ of\ the\ drum}{Area\ of\ the\ room}$ is the fraction of solar radiation absorbed by the water drum

h_w is the heat transfer coefficient of tank plate to water in W/m^{2}K

Thus, the energy balance equation for mass of water (heat capacity) can be expressed as:

$$M_w C_w \frac{dT_w}{dt} = h_w A_t (T_t - T_w)$$ (8.11)

$$\begin{bmatrix} Rate\ of\ energy \\ absorbed\ by \\ water\ in\ tank \end{bmatrix} = \begin{bmatrix} Rate\ of\ heat\ transfer \\ from\ tank\ surface \\ to\ the\ water \end{bmatrix}$$

Equation (8.11) can be explained as:

IV. Energy balance equation for the room temperature is given by:

$$A_f h_c \left(T_f - T_r\right) + U_{bcr}\left(T_c - T_r\right)A_m - 0.33NV\left(T_r - T_a\right) = M_a C_a \frac{dT_r}{dt} \tag{8.12}$$

where

M_a is the mass of the air in kg
C_a is the specific heat of the air in J/kgK

$$
\begin{bmatrix} Rate\ of\ heat \\ transfer \\ from\ floor\ to \\ the\ room\ air \end{bmatrix} + \begin{bmatrix} Rate\ of \\ heat\ transfer \\ from\ solar \\ cell\ to \\ room\ air \end{bmatrix} - \begin{bmatrix} Rate\ of\ thermal \\ energy\ withdrawn \\ from\ room\ to \\ ambient \end{bmatrix} = \begin{bmatrix} Rate\ of \\ thermal\ energy \\ absorbed\ by \\ room \end{bmatrix}
$$

Equation (8.12) can be explained as:

The term $[0.33NV(T_r - T_a)]$ in Equation (8.12), governs the heat transfer from room air to ambient in natural mode, i.e., $N \leq 10$ as explained in Chapter 1

V. Following [44], solar cell electrical efficiency can be expressed as:

$$\eta_c = \eta_0 \left[1 - \beta_0 (T_c - T_{ref})\right] \tag{8.13}$$

where

η_0 is the electrical efficiency at standard test condition (dimensionless)
β_0 is the temperature coefficient in $°C^{-1}$
T_{ref} is the reference temperature at standard test conditions in $°C$

VI. The rate of electrical energy $(\dot{E}_e)$ produced by the BiSPVT system in watts is a high-grade energy that can be evaluated using the following expression:

$$\dot{E}_e = \eta_c \tau_g \alpha_c I(t) A_m \beta \tag{8.14}$$

From Equation (8.14), it can be seen the efficiency of the solar cell and packing factor has a significant impact on the electrical energy generation.

VII. An exergy of solar radiation through non-packing area $[(1-\beta)A_m]$ of SPV module [68, 69] may be evaluated for all the cases and is expressed as:

$$\dot{E}x_{dl} = \left\{ A_m I(t)\left(1 - \beta\right) \right\} \left[1 - \frac{4}{3}\left(\frac{T_a}{T_s}\right) + \frac{1}{3}\left(\frac{T_a}{T_s}\right)^4 \right] \tag{8.15}$$

where $\dot{E}x_{dl}$ is the rate of exergy of solar energy through the non-packing area of the semi-transparent photovoltaic module in W, and T_a and T_s are the ambient and sun-surface temperature, respectively, in K.

Building Integrated Photovoltaic Thermal Systems (BiPVT)

The entrance of natural light inside the building is one of the major advantages of BiSPVT systems over BiOPVT systems, thus reducing the building's demand on artificial lighting. This natural light reduces the dependence on artificial lighting resulting into daylight savings, thus making the system sustainable. Moreover, natural light also has positive psychological effects on human behavior and performance. These savings depend on the packing factor of the PV module, and it should be minimal.

Equation (8.15) can be expressed in terms of illumination as:

$$Illumination = \left[\left\{ A_m I(t)\, (1-\beta) \right\} \left[1 - \frac{4}{3}\left(\frac{T_a}{T_s}\right) + \frac{1}{3}\left(\frac{T_a}{T_s}\right)^4 \right] \right] \times 100\,\text{lux} \qquad (8.15a)$$

As we know the illumination for different activities inside the room is different, hence the illumination level depends on level of solar radiation transmitted through non-packing factor of SPV module. For example, for packing factor equal to 1 (i.e., $\beta = 1$), the illumination will be zero as can be observed from Equation (8.15a). Further, for $\beta = 0$, there will be the highest level of illumination, which can be used for any type of activity [70]. Hence, one can optimize the packing factor for any activity of human beings inside the room.

Input exergy: An exergy of solar radiation on PV module [A_m] [68, 69] is the net input exergy, which is given in Equation (4.26)

VIII. Exergy analysis

As discussed in Chapter 4, thermal energy available from the floor and from the rear of the PV module to the room are low-grade energy and needs to be converted to high-grade energy to assess the overall exergy of the system using the second law of thermodynamics for high operating temperature or the first law of thermodynamics for medium operating temperature (Refer Section 4.4.3.1).

Now, total exergy can be determined by adding electrical power (Equation 8.14), thermal exergy (either Equation (4.27) or (4.28) and daylighting (8.15) and can be expressed as:

$$\text{Net output exergy} = \dot{E}_e + \dot{E}x_{thermal} + \dot{E}x_{dl} \qquad (8.16)$$

Thermal exergy efficiency ($\eta_{th, ex}$) and an overall exergy efficiency ($\eta_{o, ex}$) of BiSPVT can be evaluated as follows:

$$\text{Thermal exergy efficiency, } \eta_{th,ex} = \frac{\dot{E}x_{thermal}}{\dot{E}x_{sun}}\,\% \qquad (8.17)$$

$$\text{Overall exergy efficiency, } \eta_{0,ex} = \frac{\dot{E}_e + \dot{E}x_{thermal} + \dot{E}x_{dl}}{\dot{E}x_{sun}}\,\% \qquad (8.18)$$

IX. Thermal efficiency

Rate of hourly thermal energy for the BiSPVT system (Case a) in Watts can be expressed as:

$$\dot{Q}_{u,th} = A_f h_c \left(T_f - T_r \right) + A_R h_c \left(T_c - T_r \right) \qquad (8.19)$$

$$\begin{bmatrix} Total\ hourly \\ thermal\ energy \end{bmatrix} = \begin{bmatrix} Thermal\ energy \\ from\ the\ floor\ to\ room \end{bmatrix} + \begin{bmatrix} Thermal\ energy \\ from\ the\ solar\ cell \\ to\ room \end{bmatrix}$$

Equation (8.19) can be explained as:

The average monthly thermal energy (case c) in watts can be expressed as:

$$Q_u = M_w C_w \left(T_{w,max} - T_{w,min} \right) \qquad (8.20a)$$

where,

M_w is the mass of water in the drum in kg

C_w is the specific heat of the water in the drum in J/kgK

$T_{w,\,max}$ and $T_{w,\,min}$ are the maximum and minimum temperature of the water in the drum, respectively, in °C

Referring to Section 4.4.3, equivalent high-grade energy for thermal exergy in Watts can be evaluated as follows:

$$Ex_{thermal} = M_w C_w \left[\left(T_{w,max} - T_{w,min} \right) - \left(\overline{T}_a + 273 \right) \ln \left(\frac{T_{w,max} + 273}{T_{w,min} + 273} \right) \right] \qquad (8.20b)$$

$Ex_{thermal}$ is the total thermal exergy in W

$\overline{T}_a$ is the average ambient temperature in °C

Therefore, thermal exergy efficiency can be expressed as:

$$\eta_{th,ex} = \frac{\dot{Ex}_{th}}{\dot{Ex}_{sun}} \qquad (8.21a)$$

An overall hourly thermal efficiency becomes:

$$\eta_{th} = \frac{\dot{Q}_{u,th}}{I(t) A_R} \qquad (8.21b)$$

PROBLEM 8.1

Calculate the daylight savings for BiSPVT systems with design parameters as mentioned in Table 8.1 at 6:00 hours for the month of January in New Delhi, India.

Solution

Using Appendix C, Equation (8.15) and Table 8.1,

$$\dot{Ex}_{dl} = \left\{ A_m I(t) \left(1 - \beta \right) \right\} \left[1 - \frac{4}{3} \left(\frac{T_a}{T_s} \right) + \frac{1}{3} \left(\frac{T_a}{T_s} \right)^4 \right]$$

Building Integrated Photovoltaic Thermal Systems (BiPVT)

$$\dot{Ex}_{dl} = \left\{ 9.18 \times 100 \times (1-0.89) \right\} \left[1 - \frac{4}{3} \left(\frac{280.6}{6000} \right) + \frac{1}{3} \left(\frac{280.6}{6000} \right)^4 \right]$$

Therefore, $\dot{Ex}_{dl} = 94.291 \, \text{W}$

PROBLEM 8.2

Calculate the energy savings in kWh for an average level of solar radiation transmitted inside the room for the number of sunshine hours $(N') = 4$ hours for New Delhi, India, for a time interval between 10:00 to 13:00 hours. Design parameters are given in Table 8.1.

Solution

Using Equation (8.15), Appendix C and Table 8.1,

$$\dot{Ex}_{dl} = \left\{ A_m \, I(t) \, (1-\beta) \right\} \left[1 - \frac{4}{3} \left(\frac{T_a}{T_s} \right) + \frac{1}{3} \left(\frac{T_a}{T_s} \right)^4 \right]$$

$$\text{Energy savings in kWh} = \left[\left\{ A_m \, \overline{I(t)} \, (1-\beta) \right\} \left[1 - \frac{4}{3} \left(\frac{\overline{T_a}}{T_s} \right) + \frac{1}{3} \left(\frac{\overline{T_a}}{T_s} \right)^4 \right] \right] \times \frac{N'}{1000}$$

where $\overline{I(t)}$ is the average solar radiation in W/m^2, and $\overline{T_a}$ is the average ambient temperature in K.

$$\text{Energy savings} = \left[\left\{ 9.18 \times 627.25 \times (1-0.89) \right\} \left[1 - \frac{4}{3} \left(\frac{289.8}{6000} \right) + \frac{1}{3} \left(\frac{289.8}{6000} \right)^4 \right] \right] \times \frac{4}{1000}$$

Energy savings = 2.3704 kWh

8.8.4 Comparative Statement of Proposed Cases (a-d)

The above cases and Figure 8.5 are studied under same climatic conditions, i.e., hot climatic conditions and design parameters as given in Table 8.1. Hourly variation of solar intensity, $I(t)$, and ambient air temperature, T_a, for a typical summer day, New Delhi, India, are given in Appendix C.

The study has been done for a room of 9.18 m^2 integrated with semi-transparent photovoltaic modules with roof to understand the impact of every individual parameter and establish a comparison between all the four cases. This will help to conclude the selection of case as per the requirements and the best solution amongst the proposed cases for thermal heating and thermal cooling purpose.

Using the basic energy balance equations written under Section 8.8.3, Figure 8.7(a–d) shows the hourly variation of room, solar cell and ambient temperature along with solar cell electrical efficiency produced by the proposed cases for similar input parameters as described previously. The details of the results obtained can be referred from [63]. The system has also been analyzed for electrical energy produced by all cases and daylight savings with the impact of packing factor.

TABLE 8.1

Design Parameters of Building Integrated Semi-transparent Photovoltaic Thermal (BiSPVT) System

Parameters	Values	Parameters	Values	Parameters	Values	Parameters	Values
A_f	8.86 m²	h_w	100 W/m²K	M_c	64.11 kg	ρ_c	2328 kg/m³
$A_m = A_R$	9.18 m²	K_g	0.9 W/mK	T_{ref}	25°C	$\rho_{concrete}$	2400 kg/m³
b	2.5 m	K_{in}	0.035 W/mK	T_s	6000 K	ρ_{air}	1.2 kg/m³
C_a	1005 J/kgK	K_R	0.67 W/mK	U_{tcw}	50 W/m²K	ρ_w	1000 kg/m³
C_c	700 J/kgK	L_c	0.0003 m	α_c	0.9	τ_g	0.9
C_f	960 J/kgK	L_{in}	0.01 m	α_R	0.4	Room height	3 m
C_w	4190 J/kgK	L_g	0.003 m	β	0.89	Roof inclination	15°
h_o	(5.7 + 3 V) W/m²K	Lr	3.67 m	β_0	0.0045°C⁻¹		
h_c	2.8 W/m²K	L_R	0.6 m	γ	50%		
h_{cr}	5.7 W/m²K	M_a	33.6 kg	η_0	0.15		
h_i	(2.8 + 3 V) W/m²K	M_f	3190 kg	λ	2.25 × 10⁶ J/kg		

		Case specific values					
Parameters	Comparative analysis	Case a	Case b	Case c	Case d	Case e
M_w	600 kg	NA	NA	200 kg	600 kg	600 kg
$\dot{m}_{wl}$	0.01 kg/s	NA	0.01 kg/s	NA	0.01 kg/s	NA
N	1	0	1	0	1	4
T_0	20°C	37.5°C	20°C	20°C	20°C	20°C
v	1 m/s	0 m/s	1 m/s	0 m/s	1 m/s	0 m/s
Climatic conditions	New Delhi (June)	Varanasi (January)	New Delhi (June)	Srinagar (January)	New Delhi (June)	Srinagar (January)

The maximum room air temperature and solar cell temperature obtained for case (a) is achieved as 87.03°C and 102.51°C (Figure 8.7(a)) at noon hours. As discussed in the working principle, the rise in room temperature is due to the direct gain through the non-packing area and indirect gain. To reduce this increase in room temperature and make the BiSPVT system appropriate for hot climatic conditions, various measures have been studied and discussed under cases (b), (c) and (d). Referring to Figure 8.7(b), there is an increase in the solar cell electrical efficiency by about 27.83% in case (b) in comparison to the base case because of the decrease in solar cell temperature by 38.60% due to flow of water over them at peak hours. Compared to the base case (i.e., case (a)), the room air and solar cell temperature reduces by approx. 12.59% and 4.67%, respectively, with an increase in solar cell electrical efficiency by 4.12% at peak hours due to an additional thermal mass (case c) since the water mass behaves like a heat storage medium during sunshine hours and releases the same in off-sunshine hours. The thermal load leveling (Equation 3.36) reduces by 39.74% when compared with case (a) resulting in better indoor thermal conditions. The results show that water mass is not a sensitive parameter for solar cell temperature and solar cell electrical efficiency and is majorly responsible for reduction in the room air temperature. Further, it can be seen that the room temperature rises during off-sunshine hours when a storage tank is kept in the room (case c) as expected. Incase thermal heating is required during night time, a BiSPVT system with heat capacity should be considered. Case (d) is an effective technique, which incorporates the advantages associated with both cases (b) and (c). Thus, owing to the combined advantages of evaporative cooling and water mass, the drop in room and solar cell temperature at peak hours with reference to case (a) has been found to be highly significant, i.e., 44.62% and 40.06%, respectively, with an increase in solar cell electrical efficiency by 28.86%.

Building Integrated Photovoltaic Thermal Systems (BiPVT)

Using Equation (8.14), hourly variations of electrical energy produced by all the cases have been calculated and found to be 6336.49 W for case (a), 7571.89 W for case (b), 6463.51 W for case (c) and 7641.09 W for case (d). It can be seen that case (d) generates maximum electrical energy followed by cases (b), (c) and (a). Daily electrical energy produced by case (d) is approximately 1304.6 Wh greater than the daily electrical energy production by case (a) because of lower solar cell temperature and higher electrical efficiency of case (d).

The major influencing parameter for the daylight savings is the packing factor of the SPV modules. From Equation (8.15), it is clear that the daylight savings for all the proposed cases with

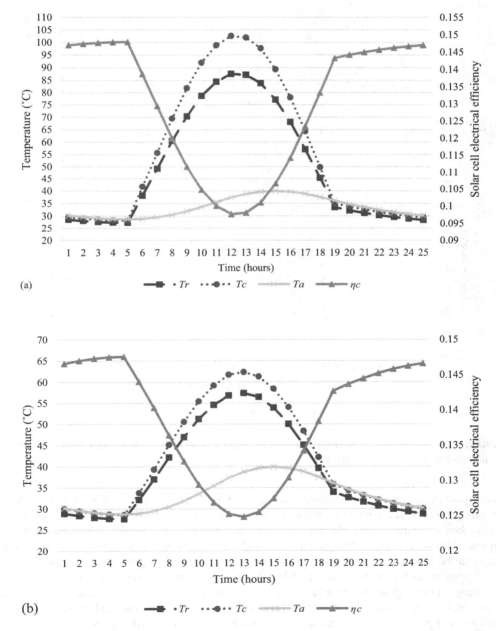

FIGURE 8.7 (a) BiSPVT system only (Case a) [63], (b) BiSPVT system with water flow (Case b) [63].

(*Continued*)

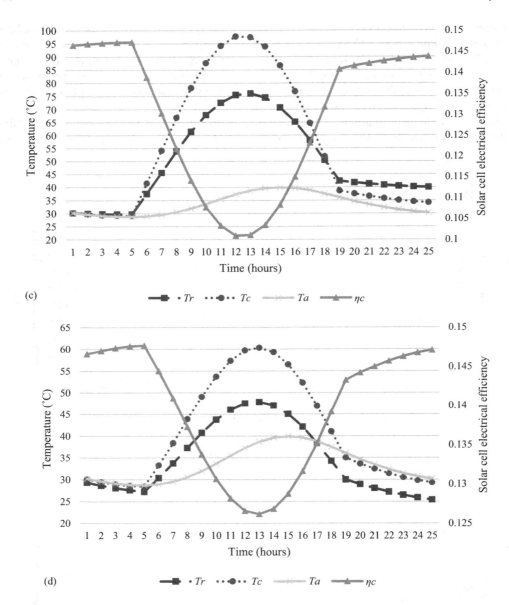

FIGURE 8.7 *(Continued)* (c) BiSPVT system with heat capacity (Case c) [63], (d) BiSPVT system with water flow and heat capacity (Case d) [63].

the same packing factor will be the same since the water mass and the water flow has no impact on the daylight savings. If the packing factor approaches the value 1, the system becomes opaque and will not admit any natural light. The non-packing factor is responsible for the solar radiation to enter the room. With a decrease in the packing area and an increase in the non-packing area, daylight savings can be increased. For packing factor equal to 0.89, the daylight savings for the proposed BiSPVT system has been estimated to be 8485 Wh with peak daylight savings equal to 979.48 Wh. With change in the packing factor from 0.89 to 0.225, the daylight savings, increased exceptionally high at peak hours by about 5921.4 Wh. Daily daylight savings can be increased by 51.29 kWh with optimization of packing factor from 0.89 to 0.225 for the proposed BiSPVT system.

Building Integrated Photovoltaic Thermal Systems (BiPVT) 241

Thus, it can be concluded that for thermal heating, case (a) should be considered, and for thermal cooling, case (d) should be used. However, in case of night heating, a BiSPVT system with heat capacity is preferred.

8.9 PERFORMANCE EVALUATION OF THE PROPOSED SYSTEMS

All the proposed cases have been assessed based on the basic energy balance equations and design parameters as given in Table 8.1 unless mentioned. The results are given in Table 8.2.

8.9.1 FOR BiSPVT SYSTEM (CASE a)

The results of the system BiSPVT system (case a) have been given in Table 8.2(a) for the climatic conditions of Varanasi in the month of January. The proposed system has been evaluated for both energy and exergy performance.

The high room temperature received is because it has been assumed that there are no thermal heat losses from the walls, and thus, the thermal energy is stored during peak hours. Such systems can be used for thermal heating of a building along with electrical energy demand for the same building. With increase in the number of air changes from 0 to 4, room air temperature was found to be within the comfortable range, and excess hot air available can be used for thermal heating of another room in the building. Electrical energy and daylight savings contribute majorly to the overall daily exergy in comparison with thermal exergy.

8.9.2 FOR BiSPVT SYSTEM WITH WATER FLOW (CASE b)

Table 8.2(a) summarizes the results for the BiSPVT system with water flow. To understand the impact of water flow over the BiSPVT system, a comparative analysis has also been done under similar design parameters for case (b). It has been found that the room and solar cell temperature drops by between 32.18% and 62.94% with an increase in the solar cell efficiency by 27.87% at noon hours due to the flow of water over the SPV modules. Annual variation of room temperature for proposed BiSPVT system with water flow has also been done, and it has been found that peak room air temperature is maximum for the month of May (58.73°C), due to high insolation levels, and minimum for the month of January (40.83°C), due to lower insolation levels. The results are based on the dynamic model. The room air temperature can be further reduced by placing heat storage inside the room. Floor material does not have a significant effect on the room air temperature due to bottom losses. Also, due to less heat capacity of the wall and roof material, no significant effect on the room air temperature can be observed.

8.9.3 FOR BiSPVT SYSTEM WITH HEAT CAPACITY (CASE c)

As discussed, heat capacity (or water mass) acts as a storage medium of heat during daytime and reduces the fluctuations of room temperature. The results (Table 8.2a) show that peak room temperature has been attained as 43.17°C at $N = 0$, and this can be reduced by increasing the number of air changes. At $N = 4$, peak room temperature according to the model will be 14.7°C. Peak water temperature has been achieved at 16:00 hours unlike the rest of the peak temperatures at noontime. This is due to the heat capacity of the water mass creating a time lag; beyond this hour, there has been no change in the water mass. An annual analysis has also been done and has been noted that TLL is more in winter months for heating purposes, which is as per our expectations. Further, TLL decreases by 35.5% for a water mass of 400 kg from December to June.

Average thermal energy and average thermal exergy is maximum in the summer months and minimum in the winter months. Electrical and daylight energy will be less in winter months

TABLE 8.2(a)

Performance Analysis for Proposed Cases (a–c)

Case	T_r* (°C)	T_f* (°C)	T_c* (°C)	η_c* (in fraction)	Daily Electrical Energy, E_e (Wh)	Daily Daylight Savings (Wh)	Daily Thermal Energy (Wh)	Total Thermal Exergy Efficiency	Overall Daily Exergy (Wh)	Overall Thermal Exergy Efficiency
Case (a)	63.5	58.78	70.75	0.119	4238	4665	7587	1.59%	9297	23.18%
Case (b)	58.73	53.96	62.94	0.1248	7592	8931	-	-	-	-
Case (c)	43.17	41.12	56.23	0.1249	3290	3520	3070	9.11%	9900	30.41%

* Denotes peak data.

Building Integrated Photovoltaic Thermal Systems (BiPVT)

compared to summers, due to lower level of insolation. Further, the average monthly thermal energy, average thermal exergy, average thermal efficiency, average thermal exergy efficiency, average overall exergy and average overall exergy efficiency increases with increases in the water mass. It has been noted that average monthly overall exergy efficiency drops in summers and increases in the winter months for all water mass. This is because thermal energy is more in summers and so is the temperature, thus there are more losses.

A comparison has also been established under the same parameters of case (c) for the BiSPVT system with and without the heat capacity to have a clear understanding of the impact of additional water mass. The room temperature reduces by 28.59% resulting in reduction of TLL by 20.39% when a water mass of 600 kg is added to the BiSPVT system for the month of June. The water mass does not have a significant impact on solar cell electrical efficiency since a drop of only 9°C in solar cell temperature (noon) can be observed with addition of water mass of 600 kg to the BiSPVT system.

8.9.4 FOR BiSPVT SYSTEM WITH HEAT CAPACITY AND WATER FLOW (CASE d)

Table 8.2(b) gives the results of the performance of a BiSPVT system with heat capacity and water flow along with a comparative analysis on similar design parameters for the system with both heat capacity and water flow and a BiSPVT system with only heat capacity. A significant increase of 23.76% in solar cell electrical efficiency can be noted. The system is sustainable in nature since the water used to flow over the photovoltaic modules can be either recirculated or used to recharge the underground water. When the water table is being recharged, it helps to achieve sustainable buildings and hence cities that do not harm the living conditions of the society. Also, with flow of water, heat capacity, natural ventilation, daylight concepts and electricity generation help achieve the thermal comfort, and such buildings are self-sustainable in nature.

8.9.5 BiSPVT SYSTEM HEAT CAPACITY WITH MOVABLE INSULATION AND SOUTH-FACING WINDOW (CASE e) [71]

Impact of movable insulation during off-sunshine hours was tested on case (c) for cold climatic conditions of Srinagar, India, as shown in Figure 8.8. The movable insulation reduces the top thermal losses from room to ambient to increase the room temperature during off-sunshine hours. This leads to a significant rise in inside temperature at desired hours for harsh cold climatic conditions. A south-facing window has been provided for direct gain, as well as daylighting inside the room if

---

TABLE 8.2(b)
Performance Analysis for Case (d)

Case	T_r^* (°C)	T_c^* (°C)	η_c^* (in fraction)	Daily Electrical Energy, E_e (Wh)	Daily Daylight Savings (Wh)
Heat capacity but no water flow (c)	76.07	97.46	0.101	6809	8931
Water flow and heat capacity (d)	**48.19**	**60.89**	**0.125**	**7987**	**8931**
Percentage change from (c) to (d)	36.65% increase	37.52% decrease	23.76% increase	17.29% increase	No change

* Denotes peak data.

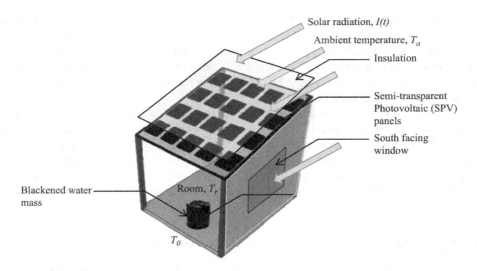

FIGURE 8.8 Schematic diagram of BiSPVT system with heat capacity, movable night insulation and south-facing window [71].

required. Otherwise this window should be covered with movable insulation (curtains/blinds, etc.) even during the daytime.

Equation (8.1) is the energy balance equation for the SPV roof integrated with the building. Considering L_g, L_c and L_{in} as thickness of glass, air conductance gap and insulation, respectively, in meters, and K_g, K_c and K_{in} as thermal conductivity of glass, air conductance gap and insulation, respectively, in W/mK.

For case (e),

$$U_{tca} = \left[\frac{1}{h_0} + \frac{L_g}{K_g}\right]^{-1} = \left[\frac{1}{5.7} + \frac{.003}{.9}\right]^{-1} = 5.59 \, \text{W}/\text{m}^2\text{K during sunshine hours, and}$$

$$U_{tca} = \left[\frac{1}{h_0} + \frac{L_g}{K_g} + \frac{L_c}{K_c} + \frac{L_g}{K_g} + \frac{L_{in}}{K_{in}}\right]^{-1} = \left[\frac{1}{5.7} + \frac{.003}{.9} + \frac{.0003}{.039} + \frac{.01}{.035}\right]^{-1}$$

$$= 2.1 \, \text{W}/\text{m}^2\text{K during off-sunshine hours with movable insulation}$$

$$U_{bcr} = \left[\frac{1}{h_i} + \frac{L_g}{K_g}\right]^{-1} = 2.77 \, \text{W}/\text{m}^2\text{K -}$$

Thus, $U\text{-}Value = U_{tca} + U_{bcr}$

For sunshine hours, $U\text{-}Value = 8.36 \, \text{W/m}^2 \, \text{K}$.
For off-sunshine hours, $U\text{-}Value = 4.87 \, \text{W/m}^2 \, \text{K}$.

To further increase the room temperature, the thickness of insulation should be *increased,* since with an increase in the thickness, there is a reduction in heat transfer, and the U-value of the system will further decrease. Further, double-glazed system also leads to reduction heat losses.

Building Integrated Photovoltaic Thermal Systems (BiPVT)

The energy balance equation for the surface of water tank (T_t) can be written as:

$$\tau_g I_{win} A_{win} + A_f h_c \left(T_f - T_r\right) + U_{bcr}\left(T_c - T_r\right)A_m - 0.33 NV\left(T_r - T_a\right) = h_0 A_t \left(T_r - T_t\right) \qquad (8.22)$$

where I_{win} is the solar intensity transmitted through window in W/m^2 and A_{win} is the area of the window opening in m².

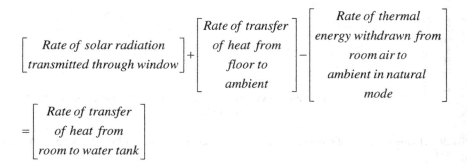

Equation (8.22) can be explained as:

A rise in room temperature of about 2°C in room temperature from −4°C as ambient temperature has been noted during off-sunshine hours due to movable insulation for harsh, cold climatic conditions at a packing factor of 0.25. To further increase the room temperature, the thickness of insulation shall be increased. Therefore, if the thickness of the insulation is about 5 cm, this rise in room temperature will be near 4°C with U-value = 3.38 W/m² K. This is as per the expectation that with a decrease in U-value, the temperature of the room will increase. To further increase the room air temperature, the numerical value of overall heat transfer coefficient from solar cell to ambient through glass cover (U_{tca}) shall be reduced with an increased level of insulation (both in terms of layer and duration).

An increase of 1°C in the room temperature at 24:00 hours was observed when BiSPVT system was proposed with night insulation with respect to BiSPVT system without night insulation at a packing factor of 0.25 and insulation of 100 mm thickness.

The peak solar cell temperature for the proposed system with night insulation is 48.68°C with solar cell electrical efficiency to be 13.4% for a packing factor of 0.89.

The impact of packing factor (for mass of water = 600 kg) was also evaluated; it was found that for packing factor equal to 0.89, TLL was minimum. It may be stated that a packing factor = 0.89 is suitable for thermal cooling in hot climatic conditions and a packing factor = 0.25 is suitable for thermal heating in cold climatic conditions. TLL has been found more for winter months (heating purposes) as per our expectations. Increasing the mass of water reduces the TLL for both summer and winter conditions. The amount of mass can be optimized by considering the summer and winter conditions.

The room temperature increases by 1.39°C (at peak hours) with a reduction in packing factor from 0.89 to 0.25. This rise in room temperature is as per our expectation since more solar insolation enters the room with an increase in the non-packing area. Due to the decrease in the non-packing area, the solar radiation absorbed by the solar cell decreases, hence the solar cell temperature reduces, thus increasing solar cell electrical efficiency by 2%. It can be observed that there is an increase in daily electrical energy generated by the said system when the packing factor changes from 0.25 to 0.89 by 2.33 kW. There is an increase in daily daylight savings by 20 kW when the packing factor reduces from 0.89 to 0.25.

It can be seen as the mass of water decreases from 600 kg to 200 kg, there is an increase in room temperature by 1.4°C. Room air temperature will drop with larger heat capacity due to more heat storage. Mass of water has negligible impact on generation of electrical power.

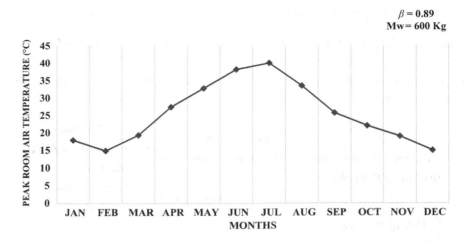

FIGURE 8.9 Hourly variation of room temperature with night insulation for BiSPVT system with water mass = 600 kg [71].

Figure 8.9 gives the annual simulation for the peak room air temperature for case (i) BiSPVT system with night insulation for packing factor 0.89, number of air changes = 4 and mass of water equal to 600 kg. The solar intensity and the number of sunshine hours for the months of August and September are low, and thus the room temperature achieved for these months is less. Generally, these months represent cloudy climatic conditions. Further, low room temperature has been noted for the winter months due to low solar intensity and thus additional heating through an active method may be used to achieve temperature within comfortable thermal conditions or the duration and level of insulation shall be increased. During the months of April to September, the attained room temperature is high, and thus amount of fossil fuel required will be comparatively less.

The number of air changes should be as few as possible, preferably zero for the winter season, and can be referred to as infiltration. For summer months, the natural ventilation should be maximum. Room temperature can further be reduced by increasing the water mass

8.10 INPUT VARIABLES OF BiSPVT SYSTEM: CASE STUDIES

Based on the thermal modeling, the impact of various input variables like number of air changes, water mass, mass flow rate, packing factor, velocity of system, relative humidity, transmissivity of glass, etc., have been assessed on the proposed cases as shown in Figure 8.5. The design parameters and climatic conditions are given in Table 8.1 and Appendix C, respectively.

8.10.1 Number of Air Changes

With an increase in the number of air changes, there will be a drop in room and floor temperature due to an increase in the transfer of thermal energy from room to outside. This results in a decrease in TLL, thus decreasing the fluctuations in room temperature. Solar cell temperature does not have much impact from N, which is very much evident from the Table 8.3; there is an insignificant effect of number in air changes on electrical power. The daylight savings is independent of the N. With increases in the number of air changes, the thermal exergy from floor to room air and from PV module to room air decreases. For winter months (thermal heating), the number of air changes shall be minimum and for summer months (thermal cooling), natural ventilation should be maximum. The effect of number of air changes has been assessed on cases (a) and (b) and summarized in Table 8.3. For case (b), with increases in number of air changes, the difference in room air temperature (ΔT_r),

TABLE 8.3

Effect of Number of Air Changes, Velocity, Packing Factor and Water Mass on the Proposed System*

Case	T_r*	T_c*	η_c*	ΔT_r*	ΔT_f*	ΔT_c*	TLL	Electrical Energy*, E_e*	Daylight Savings* (Wh)	Total Daily Thermal Exergy Efficiency (Wh)	Overall Daily Exergy Efficiency
Change when number of air changes increases from 0 to 4											
Case (a)	36.29% decrease	5.09% decrease	1.58% increase	-	-	-	9.8% decrease	1.76% increase	No impact	6.29% decrease	1.15% increase
Case (b)	-	2.81% decrease	0.97% increase	51.71% decrease	51.65% decrease	22.72% decrease	-	-	No impact	-	-
Change when velocity of system increases from 0 m/s to 4 m/s											
Case (b)	-	0.65% decrease	0.24% increase	66.46% decrease	66.49% decrease	76.75% decrease	-	-	No impact	-	-
Change when packing factor reduces from 0.89 to 0.225											
Case (b)	15.23% increase	22.74% decrease	7.61% increase	-	-	-	-	5547 Wh decrease*	5920 Wh increase	-	-
Change in water mass from 200 kg to 600 kg											
Case (c)	22.63% decrease	5.74% decrease	1.62% increase	-	-	-	6.6% decrease	1.21% increase	No impact	46.71% increase	13.96% increase

* Denotes peak data.

solar cell temperature (ΔT_c) and floor temperature (ΔT_f) with and without water flow, at noon hours reduces. This is due to a greater amount of air withdrawn from the system.

8.10.2 VELOCITY OF THE SYSTEM

With an increase in velocity of the system, the difference in room air temperature (ΔT_r), solar cell temperature (ΔT_c) and floor temperature (ΔT_f), with and without water flow, at noon hours reduces. This is due to an increase in top loss, creating a cooling effect. Similar to N, velocity also does not have much impact on the solar cell temperature and, thus, solar cell electrical efficiency. The results are given in Table 8.3.

8.10.3 PACKING FACTOR

Room and floor temperature increases with decreases in the packing factor. This is because of increases in the non-packing area, $(1-\beta)A_m$, which will permit a greater amount of solar insolation to enter the room, thus raising the temperature. Solar cell temperature will decrease with decreases in the packing factor, reducing the generation of electrical energy.

Daylight savings significantly depend on the packing factor. With decrease in the packing factor, daylight savings for the system will increase due to the increased direct gain through the non-packing area. Table 8.3 summarizes the impact of packing factor on the system.

8.10.4 RELATIVE HUMIDITY

With increases in relative humidity (γ), the room temperature increases because of the reduction in evaporative cooling and heat losses. For case (b) it has been estimated that room temperature increases by 21.58% (i.e., 12.12°C) when relative humidity increases from 30% to 90%.

8.10.5 TRANSMISSIVITY OF GLASS

The daily electrical energy and daily thermal energy generated will decrease with a decrease in the transmissivity of glass (τ_g). This is because the solar cell temperature reduces with decreases in transmissivity and so does the electrical and thermal energy. On examining case (c), the daily electrical and thermal energy drops by 61.6% and 66%, respectively, when transmissivity reduces from 0.9 to 0.3.

8.10.6 MASS OF WATER

Room and floor temperatures drop with an increase in the mass of water, resulting in decrease in the thermal load leveling. Thus, creating a comfortable indoor temperature. Mass of water is not a significant parameter for solar cell temperature and solar cell electrical efficiency. Overall exergy is large for large water mass.

Table 8.3 summarizes the impact of mass of water on case (c). Further, average monthly thermal energy, average monthly thermal exergy, average monthly thermal efficiency, and average monthly overall exergy for case (c) increases by 46.7%, 46.7%, 46.6% and 14.9%, respectively, with change in water mass from 200 kg to 600 kg.

Effect of water mass on annual overall exergy for different weather conditions has been studied by Gupta and Tiwari [67]. With an increase in water mass from 200 kg to 600 kg, yearly overall exergy increases by 15.50%. It has been found that the weather condition "hazy days" having a ratio of daily diffuse to daily global radiation ≤0.25 with sunshine hours ≥9 hours gave best results due to maximum number of days fall and comparatively larger solar intensity. It has been found for a mass

Building Integrated Photovoltaic Thermal Systems (BiPVT)   **249**

flow rate of 0.01 kg/s, there is an 18.96% decrease in peak room air temperature, a 2.89% decrease in peak solar cell temperature and a 0.95% increase in solar cell electrical efficiency when the mass of water changes from 600 kg to 4000 kg.

The impact of mass of water on the BiSPVT system with heat capacity and water flow has also been analyzed.

8.10.7 Mass Flow Rate

With an increase in the mass flow rate, the solar cell temperature reduces, thus increasing the solar cell electrical efficiency. With water flow, electrical energy generated also increases (Table 8.2b) and has no impact on the daylight savings. Due to the reduction in solar cell temperature, the indirect gain inside the room reduces, resulting in a decrease in the room and floor temperature. It has been found that for mass of water of 600 kg, there is an 8.73% decrease in peak room air temperature, a 10.80% decrease in peak solar cell temperature and a 3.57% increase in solar cell electrical efficiency when the mass flow rate changes from 0.01 kg/s to 0.10 kg/s.

8.11 BiSPVT SYSTEM BASED ON THE PV CELL TYPE

With development in BiPV technology, the need for better efficiency has also came to the surface, leading to development of different PV modules or consisting of different types of components of new buildings. Different new types of modules have been developed that promise higher efficiencies and low module temperatures. Modules include white modules, sandwiched modules (polyurethane sandwiched between top thin-film and organic color plate), dye-sensitized solar cells (DSCs), etc.; the thin film technologies are mainly based on cadmium telluride (CdTe), copper indium gallium selenide (CIGS) and amorphous silicon (a-Si). Due to the current manufacturing and material challenges, their growth in the market is limited. These are mainly related to the high processing temperatures (>500°C) and environmental concerns related to cadmium and tellurium [13]. On the other hand, DSCs emerged as flexible and lightweight with low fabrication cost [72]. Chapter 5 discusses different types of PV cells in details.

8.11.1 Literature Study

Dubey et al. [73] found that the electrical efficiency and the power output of the photovoltaic module is dependent on the operating temperature. With increases in the module temperature in a-Si modules, the electrical power increases due to the thermal recovery effect [74]. Potential of BiPV and BaPV systems were examined by Dos Santos and Rüther [75] on two systems. The first system was a 2.25 kWp c-Si; the second was a 10 kWp a-Si installed on an existing residential building. It was found that the two PV technologies selected (a-Si and c-Si) differed 6% in output performance. López and Sangiorgi [34] found that energy consumption for heating and lighting was slightly lower for a copper indium selenium (CIS) PV module, however generation of energy was higher than with a-Si PV module. They also tested a monocrystalline silicon (m-Si) PV module in the same setup instead of an a-Si PV module and found that CIS PV modules gave better results for hygro-thermal comfort analysis. The lighting requirement for m-Si was slightly higher than that of the CIS PV module. The M-Si module had better results in regard to generation of energy (0.09–1.31 kWh/day). Ordenes et al. [27] assessed the potential of six different PV technologies, namely m-Si, polycrystalline silicon (p-Si), a-Si, CdTe, copper indium diselenide (CID) and hetero-junction with thin layer (HIT). It was found that HIT had the highest efficiency of 17%, whereas a-Si accounts for the lowest efficiency of 6.3%. Salem and Kinab [76] analyzed the requirement of area for three types of PV modules, namely, m-Si, p-Si and thin films and found that m-Si and p-Si require almost the same area. In view of the economics involved, in cases where p-Si is cheaper than m-Si PV modules, p-Si should be considered. Vats et al. [77] analyzed different PV technologies (m-Si, p-Si, a-Si, CdTe, CIGS and HIT) on a roof-integrated PV system with air ducts for different packing factors. The outcome of the study was that HIT had maximum annual electrical energy production (813 kWh), while the maximum thermal output was generated by an a-Si PV module (79 kWh) for a 0.62 packing factor. Vats and Tiwari [41] have found that HIT has maximum efficiency of 16% with peak cell temperature of 42°C, and a-Si has peak efficiency of 6% with maximum cell temperature of 49°C. Hourly variation of electrical power for different PV technologies was also carried out, and it was found that a-Si produces 287 W in comparison to HIT type producing 779 W of electrical power. The reason behind HIT producing greater electrical power is its lower cell temperature. It was also found that a-Si has higher thermal energy (408 W) due to its high cell temperature due to which the room temperature rises. The thermal energy produced by HIT type was 321 W. Vats and Tiwari [41] found that HIT type has maximum electrical performance (810 kW), whereas a-Si accounts for maximum thermal energy (464 kWh). The efficiency of the module shows the actual amount of the solar radiation getting converted to electricity. With an increase in the operating temperatures, the electrical output of the modules decreases with an exception in the a-Si case due to a negligible temperature coefficient. DSCs were compared with a-Si and m-Si BiPV panels with reference to electrical power by [78]. It was found that average watt-peak value of DSC was more than a-Si and m-Si by 12% and 3%, respectively.

8.11.2 Performance of BiSPVT System Based on PV Types: A Case Study

The module selection should be based on its temperature coefficient and electrical efficiencies. Nomenclature and design parameters for the case study shall be referred from Table 8.1 with $N = 4$, $v = 0$ m/s. Values of module electrical efficiency and temperature coefficients are given in Table 8.4. In Section 8.9.1, performance analysis of case (a), i.e., BiSPVT system has been done considering the solar cell efficiency (η_c) as constant for the cell type with $\beta_0 = 0.0045°C^{-1}$ and $\eta_0 = 0.15$.

Using Equations (8.1), (8.6) and (8.13), solar cell electrical efficiency in terms of η_0 and β_0 has been derived and expressed as:

$$\eta_c = \frac{\eta_0 \left[1 - \beta_0 \left\{ \tau_g I(t) \beta \alpha_c + U_{tca} \left(T_a - T_{ref} \right) + U_{bcr} \left(T_r - T_{ref} \right) \right\} \right]}{1 - \dfrac{\eta_0 \beta_0 \tau_g I(t) \beta}{U_{tca} + U_{bcr}}} \tag{8.23}$$

And thus the expression for solar cell temperature becomes [79]

$$T_c = \frac{\tau_g I(t) \beta \left\{ \left[\alpha \tau I(t) \right]_1 - U_0 \left(T_r - T_{ref} \right) \right\} + U_{tca} T_a + U_{bcr} T_r}{U_{tca} + U_{bcr}} \tag{8.24}$$

where

$$\left[\alpha \tau I(t) \right]_1 = \frac{\alpha_c \left[1 - \dfrac{\eta_0 \beta_0 \tau_g I(t) \beta}{U_{tca} + U_{bcr}} \right] - \eta_0 \left[1 - \beta_0 \left\{ \tau_g I(t) \beta \alpha_c + U_{tca}(T_a - T_{ref}) \right\} \right]}{1 - \dfrac{\eta_0 \beta_0 \tau_g I(t) \beta}{U_{tca} + U_{bcr}}}$$

It can be seen that in Equation (8.24), solar cell temperature is dependent on the solar radiation and room air temperature. It is independent of solar cell efficiency unlike the one used in case (a) (Equation 8.6).

TABLE 8.4

Performance of BiSPVT System with Different PV Module Types [79]

	Description of Case							Daily Electrical Energy (Wh)	Overall Thermal Efficiency	Overall Exergy Efficiency (%)			
Case	Module Type	η_0	β_0 (°C^{-1})	Ref	T_r*(°C)	T_c* (°C)	η_c* (in fraction)			Electrical Energy	Day Lighting	Total Thermal Exergy	Overall Thermal Exergy
Case (i)	Mono-Si -I	0.15	0.0041	[80]	45.32	95.26	0.106	3912	21.11%	9.23	11	2.36	22.60
Case (ii)	Mono-Si- II	0.13	0.004	[81]	44.44	93.03	0.094	3427.56	20.86%	8.09	11	2.28	21.37
Case (iii)	Poly-Si (p-Si)	0.11	0.004	[81]	43.93	91.44	0.080	2911.17	20.71%	6.87	11	2.23	20.10
Case (iv)	a-Si	0.05	0.0011	[81]	40.68	81.65	0.046	1538	19.41%	3.63	11	1.85	16.48
Case (v)	Mono-Si- III	0.15	0.0045	[42]	45.99	97.71	0.102	3791.62	21.52%	8.94	11	2.48	22.43

* Denotes peak data.

252   Photovoltaic Thermal Passive House System

For floor temperature (Equations 8.7 and 8.8), room temperature (Equation 8.12), electrical energy (Equation 8.14), daylight savings (Equation 8.15), thermal energy (Equation 8.19), net output exergy (Equation 8.16), net input exergy (Equation 4.26), overall exergy efficiency (Equation 8.18), thermal exergy efficiency (Equation 8.21a) and overall thermal efficiency (Equation 8.21b), they shall be referred.

The computations have been carried out for a typical day in the month of January, New Delhi, and the results are summarized in Table 8.4.

Amongst the proposed cases (i–v), the minimum room temperature at peak hours has been achieved for case (iv) due to minimum $\beta_0 = 0.0011°C^{-1}$. However, T_r in case (i), (ii), (iii) and (v) fall very close to each other due to a similar range of β_0. Case (i) and case (v) have the same η_0 but different β_0, and there is little variation in the room air temperature. This shows that η_0 does not have much impact on the room air temperature. Solar cell electrical efficiency is dependent both on η_0 and β_0. Case (iv) has a very small variation in solar cell efficiency and, overall, the lowest solar cell efficiency when compared with other cell materials. This is attributed to lower values of module electrical efficiency and temperature coefficient. Daily electrical energy produced is found to be maximum for case (i), which equals to 3912 W, and minimum for case (iv), which equals to 1538.4 W due to lower η_0 and η_c, respectively. Amongst the proposed model, it can be seen that m-Si-I has the maximum electrical energy and overall thermal exergy, whereas the mono-Si-III system has the maximum total thermal exergy. Maximum daily thermal exergy is found to be 1050.89 W (case v) and minimum to be 787.40 W (case iv). Daylight savings is independent of the type of photovoltaic material used.

From Table 8.4, it may be concluded that in view of electrical exergy, m-Si with $\eta_0 = 0.15$ and $\beta_0 = 0.0045°C^{-1}$ is recommended.

OBJECTIVE QUESTIONS

8.1   Direct gain in a BiPVT setup is due to:
    (a)   Non-packing area
    (b)   Opaque PV modules
    (c)   Packing area
    (d)   None of the above

8.2   Thermal-load leveling can be reduced by incorporating:
    (a)   Natural light
    (b)   Heat capacity
    (c)   Water flow over PV module
    (d)   All of the above

8.3   For hot climatic conditions photovoltaic modules should be used as:
    (a)   BiSPVT systems
    (b)   BiSPVT with heat capacity
    (c)   BiSPVT with water flow
    (d)   BiSPVT with heat capacity and water flow

8.4   For night heating a _____ is preferred:
    (a)   BiSPVT system
    (b)   BiSPVT with heat capacity
    (c)   BiSPVT with water flow
    (d)   BiSPVT with heat capacity and water flow

8.5   For natural mode of ventilation, the number of air changes should be:
    (a) $N \leq 10$
    (b) $N \geq 10$
    (c) $N = 10$
    (d) $N = 0$

Building Integrated Photovoltaic Thermal Systems (BiPVT)

8.6 Production of electrical power strongly depends on:
(a) Room temperature
(b) Packing area
(c) Daylight savings
(d) None of the above

8.7 The major contribution in net output exergy is:
(a) Thermal exergy
(b) Electrical energy
(c) Thermal exergy + electrical energy
(d) Electrical energy + daylight savings

8.8 Movable insulation helps in
(a) Reducing top thermal loss
(b) Decreasing room temperature
(c) Increasing top thermal loss
(d) Increasing bottom thermal loss

8.9 Daylight savings is affected by:
(a) Mass of water
(b) Non-packing area
(c) Water flow over PV modules
(d) Electrical energy

8.10 With increase in number of air changes:
(a) TLL reduces
(b) Solar cell temperature reduces
(c) Electrical power reduces
(d) Daylight savings reduces

8.11 With increase in velocity of the system:
(a) Room temperature reduces
(b) Solar cell temperature reduces
(c) Electrical power reduces
(d) Daylight savings reduces

8.12 With increase in packing factor:
(a) Room temperature reduces
(b) Solar cell temperature reduces
(c) Electrical power reduces
(d) Daylight savings increases

ANSWERS

8.1 (a)
8.2 (b)
8.3 (d)
8.4 (b)
8.5 (a)
8.6 (b)
8.7 (d)
8.8 (a)
8.9 (b)
8.10 (a)
8.11 (a)
8.12 (a)

REFERENCES

[1] N. Gupta and G. N. Tiwari, "Energy matrices of building integrated semitransparent photovoltaic thermal systems: A case study," *Journal of Architectural Engineering*, 23(4), p. 05017006, 2017a.

[2] Y. B. Assoa, L. Gaillard, C. Ménézo, N. Negri and F. Sauzedde, "Dynamic prediction of a building integrated photovoltaic system thermal behaviour," *Applied Energy*, 214, pp. 73–82, 2018.

[3] H. Yang, G. Zheng, C. Lou, D. An and J. Burnett, "Grid-connected building-integrated photovoltaics: A Hong Kong case study," *Solar Energy*, vol. 76, pp. 55–59, 2004.

[4] T. Yang and A. K. Athienitis, "A review of research and developments of building-integrated photovoltaic/ thermal (BIPV/T) systems," *Renewable and Sustainable Energy Reviews*, vol. 66, pp. 886–912, 2016.

[5] B. P. Jelle and C. Breivik, "The path to the building integrated photovoltaics of tomorrow," *Energy Procedia*, vol. 20, pp. 78–87, 2012.

[6] C. Peng, Y. Huang and Z. Wu, "Building-integrated photovoltaics (BIPV) in architectural design in China," *Energy and Buildings*, vol. 43, pp. 3592–3598, 2011.

[7] S. Pantic, L. Candanedo and A. K. Athienitis, "Modeling of energy performance of a house with three configurations of building-integrated photovoltaic/ thermal systems," *Energy and Buildings*, vol. 42, pp. 1779–1789, 2010.

[8] S. Agrawal and G. N. Tiwari, "Performance analysis in terms of carbon credit earned on annualised uniform cost of glazed hybrid photovoltaic thermal air collector," *Solar Energy*, vol. 115, pp. 329–340, 2015.

[9] C. Li and R. Wang, "Building integrated energy storage opportunities in China," *Renewable and Sustainable Energy Reviews*, vol. 16, pp. 6191–6211, 2012.

[10] C. D. Corbin and Z. J. Zhai, "Experimental and numerical investigation on thermal and electrical performance of a building integrated photovoltaic-thermal collector system," *Energy and Buildings*, vol. 42, pp. 76–82, 2010.

[11] B. Koyunbaba, Z. Yilmaz and K. Ulgen, "An approach for energy modeling of a building integrated photovoltaic (BiPV) Trombe wall system," *Energy and Buildings*, vol. 67, pp. 680–688, 2013.

[12] R. A. Agathokleous and S. A. Kalogirou, "Double skin facades (DSF) and building integrated photovoltaics (BIPV): A," review of configurations and heat transfer characteristics," *Renewable Energy*, vol. 89, pp. 743–756, 2016.

[13] E. Biyik, M. Araz, A. Hepbasli, M. Shahrestani, R. Yao, L. Shao, E. Essah, A. C. Oliveira, T. d. Caño, E. Rico, J. L. Lechón, L. Andrade, A. Mendes and Y. B. Atlı, "A key review of building integrated photovoltaic (BIPV) systems," *Engineering Science and Technology, An International Journal*, 20, pp. 833–858, 2017.

[14] "Architecture and design," [Online]. Available: https://www.architectureanddesign.com.au/suppliers/onyx-solar-australia/onyx-to-supply-glass-for-the-world-s-largest-photo [Accessed March 30 2019].

[15] "Polysolar," [Online]. Available: http://www.polysolar.co.uk/solar-pv-case-studies/polysolar-curtain-walling-system [Accessed March 30 2019].

[16] Esmail7, 2013. *Wikipedia.* [Online] Available at: https://en.wikipedia.org/wiki/Photovoltaic_mounting_system#/media/File:PV_external_shading_device_in_zero_energy_building_of_Singapore.jpg [Accessed 2020].

[17] Tatmouss, 2010. *Wikipedia.* [Online] Available at: https://en.wikipedia.org/wiki/Photovoltaics#/media/File:Ombri%C3%A8re_SUDI_-_Sustainable_Urban_Design_%26_Innovation.jpg [Accessed March 30 2019].

[18] Bolton, H., 2007. *Wikipedia.* [Online] Available at: https://en.wikipedia.org/wiki/Solar_street_light#/media/File:Church_and_bus_stop,_Llandissiliogogo_-_geograph.org.uk_-_581087.jpg [Accessed 2020].

[19] Saurenergy, 2017. *Saur Energy International.* [Online] Available at: https://www.saurenergy.com/solar-energy-news/delhi-metro-to-set-up-solar-panels-on-pedestrian-bridges-to-generate-power [Accessed 2020].

[20] "ONYX," [Online]. Available: https://www.onyxsolar.com/tanjong-pagar [Accessed 30 March 2018].

[21] E. Akata, A. Martial, D. Njomo and B. Agrawal, "Thermal energy optimization of building integrated semi-transparent photovoltaic thermal systems," *International Journal of Renewable Energy Developments*, vol. 4, pp. 113–123, 2015.

[22] K. Mainzer, K. Fath, R. McKenna, J. Stengel, W. Fichtner and F. Schultmann, "A high-resolution determination of the technical potential for residential-roof-mounted photovoltaic systems in Germany," *Solar Energy*, vol. 105, pp. 715–731, 2014.

[23] H. B. Madessa, "Performance analysis of roof-mounted photovoltaic systems- The case of a Norwegian residential building," *Energy Procedia*, vol. 83, pp. 474–483, 2015.

[24] S. B. Sadineni, F. Atallah and R. F. Boehm, "Impact of roof integrated PV orientation on the residential electricity peak demand," *Applied Energy*, vol. 92, pp. 204–210, 2012.

[25] Wikipedia, 2020. [Online] Available at: https://en.wikipedia.org/wiki/Photovoltaics [Accessed November 30 2020].

[26] A. R. Othman and A. T. Rushdi, "Potential of building integrated photovoltaic application on rooftop of residential development in Shah Alam," *Procedia – Social and Behavioral Sciences*, vol. 153, pp. 491–500, October 2014.

[27] M. Ordenes, D. L. Marinoski, P. Braun and R. Rüther, "The impact of building-integrated photovoltaics on the energy demand of multi-family dwellings in Brazil," *Energy and Buildings*, vol. 39, no. 6, pp. 629–642, June 2007.

[28] "Wikipedia," [Online]. Available: https://upload.wikimedia.org/wikipedia/commons/6/6e/CIS_Tower.jpg [Accessed April 1 2019].

[29] S. H. Yoo and E. T. Lee, "Efficiency characteristic of building integrated photovoltaics as a shading device," *Building and Environment*, vol. 37, pp. 615–623, 2002.

[30] A. M. Memari, L. D. Iulo, R. L. Solnosky and C. R. Stultz, "Building integrated photovoltaic systems for single family dwellings: Innovation concepts," *Open Journal of Civil Engineering*, vol. 4, pp. 102–119, 2014.

[31] A. Saranti, T. Tsoutsos and M. Mandalaki, "Sustainable energy planning. Design shading devices with integrated photovoltaic systems for residential housing units," *Procedia Engineering*, vol. 123, pp. 479–487, 2015.

[32] Y. C. Huang, C. C. Chan, S. J. Wang and S. K. Lee, "Development of building integrated photovoltaic (BIPV) system with PV ceramic tile and its application for building facade," *Energy Procedia*, vol. 61, pp. 1874–1878, 2014.

[33] G. Quesada, D. Rousse, Y. Dutil, M. Badache and S. Hallé, "A comprehensive review solar facades. Opaque solar facades," *Renewable Sustainable Energy Reviews*, vol. 16, pp. 2820–2832, 2012a.

[34] C. S. P. López and M. Sangiorgi, "Comparison assessment of BIPV facade, semi-transparent modules: Further insights on human comfort conditions," *Energy Procedia*, vol. 48, pp. 1419–1428, 2014.

[35] T. Y. Y. Fung and H. Yang, "Study on thermal performance of semi-transparent building integrated photovoltaic glazings," *Energy and Buildings*, vol. 40, pp. 341–350, 2008.

[36] J. H. Song, Y. S. An, S. G. Kim, S. J. Lee, J. H. Yoon and Y. K. Choung, "Power output analysis of transparent thin-film module in building integrated photovoltaic system (BiPV)," *Energy and Buildings*, vol. 40, pp. 2067–2075.

[37] K. E. Park, G. H. Kang, H. I. Kim, G. J. Yu and J. T. Kim, "Analysis of thermal and electrical performance of semi-transparent photovoltaic (PV) module," *Energy*, vol. 35, pp. 2681–2687, 2010.

[38] N. Gupta and G. N. Tiwari, "Review of passive heating/ cooling systems of buildings," *Energy Science & Engineering*, vol. 4, no. 5, pp. 305–333, 2016.

[39] Solar Constructions [Online] Available at: http://www.solar-constructions.com/wordpress/transparent-solar-panels/ [Accessed April 1 2019].

[40] K. Vats and G. N. Tiwari, "Performance evaluation of a building integrated semitransparent photovoltaic thermal system for roof and facade," *Energy and Buildings*, vol. 45, pp. 211–218, February 2012a.

[41] K. Vats and G. N. Tiwari, "Energy and exergy analysis of a building integrated semitransparent photovoltaic thermal (BiSPVT) system," *Applied Energy*, vol. 96, pp. 409–416, 2012b.

[42] N. Gupta, A. Tiwari and G.N. Tiwari, "Exergy analysis of building integrated semitransparent photovoltaic thermal (BiSPVT) system," *Engineering Science and Technology, an International Journal*, 20, 41–50, 2016.

[43] D. H. W. Li, T. N. T. Lam, W. W. H. Chan and A. H. L. Mak, "Energy and cost analysis of semi-transparent photovoltaic in office buildings," *Applied Energy*, vol. 86, pp. 722–729, 2009.

[44] G. N. Tiwari, H. Saini, A. Tiwari, N. Gupta, P. S. Saini and A. Deo, "Periodic theory of building integrated photovoltaic thermal (BiPVT) system," *Solar Energy*, vol. 125, pp. 373–380, 2016.

[45] D. Kamthania, S. Nayak and G. N. Tiwari, "Performance evaluation of a hybrid photovoltaic thermal double pass facade for space heating," *Energy and Buildings*, vol. 43, no. 9, pp. 2274–2281, September 2011.

[46] G. Quesada, D. Rousse, Y. Dutil, M. Badache and S. Hallé, "A comprehensive review of solar facades. Transparent and translucent solar facades," *Renewable Sustainable Energy Reviews*, vol. 16, pp. 2643–2651, 2012b.

[47] L. V. Mercaldo, M. L. Addonizio, M. Della Noce, P. D. Veneri, A. Scognamiglio and C. Privato, "Thin film silicon photovoltaics: Architectural perspectives and technological issues," *Applied Energy*, vol. 86, pp. 1836–1844, 2009.

[48] T. T. Chow, Z. Qui and C. Li, "Potential application of effect of air gap on the performance of building-integrated photovoltaics 'see-through' solar cells in ventilated glazing in Hong Kong," *Solar Energy Materials and Solar Cells*, vol. 93, pp. 230–238, 2009.

[49] G. Gan, "Effect of air gap on the performance of building-integrated photovoltaics," *Energy*, vol. 34, pp. 913–921, 2009a.

[50] G. Gan, "Numerical determination of adequate air gaps for building-integrated photovoltaics," *Solar Energy*, vol. 83, pp. 1253–1273, 2009b.

[51] G. E. Lau, E. Sanvicente, G. H. Yeoh, V. Timchenko, M. Fossa, C. Ménézo and S. Giroux-Julien, "Modeling of natural convection in vertical and tilted photovoltaic applications," *Energy and Buildings*, vol. 55, pp. 810–822, December 2012.

[52] S. S. S. Baljit, H. Y. Chan and K. Sopian, "Review of building integrated applications of photovoltaic and solar thermal systems," *Journal of Cleaner Production*, vol. 137, pp. 677–689, 2016.

[53] K. Nagano, T. Mochida, K. Shimakura, K. Murashita and S. Takeda, "Development of thermal-photovoltaic hybrid exterior wallboards incorporating PV cells in and their winter performances," *Solar Energy Materials and Solar Cells*, vol. 77, no. 3, pp. 265–282, May 2003.

[54] A. Buonomano, F. Calise, A. Palombo and M. Vicidomini, "BIPVT systems for residential applications: An energy and economic analysis for European climates," *Applied Energy*, 184, pp. 1411–1431, 2016.

[55] A. D. Ondeck, T. F. Edgar and M. Baldea, "Optimal operation of a residential district-level combined photovoltaic/natural gas power and cooling system," *Applied Energy*, vol. 156, pp. 593–606, 2015.

[56] G. A. Dávi, E. Caamaño-Martín, R. Rüther and J. Solano, "Energy performance evaluation of a net plus-energy residential building with grid-connected photovoltaic system in Brazil," *Energy and Buildings*, vol. 120, pp. 19–29, 2016.

[57] A. Franco and F. Fantozzi, "Experimental analysis of a self consumption strategy for residential building: The integration of PV system and geothermal heat pump," *Renewable Energy*, vol. 86, pp. 1075–1085, 2016.

[58] F. Cucchiella, I. D'adamo and M. Gastaldi, "Photovoltaic energy systems with battery storage for residential areas: An economic analysis," *Journal of Cleaner Production*, vol. 131, pp. 460–474, 2016.

[59] N. R. Darghouth, G. Barbose and R. H. Wiser, "Customer-economics of residential photovoltaic systems (Part 1): The impact of high renewable energy penetrations on electricity bill savings with net metering," *Energy Policy*, vol. 67, pp. 290–300, 2014.

[60] D. F. A. Riza, S. I. U. H. Gilani and M. S. Aris, "Standalone photovoltaic systems sizing optimization using design space approach: Case study for residential lighting load," *Journal of Engineering Science and Technology*, vol. 10, no. 7, pp. 943–957, 2015.

[61] T. Lang, E. Gloerfeld and B. Girod, "Don't just follow the sun—A global assessment of economic performance for residential building photovoltaics," *Renewable and Sustainable Energy Reviews*, vol. 42, pp. 932–951, 2015.

[62] S. Wittkopf, S. Valliappan, L. Liu, K. Ang and S. Cheng, "Analytical performance monitoring of a 142.5 kWp grid- connected rooftop BIPV system in Singapore," *Renewable Energy*, vol. 47, pp. 9–20, 2012.

[63] N. Gupta and G. N. Tiwari. Parametric study to understand the effect of various passive cooling concepts on building integrated semitransparent photovoltaic thermal system. *Solar Energy*, vol. 180, pp. 391–400, 2019.

[64] G. N. Tiwari and R. K. Mishra, *Advanced Renewable Energy Sources*, London: Royal Society of Chemistry, 2011.

[65] N. Gupta and G. N. Tiwari, "A thermal model of hybrid cooling systems for building integrated semitransparent photovoltaic thermal system," *Solar Energy*, vol. 153, pp. 486–498, 2017b.

[66] N. Gupta and G. N. Tiwari, "Effect of water flow on building integrated semitransparent photovoltaic thermal system with heat capacity," *Sustainable Cities and Society*, 39 , 708–718, 2018.

[67] N. Gupta and G. N. Tiwari, "Effect of heat capacity on monthly and yearly exergy performance of building integrated semitransparent photovoltaic thermal system," *Journal of Renewable and Sustainable Energy*, vol. 9, 023506, 2017c.

Building Integrated Photovoltaic Thermal Systems (BiPVT)

[68] R. Petela, "Exergy of undiluted thermal radiation," *Solar Energy*, vol. 74, pp. 469–488, 2003.

[69] A. Bejan, "General criterion for rating heat-exchanger performance," *Heat Mass Transfer*, vol. 21, pp. 655–658, 1978.

[70] G. N. Tiwari, A. Tiwari and Shyam, *Handbook of Solar Radiation: Theory, Analysis and Applications*, Singapore: Springer, 2016.

[71] N. Gupta, P. Rani and G. N. Tiwari, "Effect of movable insulation on performance of building integrated Semi-transparent Photovoltaic Thermal (BiSPVT) System for harsh cold climatic conditions: A case study," *International Journal of Ambient Energy*, pp. 1–15, 2021.

[72] D. S. Ginley and D. Cahen. (Eds.), "*Fundamentals of materials for energy and environmental sustainability*," Cambridge University Press, 2012.

[73] S. Dubey, G. N. Sandhu and G. N. Tiwari, "Analytical expression for electrical efficiency of PV/T hybrid air collector," *Applied Energy*, vol. 86, pp. 697–705, 2009.

[74] T. Yamawaki, S. Mizukami, T. Masui and H. Takahashi, "Experimental investigation on generated power of amorphous PV module for roof azimuth," *Solar Energy Materials & Solar Cells*, vol. 67, pp. 369–377, 2001.

[75] Í. P. Dos Santos and R. Rüther, "The potential of building-integrated (BIPV) and building-applied photovoltaics (BAPV) in single- family, urban residences at low latitudes in Brazil," *Energy and Buildings*, vol. 50, pp. 290–297, 2012.

[76] T. Salem and E. Kinab, "Analysis of building- integrated photovoltaic systems: A case study of commercial buildings under Mediterranean climate," *Procedia Engineering*, vol. 118, pp. 538–545, 2015.

[77] K. Vats, V. Tomar and G. N. Tiwari, "Effect of packing factor on the performance of building integrated semitransparent photovoltaic thermal (BISPVT) system with air duct," *Energy and Buildings*, vol. 53, pp. 159–165, 2012.

[78] C. Cornaro, S. Bartocci, D. Musella, C. Strati, A. Lanuti, S. Mastroianni, S. Penna, A. Guidobaldi, F. Giordano, E. Petrolati, T. Brown, A. Reale and A. D. Carlo, "Comparative analysis of the outdoor performance of a dye solar cell mini-panel for building integrated photovoltaics applications," *Progress in Photovoltaics: Research and Applications*, vol. 23, no. 2, pp. 1–11, 2013.

[79] N. Gupta, G. N. Tiwari, A. Tiwari and V. Gupta, "New model for building-integrated semitransparent photovoltaic thermal system," *Journal of Renewable and Sustainable Energy* 9, 043504 (2017); doi: 10.1063/1.4999556.

[80] D. L. Evans and L. Florschuetz, "Cost studies on terrestrial photovoltaic power systems with sunlight concentration," *Solar Energy*, vol. 19, pp. 255–262, 1977.

[81] RETScreen© International, "Photovoltaic Project Analysis," vol. PV.22., 2001.

9 Environmental Aspects

9.1 INTRODUCTION

The environment is responsible for sustaining human life on the planet Earth and is widely referred to as the biosphere. The biosphere is a shallow layer compared to the total size of the earth. It extends to about 20 km from the bottom of the ocean to the highest point in the atmosphere at which life can survive without manmade protective devices.

Among the most dangerous human interferences with nature is the emission of greenhouse gases (GHGs) into the atmosphere, directly causing the climatic changes like the rise in global surface temperature. These gases absorb and trap the solar energy reflected from the Earth's surface resulting in global warning. Carbon dioxide (CO_2), methane (CH_4), nitrous oxide (NO_2) and fluorinated gases are the principal anthropogenic GHGs emitted into the atmosphere.

The emission of CO_2 is considered to have the most significant impact on the environment. Activities like combustion of fossil fuels and deforestation have resulted in a 35% increase in the concentration of carbon dioxide in the atmosphere since the beginning of the age of industrialization leading to imbalance in the carbon cycle. Burning of fossil fuels releases approximately 5.5 billion tonnes of carbon per annum in the atmosphere while deforestation accounts for approximately 1.6 billion tonnes of annual carbon release. The figures turn out to be about 7.1 billion tonnes of carbon per annum from human activities. About 3.2 billion tonnes of carbon remains in the atmosphere, thus atmospheric carbon dioxide levels increase. About 2 billion tonnes of carbon diffuses into the world's oceans, leaving 1.9 billion tonnes of carbon unaccounted for [1].

Increases in the CO_2 level lead to various important changes in the global climate. Many researchers estimate that the global mean temperature will rise in the range of 1.4°C–5.8°C over the next century as a result of increased levels of GHGs and CO_2 in the atmosphere. This rise in global temperature will cause a significant increase in average sea level (0.09–0.88 m), thus exposing low-lying coastal cities or cities located by tidal rivers such as New Orleans, Portland, Washington and Philadelphia to increasingly frequent and severe floods. Glacial retreat and species range shifts are also likely to result from global warming, and it remains to be seen whether relatively immobile species such as trees can shift their ranges fast enough to keep pace with warming.

Concentration of CO_2 in the atmosphere is indicated in Figure 9.1. The carbon dioxide in the graphs (indicated by the red curve) is measured as the mole fraction in dry air. Measured on Mauna Loa, the graphs constitute the longest record of direct measurements of CO_2 in the atmosphere. The monthly mean values, cantered on the middle of each month are given by the red-dashed lines with diamond symbol. The black lines with symbol of the square signify the same, after correction for the average seasonal cycle. The data is measured as the mole fraction, expressed as parts per million (ppm), defined as the number of molecules of carbon dioxide divided by the number of all molecules in the air, including CO_2 itself, after water vapor has been removed. The estimated uncertainty in the Mauna Loa annual mean growth rate is 0.11 ppm/year [2]. This concentration varies with seasons and different regions, especially near the ground levels with higher levels found in urban areas.

9.2 LIFE CYCLE ASSESSMENT

The models that are used to estimate the environmental benefits of retrofitting and renovation of buildings are mostly based on the viewpoint of energy use. However, it may be noted that energy use is not an environmental problem in itself, but causes several environmental impacts, which are not

DOI: 10.1201/9780429445903-9

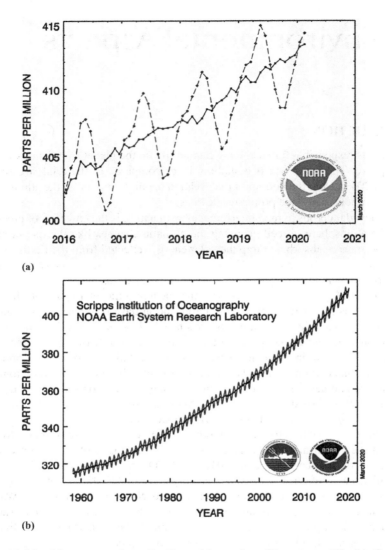

FIGURE 9.1 (a) Monthly mean carbon dioxide at Mauna Loa Observatory [2]. (b) Concentration of atmospheric CO_2 measured at Mauna Loa Observatory [2].

necessarily linearly related to the energy use. For building construction, following are the different perspectives from the energy point of view [3]:

- The energy required to construct a building
- The energy required for maintaining the comfortable conditions inside the building for the occupants
- The reduction in emission of carbon dioxide in the environment to some extent

The environmental impact of buildings during their lifetime is evaluated by life cycle assessment.

The utility of embodied energy is as an indicator of environmental impact through life cycle assessment (LCA). Cabeza et al. [4] synonym that most of the literatures available in the field of LCA and embodied energy (EE) have been carried out in developing countries and no case study has been conducted in African countries. Ramesh et al. [5] said that most of the studies with respect to LCA are from cold countries where space heating is the main concern. Increase in the use of

Environmental Aspects

energy-intensive materials like burned bricks and a shift from environmentally-friendly materials (like thatched roofs, adobe, etc.) to environmentally harmful materials (like concrete, iron sheet roofing, etc.) are the issues leading to an alarming stage [6].

9.2.1 Basic Definitions of Life Cycle Assessment [7]

Life cycle assessment is a method for analyzing and determining the environmental impact along the product chain of (technical) systems. It includes various types of technical conversions that occur in the manufacturing processes. These consist of (i) changes of material chemistry, material formulation or material structure, (ii) the removal of material resulting in an increase of (primary) outputs over the inputs, and (iii) joining and assembly of materials resulting in a decrease of (primary) outputs over the inputs.

According to ISO 14040 [8], LCA is a technique for assessing the environmental impacts and potential impacts associated with a product by:

A. Compiling an inventory of relevant inputs and outputs of a product system
B. Evaluating the potential environmental impacts associated with those inputs and outputs
C. Interpreting the results of the inventory analysis and impact assessment phases in relation to the objectives of the study

9.2.2 The Main Stages of Life Cycle Assessment

The main stages in the LCA are:

1. Production
2. Construction
3. Usage
4. End of life stage

The production stage includes the material supplies, transportation and manufacturing processes [9]. The extraction of raw materials and its impact on the environment, as well as manufacturing and waste, are some of the key assessment criteria in LCA. The carbon dioxide emissions from fossil fuels during extraction and the manufacturing process of construction materials form the majority of the embodied energy/carbon of building products [10]. Embodied energy is the key factor in evaluating the sustainability of buildings.

9.3 EMBODIED ENERGY

Embodied energy has been explained in Section 4.3.3 as the total energy required to produce any goods or services. Embodied energy can be defined as: "the quantity of energy required by all of the activities associated with a production process, including the relative proportions consumed in all activities upstream to the acquisition of natural resources and the share of energy used in making energy equipment and in other supporting functions, i.e., direct energy plus indirect energy" [11].

Therefore, EE analysis is done in order to quantify the total amount of energy used to manufacture a product/material or element. This means, the overall energy spent to extract the raw materials, manufacture the product and its components, and construct and maintain the component/product is required. Like operational energy, EE is an indicator of the level of energy use. Thus, the goal of the designer/architect should be to reduce the energy consumption through better design with high efficiency. Usually, the portion of assessment of EE is ignored due to lack to data, lack of understanding and a common belief that the embodied energy portion of assets energy consumption is insignificant. However, over recent years, the methodologies for assessment have improved, and data reliability and access have increased. Recent reports have indicated that the EE portion may be as high as 20 times the annual operational energy of an office building [11].

In addition to the EE value, other environmental indicators can also be calculated, such as CO_2 emissions. This is the basis of life cycle cost analysis (LCA) work.

The assessment of EE is based on the amount of energy consumed during the production process and depends upon the maintenance cost of building and its transportation. Reduction in EE can be brought by replacing the conventional high-energy materials (like brick, cement, concrete, etc.) with local alternate materials. Embodied energy can be reduced by 37% when local materials are used for building construction [3].

9.3.1 EMBODIED ENERGY OF DIFFERENT MATERIALS

The SBTool, UK Code for Sustainable Homes and the U.S.'s LEED are different methods to assess a building's environmental impact. In these methods, the EE of the good or service in question is rated along with other factors. In order to improve the energy efficiency of any building, various research projects are carried on. But there is a clear gap in the available literatures regarding the environmental impacts and the embodied energy.

To estimate the EE, the volume of total material used for construction is calculated and then multiplied with volume density. Then the EE per unit mass of material is multiplied by mass density of materials. Energy consumed by basic building materials is given in Appendix F.

Mud (site excavation): 0 MJ/kg; Cement: 7.8 MJ/kg; Steel: 32.0 MJ/kg; Brick: 5.0 MJ/kg; Glass: 15.9 MJ/kg; Aluminum 227 MJ/kg. It is clear that aluminum consumes about 7 times more energy than that of steel and about 14 times than glass. Thus, the amount of aluminum used in the building should be less.

9.3.2 EMBODIED ENERGY OF DIFFERENT CONSTRUCTION MATERIALS

Different materials based on the embodied energy and other factors used in the construction are discussed below:

A. **Adobe (stabilized soil block)**

Dried brick is referred to as adobe. Stabilized mud blocks are made from a mix of soil, sand, cement, lime and water. Thick adobe walls act as a thermal mass. This material is mostly used in rural areas and is capable of keeping the interiors warm in winter months and cool during the summer season. The energy content of adobe block, straw stabilized is 0.47 MJ/kg. Due to very low EE, it is considered an environmentally-friendly material. However, adobe is structurally not stable and durable when compared to burned brick and requires frequent maintenance. Also, it is vulnerable to water and rain.

B. **Hollow concrete masonry**

This is a modern building material with low mass and high compressive strength commonly used for the construction of non-load-bearing filler walls in multi-storied buildings. The EE of block greatly depends on the percentage of cement used. The basic composition of the blocks consists of cement, sand and coarse aggregates (~6 mm size). The energy content of the block will mainly depend upon the cement percentage. Energy spent for crushing of coarse aggregate will also contribute to the block energy. This percentage varies from 8%–10% by weight. Energy content of hollow blocks (200 mm × 100 mm × 100 mm) ranges between 5–7.5 MJ.

C. **Burnt clay brick**

Burnt clay bricks are frequently used in Indian construction. In case the soil type does not have sufficient clay, a binder in form of lime or cement is added. A major energy requirement in the production process is during the firing of bricks. The amount of

Environmental Aspects 263

energy required to produce each brick is approximate 5 MJ for a brick size of 100 mm × 50 mm × 50 mm as given by [3]. This material is environmentally harmful due to its low quality, very inefficient production processes and the use of local wood in brick kilns, which contribute to deforestation and air pollution. Another issue associated with this building material is excessive use of mortar due to reasons like uneven sizes of the bricks, poor construction quality, etc. The brick walls also require plastering resulting in an overall increase in the energy consumption of the brick wall. Therefore, it is important to improve the quality, production and construction processes to reduce the overall energy consumption, cost of the building and also mitigate their environmental impacts. Alternative construction methods and materials like interlocking blocks/bricks have been developed to reduce/eliminate the use of mortar in the construction processes of brick walls.

D. **Soil cement blocks**

They are manufactured by pressing a wetted soil–cement mixture into a solid block using a manually operated or mechanized machine, and then cured. Energy content of the blocks mainly depends upon the cement content. Soil–cement blocks used for the load-bearing masonry buildings will have cement content of about 6%–8%. Such blocks will have an energy content of 2.75–3.75 MJ per block of size 230 mm × 190 mm × 100 mm [1].

E. **Portland cement**

This is one of the major materials used in the construction industry. It is manufactured by two processes: the wet process and dry process. The wet process used in earlier cement plants leads to an energy consumption of 7.5 MJ/kg of cement whereas modern plants employing precalcination and dry process consume 4.2 MJ/kg of cement. The average value of 5.85 MJ/kg of cement has been used in the computation of energy in various components and systems.

F. **Hydrated lime**

The thermal energy consumed by hydrated lime is about 5.63 MJ/kg, which is equal to that of cement. This high energy content of lime is due to low thermal efficiency of small-scale kilns that are used for lime burning. Lime-pozzolana (LP) cements can provide very effective alternatives to Portland cement mainly for secondary applications such as masonry mortar, plastering, base/sub-base for flooring, etc. A typical LP cement consists of 30% lime, 60% pozzolana and 10% calcined gypsum, all three being inter-ground in a ball-mill. Such cement will have an energy content of 2.33 MJ/kg.

G. **Concrete**

Concrete is strong and durable material but is environmentally harmful because of high consumption of energy during the production of cement. When compared to burned brick, it is considered to be sustainable because of the energy-wasteful process in case of burnt brick manufacturing. The durability and long lifetime of concrete compensates for its high embodied energy

9.3.3 Embodied Energy in Floor/Roofing Systems

A. **Stabilized mud block (SMB) filler slab roof**

Commonly, RCC (reinforced cement concrete) slab is used for construction of slabs. A small portion of the bottom of the slab, below neutral axis, can be replaced by a filler material like SBM. This replacement leads to reduction in dead load of RCC slab resulting in savings of cost and energy. The total energy content of the materials constituting SMB filler slab is 590 MJ/m² of plan area of the slab. This is a floor slab designed as per IS456 code for a span of 3.6 m. There will be variations in energy content for different spans of the slab [1].

B. **Composite brick panel slab**

This consists of reinforced brickwork panel supported on RCC beams. The dimensions of the beam depend on the span and spacing of the beams. The energy content of such a slab for 3.6 m span is about 560 MJ/m² of projected plan area of the slab [1].

C. **Reinforced concrete ribbed slab roof**

The thickness of this slab is 50–60 mm and has small RCC beams. The spacing of beams ranges between 0.75–1 m, and the dimensions are dependent on the span of the space. This type of roof/floor slab can have an energy content of 491 MJ/m^2 of slab area, for a 3.6 m span [1].

D. **Masonry vault roof**

Unreinforced masonry vault roof consists of a thin masonry vault supported on ring beams with tie rods. Vault can be constructed using burnt clay bricks or SMB's. Total energy of the roof will be 575 MJ/m^2 and 418 MJ/m^2 for brick masonry and SMB masonry vault roofs respectively [1].

9.3.4 EMBODIED ENERGY IN TRANSPORTATION OF BUILDING MATERIALS

Transportation of materials also contributes majorly in the cost and energy of a building. Natural sand and crushed stone aggregate consume about 1.75 MJ/m^3 for every 1 km of transportation distance. Similarly, bricks require about 2.0 MJ/m^3 per km travel. Assuming steel and cement are also transported using trucks, diesel energy of 1 MJ/tonne/km is spent during transportation. Thermal energy spent for natural sand production is nil, but it requires about 175 MJ of diesel energy/m^3 for transporting it over a 100 km distance. Crushed aggregate consumes about 20 MJ/m^3 during its production and an additional 400%–800% more during transportation for distances of 50–100 km. The energy spent during transportation of bricks is about 4%–8% of its energy in production, for distances of 50–100 km. Transportation energy required for hauling high-energy materials, such as steel and cement, is marginal when compared to the energy spent during production. Table 9.1 gives diesel energy spent during transportation of various building materials along with the energy consumed in production [1].

9.3.5 EMBODIED ENERGY OF PV MODULE [12]

As discussed in Chapter 4, the calculations of total embodied energy for PV module per m^2 is not an easy task since it requires the energy spent in the manufacturing of each and every component.

The life cycle stages of PV, as shown in Figure 9.2, involve (1) the production of raw materials, (2) their processing and purification, (3) the manufacture of modules and balance of system (BOS) components, (4) the installation and use of the systems and (5) their decommissioning and disposal or recycling.

Tiwari and Ghosal [13] estimated that the material (i.e., silicon) required for manufacturing a PV module is 0.724 kg/m^2. The energy required to produce 1 kg of (MG-Si) by carbothermic reduction of silicon dioxide (SiO$_2$) is 20 kWh. The energy required to produce EG-Si from metallurgical-grade silicon (MG-Si) is 100 kWh/kg and there is a 90% yield. The EG-Si is melted in a Czochralski

TABLE 9.1
Specific Energy Spent in Transportation of Building Materials [1]

Material	Energy in Production	Specific Energy Spent in Transportation		Units
		50 km	75 km	
Sand	0	87.5	131.25	MJ/m^3
Crushed aggregate	20.5	87.5	131.25	MJ/m^3
Burnt clay bricks	2550	100	150	MJ/m^3
Portland cement	5858	50	75	MJ/tonnes
Steel	42,000	50	75	MJ/tonnes

Environmental Aspects

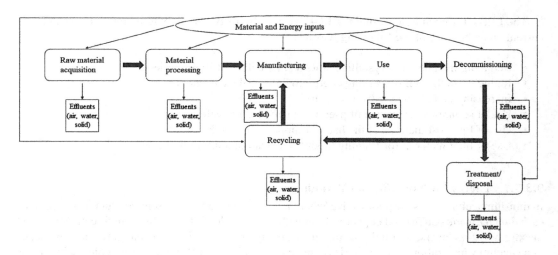

FIGURE 9.2 Flow of the life cycle stages, energy, materials and wastes for PV systems [14].

crystal puller at 1400°C and slowly crystallizes the silicon to form a single crystal ingot of silicon. The energy required is 210 kWh/kg with a total yield of 72%. The ingot is typically sliced with a thickness of 0.2–0.5 mm. Assuming the silicon wafer thickness to be 0.350 mm and the losses equivalent to 0.300 mm, giving a total yield of 54%. After trimming 156.25 cm^2 PV wafer, a pseudo square cell with an effective area of 142 cm^2 is prepared. Thus, the mass of each PV cell is 11.43 gm (142 × 0.035 × 2.3) with 91% yield and density $2.3\,\text{g}/\text{cm}^3$. The energy required to prepare 1 m^2 of silicon cell is 120 kWh. These solar cells are used to make a PV module with a packing factor of 0.82 (i.e., 82%) silicon and 18% open space between cells. The energy requirement to prepare a module is 190 kWh/m^2 [1].

For open field installation, the concrete, cement and steel are the main components used for foundation and frame, which requires maximum energy. The energy requirement for open field installation is 500 kWh/m^2 of panel. For rooftop-integrated PV systems, the energy requirement is 200 kWh/m^2 due to the absence of foundation and structure for the frame. The requirements for the BOS (i.e., all components that are a part of modules) will depend largely on the desired application. The energy requirement in different processes for production of PV module are given in Table 4.2. The EE of a PV module for 1 m^2 is given in Table 4.3.

Energy required for different processes for a unit panel are as follows:

A. Silicon purification and processing

- Production of 2.334 kg of MG-Si = 2.334 × 20 = 46.68 kWh
- Production of 2.011 kg of EG-Si = 2.011 × 100 = 201.10 kWh
- Production of 1.448 kg of EG-Si for Cz-Si = 1.448 × 210 = 304.08 kWh

B. Solar cell fabrication = 120 × (0.60534 × 0.83) = 60.29 kWh
C. PV module assembly of size 1.2 × 0.55 × 0.01 = 190 × 0.66 = 125.40 kWh
D. Installation/integration = 200 × 0.66 = 132 kWh

Hence, the total embodied energy required for installation/integration of a PV module (glass-to-glass) with PV/T systems = 869.55 ≈ 870 kWh.

The total embodied energy with BOS for open field is 870 + 500 = 1370 kWh and for rooftop is 870 + 200 = 1070 kWh.

The EPBT for open field in India is 12.45 years. The EPBT is likely to be much lower than the current value because of the following reasons:

- Reducing thickness of crystalline silicon solar cells to 150 μm or less.
- Reducing in the cell processing energy to 75% of the current value.
- Increasing cell efficiency from 14% to 23%.
- Use of square wafer instead of pseudo square-shaped wafers.
- Use of low-cost material at the back to support solar cells.
- Use of fiber material for framing modules instead of aluminum frame.

9.3.5.1 Energy for Non-Silicon PV Modules

In thin-film technologies the photoactive P/N junction is made up of two semiconductor compounds, *cadmium telluride* (CdTe) and copper indium gallium selenide (CIGS), which are directly deposited in extremely thin layers, ~10 and ~0.1 μm, respectively, on a treated transparent glass pane by means of a vacuum vaporization process. Series connection of adjacent P/N junctions is achieved by means of a series of automated laser and mechanical scribing processes, and then a second protective glass pane is added on top to form the finished module. Table 9.2 gives the quantities of inventory used in manufacturing of the CdTe and CIGS modules.

For the purposes of this analysis, the system boundaries were drawn around the module production facility, and thorough inventories of the necessary inputs for the production process of 1 m² of frameless modules were made. Table 4.8 gives the typical values concerning the energy content coefficient (ε_{PV}) of manufacturing of popular silicon- made and non-silicon-made PV modules.

Section 4.5 gives an example of calculation of embodied energy with a roof-mounted PV system with Table 4.7 having the total embodied energy calculations for the materials used for structure and frame.

9.3.5.2 Energy for Balance of System (BOS)

Building integrated photovoltaic systems are either rooftop- or façade-integrated (explained in Chapter 8), both operating with a proper balance of system (BOS). For a BiPVT application, the BOS typically includes charge controllers, inverters, battery bank, mounting structures and frames, cables and connectors. Large-scale ground-mounted PV installations require additional equipment and facilities, such as grid connections, office facilities and concrete.

TABLE 9.2

Inventory of Main Input Flows to the Cadmium Telluride (CdTe) and Copper Indium Gallium Selenide (CIGS) Module Manufacturing Process [1]

	Quantity	
Inventory	CdTe Module	CIGS Module
Glass	24,960 g/m²	24,960 g/m²
Water	1250 g/m²	1250 g/m²
Ethylene Vinyl Acetate (EVA)	630 g/m²	880 g/m²
(CdTe+CdS+CdCl$_2$+Sn+Ni/V+ITO+Sb$_2$Te$_{3)}$[1]	230 g/m²	-
(Mo+Cu+In+Ga+Se+CdS+ZnO+CuSn)[2]	-	0 g/m²
Electricity	236 kWh/m²	24.3 kWh/m²

[1] Cadmium Telluride + Cadmium sulfide + Cadmium chloride + Tin + Nickle/Vanadium + Indium tin oxide + Antimony telluride

[2] Molybdenum + Copper + Indium + Gallium + Selenium + Cadmium sulfide + Zinc oxide + Bronze

Environmental Aspects 267

Table 4.9 gives the typical values concerning the energy content coefficients and service period considered for the BOS components, the system installation, maintenance and operation stages, and the diesel–electric generator.

9.3.6 EMBODIED ENERGY AND ANNUAL OUTPUT OF RENEWABLE ENERGY TECHNOLOGIES

The estimation of embodied energy of a given technology is simple process. The estimation of embodied energy of a given renewable energy technology except photovoltaic modules can be done as follows:

a) The mass of materials (m_i) involved in the manufacturing process to be multiplied by its corresponding energy density (e_i), we get

$$m_i \times e_i$$

b) $\sum (m_i \times e_i)$

9.3.7 GUIDELINES FOR REDUCING EMBODIED ENERGY

Material selection and construction methods can significantly change the total amount of embodied energy in the building. True low-energy building design will consider this important aspect and take a broader life cycle approach to energy assessment. Merely looking at the energy used to operate the building is not really acceptable. Operational energy consumption depends on the occupants. Embodied energy content is incurred once (apart from maintenance and renovation), whereas operational energy accumulates over time and can be influenced throughout the life of the building. Reuse of building materials commonly saves about 95% of embodied energy that would otherwise be wasted. The building should be selected for the best combination for its application based on climate, transport distances, availability of materials and budget, balanced against known embodied energy content. General guidelines for reducing embodied energy are as follows [13]:

- Design for long life and adaptability, using durable low-maintenance materials.
- Ensure materials can be easily separated.
- Avoid building a bigger house than needed. This will save materials.
- Modify or refurbish instead of demolishing or adding.
- Ensure materials from demolition of existing buildings, and construction wastes are reused or recycled.
- Use locally sourced materials (including materials salvaged on site) to reduce transport.
- Select low-embodied energy materials (which may include materials with a high recycled content) preferably based on supplier-specific data.
- Avoid wasteful material use.
- Specify standard sizes and don't use energy-intensive materials as fillers.
- Ensure off-cuts are recycled and avoid redundant structure, etc. Some very energy-intensive finishes, such as paints, often have high wastage levels.
- Select materials that can be reused or recycled easily at the end of their lives using existing recycling systems.
- Give preference to materials manufactured using renewable energy sources.
- Use efficient building envelope design and fittings to minimize materials (e.g., an energy-efficient building envelope can downsize or eliminate the need for heaters and coolers, water-efficient taps allow downsizing of water pipes).
- Ask suppliers for information on their products and share this information.

9.4 MODELING OF EMBODIED ENERGY FOR BiPVT SYSTEMS [1]

As already discussed, the embodied energy in building-integrated systems involve various stages like manufacturing, installation, maintenance and operation (M&O), final decommissioning and recycling.

Now, considering the building and all the components the total embodied energy of the configuration will be

$$E_{tot} = E_{in} = E_{building} + E_{support} + E_{BIPVT} + E_{cc} + E_{inv} + E_{bat} + E_{inst+M\&O} + E_{dec} - E_{rec} \qquad (9.1)$$

where $E_{building}$, $E_{support}$, E_{BIPVT}, E_{cc}, E_{inv}, E_{bat}, $E_{inst+M\&O}$, E_{dec} and E_{rec} are the embodied energy of the building, support for the BiPVT system, BiPVT system, charge controller, inverter, battery, installation M&O, decommissioning and recycling, respectively.

9.4.1 MASONRY BUILDING

The embodied energy of the building is the sum of the embodied energy of all the construction material used. Mathematically,

$$E_{building} = E_{bricks} + E_{cement} + E_{sand} + E_{concrete} + E_{lime} + E_{MS} + E_{glass} + E_{paint} + E_{plywood} + E_{floor} \qquad (9.2)$$

$$\begin{aligned} E_{building} = {} & m_{bricks} \times \xi_{bricks} + m_{cement} \times \xi_{cement} + m_{sand} \times \xi_{sand} + m_{concrete} \times \xi_{concrete} + m_{lime} \times \xi_{lime} \\ & + m_{MS} \times \xi_{MS} + m_{glass} \times \xi_{glass} + m_{paint} \times \xi_{paint} + m_{plywood} \times \xi_{plywood} + A_{floor} \times \varepsilon_{floor} \end{aligned} \qquad (9.3)$$

where m is the mass and ξ is the embodied energy per unit mass.

Let us assume that the support for mounting BiPVT is made up of made up of aluminum having a service time equal to n_{sys}. Its embodied energy is

$$E_{support} = m_{Al} \times \xi_{Al} \qquad (9.4)$$

where m_{Al} is the mass of aluminum used in construction of support and ξ_{Al} is the embodied energy of the aluminum per unit mass.

9.4.2 PHOTOVOLTAIC THERMAL (PVT) SYSTEM

The embodied energy of the PVT is the sum of the embodied energy of the PV array, frame and duct. Therefore,

$$E_{BIPVT} = E_{PV} + E_{frame} + E_{duct} \qquad (9.5)$$

Usually, frame is made up of aluminum alloy, and the duct is made of plywood. Thus, for a PVT system having PV array area A_{PV} and service time n_{sys}, the embodied energy is

$$\begin{aligned} E_{BIPVT} = {} & A_{PV} \cdot \varepsilon_{PV} \left[1 + \text{int} \left(\frac{n_{sys} - 1}{n_{PV}} \right) \right] + A_{frame} \cdot \varepsilon_{frame} \left[1 + \text{int} \left(\frac{n_{sys} - 1}{n_{frame}} \right) \right] \\ & + A_{PV} \cdot \varepsilon_{plywood} \left[1 + \text{int} \left(\frac{n_{sys} - 1}{n_{plywood}} \right) \right] \end{aligned} \qquad (9.6)$$

Environmental Aspects 269

where ε_{PV}, ε_{frame} and $\varepsilon_{plywood}$ is the embodied energy per unit area for PV module, PV frame and the plywood used for duct manufacture, respectively. n_{PV}, n_{frame} and $n_{plywood}$ is the service time for PV module, PV frame and the duct plywood, respectively.

The panel area A_{PV} results from the peak power of the PV generator N_{PV} and the corresponding efficiency η_{PV}

$$A_{PV} = \frac{N_{PV}}{\eta_{PV}.G} \tag{9.7}$$

where G is the solar radiation at STC and its value is considered as 1000 W/m^2.

9.4.3 BALANCE OF SYSTEM (BOS)

The embodied energy of charge controller with rated power P_{cc} is

$$E_{cc} = \zeta_{cc} \cdot P_{cc} \left[1 + \text{int} \left(\frac{n_{sys} - 1}{n_{cc}} \right) \right] \tag{9.8}$$

where P_{cc} is the rated power of the charge controller, which is slightly higher than the peak power generated by the PV array.

$$P_{cc} = P_{PV-peak} + \delta P_{cc} \approx P_{PV-peak} \tag{9.9}$$

The inverter should have its rated capacity (P_{inv}) such that it fulfils the peak load demand (P_{LD}) of the consumer. Its embodied energy during n_{sys} is

$$E_{inv} = \zeta_{inv} \cdot P_{inv} \left[1 + \text{int} \left(\frac{n_{sys} - 1}{n_{inv}} \right) \right] \tag{9.10}$$

The rated capacity of the inverter is given by

$$P_{inv} = P_{LD} \left(1 + SF \right) \tag{9.11}$$

where SF is the safety factor, which is usually considered as 0.3.

The rated capacity of the battery C_{bat} should fulfil the necessary demand of the consumer during off-sunshine hours. Its embodied energy during n_{sys} is

$$E_{bat} = K_{bat} \cdot C_{bat} \left[1 + \text{int} \left(\frac{n_{sys} - 1}{n_{inv}} \right) \right] \tag{9.12}$$

Minor energy is required during the activities of installation of the system and waterproofing. Components such as junction boxes, cables, screws, etc., may be reused or changed with a new one. However, the BiPVT system with shorter lifespan of PV needs be dismantled and installed more frequently, which increases the cost of installation while a system with a longer lifespan of PV has a relatively low cost of installation. In addition to this, energy is also required for periodic maintenance and operation of the system. Therefore,

$$E_{inst+M\&O} = E_{inst} + E_{repl} + E_{M\&O} \tag{9.13}$$

9.5 EMBODIED CARBON

Embodied carbon can be referred to as the carbon dioxide or GHG emissions associated with the manufacture and use of a service or product. In simpler words, it is the CO_2 or GHG emission associated with extraction, fabrication, transporting, installing, maintaining and disposing of construction materials and products. The major part of the embodied carbon for a construction activity/product is the CO_2 emitted from the use of fossil fuel during the process of extraction and manufacturing [15].

It can be said buildings that are efficient in terms of the amount of materials used are also efficient in terms of EE and carbon. The quantity of material used in the construction process can be optimized by improving overall efficiency of the designs. This can be achieved by using better specifications, optimal structural designs, designing while keeping future requirements in mind, etc. Also, reduction in wastage of materials is important by having skilled labor, etc. Another important way to reduce the impact of embodied energy/carbon is to recycle and reuse the materials [16].

The embodied energy and embodied carbon are directly related, but the impact on deletion of resources and GHG emissions of any material is different. It depends on the primary fuel used and means of electricity generation. Renewable energy can be considered to have zero emissions if the embodied energy of the collectors and generators are neglected. Therefore, it can be stated that the embodied carbon is dependent on the fuel mix. Some materials also have negative embodied carbon when carbon sequestered during their growth has been accounted for. Trees and short-term crops used for building materials sequester atmospheric carbon dioxide during their growing period, the weight of which may be greater than the emissions produced during manufacture [17].

As given by Achintha [16], up to 5% of the energy required in the original production using the raw materials can be saved by using recycled glass. This reduction is comparatively much lower than that attained by recycling aluminum and plastic, where the potential energy savings are 95% and 88%, respectively.

Stabilized soil blocks are energy efficient and also have low embodied carbon associated with them when compared to conventional fired clay bricks and concrete blocks, and they provide nearly three times better thermal resistance. When the layer of insulation is added to conventional walls to make the comparison on an equal thermal efficiency basis, larger differences in embodied burdens are depicted. Thus, they can be used as an alternative to structural masonry materials [18]. There is an upsurge of interest among building professionals in utilizing low-embodied carbon materials.

Concrete is the predominant construction material, with global production approaching 20×10^{12} kg per annum, which accounts for much more than all other construction materials combined, and increasing at several percentage points annually as large developing nations upgrade and install infrastructure [19]. Concrete is a complex composite, which results in number of concrete ranges. Cement is responsible for concrete's large carbon footprint; a tonne of cement represents about a tonne of greenhouse gas emissions [20]. Energy of cement arises from the use of coal in the rotary kilns and energy needed for crushing and grinding the clinker.

9.5.1 EXAMPLE OF ESTIMATION OF EMBODIED CARBON DIOXIDE FOR CONCRETE [21]

Estimation of embodied CO_2 can be done for a cubic meter of in situ concrete for any mix using either natural (conventional) or lightweight aggregates when the production of CO_2 during the manufacturing process is known for that constituent and before adding the energy necessary for mixing, transporting and placing. Embodied CO_2 is based on the total amount of consumption of fuel and the production of CO_2 in kilograms per metric tonne of product; thus embodied CO_2 can be expressed in (kg CO_2/tonne) [21].

The datum used for estimation of embodied CO_2 for C30 concrete based on Portland cement, water and fine aggregates (silica sand) and coarse aggregates (granite) is given in Table 9.3(a).

Environmental Aspects

TABLE 9.3

Calculation of Embodied Carbon Dioxide for (a) a Traditional Concrete Mix [21], (b) Concrete Containing Fly Ash and Coarse Aggregate Combination [21] and (c) Concrete Containing Fly Ash and Natural Coarse Aggregate [21]

Constituent	Mix Concrete (kg/m³)			Embodied CO_2 (kg CO_2/tonne)			Embodied CO_2 (kg CO_2/m³)		
	(a)	(b)	(c)	(a)	(b)	(c)	(a)	(b)	(c)
Water	190	170	170	2	2	2	0.4	0.4	0.4
Portland cement	380	260	260	930	930	930	353.4	241.8	241.8
EN 450 fly ash (Class S)	-	130	120	-	150	50	-	2.6	6
Sand	630	640	650	8	8	6	5	5.2	3.9
5–20 mm granite	1200	300	1200	25	25	25	30	7.5	29.6
15 mm lightweight concrete	-	300	-	-	220	-	-	66	-
Total fresh concrete density	2400	1800	2400						
Total embodied CO_2							388	298	282

Table 9.3(a) includes the delivery of carbon dioxide for the Portland cement and water but excludes the delivery of carbon dioxide for the water contained with the washed fine and coarse natural aggregates as given in Table 9.3(b).

In order to reduce the embodied CO_2 obtained in Table 9.3(a), one needs to reduce the density of fresh concrete. Thus, a mix of 66.7% Portland cement, 33.3% of BS EN 450 class fly ash together with coarse aggregate combination where 60% of natural granite volume aggregate is replaced with lightweight aggregates. By doing this, a 25% reduction of density of fresh concrete can be obtained, as given in Table 9.3(b).

Reduction in the weight of the building of at least 12.5% is found due to reduction in mass of 25% with lightweight concrete in most of the cases. By doing this, it in turn allows for options either for reduced foundations or taller and larger buildings for the same foundation. This has a significant effect on the quantity of steel reinforcement used, which if reduced will have a substantial reduction in the embodied CO_2, irrespective of the source of steel.

Table 9.3(c) completes the analysis by necessarily designing a further mix that optimizes the fly ash and natural coarse aggregate.

The carbon dioxide from transport also needs to be taken under consideration. For this, the distance analysis to transport the conventional concrete materials was calculated with an assumption that typical delivery distance equals 50 km for the natural aggregates. The CO_2 arising from transport is taken as 63 g CO_2/tonne/km .

9.6 CARBON DIOXIDE EMISSIONS

The conservation of conventional energy resources reduces the conventional energy running costs, which leads to economic, political and environmental sustainability by incorporating CO_2 credit. Energy conservation is also an important method to minimize climate change due to reduction of CO_2 emission, which is produced during conventional power generation by using fossil fuels. In the current scenario, the reduction of CO_2 emissions in the energy sector by using energy-efficient equipment is the most important area of study. Energy conservation is the most economical solution to energy shortages in under-developing and developing countries. Energy conservation reduces stables energy cost and energy import and eliminates the requirement of new power plants [22]. Thus, one of the key indicators of any country's socioeconomic aspect is energy consumption. CO_2 emissions into the atmosphere also will exhaust the conventional sources of energy [23].

The GDP growth rate of India for 2019 was 4.18% (which was a 1.94% decline from the year 2018). The GDP growth rate of India for 2018 was 6.12% (which was a 0.92% decline from the year 2017). For the year 2017, GDP growth rate of India was 7.04% (which was again a decline of 1.21% from 2016) [24]. Thus, we need to sustain this growth rate. For this, we definitely need an additional secured and reliable energy source. India accounts for more than a quarter of net global primary energy demand growth between 2017–2040 [25].

In India, 42% of its energy demand is met through coal resulting double the emissions of carbon dioxide by the year 2040 [26]. Gas production is growing but is still unable to meet the demand, resulting in a significant amount of gas imports. Currently, renewable energy consumption is expected to surge from about 20 Mtoe to about 300 Mtoe by 2040. This is mainly concentrated in the power sector driven by the growth in solar capacity. Despite the growth in the renewable energy, coal still dominates the Indian power generation mix, accounting for about 80% of output by 2040. As a result, although the carbon intensity of India's power grid is projected to decline by 29% by 2040, it remains 58% above the global average. India's total net carbon dioxide emissions roughly double to 5 Gt by 2040, meaning India's share of global emissions increases from 7% today to 14% by 2040 [26].

India will exhaust its oil reserves in 22 years, its gas reserves in 30 years and its coal reserves in 80 years [27]. More alarming, the coal reserves might disappear in less than 40 years if India continues to grow at 8% a year, which reduce the loss of heat into space and therefore contribute to global temperatures through the greenhouse effect [28, 29].

Greenhouse gases (GHG) are the gases present in the earth's atmosphere. GHG are essential to maintaining the temperature of the earth; without them the planet would be so cold as to be uninhabitable. If there were absolutely no greenhouse gases (GHGs), the average surface temperature of the earth would be about $-18°C$ [30]. However, an excess of greenhouse gases can raise the temperature of a planet to lethal levels, as on Venus where the 90 bar partial pressure of carbon dioxide contributes to a surface temperature of about $467°C$ ($872°F$) [29]. Today, climatic changes is the major challenge of the world. Human emissions of GHG have increased global temperature by $1°C$ since pre-ancient times [30].

Greenhouse gases are produced by many natural and industrial processes, which currently result in carbon dioxide levels of 380 ppm in the atmosphere. Based on ice-core samples and records current levels of CO_2 that is approximately 100 ppm higher than during immediately pre-industrial times, when direct human influence was negligible [29]. Carbon dioxide concentration in the atmosphere is currently over 400 ppm, which constitutes to the highest level in over 800,000 years [31].

With reference to Figure 9.3, over 36 billion tonnes of carbon dioxide is emitted per annum globally, and this still continues to increase. When compare the CO_2 emissions country-wise, there are large-fold differences, ranging more than 100-fold in per capita CO_2 emissions. Today, China is the world's largest carbon dioxide emitter with more than one-quarter of global emissions. This is followed by the United States with 15%, Europe with 7% and Russia with 5%.

The United States has contributed most to global carbon dioxide emission and accounts for 25% of cumulative emissions followed by Europe with 22%, China with 13 %, Russia with 6% and Japan with 4% [31].

Annual total Carbon dioxide emissions from various global regions is given in Figure 9.3.

It can be seen from Figure 9.3 that global emissions increased from 2 billion tonnes of carbon dioxide in 1900 to over 36 billion tonnes 115 years later. While data from 2014 to 2017 suggested global annual emissions of CO_2 had approximately stabilized, data from the Global Carbon Project [32] reported a further annual increase of 2.7% and 0.6% in 2018 and 2019, respectively.

Environmental Aspects

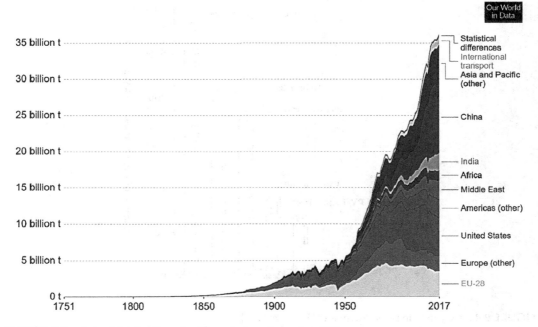

FIGURE 9.3 Annual total carbon dioxide emissions, by region [31].

The average carbon dioxide (CO_2) equivalent intensity for electricity generation from coal is approximately 0.98 kg of CO_2/kWh [33]. If the PV system has lifetime of 35 years, the CO_2 emissions per year by each component can be calculated as

$$CO_2 \text{ emissions per year} = \frac{Embodied\ energy \times 0.98}{Lifetime} \quad (9.14)$$

The CO_2 emissions per year for a PV module (glass to glass) (effective area = 0.60534 m² and size =1.20 m × 0.55 m × 0.01 m) is given in Table 9.4. The CO_2 emissions for different PVT systems are shown in Figure 9.4 [29].

TABLE 9.4
CO_2 Emissions per Year from a PV Module (Glass to Glass) (Effective Area = 0.60534 m²) [29]

S.No.	Components	Embodied Energy (kWh)	CO_2 Emissions (kg)
1	Metallurgical-grade silicon (MG-Si)	26.54	0.74
2	Electronic-grade silicon (EG-Si)	127.30	3.56
3	Czochralski silicon (Cz-Si)	267.33	7.49
4	Solar cell fabrication	60.29	1.69
5	PV module assembly	125.40	3.51
Total		**606.86**	**16.99**

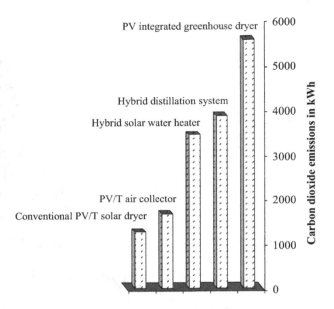

FIGURE 9.4 CO_2 emissions for different PVT systems [29].

EXAMPLE 9.1

Calculate carbon dioxide emission per year from a solar water heater in a lifetime of 10, 20 and 30 years, when the total embodied energy required for manufacturing the system is 3550 kWh.

Solution

Using Equation (9.14), we have:

For lifetime = 10 years

$$CO_2 \text{ emissions per year} = \frac{3550 \times 0.98}{10} = 347.9 \text{ kg of } CO_2$$

Similarly, for lifetime = 20 and 30 years CO_2 emissions per year is 173.9 and 115.9 kg of CO_2, respectively.

9.7 EARNED CARBON CREDITS AND CARBON DIOXIDE MITIGATION [1]

Trading of carbon credits was established to curb the greenhouse gases by reduction in the carbon footprint. Carbon credits can be defined as "a key component of national and international emissions trading schemes that have been implemented to mitigate global warming" [1]. This helps to reduce the emissions of GHG on an industrial scale by capping the total annual emissions and letting the market assign a monetary value to any shortfall through trading. Credits can be exchanged between businesses or bought and sold in international markets at the prevailing market price. Credits can be used to finance carbon reduction schemes between trading partners around the world. There are also many companies that sell carbon credits to commercial and individual customers who are interested in lowering their carbon footprint on a voluntary basis. These carbon off-setters purchase the

Environmental Aspects

275

credits from an investment fund or a carbon development company that has aggregated the credits from individual projects. The quality of the credits is based in part on the validation process and sophistication of the fund or development company that acted as the sponsor to the carbon project.

A credit means one-tonne carbon dioxide (1 credit = 1 tonne CO_2e) [34]. Kyoto Protocol is an international treaty that sets quotas for greenhouse gases for each participating country.

9.7.1 FORMULATION

If unit power is used by a consumer and the losses due to poor domestic appliances is L_a, then the transmitted power should be $\dfrac{1}{1-L_a}$ units.

If the transmission and distribution losses is L_{td}, the power that has to be generated in the power plant is $\dfrac{1}{1-L_a} \times \dfrac{1}{1-L_{td}}$ units.

The average carbon dioxide equivalent intensity for electricity generation from coal is approximately 0.98 kg of CO_2/kWh at the source. Thus, for unit power consumption by the consumer, the amount of CO_2 emission $= \dfrac{1}{1-L_a} \times \dfrac{1}{1-L_{td}} \times 0.98$

The annual CO_2 emission can be expressed as:

$$CO_2\ emission\ per\ year = \frac{E_{in}}{n_{sys}} \times \frac{1}{1-L_a} \times \frac{1}{1-L_{td}} \times 0.98\,kg \tag{9.15}$$

where E_{in} is the embodied energy input and n_{sys} is the lifetime of the system.

The CO_2 emission over the lifetime of the system is:

$$CO_2\ emission\ over\ the\ lifetime = E_{in} \times \frac{1}{1-L_a} \times \frac{1}{1-L_{td}} \times 0.98\,kg \tag{9.16}$$

The net CO_2 mitigation over the lifetime of the system is:

$$Net\ CO_2\ mitigation\ over\ the\ lifetime = Total\ CO_2\ mitigation - total\ CO_2\ emission$$

or

$$= \left(E_{aout} \times T_{LS} - E_{in}\right) \times \frac{1}{1-L_a} \times \frac{1}{1-L_{td}} \times 0.98\,kg \tag{9.17}$$

where E_{aout} is the overall exergy gain, which is the sum of the annual electrical exergy ($\dot{E}x_{el}$) and the annual thermal exergy equivalent ($\dot{E}x_{th}$). Substituting its value we have, annual exergy output as

$$E_{aout} = \left[\eta_{ca} \times I\left(t\right) \times bL \times n_s \cdot n_p\right] + \left[\dot{Q}_u \times \left(1 - \frac{T_a}{T_{airout}}\right)\right] \tag{9.18}$$

Net CO_2 mitigation over the lifetime in tonnes of CO_2 is given by

$$= \left(E_{aout} \times n_{sys} - E_{in}\right) \times \frac{1}{1-L_a} \times \frac{1}{1-L_{td}} \times 0.98 \times 10^{-3} \tag{9.19}$$

If CO_2 emission is being traded at x US\$ C/tonnes of CO_2 mitigation then the carbon credit earned by the system is:

$$Earned\ carbon\ credit = x\,\$C\times\left(E_{aout}\times n_{sys}-E_{in}\right)\times\frac{1}{1-L_a}\times\frac{1}{1-L_{td}}\times0.98 \qquad (9.20)$$

9.8 CASE STUDY WITH THE BiPVT SYSTEM

If unit power is used by a consumer and the losses due to poor domestic appliances are around 20%, then the transmitted power should be $\frac{1}{1-0.2}=1.25$ units. If the transmission and distribution losses are 40%, common in Indian conditions, then the power that has to be generated in the power plant is $\frac{1.25}{1-0.4}=2.08$ units. The average CO_2 equivalent intensity for electricity generation from coal is approximately 0.98 kg of CO_2 per kWh at the source. Thus, for unit power consumption by the consumer the amount of CO_2 emission is $2.08\times0.98=2.04$ kg. For the BiPVT system, the annual CO_2 mitigation in tonnes of CO_2 is given by:

$$\left(E_{aout}-\frac{E_{in}}{n_{Sys}}\right)\times2.04\times10^{-3} \qquad (9.21)$$

Assume that the overall embodied energy for following BiPVT systems

monocrystalline silicon (c-Si, $n_{sys}=30$ years) is 607,613 MJ,
multi-crystalline silicon (p-Si, $n_{sys}=30$ years) is 540,628 MJ,
ribbon silicon (r-Si, $n_{sys}=25$ years) is 409,716 MJ,
amorphous silicon (a-Si, $n_{sys}=20$ years) is 272,324 MJ,
cadmium telluride (CdTe, $n_{sys}=15$ years) is 211,984 and
copper indium gallium selenide (CIGS, $n_{sys}=5$ years) is 63,937 MJ.

The overall exergy calculations for the climatic conditions of New Delhi show that c-Si, p-Si, r-Si, a-Si, CdTe and CIGS BiPVT systems covering 45 m^2 of roof area generate 16,224; 14,352;12,512; 7790; 9547 and 11,037 kW of overall exergy output, respectively. Thus, the annual CO_2 mitigation for c-Si, p-Si, r-Si, a-Si, CdTe and CIGS systems are 77.83, 68.64, 58.45, 29.44, 41.29 and 54.97 tonnes, respectively. If CO_2 emissions are being traded at US\$20 per tonnes of CO_2 mitigation, then the carbon credit earned by the BiPVT system with the c-Si, p-Si, r-Si, a-Si, CdTe and CIGS technologies are US\$1557, 1373, 1169, 589, 826 and 1099, respectively. This shows that the monocrystalline silicon BiPVT system gives the highest earnings through carbon credit trading.

9.9 KYOTO PROTOCOL AND THE UNITED NATIONS FRAMEWORK CONVENTION ON CLIMATE CHANGE

The Kyoto Protocol is an international treaty that extends the 1992 United Nations Framework Convention on Climate Change (UNFCCC) that commits state parties to reduce the emissions of GHG, based on fact that global warming is occurring due to human-made emissions of carbon dioxide. The Protocol was initially adopted on December 11, 1997 in Kyoto, Japan, and came into force on February 16, 2005. The most notable non-member of the Protocol is the United States, which is a signatory of UNFCCC and was responsible for 36.1% of the 1990 emission levels [1].

Environmental Aspects

The following were agreed upon by the parties:

- Developed counties contributed to the largest share of historical and current global emissions of GHGs.
- Per capita emissions in developing countries are still relatively low.
- The share of global emissions originating in developing countries will grow to meet social and development needs.

United Nations Framework Convention on Climate Change (UNFCCC or FCCC) is an international environmental treaty produced at the United Nations Conference on Environment and Development (UNCED), also referred to as the Earth Summit, which was held in Rio de Janeiro in 1992 and had the classification of parties as Annex I Countries, Annex II Countries and Non-Annex I Counties.

Annex I countries are basically the industrialized countries that were members of the Organization for Economic Co-operation and Development (OECD). 1992. In addition to these, countries with economies in a transition stage, including the Russian Federation, the Baltic States and several Central and Eastern European States are part of this Annex I.

Annex I countries pledge to a 5.2% reduction in the production of their collective shares of four greenhouse gases [i.e., carbon dioxide (CO_2), nitrous oxide (N_2O), methane (CH_4) and sulfur hexafluoride (SF_6)] and two industrial gases, namely hydrofluorocarbons and perfluorocarbons. All member countries gave general commitments. The reduction of CO_2, N_2O and CH_4 will be measured against a base year of 1990, while that of hydrofluorocarbons (HFCs), perfluorocarbons (PFCs) and SF_6 can be measured against either a 1990 or 1995 baseline. These reduction targets are in addition to the industrial gases and chlorofluorocarbons (CFCs), which are dealt with under the 1987 Montreal Protocol on substances that deplete the ozone layer.

National or joint targets for the reduction ranges from

- 8% for the European Union (EU) and others
- 7% for the US (non-binding as the US is not a signatory)
- 6% for Japan
- 0% for Russia

Emission limits do not include emissions by international aviation and shipping.

There is no immediate restriction for the Non-Annex I countries to serve the purposes as follows:

- It avoids restrictions on their development because emissions are strongly linked to industrial capacity.
- They can sell emissions credits to nations whose operators have difficulty meeting their emissions targets.
- They get money and technologies for low-carbon investments from Annex II countries.

Annex II Countries consist of 24 original members of OECD and EU but not the economies in transition parties. They have to provide the financial resources to enable developing counties to undertake reduction in emission activities under the convention and to help them adapt to adverse effects of climate change. In addition to above, they have to promote the development and transfer of environmentally-friendly technologies to economies in transition parties and developing countries.

Non-Annex I countries are mostly the developing countries that reflect their less advanced economic development and their lower GHG emissions to date. The overall emissions of these countries do not grow as fast as those of Annex I countries. Non-Annex I countries may volunteer to become Annex I countries when they are sufficiently developed. Some opponents of the convention argue that the split between Annex I and Non-Annex I countries is unfair and both need to reduce their emissions unilaterally. Some countries claim that their costs of following the convention requirements will stress their economy.

9.9.1 THE PROTOCOL AND THE GREEN GROWTH

Various flexible mechanisms are provided by the protocol referred to as Kyoto Mechanisms and allow the Annex I Countries to meet their targets of GHGs emissions by acquiring GHGs emission reduction credits. These credits are earned by the Annex I countries via financial exchanges and projects that reduce the emissions in non-Annex I countries, from other Annex I countries or from Annex I countries with excess credits. Following are the flexible mechanisms in the Kyoto Protocol:

- Emissions trading (ET): This mechanism enables the Annex I countries to buy emission permits from the Annex I countries to meet their domestic emission reduction targets.
- Joint Implementation (JI): Annex I countries can offset their emissions through the purchase of Emission Reduction Units (ERU) from offset projects from any other Annex I countries as an alternative to meet their domestic emission reduction targets.
- Clean Development Mechanism (CDM): Annex I countries can meet their domestic emission reduction targets through the purchase of Certified Emission Reductions (CERs) from offset projects non-Annex I countries that are signatory to the UNFCCC.

Meaningful emission reductions within a trading system can only occur if they can be measured at the level of operator or installation and reported to a regulator, there is an open source tool to measure and plan the emissions accordingly. For greenhouse gases all trading countries maintain an inventory of emissions at the national and installation level; in addition, the trading groups within North America maintain inventories at the state level through The Climate Registry. For trading between regions these inventories must be consistent, with equivalent units and measurement techniques. In some industrial processes emissions can be physically measured by inserting sensors and flowmeters in chimneys and stacks, but many types of activity rely on theoretical calculations for measurement. Depending on local legislation, these measurements may require additional checks and verification by government or third-party auditors, prior or post submission to the local regulator [1, 35].

9.10 CARBON DIOXIDE MITIGATION WITH USE OF PHOTOVOLTAICS

The countries are moving towards non-renewable sources of energy to reduce the emission of carbon dioxide and follow Kyoto Protocol. Among the renewable sources of energy, solar energy is considered to be economical and pollution-free. Therefore, photovoltaic (PV) systems are best suited for maintaining the increase in demand of energy and keeping the environment free from pollution. PV electricity use replaces the power from fossil fuel plants, reduces carbon emissions and thus slows down the greenhouse effect.

The estimation of embodied energy is important for renewable energy system to assess the amount of energy required in construction and fabrication of the complete system. In case of PV systems, embodied energy and energy payback time (refer to Chapter 4) is evaluated. This is important to compare the energy produced by the system with the energy required by the system for its manufacturing. This analysis is done by life cycle analysis of the system. LCA is discussed in Chapter 10.

Carbon dioxide emissions for PV systems in India has been assessed by Chaurey and Kandpal [36]. It was found that that carbon credits reduce the financial burden on users by 19% and reduce pollutant emissions. CO_2 mitigation potential for a PV system was also analyzed by Purohit [37] for residential buildings. This was done under the clean development mechanism (CDM) in India, and the theoretical potential number of SPV came out to be 97 million. In an experimental setup [34],

Environmental Aspects

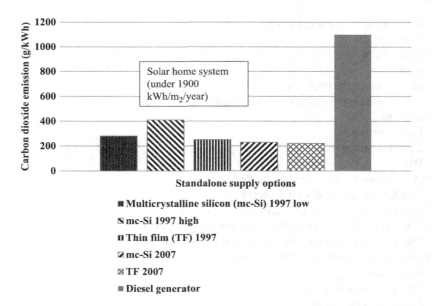

FIGURE 9.5 Life cycle carbon dioxide emissions from solar home systems and from a diesel generator [38].

it was found that the reduction in emissions of CO_2 with use of a PV system against a coal-based plant for the generation of electricity is 2.525 tonnes CO_2e with monitoring savings of about Rs.143,162.19 for a lifetime of 35 years. Thus, the monitory savings due to carbon credit should also be considered during economic analysis of the PV system. Alsema [38] found out the potential for CO_2 mitigation for a solar home system for a setup in Indonesia comprised of a 50 Wp module and a 70 Ah battery yielding 1.30 kWh/Wp/year under a 1900 kWh/m²/year irradiation. Life cycle carbon dioxide emissions from solar home systems and from a diesel generator are given in Figure 9.5 based on the setup by [38]. It can be seen that the CO_2 emissions from the PV installation are significantly lower than that of diesel generator. One consequence of this result is that one should be careful when attributing a large CO_2 mitigation potential to solar home systems.

Projections show that to achieve its target of a carbon neutral and sustainable energy supply system, Germany requires 200 GW of installed PV in total [39].

Chapter 11 gives examples of application of photovoltaics worldwide and their respective reduction in carbon dioxide emissions.

OBJECTIVE QUESTIONS

9.1 For zero embodied energy, CO_2 emission is
 (a) < 1
 (b) > 1
 (c) = 1
 (d) = 0

9.2 1 kWh is equivalent to how many kg CO_2 emission from the following at the source of coal base power generation
 (a) 0.98 kg CO_2 emission
 (b) 9.8 kg CO_2 emission
 (c) 0.098 kg CO_2 emission
 (d) none of them

9.3 The lowest embodied energy is of _____
  (a) Concrete
  (b) Burnt brick
  (c) Adobe
  (d) Hollow concrete masonry
9.4 _____ is responsible for concrete's large carbon footprint
  (a) Cement
  (b) Fine aggregates
  (c) Coarse aggregates
  (d) Water
9.5 Energy-efficient material can be referred to as
  (a) One having high embodied energy
  (b) One having low embodied carbon
  (c) One having high embodied energy and embodied carbon
  (d) One having low embodied energy and embodied carbon
9.6 Carbon dioxide emissions are based upon
  (a) Embodied energy
  (b) Energy payback period
  (c) Energy production factor
  (d) None of the above

ANSWERS

9.1 (d)
9.2 (a)
9.3 (c)
9.4 (a)
9.5 (d)
9.6 (a)

REFERENCES

[1] B. Agrawal and G. Tiwari, *Building Integrated Photovoltaic Thermal Systems: For Sustainable Developments*, United Kingdom: RSC Publications, 2010.

[2] NOAA, "Earth system research laboratory: Global monitoring system," *NOAA Research*, [Online]. Available: https://www.esrl.noaa.gov/gmd/ccgg/trends/. [Accessed 2020].

[3] S. Mishra and J. A. Usmani, "Comparison of embodied energy in different masonry wall materials," *International Journal of Advanced Engineering Technology*, pp. 90–92, 2013.

[4] L. F. Cabeza, L. Rincón, V. Vilariño, G. Pérez and A. Castell, "Life cycle assessment (LCA) and life cycle energy analysis (LCEA) of buildings and the building sector: A review," *Renewable and Sustainable Energy Reviews*, vol. 29, pp. 394–416, 2014.

[5] T. Ramesh, R. Prakash and K. K. Shukla, "Life cycle energy analysis of buildings: An overview," *Energy and Buildings*, vol. 42, pp. 1592–1600, 2010.

[6] A. Hashemi, H. Cruickshank and A. Cheshmehzangi, "Environmental impacts and embodied energy of construction methods and materials in low-income tropical housing," *Sustainability*, vol. 7, pp. 7866–7883, 2015.

[7] CARPEessence, "Life Cycle Analysis," *Consortium on Applied Research and Professional Education*, Universitat Politecnica de Valencia.

[8] ISO. 14040:2006, "Environmental management — Life cycle assessment — Principles and framework," *International Organization for Standardization*, Geneva, 2006.

[9] The British Standards Institution, "*Sustainability of Construction Works—Assessment of Environmental Performance of Buildings—Calculation Method*," London, UK: The British Standards Institution, 2012.

Environmental Aspects

[10] J. Anderson and J. Thornback, *A Guide to Understanding the Embodied Impacts of 823 Construction Products*. London, UK: Construction Products Association, 2012.

[11] G. J. Treloar, *"Energy analysis of the construction of office buildings,"* Deakin University, Geelong: Master of Architecture thesis, 1994.

[12] W. Palz and H. Zibetta, "Energy pay-back time of photovoltaic modules," *International Journal of Solar Energy*, vol. 10, pp. 211–216, 1991.

[13] G. N. Tiwari and M. K. Ghosal, *Renewable Energy Resources: Basic Principles and Applications*, New Delhi: Narosa Publishing House, 2005.

[14] V. M. Fthenakis and H. C. Kim, "Photovoltaics: Life-cycle analyses," *Solar Energy*, vol. 85, no. 8, pp. 1609–1628, 2011.

[15] C. Cao, "Embodied carbon is the carbon dioxide (CO_2) or greenhouse gas (GHG) emissions associated with the manufacture and use of a product or service," in *Advanced High Strength Natural Fibre Composites in Construction*, Woodhead Publishing, 2017.

[16] M. Achintha, "Sustainability of glass in construction," in *Sustainability of Construction Materials* (2nd ed.), Woodhead Publishing Series in Civil and Structural Engineering, 2016, pp. 79–104.

[17] A. Miller and K. Ip, "Sustainable construction materials," in *Sustainability, Energy and Architecture: Case Studies in Realizing Green Buildings*, United States of America: Elsevier, 2013, pp. 79–92.

[18] A. D. González, "Assessment of the energy and carbon embodied in straw and clay masonry blocks," in *Eco-Efficient Masonry Bricks and Blocks: Design, Properties and Durability*, Woodhead Publishing, 2015, pp. 461–480.

[19] F. Krausmann, S. Gingrich, N. Eisenmenger, K. H. Erb, H. Haberl and M. Fischer-Kowalski, "Growth in global materials use, GDP and population during the 20th century," *Ecological Economists*, vol. 68, pp. 2696–2705, 2009.

[20] P. Melton, "The urgency of embodied carbon and what you can do about ItG," *Building Green*, [Online]. Available: https://www.buildinggreen.com/feature/urgency-embodied-carbon-and-what-you-can-do-about-it. [Accessed 2020].

[21] R. F. W. Boarder, P. L. Owens and J. M. Khatib "The sustainability of lightweight aggregates manufactured from clay wastes for reducing the carbon footprint of structural and foundation concrete," in *Sustainability of Construction Materials*, 2nd ed. Woodhead Publishing Series in Civil and Structural Engineering, 2016.

[22] G. N. Tiwari, A. Tiwari and Shyam, *Handbook of Solar Energy*, Springer, 2016.

[23] G. N. Tiwari, *Advances in Solar Energy Technology*, New Delhi: Nova Science Publishers., 2005.

[24] "Macrotrends," [Online]. Available: https://www.macrotrends.net/countries/IND/india/gdp-growth-rate. [Accessed 2020].

[25] Reuters, The Hindu: India to become largest source of energy demand growth to 2040, says International Energy Agency. https://www.thehindu.com/business/india-to-be-largest-source-of-energy-demand-growth-to-2040-says-international-energy-agency/article33790095.ece. [Accessed May 2021].

[26] B. E. Economics, "BP Energy outlook 2019," [Online]. Available: Gas production grows but fails to keep pace with demand implying a significant. [Accessed 2020].

[27] R. Kalshian "Energy versus emissions: The big challenge of the new *millennium*," 2006, By info change news and features [Online]. Available: www.infochangeindia.org/agenda5_01.jsp

[28] Wikipedia, "Per capita greenhouse gas emissions on world," [Online]. Available: http://en.wikipedia.org.

[29] G. N. Tiwari and R. K. Mishra, *Advanced Renewable Energy Sources*, UK: RSC Publishing, 2012.

[30] Q. Ma, "Greenhouse gases: refining the role of carbon dioxide," NASA Science Briefs, 1998.

[31] H. Ritchie and M. Roser, "CO_2 and greenhouse gas emissions," *Our World in Data,* 2020. [Online]. Available: https://ourworldindata.org/co2-and-other-greenhouse-gas-emissions. [Accessed 2020].

[32] G. C. Project [Online], Available: https://www.globalcarbonproject.org/. [Accessed 2020].

[33] M. Watt, A. Johnson, M. Ellis and N. Outhred "Life cycle air emission from PV power systems," *Progress in Photovoltaic Research Applications*, vol. 6, no. 2, pp. 127–137, 1998.

[34] S. K. Rajput and O. Singh, "Reduction in CO_2 emission through photovoltaic system: A case study," in: *3rd IEEE International Conference on Nanotechnology for Instrumentation and Measurement*, GBU, Greater Noida, India, 2017.

[35] B. Agrawal and G. N. Tiwari, *Developments in Environmental Durability for Photovoltaics*, UK: Pira International Ltd., 2008.

[36] A. Chaurey and T. C. Kandpal, "Carbon abatement potential of solar home systems in India and their cost reduction due to carbon finance," *Energy Policy*, vol. 37, pp. 115–125, 2009.

[37] P. Purohit, "CO_2 emission mitigation potential of solar home systems under clean development mechanism in India," *Energy*, vol. 34, pp. 1014–1023, 2009.

[38] E. A. Alsema, "Energy requirements and CO_2 mitigation potential of PV systems," in *BNL/NREL Workshop "PV and the Environment*, Keystone, CO, 1998.

[39] H. Wirth, "Recent facts about photovoltaics in Germany," Fraunhofer, ISE 2018.

10 Life Cycle Analysis

10.1 INTRODUCTION

Techno-economic analysis is an engineering subject wherein the engineering skills like judgement and experience are used along with scientific principles and techniques and applied to problems of project cost control, profitability analysis, planning, scheduling and optimization of operational research, etc. Thus, it is fundamental for life cycle cost analysis. This study involves a wide range of subjects like quality and resource management, time value of money, maintenance, integrated projects control, organizational structures, life cycle and risk analysis, etc.

Economic analysis is the process in which we analyze the strength and weakness of any economy. Instead of merely looking at the facility in terms of cost to design and build, owners can broaden their perspective to include operations, maintenance, repair, replacement and disposal costs. The sum of initial and future costs associated with the construction and operation of a building over a period of time is called the life cycle cost (LCC) of a facility. Thus, life cycle cost assessment (LCCA) is an economic evaluation technique, which is used to determine the total cost of owning and operating a facility over a period of time. LCC is associated with an energy delivery system over its lifetime or over a selected period of analysis and takes into account the time value of money [1].

Today, interest in the development of and dissemination of technologies based out of renewable energy has again rehabilitated in the view of increasing global climate change concerns. In addition to this development, the matter related to their financial and economic viability and financing of renewable energy system should be being given importance.

Cost analysis is used for project cost control through benefit-cost analysis, profitability analysis, planning, scheduling and optimization of operational research, etc. For technology based on solar energy, the economic viability of the same needs to be well-understood for the users to understand the importance of the technology importance and they can utilize the area under their command to their best advantage.

Close estimation becomes crucial and critical with the advancement of technology and society to remain competitive. An estimate based on over-design may be too high to sustain, whereas that based on under-design may be successful for a while but again it is not possible to sustain. An effective economic analysis can be made by the knowledge of cost analysis, which can be done by the aid of cash flow diagrams and some other methods [2].

LCCA depends on critical parameters as follows:

- Initial investment (Present value/First cost) for construction of renewable energy system (P)
- Annual operating cost (O)
- Annual maintenance cost (M)
- Annual energy output either in term of thermal energy or exergy (AEC)
- The rate of interest (i)
- Overhauling cost of renewable energy system, if any, during life of the system (OC)
- Life of the system (n) and its salvage value (S)
- Environmental impact due to carbon dioxide emission by embodied energy
- The energy used to operate it (annually) and pre-treatments
- CO_2 credit (CC) earned due to use of renewable energy system
- Energy payback time (EPBT)

DOI: 10.1201/9780429445903-10

For effective economic analysis of renewable energy systems, the subsequent sections deal with the knowledge of cost analysis, cash-flow diagram, payback time and benefit-cost (B/C) analysis etc.

10.2 CASH FLOW DIAGRAM

Cash flow diagrams are the pictorial representation on time scale of the transactions taking place during the course of a given project. These diagrams are useful in solving the economic analysis, giving the insight of inflow and outflow of the money from the system. These transactions can be anything like initial investments, maintenance costs, projected earnings or savings resulting from the project, as well as salvage and resale value of equipment at the end of the project.

$$Net\ cash\ flow = Receipts\,(credits) - Disbursements\,(Debits)$$

For economic viability, the net cash flow should be positive for any system.

In a cash flow diagram, there is a horizontal axis, which represents the duration of the project (with equidistant markings, one per period up to the duration of the project) and arrows in upward and downward directions. The horizontal lines also address payment and receipt of funds at the end of the period of occurrence. Funds that are expenses are referred to as negative cash flow and are represented by a downward arrow, downward from the time line with their bases at the appropriate positions along the line. Funds that are receipts are referred to as positive cash flows and are represented by arrows extending upward from the line. Arrow lengths are approximately proportional to the magnitude of the cash flow. Figure 10.1(a) shows the positive cash flow diagram. Figure 10.1(b) shows the negative cash flow diagrams.

In the cash flow diagrams in Figures 10.1(a) and (b), a uniform end-of-year annual amount will be considered at the end of each year of time scale.

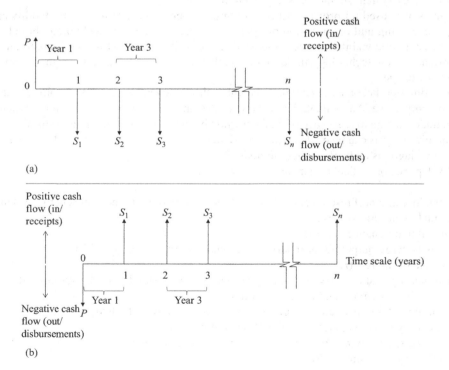

FIGURE 10.1 Cash flow diagram (a) loan transaction, (b) investment transaction.

Life Cycle Analysis   **285**

Cash flow diagrams can occur at the beginning or in the middle of an interest period or at practically any point in time. To simplify the calculations in economic analysis, we generally assume the end-of-period convention, which is the practice of placing all cash flow transactions at the end of an interest period. This relieves us from dealing with the effects of interest within an interest period.

10.3 COST ANALYSIS

10.3.1 CAPITAL RECOVERY FACTOR

- **Single payment future value factor**

Let P be the initial investment (Present value/First cost) at zero time ($n = 0$) at the interest rate of i per year. If future value at the nth years is S_n, then based on the following cash flow (Figure 10.1b), the future value (S_n) in the term of present value at the nth years will be obtained.

At the end of one year, the time value of investment P is given by

$$S_1 = P + iP = P(1+i) \tag{10.1a}$$

Similarly, at the end of second year, the value of P becomes

$$S_2 = S_1 + iS_1 = P + iP + i(P + iP) = P(1+i) + Pi(1+i) = P(1+i)^2 \tag{10.1b}$$

Similarly, at the end of third year, the value of P becomes

$$S_3 = P(1+i)^3 \tag{10.1c}$$

Therefore, the value of P at n^{th} will be

$$S_n = P(1+i)^n \tag{10.1d}$$

For simplification, let us assume $S_n = S$, Equation (10.1d) becomes

$$S = P(1+i)^n \tag{10.2a}$$

It may be noted that for the positive rate of interest, the future value will be greater than the present value invested, i.e., for $i > 0$, $S > P$. Thus, considering compound interest law, Equation (10.2a) becomes

$$S = PF_{PS} \tag{10.2b}$$

where $F_{PS} = (1+i)^n$

The future value (S_n) in term of present value at the nth years as obtained in Equation (10.2b) can be explained as under

$$\left[Future\ value(S_n) \right] = \left[Present\ value(P) \right] \left[Future\ value\ factor(F_{PS}) \right]$$

or

$$\left[Future\ value(S_n) \right] = \left[Present\ value(P) \right] \left[Compound\ interest\ factor \right]$$

F_{PS} is conversion factor for present value (P) to future value (S), refer Table 10.1. Thus, compound interest factor when multiplied with present value gives the future value.

F_{PS} more completely designated as $F_{PS,\,i,\,n}$, where i is the rate of interest and n is the number of years, hence

$$F_{PS,i,n} = \left(1+i\right)^n \tag{10.2c}$$

Table 10.1 gives the values of conversion factors with number of years for a given rate of interest.

If the analysis has been done in less than a year's time, then let's assume that if one year is divided into p equal units of time period, then n becomes np and i becomes $\dfrac{i}{p}$, which is the rate of return per unit period. Then the expression for the future value using Equation (10.2a) becomes,

$$S = P\left(1+\frac{i}{p}\right)^{np} \tag{10.3a}$$

Using spreadsheet computer software, MS-Excel, the future value can be computed by writing the expression as $= FV(i,n,-P)$ [1]

Equation (10.3a) can be written as:

$$S = P\left[\left(1+\frac{i}{p}\right)^p\right]^n \tag{10.3b}$$

TABLE 10.1

The Values of Conversion Factors with Number of Years for a Given Rate of Interest

n	F_{PS}	F_{SP}	F_{RP}	F_{PR}	F_{RS}	F_{SR}	F_{PK}	F_{GR}
				$i = 0.02$				
1	1.02	0.980392	0.980392	1.02	1	1	51	1
2	1.0404	0.961169	1.941561	0.51505	2.02	0.49505	25.75248	1.01
3	1.061208	0.942322	2.883883	0.346755	3.0604	0.326755	17.33773	1.020133
4	1.082432	0.923845	3.807729	0.262624	4.121608	0.242624	13.13119	1.030402
5	1.104081	0.905731	4.71346	0.212158	5.20404	0.192158	10.60792	1.040808
6	1.126162	0.887971	5.601431	0.178526	6.308121	0.158526	8.926291	1.051353
7	1.148686	0.87056	6.471991	0.154512	7.434283	0.134512	7.725598	1.06204
8	1.171659	0.85349	7.325481	0.13651	8.582969	0.11651	6.82549	1.072871
9	1.195093	0.836755	8.162237	0.122515	9.754628	0.102515	6.125772	1.083848
10	1.218994	0.820348	8.982585	0.111327	10.94972	0.091327	5.566326	1.094972
11	1.243374	0.804263	9.786848	0.102178	12.16872	0.082178	5.108897	1.106247
12	1.268242	0.788493	10.57534	0.09456	13.41209	0.07456	4.72798	1.117674
13	1.293607	0.773033	11.34837	0.088118	14.68033	0.068118	4.405918	1.129256
14	1.319479	0.757875	12.10625	0.082602	15.97394	0.062602	4.130099	1.140996
15	1.345868	0.743015	12.84926	0.077825	17.29342	0.057825	3.891274	1.152894
16	1.372786	0.728446	13.57771	0.07365	18.63929	0.05365	3.682506	1.164955
17	1.400241	0.714163	14.29187	0.06997	20.01207	0.04997	3.498492	1.177181
18	1.428246	0.700159	14.99203	0.066702	21.41231	0.046702	3.335105	1.189573
19	1.456811	0.686431	15.67846	0.063782	22.84056	0.043782	3.189088	1.202135
20	1.485947	0.672971	16.35143	0.061157	24.29737	0.041157	3.057836	1.214868

(Continued)

Life Cycle Analysis

TABLE 10.1 (Continued)
The Values of Conversion Factors with Number of Years for a Given Rate of Interest

n	F_{PS}	F_{SP}	F_{RP}	F_{PR}	F_{RS}	F_{SR}	F_{PK}	F_{GR}
				$i = 0.04$				
1	1.04	0.961538	0.961538	1.04	1	1	26	1
2	1.0816	0.924556	1.886095	0.530196	2.04	0.490196	13.2549	1.02
3	1.124864	0.888996	2.775091	0.360349	3.1216	0.320349	9.008713	1.040533
4	1.169859	0.854804	3.629895	0.27549	4.246464	0.23549	6.887251	1.061616
5	1.216653	0.821927	4.451822	0.224627	5.416323	0.184627	5.615678	1.083265
6	1.265319	0.790315	5.242137	0.190762	6.632975	0.150762	4.769048	1.105496
7	1.315932	0.759918	6.002055	0.16661	7.898294	0.12661	4.16524	1.128328
8	1.368569	0.73069	6.732745	0.148528	9.214226	0.108528	3.713196	1.151778
9	1.423312	0.702587	7.435332	0.134493	10.5828	0.094493	3.362325	1.175866
10	1.480244	0.675564	8.110896	0.123291	12.00611	0.083291	3.082274	1.200611
11	1.539454	0.649581	8.760477	0.114149	13.48635	0.074149	2.853726	1.226032
12	1.601032	0.624597	9.385074	0.106552	15.02581	0.066552	2.663804	1.25215
13	1.665074	0.600574	9.985648	0.100144	16.62684	0.060144	2.503593	1.278988
14	1.731676	0.577475	10.56312	0.094669	18.29191	0.054669	2.366724	1.306565
15	1.800944	0.555265	11.11839	0.089941	20.02359	0.049941	2.248528	1.334906
16	1.872981	0.533908	11.6523	0.08582	21.82453	0.04582	2.1455	1.364033
17	1.9479	0.513373	12.16567	0.082199	23.69751	0.042199	2.054963	1.393971
18	2.025817	0.493628	12.6593	0.078993	25.64541	0.038993	1.974833	1.424745
19	2.106849	0.474642	13.13394	0.076139	27.67123	0.036139	1.903465	1.45638
20	2.191123	0.456387	13.59033	0.073582	29.77808	0.033582	1.839544	1.488904
				$i = 0.06$				
1	1.06	0.943396	0.943396	1.06	1	1	17.66667	1
2	1.1236	0.889996	1.833393	0.545437	2.06	0.485437	9.090615	1.03
3	1.191016	0.839619	2.673012	0.37411	3.1836	0.31411	6.235164	1.0612
4	1.262477	0.792094	3.465106	0.288591	4.374616	0.228591	4.809858	1.093654
5	1.338226	0.747258	4.212364	0.237396	5.637093	0.177396	3.956607	1.127419
6	1.418519	0.704961	4.917324	0.203363	6.975319	0.143363	3.389377	1.162553
7	1.50363	0.665057	5.582381	0.179135	8.393838	0.119135	2.985584	1.19912
8	1.593848	0.627412	6.209794	0.161036	9.897468	0.101036	2.683932	1.237183
9	1.689479	0.591898	6.801692	0.147022	11.49132	0.087022	2.450371	1.276813
10	1.790848	0.558395	7.360087	0.135868	13.18079	0.075868	2.264466	1.318079
11	1.898299	0.526788	7.886875	0.126793	14.97164	0.066793	2.113216	1.361058
12	2.012196	0.496969	8.383844	0.119277	16.86994	0.059277	1.98795	1.405828
13	2.132928	0.468839	8.852683	0.11296	18.88214	0.05296	1.882668	1.452472
14	2.260904	0.442301	9.294984	0.107585	21.01507	0.047585	1.793082	1.501076
15	2.396558	0.417265	9.712249	0.102963	23.27597	0.042963	1.716046	1.551731
16	2.540352	0.393646	10.1059	0.098952	25.67253	0.038952	1.649202	1.604533
17	2.692773	0.371364	10.47726	0.095445	28.21288	0.035445	1.590747	1.659581
18	2.854339	0.350344	10.8276	0.092357	30.90565	0.032357	1.539276	1.716981
19	3.0256	0.330513	11.15812	0.089621	33.75999	0.029621	1.493681	1.776842
20	3.207135	0.311805	11.46992	0.087185	36.78559	0.027185	1.453076	1.83928
				$i = 0.08$				
1	1.08	0.925926	0.925926	1.08	1	1	13.5	1
2	1.1664	0.857339	1.783265	0.560769	2.08	0.480769	7.009615	1.04
3	1.259712	0.793832	2.577097	0.388034	3.2464	0.308034	4.850419	1.082133
4	1.360489	0.73503	3.312127	0.301921	4.506112	0.221921	3.77401	1.126528
5	1.469328	0.680583	3.99271	0.250456	5.866601	0.170456	3.130706	1.17332

(Continued)

TABLE 10.1 (Continued)
The Values of Conversion Factors with Number of Years for a Given Rate of Interest

n	F_{PS}	F_{SP}	F_{RP}	F_{PR}	F_{RS}	F_{SR}	F_{PK}	F_{GR}
6	1.586874	0.63017	4.62288	0.216315	7.335929	0.136315	2.703942	1.222655
7	1.713824	0.58349	5.20637	0.192072	8.922803	0.112072	2.400905	1.274686
8	1.85093	0.540269	5.746639	0.174015	10.63663	0.094015	2.175185	1.329578
9	1.999005	0.500249	6.246888	0.16008	12.48756	0.08008	2.000996	1.387506
10	2.158925	0.463193	6.710081	0.149029	14.48656	0.069029	1.862869	1.448656
11	2.331639	0.428883	7.138964	0.140076	16.64549	0.060076	1.750954	1.513226
12	2.51817	0.397114	7.536078	0.132695	18.97713	0.052695	1.658688	1.581427
13	2.719624	0.367698	7.903776	0.126522	21.4953	0.046522	1.581523	1.653484
14	2.937194	0.340461	8.244237	0.121297	24.21492	0.041297	1.516211	1.729637
15	3.172169	0.315242	8.559479	0.11683	27.15211	0.03683	1.460369	1.810141
16	3.425943	0.29189	8.851369	0.112977	30.32428	0.032977	1.412211	1.895268
17	3.700018	0.270269	9.121638	0.109629	33.75023	0.029629	1.370368	1.985307
18	3.996019	0.250249	9.371887	0.106702	37.45024	0.026702	1.333776	2.080569
19	4.315701	0.231712	9.603599	0.104128	41.44626	0.024128	1.301595	2.181382
20	4.660957	0.214548	9.818147	0.101852	45.76196	0.021852	1.273153	2.288098
				$i = 0.10$				
1	1.1	0.909091	0.909091	1.1	1	1	11	1
2	1.21	0.826446	1.735537	0.57619	2.1	0.47619	5.761905	1.05
3	1.331	0.751315	2.486852	0.402115	3.31	0.302115	4.021148	1.103333
4	1.4641	0.683013	3.169865	0.315471	4.641	0.215471	3.154708	1.16025
5	1.61051	0.620921	3.790787	0.263797	6.1051	0.163797	2.637975	1.22102
6	1.771561	0.564474	4.355261	0.229607	7.71561	0.129607	2.296074	1.285935
7	1.948717	0.513158	4.868419	0.205405	9.487171	0.105405	2.054055	1.35531
8	2.143589	0.466507	5.334926	0.187444	11.43589	0.087444	1.87444	1.429486
9	2.357948	0.424098	5.759024	0.173641	13.57948	0.073641	1.736405	1.508831
10	2.593742	0.385543	6.144567	0.162745	15.93742	0.062745	1.627454	1.593742
11	2.853117	0.350494	6.495061	0.153963	18.53117	0.053963	1.539631	1.684652
12	3.138428	0.318631	6.813692	0.146763	21.38428	0.046763	1.467633	1.782024
13	3.452271	0.289664	7.103356	0.140779	24.52271	0.040779	1.407785	1.886362
14	3.797498	0.263331	7.366687	0.135746	27.97498	0.035746	1.357462	1.998213
15	4.177248	0.239392	7.60608	0.131474	31.77248	0.031474	1.314738	2.118165
16	4.594973	0.217629	7.823709	0.127817	35.94973	0.027817	1.278166	2.246858
17	5.05447	0.197845	8.021553	0.124664	40.5447	0.024664	1.246641	2.384983
18	5.559917	0.179859	8.201412	0.12193	45.59917	0.02193	1.219302	2.533287
19	6.115909	0.163508	8.36492	0.119547	51.15909	0.019547	1.195469	2.692584
20	6.7275	0.148644	8.513564	0.11746	57.275	0.01746	1.174596	2.86375
				$i = 0.12$				
1	1.12	0.892857	0.892857	1.12	1	1	9.333333	1
2	1.2544	0.797194	1.690051	0.591698	2.12	0.471698	4.930818	1.06
3	1.404928	0.71178	2.401831	0.416349	3.3744	0.296349	3.469575	1.1248
4	1.573519	0.635518	3.037349	0.329234	4.779328	0.209234	2.74362	1.194832
5	1.762342	0.567427	3.604776	0.27741	6.352847	0.15741	2.311748	1.270569
6	1.973823	0.506631	4.111407	0.243226	8.115189	0.123226	2.026881	1.352532
7	2.210681	0.452349	4.563757	0.219118	10.08901	0.099118	1.825981	1.441287
8	2.475963	0.403883	4.96764	0.201303	12.29969	0.081303	1.677524	1.537462
9	2.773079	0.36061	5.32825	0.187679	14.77566	0.067679	1.563991	1.64174
10	3.105848	0.321973	5.650223	0.176984	17.54874	0.056984	1.474868	1.754874

(*Continued*)

Life Cycle Analysis

TABLE 10.1 (Continued)
The Values of Conversion Factors with Number of Years for a Given Rate of Interest

n	F_{PS}	F_{SP}	F_{RP}	F_{PR}	F_{RS}	F_{SR}	F_{PK}	F_{GR}
11	3.47855	0.287476	5.937699	0.168415	20.65458	0.048415	1.403462	1.877689
12	3.895976	0.256675	6.194374	0.161437	24.13313	0.041437	1.345307	2.011094
13	4.363493	0.229174	6.423548	0.155677	28.02911	0.035677	1.29731	2.156085
14	4.887112	0.20462	6.628168	0.150871	32.3926	0.030871	1.25726	2.313757
15	5.473566	0.182696	6.810864	0.146824	37.27971	0.026824	1.223535	2.485314
16	6.130394	0.163122	6.973986	0.14339	42.75328	0.02339	1.194917	2.67208
17	6.866041	0.145644	7.11963	0.140457	48.88367	0.020457	1.170473	2.87551
18	7.689966	0.13004	7.24967	0.137937	55.74971	0.017937	1.149478	3.097206
19	8.612762	0.116107	7.365777	0.135763	63.43968	0.015763	1.131358	3.338931
20	9.646293	0.103667	7.469444	0.133879	72.05244	0.013879	1.115657	3.602622
				$i = 0.14$				
1	1.14	0.877193	0.877193	1.14	1	1	8.142857	1
2	1.2996	0.769468	1.646661	0.60729	2.14	0.46729	4.337784	1.07
3	1.481544	0.674972	2.321632	0.430731	3.4396	0.290731	3.076653	1.146533
4	1.68896	0.59208	2.913712	0.343205	4.921144	0.203205	2.451463	1.230286
5	1.925415	0.519369	3.433081	0.291284	6.610104	0.151284	2.080597	1.322021
6	2.194973	0.455587	3.888668	0.257157	8.535519	0.117157	1.836839	1.422586
7	2.502269	0.399637	4.288305	0.233192	10.73049	0.093192	1.66566	1.532927
8	2.852586	0.350559	4.638864	0.21557	13.23276	0.07557	1.539786	1.654095
9	3.251949	0.307508	4.946372	0.202168	16.08535	0.062168	1.44406	1.787261
10	3.707221	0.269744	5.216116	0.191714	19.3373	0.051714	1.369382	1.93373
11	4.226232	0.236617	5.452733	0.183394	23.04452	0.043394	1.309959	2.094956
12	4.817905	0.207559	5.660292	0.176669	27.27075	0.036669	1.261924	2.272562
13	5.492411	0.182069	5.842362	0.171164	32.08865	0.031164	1.222598	2.468358
14	6.261349	0.15971	6.002072	0.166609	37.58107	0.026609	1.190065	2.684362
15	7.137938	0.140096	6.142168	0.162809	43.84241	0.022809	1.162921	2.922828
16	8.137249	0.122892	6.26506	0.159615	50.98035	0.019615	1.14011	3.186272
17	9.276464	0.1078	6.372859	0.156915	59.1176	0.016915	1.120825	3.477506
18	10.57517	0.094561	6.46742	0.154621	68.39407	0.014621	1.104437	3.79967
19	12.05569	0.082948	6.550369	0.152663	78.96923	0.012663	1.090451	4.156276
20	13.74349	0.072762	6.623131	0.150986	91.02493	0.010986	1.078471	4.551246
				$i = 0.16$				
1	1.16	0.862069	0.862069	1.16	1	1	7.25	1
2	1.3456	0.743163	1.605232	0.622963	2.16	0.462963	3.893519	1.08
3	1.560896	0.640658	2.24589	0.445258	3.5056	0.285258	2.782862	1.168533
4	1.810639	0.552291	2.798181	0.357375	5.066496	0.197375	2.233594	1.266624
5	2.100342	0.476113	3.274294	0.305409	6.877135	0.145409	1.908809	1.375427
6	2.436396	0.410442	3.684736	0.27139	8.977477	0.11139	1.696187	1.496246
7	2.82622	0.35383	4.038565	0.247613	11.41387	0.087613	1.547579	1.630553
8	3.278415	0.305025	4.343591	0.230224	14.24009	0.070224	1.438902	1.780012
9	3.802961	0.262953	4.606544	0.217082	17.51851	0.057082	1.356766	1.946501
10	4.411435	0.226684	4.833227	0.206901	21.32147	0.046901	1.293132	2.132147
11	5.117265	0.195417	5.028644	0.198861	25.7329	0.038861	1.24288	2.339355
12	5.936027	0.168463	5.197107	0.192415	30.85017	0.032415	1.202592	2.570847
13	6.885791	0.145227	5.342334	0.187184	36.7862	0.027184	1.169901	2.829707
14	7.987518	0.125195	5.467529	0.182898	43.67199	0.022898	1.143112	3.119428
15	9.265521	0.107927	5.575456	0.179358	51.65951	0.019358	1.120985	3.443967

(Continued)

TABLE 10.1 (Continued)
The Values of Conversion Factors with Number of Years for a Given Rate of Interest

n	F_{PS}	F_{SP}	F_{RP}	F_{PR}	F_{RS}	F_{SR}	F_{PK}	F_{GR}
16	10.748	0.093041	5.668497	0.176414	60.92503	0.016414	1.102585	3.807814
17	12.46768	0.080207	5.748704	0.173952	71.67303	0.013952	1.087202	4.216061
18	14.46251	0.069144	5.817848	0.171885	84.14072	0.011885	1.07428	4.674484
19	16.77652	0.059607	5.877455	0.170142	98.60323	0.010142	1.063385	5.189644
20	19.46076	0.051385	5.928841	0.168667	115.3797	0.008667	1.054169	5.768987

$i = 0.18$

n	F_{PS}	F_{SP}	F_{RP}	F_{PR}	F_{RS}	F_{SR}	F_{PK}	F_{GR}
1	1.18	0.847458	0.847458	1.18	1	1	6.555556	1
2	1.3924	0.718184	1.565642	0.638716	2.18	0.458716	3.54842	1.09
3	1.643032	0.608631	2.174273	0.459924	3.5724	0.279924	2.555133	1.1908
4	1.938778	0.515789	2.690062	0.371739	5.215432	0.191739	2.065215	1.303858
5	2.287758	0.437109	3.127171	0.319778	7.15421	0.139778	1.776544	1.430842
6	2.699554	0.370432	3.497603	0.28591	9.441968	0.10591	1.58839	1.573661
7	3.185474	0.313925	3.811528	0.262362	12.14152	0.082362	1.457567	1.734503
8	3.758859	0.266038	4.077566	0.245244	15.327	0.065244	1.362469	1.915874
9	4.435454	0.225456	4.303022	0.232395	19.08585	0.052395	1.291082	2.120651
10	5.233836	0.191064	4.494086	0.222515	23.52131	0.042515	1.236192	2.352131
11	6.175926	0.161919	4.656005	0.214776	28.75514	0.034776	1.193202	2.614104
12	7.287593	0.13722	4.793225	0.208628	34.93107	0.028628	1.159043	2.910923
13	8.599359	0.116288	4.909513	0.203686	42.21866	0.023686	1.13159	3.247589
14	10.14724	0.098549	5.008062	0.199678	50.81802	0.019678	1.109323	3.629859
15	11.97375	0.083516	5.091578	0.196403	60.96527	0.016403	1.091127	4.064351
16	14.12902	0.070776	5.162354	0.19371	72.93901	0.01371	1.076167	4.558688
17	16.67225	0.05998	5.222334	0.191485	87.06804	0.011485	1.063807	5.121649
18	19.67325	0.05083	5.273164	0.189639	103.7403	0.009639	1.053553	5.763349
19	23.21444	0.043077	5.316241	0.188103	123.4135	0.008103	1.045016	6.495449
20	27.39303	0.036506	5.352746	0.18682	146.628	0.00682	1.037889	7.331399

$i = 0.20$

n	F_{PS}	F_{SP}	F_{RP}	F_{PR}	F_{RS}	F_{SR}	F_{PK}	F_{GR}
1	1.2	0.833333	0.833333	1.2	1	1	6	1
2	1.44	0.694444	1.527778	0.654545	2.2	0.454545	3.272727	1.1
3	1.728	0.578704	2.106481	0.474725	3.64	0.274725	2.373626	1.213333
4	2.0736	0.482253	2.588735	0.386289	5.368	0.186289	1.931446	1.342
5	2.48832	0.401878	2.990612	0.33438	7.4416	0.13438	1.671899	1.48832
6	2.985984	0.334898	3.32551	0.300706	9.92992	0.100706	1.503529	1.654987
7	3.583181	0.279082	3.604592	0.277424	12.9159	0.077424	1.38712	1.845129
8	4.299817	0.232568	3.83716	0.260609	16.49908	0.060609	1.303047	2.062386
9	5.15978	0.193807	4.030967	0.248079	20.7989	0.048079	1.240397	2.310989
10	6.191736	0.161506	4.192472	0.238523	25.95868	0.038523	1.192614	2.595868
11	7.430084	0.134588	4.32706	0.231104	32.15042	0.031104	1.155519	2.922765
12	8.9161	0.112157	4.439217	0.225265	39.5805	0.025265	1.126325	3.298375
13	10.69932	0.093464	4.532681	0.22062	48.4966	0.02062	1.1031	3.730508
14	12.83918	0.077887	4.610567	0.216893	59.19592	0.016893	1.084465	4.22828
15	15.40702	0.064905	4.675473	0.213882	72.03511	0.013882	1.069411	4.802341
16	18.48843	0.054088	4.729561	0.211436	87.44213	0.011436	1.057181	5.465133
17	22.18611	0.045073	4.774634	0.20944	105.9306	0.00944	1.047201	6.231209
18	26.62333	0.037561	4.812195	0.207805	128.1167	0.007805	1.039027	7.117593
19	31.948	0.031301	4.843496	0.206462	154.74	0.006462	1.032312	8.144211
20	38.3376	0.026084	4.86958	0.205357	186.688	0.005357	1.026783	9.3344

Life Cycle Analysis

where the expression $\left(1+\dfrac{i}{p}\right)^{p}$ can be expressed as:

$$\left(1+\frac{i}{p}\right)^{p}=1+\textit{effective rate of return} \tag{10.3c}$$

or

$$\textit{Effective rate of return}=\left(1+\frac{i}{p}\right)^{p}-1 \tag{10.3d}$$

$$=i \text{ for } p=1$$

$$>i \text{ for } p>1$$

For simple interest

$$S=P(1+ni)=P+(iP)n \tag{10.4}$$

- **Single payment future value factor**

Discounting is a process for calculating the present value for an amount in the future value using a specified discount rate.

Let S be the future value, at the end of n years, then the present value (P) at an annual discount (or interest) rate i can be written as following expression.

Thus, Equation (10.2a) can be rewritten as:

$$P=\frac{S}{(1+i)^{n}}=S(1+i)^{-n} \tag{10.5a}$$

It can be concluded that the future amount (at n^{th} $year$) is reduced when converted against the calendar to the present value (at zeroth time), assuming i to be positive.

Equation (10.5a) can be rewritten as:

$$P=SF_{SP} \tag{10.5b}$$

Equation (10.5b) can be explained as:

$$\left[\textit{Present value}(P)\right]=\left[\textit{Future value}(S_n)\right]\left[\textit{Present value factor}(F_{SP})\right]$$

F_{SP} is conversion factor for future value (S) to present value (P) and is referred to as compound interest factor (Table 10.1) and it is expressed as:

$$F_{SP,i,n}=(1+i)^{-n} \tag{10.6}$$

The numerical value of F_{SP} is always less than one because the denominator is always higher than one. For this reason, present-worth calculations are generally referred to as discounted cash flow (DCF) method. Other terms generally used in reference to present-worth (PW) calculations are present value (PV) and net present value (NPV) [3].

From Equation (10.2b and 10.6), we have

$$F_{PS} = \frac{1}{F_{SP}}$$

or

$$F_{PS} \times F_{SP} = 1$$

Using MS-Excel, the present value can be computed by writing the expression as =PV(i,n,,-S) [1].

From Equation (10.2a and 10.5b), the initial investment (P) and the future value (S) of investment P can be calculated.

$$S = P(1+i)^{n}, \textit{moving with the calendar} \qquad (10.7a)$$

$$S = P(1+i)^{-n}, \textit{moving against the calendar} \qquad (10.7b)$$

Combining the Equations (10.7a and 10.7b), we have

$$A_{t2} = A_{t1}(1+i)^{n} \qquad (10.7c)$$

where n is referred to as the time-value conversion relationship, which is positive with the calendar and negative against the calendar.

Equation (10.7c) can be explained as:

$$\left[Amount\ at\ time\ 2 \right] = \left[Amount\ at\ time\ 1 \right]\left[Compound\ interest\ operator \right]$$

EXAMPLE 10.1

If USD$20,000 compounds to USD$28,240 (Rs. 1,448,800) in 4 years what will be the rate of return? (Consider USD$1= Rs. 72.44)

Solution

From Equation (10.2a), using S = USD$28,240; P = USD$20,000 and n = 4 years, we get

$$28,240 = 20,000(1+i)^{4}$$

Solving the above, we get

$$i = 0.09 \text{ or } 9\% \text{ per annum.}$$

Using MS-Excel, the above can be solved by using the following
=RATE(4,-20000,28232)
= 9%
or,

Life Cycle Analysis   293

From Equation (10.2a), using S = Rs. 2,045,706; P = Rs. 1,448,800 and n = 4 years, we get

$$2{,}045{,}706 = 1{,}448{,}800\left(1+i\right)^4$$

Solving the above, we get

$$i = 0.09 \text{ or } 9\% \, per \, annum.$$

Using MS-Excel, the above can be solved by using the following
=RATE(4,-1448800,2045706)
= 9%

EXAMPLE 10.2

Estimate the time required to double the amount if compounded annually at 10% per annum.

Solution

Let us assume that the amount will double in n years, then $S = 2P$
From Equation (10.2a), we get

$$2P = P\left(1+0.10\right)^n$$

Solving the above, we get

$$\log 2 = n\log 1.1, \text{ i.e., } n = 7.3 \, years$$

Therefore, the amount will double in 7.3 years.

EXAMPLE 10.3

Estimate the effective rate of return for 10% interest for p = 5 and p = 12.

Solution

Using Equation (10.3d),
For p = 5;

$$Effective \, rate \, of \, return = \left(1+\frac{i}{p}\right)^p - 1 = \left(1+\frac{0.10}{5}\right)^5 - 1 = 0.104$$

Similarly, for p = 12;

$$Effective \, rate \, of \, return = \left(1+\frac{i}{p}\right)^p - 1 = \left(1+\frac{0.10}{12}\right)^{12} - 1 = 0.1047$$

EXAMPLE 10.4

A student borrows USD\$20,000 and returns USD\$21,000 at the end of six months. What was the rate of interest paid by the student? (Consider USD\$1= Rs. 72.44)

Solution

In the given problem $S = 21,000$ (Rs. 1,521,240); $P = 20,000$ (Rs. 1,448,800) and $n = 6/12$
  Substituting these values in Equation (10.4), we get,
  Calculations in USD

$$21,000 = 20,000\left(1+\frac{6}{12}i\right)$$

Or (calculation in INR)

$$1,521,240 = 1,448,800\left(1+\frac{6}{12}i\right)$$

$$1.05 = 1+0.5i, i = 0.10 \, or \, 10\%$$

EXAMPLE 10.5

A person borrows USD\$10,000 at 10% for 4 years and 4 months. Calculate the money paid by considering compound interest. (Consider USD\$1= Rs. 72.44)

Solution

Using Equation (10.4)

$$S = 10,000\left(1+\frac{4}{12}(0.10)\right) = USD\$10,333\left(Rs.\,748,522.5\right)$$

For 4 months and USD\$10,333 become P for another 4 years. For compound interest

$$S = PF_{PS,10\%,4}$$

Using Equation (10.2b), we have

$$S = 10,333(1.4641) = USD\$15,129 \cdot \left(Rs.\,1,095,945\right)$$

EXAMPLE 10.6

Calculate the worth of USD\$1200 in 8 years when invested now at 10%. Also draw a cash flow diagram for the same. (Consider USD\$1= Rs. 72.44)

Solution (Figure 10.2)

Using Equation (10.7a), we have

$$S = P(1+i)^n = 1200(1+0.10)^8 = USD\$2572.31\left(Rs.\,186,338.1\right)$$

Life Cycle Analysis

FIGURE 10.2 Solution to Example 10.6.

Using MS-Excel, the above can be solved by using the following
=FV(10%,8,-1200)
= USD$2572.31 (Rs. 186,338.1)

EXAMPLE 10.7

A solar house has been purchased at USD$36,000. He assumed that the cost of the property will increase 14% per annum, how long will it take to get USD$50,000 market price?

Solution

Referring to above example,

$$50,000 = 36,000(1+0.14)^n$$

$$n = \frac{\ln\left(\frac{50,000}{36,000}\right)}{\ln(1.14)} = 2.5 \text{ years}$$

Using MS-Excel, the above can be solved by using the following
=NPER(14%,0,-36000,50000)
= 2.5 years

EXAMPLE 10.8

The electrification of houses in a certain city with a PV grid is targeted to increase from the present 12 million to 59 million in 5 years. Calculate the growth rate required to achieve the target.

Solution

Using Equation (10.7a), we have

$$59 \times 10^6 = 12 \times 10^6 (1+i)^5$$

$$i = 37.51\%$$

296 Photovoltaic Thermal Passive House System

Thus, 27.51% growth rate is required to achieve the target.
Using MS-Excel, the above can be solved by using the following
=RATE(5,-12E06,59E06)
=37.51%

EXAMPLE 10.9

It is estimated that about 120 million households in the country can benefit from the use of improved PV/T drying techniques. What is the required growth rate to achieve the potential in the next 20 years if the number of improved drying techniques disseminated so far is 30 million?

Solution

Using Equation (10.7a), we have after taking log of both sides

$$\log(1+i) = 1/n\log(S/P)$$

$$\text{or} \log(1+i) = 1/20\log(120/30)$$

$$\text{or} \log(1+i) = 0.05\log 4$$

$$\text{or} = 0.030103$$

which may be solved to give

$$i \approx 0.07177 \left(\text{or} \approx 7.18\%\right)$$

Thus a compound rate growth rate of more than 7% could be required to achieve the estimated potential of improved drying techniques utilization in the country in the next 20 years.

EXAMPLE 10.10

Suppose a person is offered the alternative of receiving either USD\$15,000 at the end of 10 years or P USD\$ today. Because he does not have any money requirement today, he deposits the P USD\$ in an account that pays 8% interest compounded annually. What value of P would make him indifferent to his choice between the P USD\$ today and the promise USD\$15,000 at the end of 10 years? (Consider USD\$1= Rs. 72.44)

Solution

Using Equation (10.5a and 10.5b), we have

$$P = \frac{S}{(1+i)^n} = \frac{15,000}{(1+0.08)^{10}} = \text{USD\$}6947.90 \,(\text{Rs.}503,305.9)$$

Using MS-Excel, the above can be solved by using the following
=PV(8%,10,-15000)
=USD\$6947.90 (Rs. 503,305.9)

10.3.2 Uniform Annual Cost

To solve the engineering economic problems, the cash flow diagram is drawn with expenditures (i.e., debits) and receipts (i.e., credits) in the opposite direction along a horizontal line. Based on this, a cash flow diagram as shown in Figure 10.3 is drawn to discuss the uniform annual cost.

Referring to Figure 10.3, let R be the amount to be recovered in equal payments (i.e., unacost) for a period of n years at an annual interest rate i. The smallest unit of time considered normally is a year. Unacost can be expenditures (debits), such as annual operating and maintenance for a solar system, and it can be shown below line. If R is receipts (credits), such as cost of annual energy savings and carbon credit (CC) from the solar system then it can be shown above line of cash flow [4]. Now, the total collected amount in terms of present value P is simply the sum of R payments multiplied by the appropriate single payment present value factor for years 1 through n. This summation is a geometric series as follows:

$$P = R\left[\frac{1}{1+i} + \frac{1}{(1+i)^2} + \ldots + \frac{1}{(1+i)^n}\right] \tag{10.8a}$$

$$P = R\sum_{1}^{n}\frac{1}{(1+i)^n} \tag{10.8b}$$

Here, $\frac{1}{(1+i)^n}$ is the present worth factor (PWF).

Equation (10.8a) represents a geometric series, which has $1/(1 + i)$, as the first term and $1/(1 + i)$ as the ratio of n successive terms. The summation of Equation (10.8a) can be calculated using the following expression:

$$\sum_{1}^{n}\frac{1}{(1+i)^n} = \frac{1}{1+i} + \frac{1-\left(\frac{1}{1+i}\right)^n}{1-\left(\frac{1}{1+i}\right)} = \frac{(1+i)^n - 1}{i(1+i)^n} \tag{10.8c}$$

Equation (10.8b) becomes

$$P = R\frac{(1+i)^n - 1}{i(1+i)^n} \tag{10.9a}$$

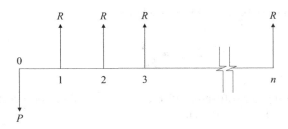

FIGURE 10.3 Cash flow diagram for unacost.

$$P = RF_{RP,i,n} \tag{10.9b}$$

Equation (10.9b) can be explained as under:

$$\left[\,Present\ value\,\right] = \left[\,unacost\,\right]\left[\,unacost\ present\ value\,\right]$$

where

$$F_{RP,i,n} = \frac{(1+i)^n - 1}{i(1+i)^n} \tag{10.9c}$$

and is referred to as the equal payment series, present value factor or annuity present value factor (Table 10.1).

Using spreadsheet computer software, MS-Excel, the present value can be computed by writing the expression as $=PV(i,n,-R)$ [1].

If R is considered as cost of annual energy saving and CC is carbon credit due to use of renewable energy technology, then Equations (10.9a and 10.9b) becomes

$$P = (R + CC)F_{RP,i,n} \tag{10.10}$$

Equations (10.9a) can be rewritten as:

$$R = P\frac{i(1+i)^n}{(1+i)^n - 1} \tag{10.11a}$$

$$R = PF_{PR} \tag{10.11b}$$

Equation (10.11b) can be explained as under:

$$\left[\,unacost\,\right] = \left[\,present\ value\,\right]\left[\,capital\ recovery\ factor\,(\text{CRF})\,\right]$$

where F_{PR} (Table 10.1) is designated as

$$F_{PR,i,n} = \frac{i(1+i)^n}{(1+i)^n - 1} \tag{10.11c}$$

Referring to Equations (10.9c and 10.11c), we have

$$F_{RP,i,n} = \frac{1}{F_{PR,i,n}} \tag{10.11d}$$

Using spreadsheet computer software, MS-Excel, the annual uniform cost can be computed by writing the expression as $=PMT(i,n,P)$ [1].

Life Cycle Analysis 299

EXAMPLE 10.11

A large-capacity water heating system is expected to save USD$4000 every year in terms of fuel savings. If the effect of escalation in the price of fuel saved is neglected, what is the present worth of fuel saving in the 5th, 10th, 15th, 20th, 25th and 30th years for a discount rate of 12%?

Solution

Given that amount of fuel saving is USD$4000/year and $i = 0.12$. The values of the present worth factors and the corresponding present worth of annual fuel saving for the desired years are tabulated below:

Year (n)	Present Worth Factor $(PWF) = \left[\dfrac{1}{(1+i)^n} \right]$	Present Worth of Fuel Savings (4000 × PWF)
5	$1/(1.2)^5 = 0.5670$	2269.7
10	$1/(1.2)^{10} = 0.3220$	1287.8
15	$1/(1.2)^{15} = 0.1827$	730.7
20	$1/(1.2)^{20} = 0.1037$	414.6
25	$1/(1.2)^{25} = 0.0588$	235.2
30	$1/(1.2)^{30} = 0.0334$	133.5

It may be noted that the present worth of fuel savings in later years of the useful life of the domestic solar water heating system is rather small. Thus, the present value analysis of a renewable energy system with longer useful life may not be representative of its actual usefulness to the user.

EXAMPLE 10.12

A man borrows Rs. 150,000 for constructing a room having a roof integrated with a PVT system. The loan carries an interest rate of 10% year and is to be repaid in equal instalments over the next 7 years. Calculate the amount of this annual instalment. (Consider USD$1= Rs. 72.44)

Solution

Using Equation (10.11a), we have

$$R = P \frac{i(1+i)^n}{(1+i)^n - 1} = 150{,}000 \times \frac{0.10 \times (1+0.10)^7}{(1+0.10)^7 - 1} = \text{Rs. } 30{,}810.82 \ (\text{USD} \$439.13)$$

Using MS-Excel, the above can be solved by using the following
=PMT(10%,7,-150000)
=Rs. 30,810.82 (USD$439.13)

10.3.3 SINKING FUND FACTOR

The future value (S), which is commonly also referred to as salvage value, can also be converted into a uniform end-of-year annual amount (R) at the end of n years as given in Figure 10.4. The salvage value is basically the scrap or residual value.

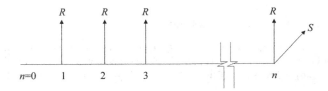

FIGURE 10.4 Cash flow diagram for sinking fund factor.

Let R be the amount invested at the end of every year for n period (years) at an annual interest rate i, as shown in Figure 10.4, then the total accumulated fund (S) at the end of duration is simply the sum of R payments multiplied by the appropriate single-payment future value factor for years 1 through n. This summation is a geometric series as follows:

$$S = \sum_{j=n-1}^{0} R(1+i)^j = R\left[(1+i)^{n-1} + (1+i)^{n-2} + (1+i)^{n-3} + \ldots + (1+i)^2 + (1+i)^1 + 1\right] \quad (10.12a)$$

Expressing Equation (10.11a) in terms of S with the help of Equation (10.5a), we have

$$R = \left[S(1+i)^{-n}\right]\frac{i(1+i)^n}{(1+i)^n - 1} \quad (10.12b)$$

$$R = S\frac{i}{(1+i)^n - 1} \quad (10.12c)$$

or

$$S = R\frac{(1+i)^n - 1}{i} \quad (10.12d)$$

$$R = SF_{SR,i,n} \quad (10.13a)$$

or

$$S = RF_{RS,i,n} \quad (10.13b)$$

Equation (10.13a) can be explained as:

$$[unacost] = [Future\ amount][sinking\ fund\ factor\,(SFF)]$$

Equation (10.13b) can be explained as:

$$[future\ amount] = [unacost][equal\ payment\ series\ future\ value\ factor]$$

where

$$F_{SR,i,n} = \frac{i}{(1+i)^n - 1} \quad (\text{Table}\,10.1) \quad (10.14a)$$

Life Cycle Analysis

SFF is a factor used to calculate the uniform end-of-year annual amount for the salvage value of the system.

$$F_{RS,i,n} = \frac{(1+i)^n - 1}{i} \text{ is the equal payment series future value factor} \quad (10.14b)$$
$$\text{or annuity future value factor}$$

Using MS-Excel, the future value can be computed by writing the expression as = $FV(i,n,-R)$ [1]. As seen from Equations (10.14a) and (10.14b),

$$SFF = \frac{1}{\text{equal payment series future value factor}}$$

i.e.,

$$F_{SR,i,n} = \frac{1}{F_{RS,i,n}} \quad (10.14c)$$

Using MS-Excel, the annual uniform cost can be computed by writing the expression as =$PMT(i,n,,-S)$ [1].

From Equations (10.9b) and (10.13b), it is clear that R is related to P and S. Similarly, a uniform beginning-of-year annual amount (R_b) may also be derived in terms of P and S.

Figures 10.5(a) and (b) give the cash flow diagram for R_b and R, respectively.

Referring to Equation (10.5b), the relation between R_b and R can be established as follows:

$$R_b = \frac{R}{(1+i)} \quad (10.15a)$$

or

$$R = R_b(1+i) \quad (10.15b)$$

Now, expression to show the relationship between P, S and R_b can be derived as follows.
Using Equations (10.15b) and (10.11b), we have

$$R_b(1+i) = PF_{PR,i,n} \quad (10.16a)$$

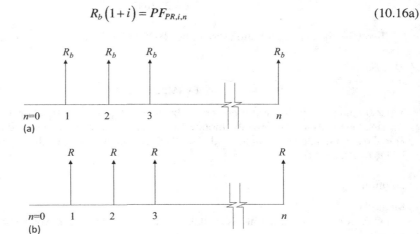

FIGURE 10.5 Cash flow diagram for uniform (a) end of year amount and (b) beginning of year amount.

$$R_b = \frac{P}{(1+i)} \times F_{PR,i,n} \qquad (10.16b)$$

Similarly, from Equation (10.13a)

$$R_b = \frac{S}{(1+i)} \times F_{SR,i,n} \qquad (10.16c)$$

EXAMPLE 10.13

A person decides to spend USD$4000 in the first, second, third and fourth year on energy-efficient equipment. They agree to set aside a certain amount now and each year thereafter until the fourth year. If the contribution forms an arithmetical progression for all year, increasing by 20% after the first year; calculate their first contribution if money is worth 10%. (Consider USD$1= Rs. 72.44)

Solution

Let the first contribution be x as given in the cash flow diagram, Figure 10.6.

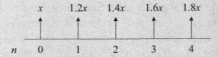

FIGURE 10.6 Solution to Example 10.13.

Consider 2 years from now as the focal point. Now using time-value conversion relation in a cash-flow diagram, we get,

$$x(1.10)^{-2} + 1.2x(1.10)^{-1} + 1.4x(1.10)^0 + 1.6x(1.10)^1 + 1.8x(1.10)^2$$
$$= 4000(1.10)^{-1} + 4000(1.10)^0 + 4000(1.10)^1 + 4000(1.10)^2$$

$$7.2553x = 16,876.36,723.1$$

$$x = 2326.05$$

Therefore, the first contribution should be USD$2326.05 (Rs. 168,499.1)

EXAMPLE 10.14

A person plans to create a forborne annuity by depositing Rs. 100,000 at the end of each year for 8 years. He wants to withdraw the money 14 years from now to buy a hybrid solar water heater. Find the accumulated value at the end of the 14th year, if money is worth 10% per year. (Consider USD$1= Rs. 72.44)

Solution

Method I
   Let us consider that amount x is available at the end of 14th year, which can be considered as receipt.

Life Cycle Analysis

Using Equation (10.9b), the present value (zero time) can be estimated

$$P = 1,00,000 \times F_{RP,10\%,14} = \text{Rs. } 533,490 \ (\text{USD}\$7364.57)$$

Rs. 533,490 is deposited for 14 years, then the future value at the end of 14 years can be calculated using Equation (10.2b)

$$S = 533,490 \times F_{PS,10\%,14} = 533,490 \times 3.79 = \text{Rs. } 20,25,900 \ (\text{USD}\$27,966.59)$$

Method II

The problem can also be solved by considering Rs. 100,000 paid for 14 years less Rs. 100,000 paid as annuity for the last 16 years. Using Equation (10.12d), we get

$$S = 1,00,000 \left[\frac{(1.10)^{14} - 1}{0.10} - \frac{(1.10)^6 - 1}{0.10} \right] = \text{Rs. } 2,025,900 \ (\text{USD}\$27,966.59)$$

EXAMPLE 10.15

A person wants a down payment of Rs. 2000 on a water-heating system of amount Rs. 10,000. An annual end-of-year payment (R) of Rs. 1174.11 is required for 12 years. However, the person elects to pay Rs. 1000 yearly and a balance payment at the end. Find the balance payment if money is worth 10% interest. (Consider USD$1= Rs. 72.44)

Solution

Using Equations (10.5b) and (10.9b), we have

$$10,000 = 2000 + 1000 \times F_{RP,10\%,12} + x F_{SP,10\%,12}$$

Solving the above, we get the balance payment as

$$x = \text{Rs. } 3723.10 \ (\text{USD}\$51.39)$$

EXAMPLE 10.16

Find the equivalent present value (P) of the following series of receipt (USD$) as of the end of the fourth year if money is worth 10% per year for the cash flow diagram given in Figure 10.7(a). (Consider USD$1= Rs. 72.44)

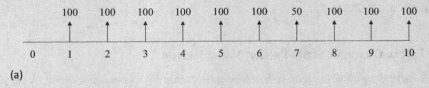

(a)

FIGURE 10.7 Cash flow diagram for (a)

Solution

After adding and subtracting 50 at the 7th year, the cash flow diagram becomes as given in Figure 10.7(b), and the solution to the problem by using Equations (10.5b) and (10.9b) becomes

$$P = 100 \times F_{RP,10\%,10} - 50 \times F_{SP,10\%,7} = 100(6.1438) - 50(0.3855) = 595.10 \ (\text{zero time})$$

Move, 4 years with the calendar and using Equation (10.2b), the equivalent present value at the 4th year becomes

$$595.10 \times F_{PS,6\%,3} = 595.10 \times 2.5937 = \text{USD}\$1543.52 \ (\text{Rs.}111{,}812.6)$$

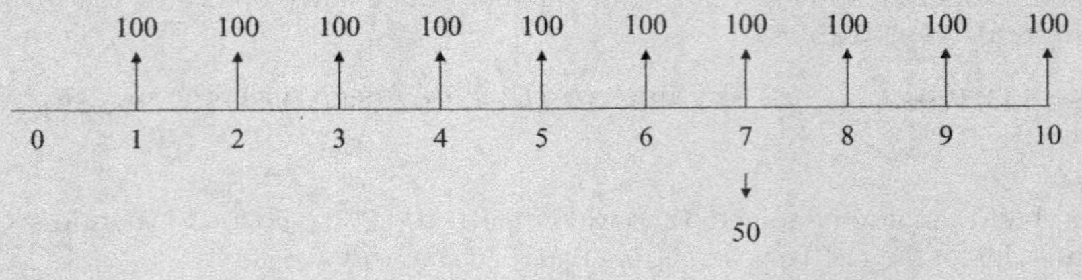

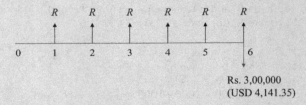

(b)

FIGURE 10.7 (b) Solution to Example 10.16.

EXAMPLE 10.17

A person wants to purchase energy-efficient equipment after 6 years whose expected cost at that time is Rs. 300,000. For this they open a recurring deposit account of 6 years in which they deposit amount R at the end of each year at an interest rate of 10%, as shown by cash flow diagram in Figure 10.8. Determine the value of R. (Consider USD$1= Rs. 72.44)

FIGURE 10.8 Cash flow diagram.

Solution

Using Equation (10.12c)

$$R = S\frac{i}{(1+i)^n - 1} = 300{,}000\frac{0.10}{(1+0.10)^6 - 1} = \text{Rs.}\,38{,}882.2 \ (\text{USD}\$536.75)$$

Using MS-Excel, the above can be solved by using the following
=PMT(10%,6,-300000)
= Rs. 38,882.2 (USD$536.75)

10.3.4 Linear Gradient Series Present Value Factor

Sometimes, periodic payments tend to increase or decrease by a constant amount (let say, G) from time to time as given in Figure 10.9. The gradient (G) can be either positive or negative. In the former case, i.e., $G > 0$ (positive), it is referred to as an increasing gradient series, and in the latter case when $G < 0$ (negative), it is then referred to as a decreasing gradient series.

The payment in terms of present value is given by the series as follows

$$P = \sum_{j=1}^{n} \frac{(n-1)G}{(1+i)^j} = G\left[0 + \frac{1}{(1+i)^2} + \frac{2}{(1+i)^3} + \frac{3}{(1+i)^4} + \ldots + \frac{n-2}{(1+i)^{n-1}} + \frac{n-1}{(1+i)^n}\right] \quad (10.17a)$$

Life Cycle Analysis

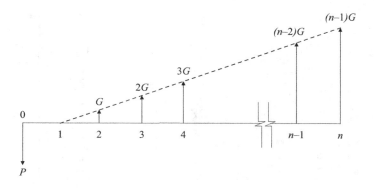

FIGURE 10.9 Linear gradient series present value factor.

$$P = G\left[\frac{(1+i)^n - i \cdot n - 1}{i^2(1+i)^n}\right] = G\left[\frac{(1+i)^n - 1}{i^2(1+i)^n} - \frac{n}{i(1+i)^n}\right] \quad (10.17b)$$

$$P = G \cdot F_{GP} \quad (10.17c)$$

where

$$F_{GP} = \left[\frac{(1+i)^n - i \cdot n - 1}{i^2(1+i)^n}\right] \text{ is the gradient series present value factor} \quad (10.17d)$$

F_{GP} is more completely written as $F_{GP,\,i,\,n}$.

Equation (10.17c) can be explained as:

$$[\text{Present value}(P)] = [\text{linear gradient value}][\text{gradient series present value factor}]$$

EXAMPLE 10.18

A company installs PV with a useful life of 15 years on the roofs of all buildings. It is estimated that the maintenance costs will increase from the first year of USD$100 at the rate of USD$25 per year over the lifetime. Assuming the maintenance costs occur at the end of each year. The firm set up a maintenance account that earns 10% annual interest through which all the expenses of maintenance will be paid out. How much does the firm have to deposit now? (Consider USD$1= Rs. 72.44)

Solution

Using Equation (10.9b) and (10.17c), the solution to above problem with reference to Figure 10.10 becomes,

$$P = R \cdot F_{RP} + G \cdot F_{GP} = R \cdot \frac{(1+i)^n - 1}{i(1+i)^n} + G \cdot \frac{(1+i)^n - i \cdot n - 1}{i^2(1+i)^n}$$

$$= 100 \times \frac{(1+0.10)^{15} - 1}{0.10(1+0.10)^{15}} + 25 \times \frac{(1+0.10)^{15} - 0.10 \times 15 - 1}{(0.10)^2(1+0.10)^{15}}$$

$$= \text{USD}\$760.61 + 1003.80 = \text{USD}\$1764.41\ (\text{Rs}.127{,}813.9)$$

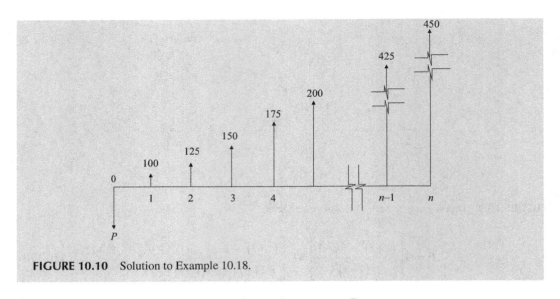

FIGURE 10.10 Solution to Example 10.18.

10.3.5 Gradient to Equal Payment Series Conversion Factor

Figure 10.11 shows that the gradient series may be converted into an equivalent equal payment series using the empirical relationship can be obtained by substituting value of P from Equation (10.17b) into Equation (10.11a) as follows:

$$R = G\left[\frac{(1+i)^n - i \cdot n - 1}{i^2(1+i)^n} \cdot \frac{i(1+i)^n}{(1+i)^n - 1}\right] \quad (10.18a)$$

$$R = G\left[\frac{(1+i)^n - i \cdot n - 1}{i\{(1+i)^n - 1\}}\right] = G\left[\frac{1}{i} - \frac{n}{(1+i)^n - 1}\right] \quad (10.18b)$$

$$\text{or } R = G \cdot F_{GR} \quad (10.18c)$$

where

$$F_{GR} = \left[\frac{1}{i} - \frac{n}{(1+i)^n - 1}\right] \text{ is called the gradient to equal payment series conversion factor.} \quad (10.18d)$$

F_{GR} is more completely written as $F_{GR, i, n}$.

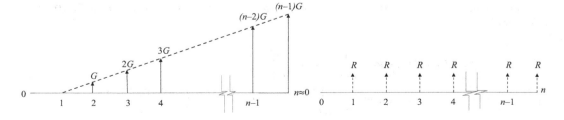

FIGURE 10.11 Converting a gradient series into an equivalent uniform series.

Life Cycle Analysis

Equation (10.18c) can be explained as:

$$[Annualized\ uniform\ cost] = [linear\ gradient\ value]$$
$$[gradient\ to\ equal\ payment\ series\ conversion\ factor]$$

EXAMPLE 10.19

The two cash flows in Figure 10.12 are equivalent at an interest rate of 10%, compounded annually. Determine the value of R. (Consider USD$1= Rs. 72.44)

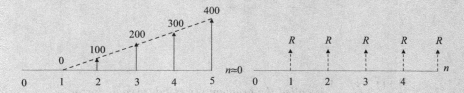

FIGURE 10.12 Cash flow diagram

Solution

From Figure 10.12, G = USD$100, i = 10%, n = 5 years.
Using Equation (10.18b) we have,

$$R = G \cdot F_{GR} = G\left[\frac{(1+i)^n - i \cdot n - 1}{i\{(1+i)^n - 1\}}\right]$$

$$R = 100 \times \left[\frac{(1+0.10)^5 - 0.10 \times 5 - 1}{0.10 \times \{(1+0.10)^5 - 1\}}\right]$$

$$R = USD\$181.01\ (Rs.13{,}112.36)$$

10.3.6 Linear Gradient Series Future Value Factor

Figure 10.13 represents a gradient series whose equivalent future value S can be obtained by substituting value of R from Equation (10.18b) into Equation (10.12d) as follows

$$S = G\left[\frac{(1+i)^n - i \cdot n - 1}{i\{(1+i)^n - 1\}} \cdot \frac{(1+i)^n - 1}{i}\right] \qquad (10.19a)$$

$$S = G\left[\frac{(1+i)^n - i \cdot n - 1}{i^2}\right] \qquad (10.19b)$$

or $\quad S = G \cdot F_{GS}$ \qquad (10.19c)

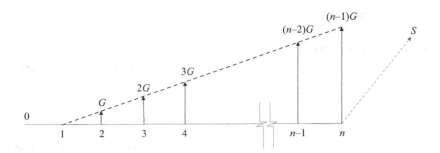

FIGURE 10.13 Linear gradient series future value factor.

where

$$F_{GS} = \frac{(1+i)^n - i \cdot n - 1}{i^2} \text{ is called the gradient series future value factor.} \quad (10.19d)$$

F_{GS} is more completely written as $F_{GS, i, n}$.

Equation (10.19c) can be explained as:

$$[Future\ value] = [linear\ gradient\ value][gradient\ series\ future\ value\ factor]$$

10.4 CAPITALIZED COST

We have already understood that cash flow is based on finite time scale (Section 10.2). However, after getting net cash flow, each net cash flow is repeated on infinite time scale.

Capitalized cost is the present value on an infinite time basis. They are used for the analysis of the systems whose service life is perpetual or the horizon is extremely long (say, 30 years or more), such as bridges, hydroelectric dams, waterway constructions, etc. as given in Figure 10.14.

For a system costing P_n and lasting n years, the present value replacing out to infinity is

$$K = P_n \left[1 + \frac{1}{(1+i)^n} + \frac{1}{(1+i)^{2n}} + \ldots \right] = P_n \sum_{x=0}^{\infty} \frac{1}{(1+i)^{xn}} \quad (10.20a)$$

Equation (10.20a) represents a geometric series with the first term as 1 and the ratio of the consecutive terms as follows.

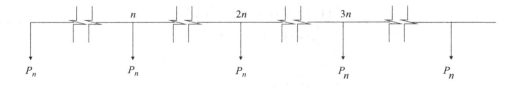

FIGURE 10.14 Capitalized cost.

Life Cycle Analysis   309

The present value, replacing the infinity series will be

$$\sum \frac{1}{\left(1+i\right)^{xn}} = \frac{1-\left(\dfrac{1}{\left(1+i\right)^{n}}\right)^{\infty}}{1-\dfrac{1}{\left(1+i\right)^{n}}} = \frac{\left(1+i\right)^{n}}{\left(1+i\right)^{n}-1} \tag{10.20b}$$

Equation (10.20a) becomes,

$$K = P_n \frac{\left(1+i\right)^{n}}{\left(1+i\right)^{n}-1} = P_n \times F_{PK,i,n} \tag{10.20c}$$

where K is the capitalized cost and $F_{PK,i,n}$ is the factor that converts a present value to capitalized cost (Table 10.1) and is given by

$$F_{PK,i,n} = \frac{\left(1+i\right)^{n}}{\left(1+i\right)^{n}-1} \text{ is the capitalized cost factor} \tag{10.20d}$$

Equation (10.20c) can be explained as:

$$\left[Capitalized\ cost\right] = \left[net\ present\ value\left(cash\ flow\right)on\ basis\ n\ years\ time\ scale\right]$$
$$\left[capitalized\ cost\ factor\right]$$

From Equation (10.11c), we have

$$F_{PR,i,n} = iF_{PK,i,n} \tag{10.20e}$$

Equation (10.20e) can be explained as

$$\left[Capitalized\ recovery\ factor\left(CRF\right)\right] = \left[rate\ of\ return\right]\left[capitalized\ cost\ factor\left(CCF\right)\right]$$

Relation between uniform end-of-year annual amount (R) and capitalized cost (K) can be understood from the following relation:

$$R = iK \tag{10.21}$$

Equation (10.21) can be explained as:

$$\left[Unacost\right] = \left[rate\ of\ return\right]\left[capitalized\ cost\right]$$

10.5   COST COMPARISONS WITH EQUAL DURATION

In this section a uniform expense is referred to as a uniform end-of-year cost.

If two energy-efficient systems have same service life but different associated costs, they can compare by submitting all the costs in terms of either present value or future. Two solar energy systems have been considered in the following solved example for cost comparison.

EXAMPLE 10.20

The cost comparison of two solar water heating systems is given in Table 10.2. Find out which system is more economical if the money is worth 10% rate of interest per year. (Consider USD$1= Rs. 72.44)

TABLE 10.2
Cost Comparison of Two Solar Water Heating Systems

Sr. No.	Economic Components	System A	System B
1	First cost	Rs. 35,000	Rs. 20,000
2	Uniform end-of-year maintenance per annum	Rs. 3000	Rs. 5500
3	Overhaul, end of the third year	-	Rs. 4000
4	Salvage value	Rs. 4500	Rs. 1000
5	Life of the system	5 years	5 years
6	Benefit from quality control as a uniform end-of-year amount per annum	Rs. 1000	-

Solution

Refer to cash flow diagrams from Figures 10.15(a) and (b)

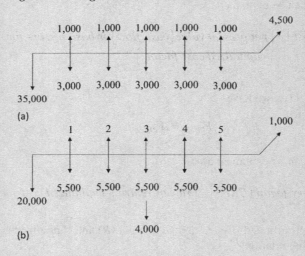

FIGURE 10.15 Solution in Rs. (a) System A and (b) System B.

The present value of the costs for system A can be obtained by using Equations (10.5b) and (10.9b), we obtain

$$P_A = 35{,}000 + (3000 - 1000) F_{RP,10\%,5} - 4000 F_{SP,10\%,5}$$

$$P_A = 35{,}000 + (3000 - 1000) \times \frac{(1+0.10)^5 - 1}{0.10 \times (1+0.10)^5} - 4500 \times \frac{1}{(1+0.10)^5}$$

$$P_A = \text{Rs. } 39{,}787.43 \ (\text{USD\$} 549.24)$$

Life Cycle Analysis

Similarly, the present value of the costs for system B can be obtained:

$$P_B = 20{,}000 + 5500 F_{RP,10\%,5} + 4000 F_{SP,10\%,3} - 1000 F_{SP,10\%,5}$$

$$P_B = 20{,}000 + 5500 \times \frac{(1+0.10)^5 - 1}{0.10 \times (1+0.10)^5} + 4000 \times \frac{1}{(1+0.10)^3} - 1000 \times \frac{1}{(1+0.10)^5}$$

$$P_B = \text{Rs. } 43{,}233.67 \ (\text{USD } \$596.82)$$

From these calculations, it is clear that system A is more economical than system B.

10.6 NET PRESENT VALUE (NPV)

The difference between the present value of the benefits and the costs resulting from an investment is the net present value (NPV) of the investment. For NPV, the cash flow as given in Figure 10.16 should be prepared based on expenditures (debits) or receipts (credits) data available as follows.

Receipts (credits) have been denoted by B_j and expenditures (debits) by C_j at n^{th} year. The numerical values of B_j include all credits such as annualized salvage value, annual C-credit, cost of annual energy savings, etc. In the same way, C_j includes annual operations and maintenance (O&M) and annual overhauling cost, if any. The initial investment (P) will always be negative. NPV of the investment is mathematically given as follows [4]:

$$NPV = -P + \sum_{j=1}^{n} \frac{B_j - C_j}{(1+i)^j} \tag{10.22a}$$

For $(B_j - C_j)$ constant and for all j except for $j = 0$

$$NPV = -P + (B-C) \sum_{j=1}^{n} \frac{1}{(1+i)^n} = -P + (B-C)\left[\frac{(1+i)^n - 1}{i(1+i)^n}\right] \tag{10.22b}$$

Equation (10.22b) is valid for the constant rate of interest per year and $(B_j - C_j)$. For unequal rate of interest and $(B_j - C_j)$, Equation (10.22b) becomes

$$NPV = P + \frac{B_1 - C_1}{(1+i_1)} + \frac{B_2 - C_2}{(1+i_1)(1+i_2)} + \ldots$$
$$+ \frac{B_j - C_j}{(1+i_1)(1+i_2)\ldots(1+i_j)} + \ldots + \frac{B_n - C_n}{(1+i_1)(1+i_2)\ldots(1+i_n)} \tag{10.22c}$$

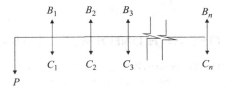

FIGURE 10.16 Net present value.

312   Photovoltaic Thermal Passive House System

The feasibility of any project is determined by NPV, whether the project shall be accepted or rejected.

- If NPV > 0, the project shall be accepted, which means that financial position of the investor will be improved by undertaking the project.
- If NPV = 0, the project remains indifferent, which means that the present value of all benefits over the useful lifetime is equal to the present value of all the costs.
- If NPV < 0, the project shall be rejected because this indicates financial loss.

10.6.1 LIMITATIONS OF THE NPV METHOD

Following are the limitations of NPV method:

- This method majorly aims at benefits and is unable to distinguish between an investment involving relatively large costs and benefits and one involving much smaller costs and benefits as long as the two projects result in equal NPV. Thus, it does not give any indication of the scale of efforts required to achieve the results.
  - (a) The results of NPV are dependent on the rate of interest/discount rate chosen. Thus failure to select an appropriate value of the interest rate used in the computation of NPV may alter or even reverse the relative ranking of different alternatives being compared using this method. This can be explained from the following example. With a very low interest rate, an alternative with benefits spread far into the future may unjustifiably appear more profitable than an alternative whose benefits are more quickly realized but is of a lower amount in undiscounted terms.

EXAMPLE 10.21

The life of a photovoltaic system is 30 years with P = Rs. 100,000, B_j = Rs. 15,000 worth of annual saving of diesel; annual maintenance cost is estimated C_j = Rs. 5000. Calculate the NPV of the investment at an interest rate of 8%. (Consider USD$1= Rs. 72.44)

Solution

Net annual benefits of using a PV system $(B - C)$ = 15000 − 5000 = Rs. 1700 (USD$23.46)

For constant net annual benefits, Equation (10.22b) can be used for determining the NPV as follows:

$$NPV = -P + (B-C)\left[\frac{(1+i)^n - 1}{i(1+i)^n}\right]$$

$$= -100,000 + 1700\left[\frac{(1+0.08)^{30} - 1}{0.08(1+0.08)^{30}}\right] = \text{Rs.}\,12,577.8\,(\text{USD}\$173.63)$$

Therefore the investment in the PV system is a financially viable investment.

10.7 COST COMPARISONS WITH UNEQUAL DURATION

Let us assume that there are two energy-efficient systems with different duration of lives, then a fair comparison can be made only on the basis of equal duration. One of the methods for comparison is to compare single present value of costs on the basis of a common denominator of their service lives. Example 10.22 has been solved for different methods.

Life Cycle Analysis 313

10.7.1 SINGLE PRESENT VALUE METHOD (METHOD I)

Consider the following example in which there are two solar water heating systems, and their cost comparison is tabulated in Table 10.3. The economic viability has been found out for the systems in question.

EXAMPLE 10.22

Table 10.3 gives the comparative cost analysis of two solar water heating systems. Find out the most economical system of the two if the money is worth 10% per annum. (Consider USD$1 = Rs. 72.44)

TABLE 10.3
Cost Comparison of Two Solar Water Heating Systems

		Economic Components		
System	First Cost	Uniform End-of-Year Maintenance per Annum	Salvage Value	Service Life
System A	Rs. 20,000	Rs. 4000	Rs. 500	2 years
System B	Rs. 30,000	Rs. 3000	Rs. 1500	3 years

Solution

The cash flow diagrams for both systems are first reduced to single present value of the cost and are given in Figures 10.17(a–d).

For system A:

Present value of system A as of its time of installation = Rs. 26,529 (USD$366.22)

Present value at 10% for system A,

$$P_{A6} = 26,529 + 26,529 F_{SP,10\%,2} + 26,529 F_{SP,10\%,4}$$

$$P_{A6} = \text{Rs. } 66,573.45 \text{ (USD\$919.01)}$$

For system B:

Present value of system B at its time of installation = Rs. 36,334 (USD$501.57)

Similarly, P_{B6} = Rs. 63,632.27 (USD$878.41)

The ratio of the cost becomes

$$\frac{P_A}{P_B} = \frac{66,573.45}{63,632.27} = 1.0462$$

Thus, system B is comparatively more economical than system A.

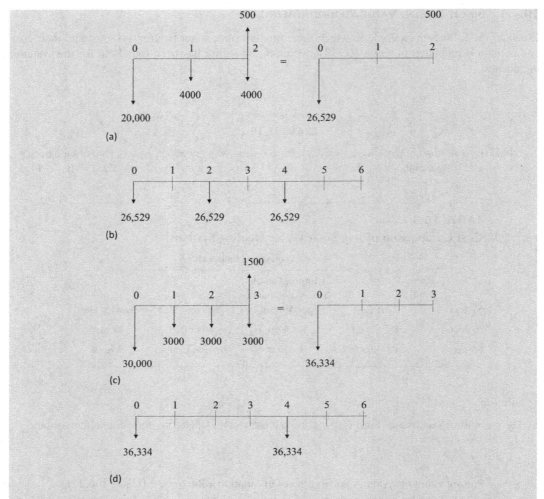

FIGURE 10.17 Solution in Rs. for System A (a) cash flow diagram reduced to single present value of the cost and (b) simplified diagram repeated to obtain 6-year duration. (c) cash flow diagram reduced to single present value of the cost and (d) simplified diagram repeated to obtain 6-year duration.

10.7.2 Annual Cost Method (Method II)

Using Equation (10.11b) uniform-end-of-year cost will be calculated for both the systems for Example 10.22 and taking:

$$Present\ value\ for\ system\ A = P_{A2} = Rs.\ 26,529\ \left(USD\$366.22\right)$$

$$Present\ value\ for\ system\ B = P_{B3} = Rs.\ 36,334\ \left(USD\$501.57\right)$$

$$R_A = P_{A2} \times F_{PR,10\%,2} = 26,529 \times 0.57619 = Rs.\ 15,285.74\ \left(USD\$211.01\right)$$

$$R_B = P_{B3} \times F_{PR,10\%,3} = 36,334 \times 0.40211 = Rs.\ 14,610.26\ \left(USD\$201.68\right)$$

Life Cycle Analysis   315

The ratio of the cost becomes,

$$\frac{R_A}{R_B} = \frac{15,285.74}{14,610.26} = 1.0462$$

Thus, system B is comparatively more economical than system A.

10.7.3 Capitalized Cost Method (Method III)

Solving Example 10.22 using capitalized cost method:

$$P_{A2} \text{ for system } A = \text{Rs. } 26,529 \left(\text{USD}\$366.22\right)$$

$$P_{B3} \text{ for system } B = \text{Rs. } 36,334 \left(\text{USD}\$501.57\right)$$

For system A, using the capitalized cost method Equation (10.20d), we have

$$F_{PK,10\%,2} = \frac{(1+0.1)^2}{(1+0.1)^2 - 1} = 5.7619$$

From Equation 10.20c, we have

$$K_A = P_{A2} \times F_{PK,10\%,2} = 26,529 \times 5.7619 = \text{Rs. } 152,857.45 \left(\text{USD}\$2110.12\right)$$

Similarly, for system B, sing the capitalized cost method (Equation 10.20d), we have

$$F_{PK,10\%,3} = \frac{(1+0.1)^3}{(1+0.1)^3 - 1} = 4.0211$$

From Equation 10.20c, we have

$$K_B = P_{B3} \times F_{PK,10\%,3} = 36,334 \times 4.0211 = \text{Rs. } 146,102.65 \left(\text{USD}\$2106.87\right)$$

The ratio of capitalized cost is obtained as

$$\frac{K_A}{K_B} = \frac{152,857.45}{146,102.65} = 1.0462$$

Thus, system B is comparatively more economical than system A.

10.7.4 Method IV

As a matter of fact, it is possible to convert a present value P_{n1}, of n_1 year's duration to an equivalent present value P_{n2} of n_2 year's duration.

From Equation (10.20c) and Equation (10.20e), we have

$$P_{n1} \times F_{PR,i,n1} = P_{n2} \times F_{PR,i,n2} \tag{10.23a}$$

$$P_{n2} = P_{n1} \frac{F_{PR,i,n1}}{F_{PR,i,n2}} \tag{10.23b}$$

Solving Example 10.22 using method IV, we have

$$P_{A2} \text{ for system } A = \text{Rs. } 26,529 \, (\text{USD\$366.22}) (\text{for 2 years})$$

$$P_{B3} \text{ for system } B = \text{Rs. } 36,334 \, (\text{USD\$501.57}) \, (\text{for 3 years})$$

Now, converting the present value of system B to an equivalent value for 2 years duration by using Equation (10.23b)

$$P_{B2} = P_{B3} \frac{F_{PR,10\%,3}}{F_{PR,10\%,2}} = 36,334 \frac{0.40211}{0.57619} = \text{Rs. } 25,356.68 \, (\text{USD\$350.03})$$

The ratio of cost is obtained as

$$\frac{P_{A2}}{P_{B2}} = \frac{26,529}{25,356} = 1.0462$$

Again, system B is comparatively more economical than system A as concluded by previous methods.

10.8 PAYBACK TIME

The amount used for the project (or the amount going into the project) is taken as negative, whereas the amount coming back from the project is always positive. Thus, the profit can be measured as the total income for a project compared to the total outlay. Payback time or payment time or payback period is one of the criteria to measure the profit.

If C_i is the initial cost on the investment, and F is the annual cash flow then the payback period is given by

$$\text{Payback period} = \frac{\text{Initial investment}}{\text{Annual cash flow}} = \frac{C_i}{F} \tag{10.24}$$

A short payback period indicates that the investment provides revenues early in its life sufficient to cover the initial outlay. Thus, an investment with a short payback period can be viewed as having a higher degree of liquidity than one with a longer payback period.

10.8.1 ANALYTICAL EXPRESSION FOR PAYBACK TIME

Payback time is the time period that is required to recover the initial investment (P) and can be computed by summing the annual cash flow values and estimating n through the relation.

Using Equation (10.22b) and making net present value equal to zero for $(B_j - C_j)$ to be constant for all j except $j = 0$ and $(B - C)$ is the net cash flow (CF) at the end of every year; we have

$$NPV = -P + (B - C)\left[\frac{(1+i)^n - 1}{i(1+i)^n}\right] = 0 \tag{10.25}$$

Life Cycle Analysis
317

Now Equation (10.22b) can be explained as:

$0 = -$initial investment $+$ sum of annual cash flows and can be rewritten as :

$$0 = -P + \sum_{t=1}^{n} CF_t \left(F_{SP,i\%,t} \right) \qquad (10.26a)$$

where CF_t is the net cash flow at the end of year t.

If cash flow is same for every year, F_{RP} factor may be used in the relation

$$0 = -P + CF_t \times F_{RP,i\%,n} \qquad (10.26b)$$

After n years the cash flow will recover the investment and a return of $i\%$. If the expected retention period of the asset/project is less than n years, then investment is not advisable.

Considering $i = 0$, Equation (10.24c) can be written as:

$$0 = -P + \sum_{t=1}^{n} CF_t \qquad (10.26c)$$

If CF_t values are assumed to be equal, then

$$n = \frac{P}{CF} \qquad (10.27a)$$

There is little value in techno-economic study for n computed from Equations (10.25d) and (10.25e). When $i\% > 0$ is used to estimate n, the results incorporate the risk considered in the project undertaken.

Using Equations (10.11d) and (10.25d), the expression for the payback period for unequal annual savings can be written as

$$n = \frac{\ln\left[\dfrac{CF}{CF - P \times i} \right]}{\ln\left[1 + i \right]} \qquad (10.27b)$$

EXAMPLE 10.23

Find the payback period for a PVT system initially costs USD\$6200 and the expected uniform annual benefit is USD\$775.

Solution

Using Equation (10.24)

$$Payback\ period = \frac{C_i}{F} = \frac{6200}{775} = 8\ years$$

10.8.2 PAYBACK PERIOD WITHOUT INTEREST

It is the time required to reduce the investment to zero. The working capital is not considered in evaluating payout time without interest.

It is the length of time required to recover the initial cost of the investment from the net cash flow produced by that investment for an interest rate of zero. If C_i is the initial cost on the investment, and F_t is the net cash flow in period t, then the payback period is defined as the smallest value of n that satisfies the equation

$$\sum_{t=0}^{n} F_t \geq 0 \tag{10.28}$$

10.8.3 PAYBACK PERIOD WITH INTEREST

This is the method that allows for a return on investment and is subject to variations. The variation includes an interest on working capital.

It is the length of time required until the investments equivalent receipts exceed the equivalent capital outlays. If C_i is the initial cost on the investment, and F_t is the net cash flow in period t, then the payback period with an annual interest rate i is the smallest value of n' that satisfies the equation

$$\sum_{t=0}^{n'} F_t (1+t)^{-t} \geq 0 \tag{10.29}$$

EXAMPLE 10.24

The data for various profit and cash flow is given in Table 10.4. Find out the payback period with interest on the remaining investment at 10% per annum.

TABLE 10.4
Data for Various Profit and Cash Flow

		0	1	2	3	4	5
A	Time (end year)	0	1	2	3	4	5
B	After tax profit (Rs.)	−10,000	2750	2000	1300	700	0
C	Tax benefit due to depreciation (Rs.)	0	2000	2000	2000	2000	2000
D	Cash flow (=B+C) (Rs.)	−10,000	4750	4000	3300	2700	2000

Solution

$$\text{Investment for the year} = \text{Investment for the previous year}$$
$$- (\text{cash flow} - \text{interest rate for the previous year}) \tag{10.30}$$

The data for cumulative cash flow with interest is given in Table 10.5.

It can be concluded that the payback period is about 2.95 years, which is higher due to the interest rate as per expectation.

Life Cycle Analysis

TABLE 10.5

Data for Cumulative Cash Flow with Interest

A	End Year	0	1	2	3
B	Investment for year (Rs.)		10,000	6250	2875
C	Interest on (B) (Rs.)		1000	625	287.5
D	Cash flow after interest (Rs.)	−10,000	4750	4000	3300
E	Cash flow net = (D-C) (Rs.)		3750	3375	3012.50
F	Cumulative cash flow (Rs.)	−10,000	−6250	−2875	137.50

EXAMPLE 10.25

The cash flow of two systems is tabulated in Table 10.6. Find the payback period for both the systems.

TABLE 10.6

Data for Cash Flow for Two Systems

System	Economic Components			
	Present Cost	Net Income per Annum	Maximum Life	Rate of Interest
System I	Rs. 16,000	Rs. 4000	7 years	15%
System II	Rs. 8000	Rs. 1000 (for year 1–5)	15 years	15%
		Rs. 3000 (for years 6–15)		

Solution

Cash flow diagram for system I and II are given in Figure 10.18 (a, b), and equivalent cash flow for n years as a payback period for System II is given in Figure 10.18(c).

System I:
   Using Equation (1.25c), we have

$$0 = -16,000 + 4000 \times F_{RP,15\%,n}$$

After solving this, we have $n = 6.57$ years, which is <7 years.

System II:
   Now $3000 \times F_{RP,15\%,n-5}$ can be converted into present worth by using Equations (10.6) and we have

$$0 = -8000 + 1000 \times F_{RP,15\%,5} + \left(3000 \times F_{RP,15\%,n-5}\right) \times F_{SP,15\%,n}$$

By solving the above, we get $n = 9.52$ years, which is less than expected life.

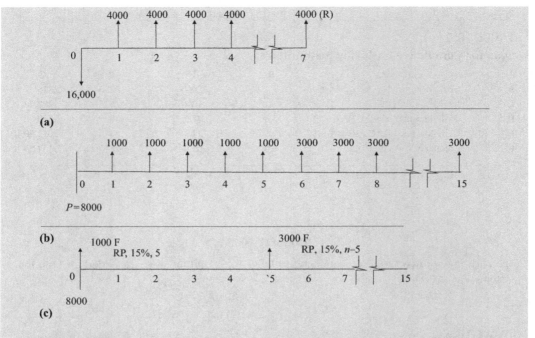

FIGURE 10.18 Solution to Example 10.25 (a) System I, (b) System II and (c) equivalent cash flow for n years as a payback period for System II.

EXAMPLE 10.26

The initial cost of a system is USD$1000, and the expected annual benefits by the end of each year are USD$400, USD$300, USD$200, USD$200, USD$165, USD$165 for six consecutive years. Find out the payback period without interest and also with 10% interest.

Solution

(a) Payback period without interest is

$$-1000 + 400 + 300 + 200 + 200 + 165 + 165 \geq 0$$

In this example, the shortest payback period is between 3–4 years, i.e., 3.5 years.

(b) Payback period with $i = 10\%$ is,

$$-1000 + 400 F_{SP,10\%,1} + 300 F_{SP,10\%,2} + 200 F_{SP,10\%,3} + 200 F_{SP,10\%,4} + 165 F_{SP,10\%,5} + 165 F_{SP,10\%,6} \geq 0$$

In this example the shortest payback period is 5 years.

10.9 BENEFIT–COST ANALYSIS

This is a method based on the ratio of benefits-to-cost to compare the investment costs of a project with its expected profits indicating the total return expected per unit of money spent. The ratio of benefits to costs as a measure of financial or economic efficiency is conceptually simple and quite versatile, and it measures cost efficiency. Benefits and costs are adjusted for the time value of money, so that all flows of benefits and flows of project costs over time are expressed on a common basis in terms of their present value.

Life Cycle Analysis 321

Thus this acts as a powerful tool to choose the right project, which is based on analysis of advantages versus disadvantages. In general, the benefits are advantages (fuel saving in the case of energy projects) expressed in monetary terms, and the disadvantages are the associated disbenefits. The costs are the anticipated expenditures for construction, installation, operation, maintenance, etc.

This analysis generally excludes consideration of factors that are not measured ultimately in economic terms. This is frequently used by government agencies for projects whose benefits are reaped by the common public, and the costs are incurred by the government, to evaluate the desirability of a given intervention.

Following terminology is used for benefit-cost analysis

- **Owner:** Public: One who incurs the costs as the government.
- **Benefits (B):** Benefits are the advantages to the owner. It will be positive in the cash flow diagram. Benefits can be classified as primary and secondary benefit to make the analysis for efficient. The former benefit is directly attributable to the project while the latter is indirectly attributable to the project. If primary benefits alone are sufficient to justify project costs, we can save time and effort by not quantifying the secondary benefits. Then the costs are identified by classifying the expenditures required and any savings (or revenues), including both capital investments and annual operating costs.

  All benefits such as salvage value, annual C-credit, cost of annual energy savings, etc. should be converted into present value.
- **Disbenefits (D):** The project involves disadvantages/loss per year to the owner. This should be also converted into present value.

  A project is considered to be attractive when the benefits derived from its execution exceed its associated costs.
- **Costs (C):** The anticipated expenditures for construction, annual operation and maintenance, and annual overhauling cost, etc. should be converted into present value and then added to construction cost as given below

The equivalent present value cost C may be split into two components – (i) the initial capital expenditure, and (ii) the annual costs accrued in each successive period. If it is assumed that the initial investment is required in the first m periods, and that the annual costs accrue in each of the following periods till the end of the useful life of n periods, the above two components of the equivalent present value cost C and may be expressed as

For bigger projects, the initial investment (C_0) is required in the first m periods.

$$C_0 = \sum_{j=0}^{m} \frac{C_j}{(1+i)^j} \tag{10.31}$$

$$C'' = \sum_{j=m+1}^{n} \frac{C_j}{(1+i)^j} \tag{10.32}$$

with $C = C_0 + C''$.

To start with this analysis, first, all the benefits and disbenefits need to be identified, and net benefit is calculated as:

$$Net\ benefit = Benefit - Disbenefit \tag{10.33a}$$

The conventional benefit-cost ratio (B/C ratio) can be calculated as:

$$B/C = (Benefits - Disbenefits)/\text{cost} = (B-D)/C \tag{10.33b}$$

322   Photovoltaic Thermal Passive House System

The modified *B/C* ratio, which is gaining support, includes O&M costs in the numerator and treats them in a manner similar to disbenefits, and this is given by

$$Modified\ B/C = (Benefits - Disbenefits - O\&M\ cost)/(Initial\ investment) \qquad (10.33c)$$

The salvage value can also be considered in the denominator.
The *B/C* ratio influences the decision on the project approval.

If $B/C > 1$, accept the project
    $B/C < 1$, reject the project.

Thus, in case of mutually exclusive projects, *B/C* ratio gives a method to compare them against each other.

10.9.1 Types of Benefit–Cost Analysis

Using the above expressions, the following three types of benefits-cost ratios are usually defined.

10.9.1.1 Aggregate *B/C* Ratio

This is the ratio of the present value of total benefits to total costs.

$$\left(\frac{B}{C}\right)_{aggregate} = \frac{B}{C} = \frac{B}{C_0 + C''}, C > 0 \ (or\ C_0 + C'' > 0) \qquad (10.34a)$$

$$or \left(\frac{B}{C}\right)_{aggregate} = \frac{\sum_{j=0}^{n} \frac{B_j}{(1+j)^n}}{\sum_{j=0}^{n} \frac{C_j}{(1+j)^n} + \sum_{j=m+1}^{n} \frac{C_j}{(1+j)^n}} \qquad (10.34b)$$

$$\left(\frac{B}{C}\right)_{aggregate} > 1 : \text{Accept the project}$$

10.9.1.2 Net *B/C* Ratio

Here, only the initial capital expenditure is considered a cash outlay, and equivalent benefits become net benefits (i.e., annual revenues minus annual outlays). This type of *B/C* ratio provides an index, which indicates the benefits expected per unit of capital investment, and can hence be used as a profitability index. Note that it simply compares the present value of net revenues with the present value of capital investment. Thus $\left(\frac{B}{C}\right)_{net}$ ensures that there is a surplus at time zero and the project is favorable.

The net benefits-cost ratio is expressed as

$$\left(\frac{B}{C}\right)_{net} = \frac{B - C'}{C_0}, C_0 > 0 \qquad (10.35)$$

$$\left(\frac{B}{C}\right)_{net} > 1 : \textit{Project is viable}$$

Life Cycle Analysis   **323**

10.9.2 ADVANTAGES AND DISADVANTAGES OF *B/C* RATIO

Benefit-cost analysis helps planners put relevant costs and benefits on a common temporal footing in order to help people make informed decisions. It provides people with an understanding as to the economic costs of decisions, and allows arguments to be made for or against a change based upon economic considerations. Following are the advantages of *B/C* ratio:

- It gives a comparison between alternatives on a common scale and permits evaluation of different-sized alternatives.
- It can be used to rank alternatives projects to determine the most profitable alternative for an investor with a limited budget.
- It gives the suitability of the project and indicates whether a project is worthwhile.
- Optimal size of the project can also be determined if it is computed for increments in the investment size.

The disadvantage of this analysis is that it is used to measure effects that may be difficult, or improper, to measure in financial terms. Following are the disadvantages of *B/C* ratio:

- This is influenced by the decision as to whether an item is classified as a cost or a disbenefit, i.e., whether it appears in the denominator or numerator of the ratio. Often it may be an arbitrary decision but can lead to inefficient ranking of investment alternatives.
- The simple benefit-cost ratio cannot be used to determine the efficient scale of a given project. Incremental analysis is required to be undertaken for this purpose.

EXAMPLE 10.27

A non-profit organization is expecting an investment of Rs. 100,000 to install a PV system. The grant would extend over a 10-year period and would create an estimated saving of Rs. 20,000 per year. The organization uses a rate of return of 6% per year on all grant investments. An estimated Rs. 5000 a year would have to be released from other sources for expenses. In order to make this program successful a Rs. 3000/year operating expense will be incurred by the organization from its regular O&M budget. Use the following analysis methods to determine whether the program is justified over a 10-year period:

(a) Conventional B/C
(b) Modified B/C

(Consider USD$1= Rs. 72.44)

Solution

$$Benefits = Rs.\ 20,000\ per\ annum\ \left(USD\$276.09\ per\ annum\right)$$

$$Investment\ cost = Rs.\ 100,000\ F_{PR,6\%,10} = Rs.\ 13587\ per\ annum\ \left(USD\$187.56\ per\ annum\right)$$

$$O\&M\ Cost = Rs.\ 3000\ per\ annum\ \left(USD\$41.41\ per\ annum\right)$$

$$Disbenefits = Rs.\ 5000\ per\ annum\ \left(USD\$69.02\ per\ annum\right)$$

(a) Conventional method:

$$B/C = \frac{20,000 - 5000}{13,587 + 3000} = 0.90$$

324   Photovoltaic Thermal Passive House System

The project is not justified since $B/C<1$

(b) Modified method:

$$Modified \ B/C = \frac{20,000-5000-3000}{13,587} = 0.88$$

Again, the project is not justified since $B/C<1$.

EXAMPLE 10.28

The building regulation in a city stipulates that all new student hostels must use solar energy hot water system. There are two options available namely:

System A: Double-glazed flat plate collectors
System B: Evacuated tubular collectors

The associated costs with the above systems are given in Table 10.7. Find out which system

TABLE 10.7
Associated Costs of Two Systems

	Cost (Rs.)	
Cost and Benefits	**System A**	**System B**
Capital cost	Rs. 3,000,000	Rs. 2,500,000
Annual maintenance cost	Rs. 55,000	Rs. 45,000
Annual benefits due to fuel savings	Rs. 700,000	Rs. 600,000
Life, Years (n)	20	20
Rate of interest	10%	10%

should be adopted on the basis of incremental net benefit-cost ratio.
(Consider USD\$1 = Rs. 72.44)

Solution

The incremental capital cost of System A over System B = 3,000,000 − 2,500,000 = Rs. 500,000 (USD\$6902.26)

> The incremental net annual benefit of System A over net annual benefit of System B
> = (Annual benefits due to fuel savings − Annual maintenance cost) of System A
> − (Annual benefits due to fuel savings − Annual maintenance cost) of System B
> = (700,000 − 55,000) − (600,000 − 45,000) = Rs. 90,000 (USD\$1242.408)

The cumulative present worth of the incremental benefits over 20 years of useful life of System A over System B is

$$= \sum_{j=1}^{n} \frac{90,000}{(1+i)^j} = 90,000 \left[\frac{(1+0.1)^{20}-1}{0.1(1+0.1)^{20}} \right] = 90,000 \times 8.51 = Rs. \ 766,220 \ (USD\$10,577.31)$$

Thus, the net incremental benefits to cost ratio = 766,220/500,000 ≈ 1.53.

A value greater than one for the ratio of net incremental benefits to incremental capital cost implies that the additional discounted benefits more than justify the extra capital cost of System A compared to System B. Therefore, System A should be selected for installation on the hostel.

Life Cycle Analysis

The computation for the net benefit to cost ratio for each alternative independent of each other are given below.

The net benefits to cost ratio for System A is

$$= \frac{(700,000 - 55,000)\left[\dfrac{(1+0.1)^{20} - 1}{0.1(1+0.1)^{20}}\right]}{3,000,000} = 1.83$$

Similarly, the net benefits to cost ratio for System B is:

$$= \frac{(600,000 - 45,000)\left[\dfrac{(1+0.1)^{20} - 1}{0.1(1+0.1)^{20}}\right]}{2,500,000} = 1.89$$

It may be noted that an appraisal of the two alternatives using their net benefits-to-cost ratios would suggest that System B is selected.

As the results obtained with the two methods do not match, the net present values of both alternatives are determined to identify the correct method.

NPV of System A:

$$NPV_A = -3,000,000 + (700,000 - 55,000)\left[\frac{(1+0.1)^{20} - 1}{0.1(1+0.1)^{20}}\right]$$

$$= Rs.\ 2,491,248.599\ (USD\$34,390.51)$$

NPV of alternative B is

$$NPV_B = -2,500,000 + (600,000 - 45,000)\left[\frac{(1+0.1)^{20} - 1}{0.1(1+0.1)^{20}}\right]$$

$$= Rs.\ 2,225,027\ (USD\$30,715.46)$$

i.e., $NPV_A > NPV_B$

Thus, the appraisal based on incremental costs and benefits is correct.

EXAMPLE 10.29

Initial cost of the material used in construction of a storage tank of water with 100 liter capacity is given in Table 10.8.

Estimated salvage value is 40% of the initial cost since the various system components can be reused

$$Salvage\ value\ of\ built\ in\ storage\ water\ heater = Rs.\ 2500$$

$$Maintenance\ cost = 0.20 \times Annual\ first\ cost$$

$$Useful\ life\ of\ the\ system\ (n) = 10\ years\ and\ r = 0.15$$

Find out the total useful energy assuming the efficiency of the system to be 70% and inclination of the absorber at $45° = 5.8768$ kWh/m^{2}d.

(Consider USD\$1= Rs. 72.44)

326   Photovoltaic Thermal Passive House System

TABLE 10.8
Cost of the Materials Used in Water Storage Tank

Materials Used	Cost (Rs.)
Steel structure (for box and cover material)	1500
Glass	190
Insulation	175
Paint	170
Stand, bucket, frame, etc.	650
Labor	700
Total	3385

Solution

Capital recovery factor from Equation (10.11c) is:

$$CRF = \frac{i(1+i)^n}{(1+i)^n - 1} = 0.199$$

Capital recovery factor from Equation (10.14a) is:

$$SFF = \frac{i}{(1+i)^n - 1} = 0.049$$

From the Table 10.8, P = Rs. 3385 (USD$46.72)

$$Salvage\ value\ (S) = \text{Rs. } 2500\ (USD\$34.51)$$

$$Annual\ first\ cost = CRF \times P = \text{Rs. } 674.46\ (USD\$9.31)$$

$$Annual\ salvage\ cost = SFF \times S = \text{Rs. } 123.13\ (USD\$1.69)$$

$$Annual\ maintenance\ cost = 0.20 \times Annual\ first\ cost = 0.20 \times 674.46 = \text{Rs. } 134.89\ (USD\$1.86)$$

Hence, annual cost/m² = (134.89−123.13) + 674.46 = Rs. 686.23 (USD$9.47)
Av. total insolation at 45° inclination of the absorber = 5.8768 × 365 = 2145.03 kWh/m²
Useful energy at efficiency of 70 % = 1501.5 kWh/m²

EXAMPLE 10.30

Initial cost of the material used in single basin solar still with yearly average yield of the still = 2 l/m²d is given in Table 10.9.

Given, salvage value to be 30% of initial cost and n = 15, r = 0.12, and maintenance rate is taken from Example 10.29. Find out the annual cost/kg.

(Consider USD$1= Rs. 72.44)

Life Cycle Analysis

TABLE 10.9
Cost of the Materials Used in Solar Still

Materials used	Cost (Rs.)
Steel and aluminum structures	1500
Glass	200
Rubber material	45
Paint	90
Insulation	45
Labor	700
Total Cost	2580

Solution

From Table 10.9, Salvage value = $0.30 \times P = 0.30 \times 2580 =$ Rs. 774 (USD$10.68)
CRF and SFF can be calculated as given in Example 10.29, and we get,

$$CRF = 0.146$$

$$SFF = 0.026$$

$$Annual\ first\ cost = CRF \times P = Rs.\ 378.80\ (USD\$5.22)$$

$$Annual\ salvage\ cost = SFF \times S = Rs.\ 20.76\ (USD\$0.28)$$

$$Annual\ maintenance\ cost = 0.20\ Annual\ first\ cost = 0.20 \times 378.80 = Rs.\ 75.76\ (USD\$1.04)$$

$$Hence, annual\ cost/m^2 = (378.80 + 75.76 - 20.76) = Rs.\ 433.80\ (USD\$5.98)$$

$$Annual\ yield\ of\ the\ still = 2 \times 365 = 730\ l$$

Using latent heat of vaporization = 0.65 kWh/kg, we have

$$Annual\ useful\ energy = 730 \times 0.65 = 474.5\ kWh$$

$$Annual\ cost/kg = 378.80/730 = Rs.\ 0.51\ (USD\$0.007)$$

10.10   INTERNAL RATE OF RETURN

The internal rate of return (IRR) is a widely-accepted discounted measure of investment worth and is used as an index of profitability for the appraisal of projects. The IRR is defined as the rate of interest that equates the present value of a series of cash flows to zero.

The IRR is widely used in the appraisal of projects due to following reasons:

- The IRR on a project is its expected rate of return.
- It employs a percentage rate of return as the decision variable, which suits the banking community.
- For situations in which IRR exceeds the cost of the funds used to finance the project—a surplus would remain after paying for the capital investment, P.

328 Photovoltaic Thermal Passive House System

The IRR of an investment, P, is the discount rate at which the NPV of costs (negative cash flows, Figure 10.1, Section 10.2) of the investment equals the NPV of the benefits (positive cash flows, Figure 10.1, Section 10.2) of the investment.

Alternatively, the internal rate of return is the interest rate that causes the discounted present value of the benefits in a cash flow to be equal to the present value of the costs.

For appraisal of projects, IRR is widely used because of the following reasons:

- The IRR on a project is its expected rate of return.
- It employs a percentage rate of return as the decision variable, which suits the banking community.
- For situations in which IRR exceeds the cost of the funds used to finance the project, a surplus would remain after paying for the capital.

10.10.1 Iterative Method to Compute IRR [1, 3]

The following steps shall be followed for computation of IRR by iterative approach:

Step 1: Make a guess at a trial rate of interest as follows:

For constant net cash flow, i.e., $(B_j - C_j) = (B - C)$ and for all j except for $j = 0$ and following Equation (10.22b)

$$\sum_{j=1}^{n} \frac{B_j}{\left(1+i_{IRR}\right)^j} - \left[P + \sum_{j=1}^{n} \frac{C_j}{\left(1+i_{IRR}\right)^j}\right] = -P + \sum_{j=1}^{n} \frac{B_j - C_j}{\left(1+i_{IRR}\right)^j} = -P + (B-C)\sum_{j=1}^{n} \frac{1}{\left(1+i_{IRR}\right)^j} = 0$$

For infinite series, i.e., $n \rightarrow \infty$, above equation becomes

$$-P + \frac{(B-C)}{i_{IRR}} = 0 \text{ or } i_{IRR} = \frac{(B-C)}{P} \tag{10.36}$$

Step 2: Using the guessed rate of interest, calculated the *NPV* of all disbursements and receipts.

$$Net\ NPV = \sum_{j=1}^{n} \frac{B_j}{\left(1+i_{IRR}\right)^j} - \left[P + \sum_{j=1}^{n} \frac{C_j}{\left(1+i_{IRR}\right)^j}\right] \tag{10.37}$$

Step 3: If the calculated value of *NPV* is positive then the receipts from the investments are worth more than the disbursements of the investments, and the actual value of IRR would be more than the trial rate. On the other hand, if *NPV* is negative the actual value of IRR would be less than the trial rate of interest. Adjust the estimate of the trial rate of return accordingly.

Step 4: Precede with steps 2 and 3 again until one value of i $(=i_1)$ is found that results in a positive (+) *NPV* and the next higher value of i $(=i_2)$ is found with a negative *NPV*.

Step 5: Solve for the value of $IRR = i_{IRR}$ by using the following

$$i_{IRR} = i_1 - NPV_1 \frac{(i_2 - i_1)}{NPV_2 - NPV_1} \tag{10.38}$$

The IRR is a measure of profitability for the assessment of the project. An important aspect of the iterative method of computing *IRR* is making the initial estimate. If the initial estimate is too far

Life Cycle Analysis

from the actual value of *IRR*, a large number of trials will have to be made to obtain the two consecutive values of interest rate (i_1 and i_2) to permit accurate interpolation. It should be noted that the initial estimate of the *IRR* will always be somewhat in error, and several iterations will normally be required to determine i_1 and i_2. A simple approach for making a guess of the first trial rate of return is given in the following.

The *NPV* of a capital investment C_i resulting in uniform net annual cash flows of amount *A* for an infinite time horizon can be expressed as

$$NPV = -C_i + \left[\frac{A}{(1+i)} + \frac{A}{(1+i)^2} + \frac{A}{(1+i)^3} + \ldots\right] \Rightarrow NPV$$

$$= -C_i + \frac{A}{(1+i)}\left[\frac{1}{1-(1+i)^{-1}}\right] = -C_i + \frac{A}{i} \quad (10.39)$$

where *i* is the interest rate.

Since *NPV* = 0 at *i* = *IRR*, we have

$$-C_i + \frac{A}{i} = 0 \Rightarrow IRR = i = \frac{A}{C_i} \quad (10.40)$$

In actual practice, for investment projects with finite life the IRR shall be less than $\frac{A}{C_i}$. However, to begin with, for cases with uniform periodic cash flows, the figure $\frac{A}{C_i}$ or a value close to it may be used as the trial rate of return in the iterative procedure used for determining *IRR*.

The above interpolation between two consecutive values of interest rates that bracket the *IRR* always overestimates its true value. This is because the linear interpolation technique makes an implicit assumption that between two interest rates i_1 and i_2 the *IRR* changes, following a straight line, whereas the true value of *IRR* follows a concave curvilinear function between the two values. However, the error introduced by interpolation is usually very small. Referring to Figure 10.19, the true value of *IRR* is that value of *i* for which the *NPV* i function intersects the horizontal axis, whereas the interpolated value of *IRR* is somewhat higher than the true value. Obviously, the interpolation error would become less and less as the incremental change in the trial values of *i* used in iteration is made smaller and smaller.

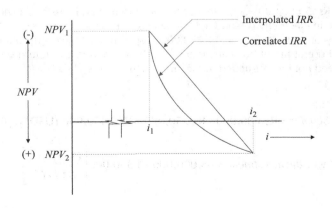

FIGURE 10.19 Interpolation of *IRR*.

EXAMPLE 10.31

Calculate the internal rate of return for the investment in a solar system, which will cost Rs. 500,000 to purchase and install, will last 10 years, and will result in fuel savings of Rs. 145,000 per year. Also assume that the salvage value of the heat exchanger at the end of 10 years is negligible.
(Consider USD\$1 = Rs. 72.44)

Solution

Let the first guess that the value of *IRR* is 25%.

$$NPV \text{ at } 25\% = 145,000 \left[\frac{(1+0.25)^{10} - 1}{0.25(1+0.25)^{10}} \right] - 500,000$$
$$= 145,000(3.57) - 500,000 = \text{Rs.} 17,722 \text{ (USD\$244.64)}$$

Since the NPV at 25% is positive, the IRR shall be greater than 25%. If the next trial value is chosen at 30%, then

$$NPV \text{ at } 30\% = 145,000 \left[\frac{(1+0.3)^{10} - 1}{0.3(1+0.3)^{10}} \right] - 500,000$$
$$= 145,000(3.09) - 500,000 = \text{Rs.} -51,724 \text{ (USD\$714.02)}$$

Obviously, the true *IRR* lies between 25% and 30%. By interpolating between the two, the *IRR* can be estimated as

$$IRR = \left(\frac{0.3 - 0.25}{17,722 + 51,724} \right) 0.25 \times 17,722 = 0.2627 = 26.27\%$$

A better estimate of the true *IRR* may be obtained by using smaller incremental changes in the interest rate.

EXAMPLE 10.32

Installation of a Rs. 50,00,000 energy management system in an industry is expected to result in a 25% reduction in electricity use and a 40% savings in process heating costs. This translates to net yearly savings of Rs. 600,000 and Rs. 750,000, respectively. If the energy management system has an expected useful life of 20 years, determine the internal rate of return on the investment. Salvage value need not be considered in the analysis. (Consider USD\$1 = Rs. 72.44)

Solution

$$Total \ annual \ benefits = \text{Rs.} 600,000 + \text{Rs.} 750,000 = \text{Rs.} 1,350,000 \text{ (USD\$18,636.11)}$$

$$NPV \ of \ the \ investment = -5,000,000 + 1,350,000 \left[\frac{(1+i)^{20} - 1}{i(1+i)^{20}} \right]$$

NPV at *i* = 27%

$$= -5{,}000{,}000 + 1{,}350{,}000 \left[\frac{(1+0.27)^{20} - 1}{0.27(1+0.27)^{20}} \right]$$

$$= -5{,}000{,}000 + 4{,}958{,}034 = \text{Rs.} - 41{,}965 \, (\text{USD} \, \$579.30)$$

NPV at *i* = 26%

$$= 5{,}000{,}000 + 1{,}350{,}000 \left[\frac{(1+0.26)^{20} - 1}{0.27(1+0.26)^{20}} \right]$$

$$= 5{,}000{,}000 + 5{,}141{,}263 = \text{Rs.} \, 141{,}263 \, (\text{USD} \, \$1950.06)$$

Thus, the IRR can be obtained by interpolating between $i = 26\%$ and $i = 27\%$ in the following manner:

$$IRR = 0.26 \left(\frac{0.27 - 0.26}{141{,}263 + 41{,}965} \right) 141{,}263 = 0.2677$$

i.e., the internal rate of return is 26.77%.

10.10.2 Multiple Values of IRR [1, 3]

The NPV of a set of cash receipts and disbursements can be expressed as an nth degree polynomial of the form

$$NPV(i_{IRR}) = 0 = F_0 + F_1 x + F_2 x^2 + F_3 x^3 + \ldots + F_n x^n \tag{10.41}$$

where $x = \dfrac{1}{1+i}$ and $F_i's$ are coefficients of the *n* terms in the polynomial.

For this polynomial, in principle, there may be *n* different roots or values of *x*, which satisfy Equation (10.41). Thus, it is possible that the NPV *i* function crosses the *i*-axis several times as shown in Figure 10.20.

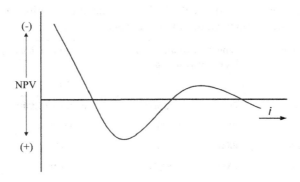

FIGURE 10.20 Multiple values of IRR.

332 Photovoltaic Thermal Passive House System

It may be noted that a unique value of IRR of special interest in applying the IRR method and consequently multiple values of IRR essentially hinder the application of the IRR criterion. In fact, in cases with multiple IRR values, use of the IRR criterion is normally not recommended.

10.11 EFFECT OF DEPRECIATION [2, 5, 6]

Let us define the following terms to be used frequently:

Initial cost (C_i): This is also referred as first cost or initial value or single amount. It is the installed cost of the system. The cost includes the purchase price, delivery and installation fee and other depreciable direct costs (defined later) incurred to ready the asset for use.

Salvage value (C_{sal}): This is the expected market value at the end of useful life of the asset. It can be negative if dismantling cost or carrying-away cost is anticipated and is more than salvage value. It can be also be equal to zero. For example, for the window glass, $C_{sal} = 0$.

Depreciation (C_d): This is the loss incurred by an expenditure that decreases in value with time. This must be apportioned over its lifetime. Depreciation can be expressed in terms of salvage and initial value as follows:

$$C_d = C_i - C_{sal} \tag{10.42}$$

Book value (B): Book value represents the remaining undepreciated investment on corporate books. It can be obtained after the total amount of annual depreciation charges to date has been subtracted from the first cost (present value/initial cost). At the beginning of the first year, book value of depreciation is the initial cost/investment at $n = 0$. The book value is usually determined at end of each year.

$$Book\ value = initial\ cost\left(first\ cost\right) - accumulated\ cost$$

$$or \quad Book\ value = salvage\ value + future\ depreciation$$

Let us understand this with following example [4].

Let us assume a PV system whose initial cost is Rs. 15,000 (USD\$207.76) with salvage value as Rs. 5000 (USD\$69.02) after 5 years. (Consider USD\$1= Rs. 72.44)

$$Thus\ the\ annual\ depreciation\ of\ the\ system = \frac{15,000 - 5000}{5} = Rs.\ 2000\left(USD\$27.60\right)$$

Book value at the end of current year equals the book value at the beginning of next year. The PV system is depreciated until the book value equals salvage/scrap value.

Annual Depreciation		Accumulated Depreciation Cost at year-end		Book Value at year-end	
				Rs.	USD\$
Rs.	USD\$	Rs.	USD\$	(Initial cost) 15,000	(Initial cost)
2000	27.60	2000	27.60	13,000	207.06
2000	27.60	4000	55.21	11,000	151.84
2000	27.60	6000	82.82	9000	124.24
2000	27.60	10,000	138.04	7000	96.63
2000	27.60	12,000	165.65	(Salvage value) 5000	(Salvage value) 69.02

Life Cycle Analysis

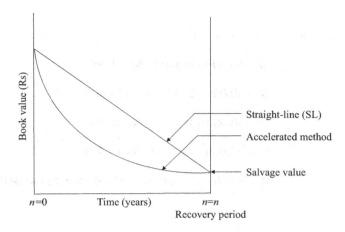

FIGURE 10.21 Variation of book value with time (n) showing depreciation rate.

Depreciation rate (D_t): This is the fraction of first cost (C_i) removed through depreciation from corporate book. This rate may be the same, i.e., straight-line (SL) rate or different for each year of the recovery period as given in Figure 10.21.

Mathematically for straight-line, it can be expressed as follows:

$$D_t = \frac{C_i - C_{sal}}{n} \qquad (10.43)$$

where n is the life of the system.

Now, the book value at n^{th} year can be expressed as

$$B_n = C_i - nD_t \qquad (10.44)$$

Recovery period (n): This is the life of the asset (in years) for depreciation and tax purpose. It is also referred to as the expected life of asset in years.

Market value: It is the actual amount that could be obtained after selling the asset in the open market. For example, (i) the market value of a commercial building tends to increase with a period in the open market, but the book value will decrease as depreciation charges are taken in to account, and (ii) an electronic equipment (computer system) may have a market value much lower than book value due to the rapid change of technology.

EXAMPLE 10.33

Calculate the depreciation rate and the book value of the asset having first cost of Rs. 50,000 and salvage value of Rs. 10,000 after 5 years. (Consider USD$1= Rs. 72.44)

Solution

Using Equation (10.43), we have

$$D_t = \frac{C_i - C_{sal}}{n} = \frac{50,000 - 10,000}{5} = Rs.\,8000$$

Book value at the end of every year will use Equation (10.44),

$$B_1 = 50,000 - 1 \times 8000 = \text{Rs. } 42,000$$

$$B_2 = 50,000 - 2 \times 8000 = \text{Rs. } 34,000$$

$$B_3 = 50,000 - 3 \times 8000 = \text{Rs. } 26,000$$

$$B_4 = 50,000 - 4 \times 8000 = \text{Rs. } 18,000$$

$$B_5 = 50,000 - 5 \times 8000 = \text{Rs. } 10,000 \ (\text{USD} \$138.04)$$

10.11.1 EXPRESSION FOR BOOK VALUE

With C_i as initial cost of an asset and C_{sal} as salvage value in n years, then total depreciation or depreciable first cost is given by,

$$C_d = C_i - C_{sal} \tag{10.45}$$

Let D_{f1}, D_{f2}, $D_{f3}...D_{fn-1}$, D_{fn} be the fractional depreciation for each year, then depreciation for m^{th} year will be,

$$D_m = D_{fm} \times C_d \tag{10.46}$$

The book value for an asset at the end of mth year can be obtained by subtracting the accumulated depreciation expense to that time from the original value of the asset as given in Equation (10.47) and Equation (10.48).

$$\left[Book\ value \right] = \left[Initial\ cost \left(first\ cost \right) \right] - \left[Accumulated\ cost \right] \tag{10.47}$$

Equation (10.47) can be explained as

$$B_m = C_i - C_d \sum D_{fi} \tag{10.48}$$

Subtracting and adding C_d to the right-hand side of the above equation, we get

$$B_m = C_i - C_d + C_d - C_d \sum D_{fi} = C_{sal} + C_d \left[1 - \sum D_{fi} \right] \tag{10.49}$$

The book value may bear no relation to the resale value.

10.11.2 STRAIGHT-LINE DEPRECIATION [2]

In case the fractional depreciation is the same for all years, i.e.,

$D_{f1} = D_{f2} = D_{f3} = ... = D_{fn-1} = D_{fn}, or\ D_f = \dfrac{1}{n},$ then depreciation for m^{th} year will be

$$D_m = C_d / n \tag{10.50a}$$

Life Cycle Analysis

Then accumulated depreciation up to m^{th} year will be

$$\Sigma D_m = C_d\, m/n \tag{10.50b}$$

The book value for an asset at the end of mth year will be using Equation (10.48),

$$B_m = C_i - C_d\, m/n \tag{10.51a}$$

Subtracting and adding C_d in the right-hand side of the above equation, we get

$$B_m = C_i - C_d + C_d - C_d\left(m/n\right) = C_{sal} + C_d\left[1-\left(\frac{m}{n}\right)\right] \tag{10.51b}$$

Equation (10.51b) can be explained as:

$$\left[Book\ value\right] = \left[Salvage\ value\right] + \left[Future\ depreciation\right]$$

Depreciation remaining for future years from m^{th} year to n^{th} year is given by

$$C_d\sum_{m}^{n} D_{f,m} = C_d\sum_{m=m}^{m=n}\frac{1}{n} = C_d\frac{n-m}{n} = C_d\left[1-\frac{m}{n}\right] \tag{10.52}$$

Present value of Re. 1 of depreciation ($C_d =$ Re. 1) is,

$$F_{SLP,i,n} = \frac{1}{n}\left[\frac{1}{1+i} + \frac{1}{\left(1+i\right)^2} + \frac{1}{\left(1+i\right)^3} + \ldots + \frac{1}{\left(1+i\right)^n}\right] \tag{10.53a}$$

$$F_{SLP,i,n} = \frac{1}{n}\left[\frac{\left(1+i\right)^n - 1}{i\left(1+i\right)^n}\right] = \frac{1}{n}F_{RP,i,n} \tag{10.53b}$$

Substituting the value of $F_{PK,i,n}$ from Equation (10.20d) in Equation (10.53b), we have

$$F_{SLP,i,n} = \frac{1}{ni F_{PK,i,n}} \tag{10.53c}$$

10.11.3 Sinking Fund Depreciation

The sinking fund depreciation (SF) has been explained in Figure 10.22 for an annual deposit (Rs.) made at end of every year to a sinking fund to restore the depreciation value at the end of n years. The annual deposit (Rs.) can be expressed for a known first cost (C_i) and salvage value (C_{sal}) as

$$\text{Rs.} = C_d F_{SP,i,n} F_{PR,i,n} = C_d F_{SR,i,n} = C_d\frac{i}{\left(1+i\right)^n - 1} = \left(C_i - C_{sal}\right)\left[\frac{i}{\left(1+i\right)^n - 1}\right] \tag{10.54}$$

The depreciation for any year is the sinking-fund increase for that year, which is the deposit for the year plus an interest earned by the fund for the year.

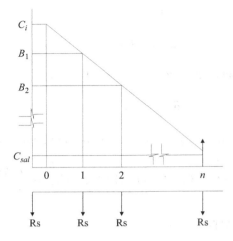

FIGURE 10.22 Cash flow diagram for sinking fund depreciation.

10.11.4 Accelerated Depreciation

Accelerated depreciation gives a higher depreciation charge in the first year of an asset's life and gradually decreasing charges in subsequent years. This is because the efficiency of asset declines, maintenance costs increase, and there is a possibility of availability of better equipment with time.

One popular accelerated method is the declining-balance method. Under this method the book value is multiplied by a fixed rate. The rate is the multiple of the straight-line rate; typically, it is twice the straight-line rate and popularly known as double-declining-balance depreciation. For n^{th} year, the depreciation charge is given by

$$D_n = \alpha C_i (1-\alpha)^{n-1} \qquad (10.55)$$

where fraction $\alpha = \dfrac{\text{Multiplier}}{N}$

and N is the estimated useful life of the asset.

The total declining-balance (TDB) depreciation by the end of n years can be estimated by

$$TDB = C_i \times \left[1 - (1-\alpha)^n\right] \qquad (10.56)$$

The book value (B_n) at the end of n years is calculated by

$$B_n = C_i - TDB = C_i \times (1-\alpha)^n \qquad (10.57)$$

Another accelerated method for allocating the cost of an asset is called sum-of-years'-digit (SOYD) depreciation. In this method, the numbers 1, 2, 3, ..., N are summed, where N is the estimated years of useful life, i.e.,

$$SOYD = 1 + 2 + 3 + \ldots + N = \dfrac{N(N+1)}{2} \qquad (10.58)$$

Life Cycle Analysis

The annual depreciation rate is a fraction in which the denominator is the SOYD, and the numerator is, for the first year N; second year $N-1$; for the third year $N-2$; and so on. Thus the depreciation charge for each year is estimated by

$$D_n = \frac{N-n+1}{SOYD} \times (C_i - C_{sal}) \tag{10.59}$$

EXAMPLE 10.34

The present cost of a system is USD\$1000, whose estimated salvage value is USD\$100, at the end of the useful life of 5 years. Compute the annual depreciation allowances and the resulting book value using:

(a) Double-declining depreciation method
(b) SOYD depreciation method

(Consider USD\$1= Rs. 72.44)

Solution

Given C_i = USD\$1000, C_{sal} = USD\$100, N = 5 years.
(a) For double-declining depreciation the multiplier is 2, therefore the fraction is

$$\alpha = \frac{2}{5} = 40\%$$

The depreciation deduction for the first year will be 40% × USD\$1000 = USD\$400.
Book value at the end of first year is USD\$1000 – USD\$400 = USD\$600.
The depreciation deduction for the second year will be 40% × USD\$600 = USD\$240.
Book value at the end of second year is USD\$600 – USD\$240 = USD\$360.
The depreciation deduction for the third year will be 40% × USD\$360 = USD\$144.
Book value at the end of third year is USD\$360 – USD\$144 = USD\$216.
The depreciation deduction for the fourth year will be 40% × USD\$216 = USD\$86.40.
Book value at the end of fourth year is USD\$216 – USD\$86.40 = USD\$129.60.
The depreciation deduction for the fifth year would be 40% × USD\$129.60 = USD\$51.84, which gives book value at the end of fifth year is USD\$129.60 – USD\$51.84 = USD\$77.76, which is less than USD\$100.00. As the book value cannot be less than scrap value, at the end of fifth year the book value is USD\$100.00 and depreciation is USD\$129.60 – USD\$100.00 = USD\$29.60.
Note: The salvage value is not considered in determining the annual depreciation in case of the declining depreciation method, but the book value of the asset is never brought below its salvage value. The process continues until the salvage value, or the end of the asset's useful life, is reached. In the last year of depreciation, a subtraction might be needed in order to prevent book value from falling below estimated scrap value.
The variation in the book value with time is shown in Figure 10.23.
(b) For the SOYD method,

$$SOYD = \frac{5(5+1)}{2} = 15$$

The depreciation deduction for the first year will be (5/15) × (1000 –100) = USD\$300.

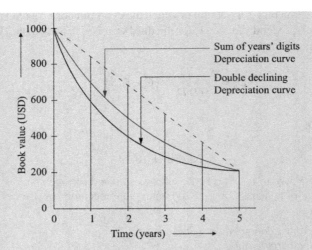

FIGURE 10.23 Variation in book value with time, accelerated depreciation.

Book value at the end of first year is USD$1000 – USD$300 = USD$700 USD$1000 – USD$300 = USD$700.
The depreciation deduction for the second year will be (4/15) × (1000–100) = USD$240.
Book value at the end of second year is USD$700 – USD$240 = USD$460.
The depreciation deduction for the third year will be (3/15) × (1000 –100) = USD$180.
Book value at the end of third year is USD$460 – USD$180 = USD$280.
The depreciation deduction for the fourth year will be (2/15) × (1000 – 100) = US$120.
Book value at the end of fourth year is USD$280 – USD$120 = USD$160
The depreciation deduction for the fifth year will be (1/15) × (1000–100) = USD$60.
Book value at the end of fifth year is USD$160 – USD$60 = USD$100.
The variation in the book value with time is shown in Figure 10.23.

10.12 COST COMPARISON AFTER TAXES

10.12.1 WITHOUT DEPRECIATION

Let

i = rate of return before tax,
t = tax rate, and
r = rate of return after taxes.

The expression of r becomes,

$$r = i(1-t) \tag{10.60}$$

The cash flow diagram before the taxes is given in Figure 10.24 in which an investment compounds at rate (i) and the first cost (P) can be expressed in terms of unacost (R) as

$$P = RF_{PR,i,n} = R\frac{(1+i)^n - 1}{i(1+i)^n} \tag{10.61a}$$

Life Cycle Analysis

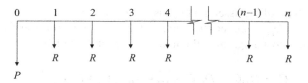

FIGURE 10.24 Cash flow diagram before taxes.

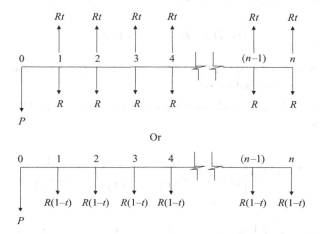

FIGURE 10.25 Cash flow diagram after taxes.

The cash flow diagram before the taxes is given in Figure 10.25
In this case, it will compound at a rate r and an expression for first cost (P) can be written as

$$P = \frac{R(1-t)}{(1+r)} + \frac{R(1-t)}{(1+r)^2} + \ldots + \frac{R(1-t)}{(1+r)^n} = R(1-t)\frac{(1+r)^n - 1}{r(1+r)^n} = R(1-t)F_{RP,i,n} \quad (10.61b)$$

From this equation, an expression for unacost after taxes can be expressed as

$$R = \frac{P}{(1-t)} \times \frac{r(1+r)^n}{(1+r)^n - 1} = \frac{P}{(1-t)} F_{PR,r,n} \quad (10.62a)$$

$$\text{Also } R = \frac{S}{(1-t)} \times \frac{r}{(1+r)^n - 1} = \frac{S}{(1-t)} F_{SR,r,n} \quad (10.62b)$$

$$\text{Here } S = P(1+r)^n \quad (10.62c)$$

10.12.2 With Depreciation

Let us consider C_i as the initial cost of an asset that lasts n years with the salvage value of C_{sal}. The depreciable cost as given in Equation (10.45) will be

$$C_d = C_i - C_{sal}$$

At the time of purchase of the asset, no tax has been considered.

Let $D_{f1}, D_{f2}, D_{f3}, \ldots = D_{fn-1}, D_{fn}$ be the fractional depreciations for every year as assumed in an earlier section. The time-cost diagram without tax is given in Figure 10.26

$$C_d = D_{f1}C_d + D_{f2}C_d + \ldots + D_{f(n-1)}C_d + D_{fn}C_d$$

$$C_d = D_f C_d + D_f C_d + C_d + \ldots + D_f C_d \; (\text{for fractional depreciation})$$

$$\text{Also } D_{f1} = D_{f2} = D_{fn} = 1/n$$

Now the taxable base is reduced to $D_{f1}C_d$ and a saving or reduction in taxes amounting to $D_{f1}C_d \cdot t$ is realized. The time-cost diagram is given in Figure 10.26.

Using these time-cost diagrams and assuming $D_{f1} = D_{f2} = D_{f3} = \ldots = D_{fn-1} = D_{fn} = D_f = \dfrac{1}{n}$, the present value becomes,

$$P = C_d - D_f C_d t \left[\dfrac{1}{(1+r)} + \dfrac{1}{(1+r)^2} + \dfrac{1}{(1+r)^3} \pm - \mp \dfrac{1}{(1+r)^n} \right]$$

$$P = C_d - C_d t \left(\dfrac{1}{n}\right) \left[\dfrac{1}{(1+r)} + \dfrac{1}{(1+r)^2} + \dfrac{1}{(1+r)^3} \pm - \mp \dfrac{1}{(1+r)^n} \right]$$

$$P = C_d - C_d t \, F_{SLP,r,n}$$

$$P = C_d \left[1 - t F_{SLP,r,n} \right] \tag{10.63a}$$

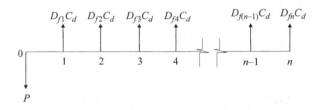

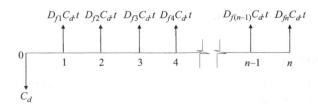

FIGURE 10.26 The cash flow of unacost with depreciation and tax.

Life Cycle Analysis

The expression for unacost (R) and capitalized cost (K) are given here

$$R = PF_{PR,r,n} = C_d \left[1 - tF_{SLP,r,n} \right] F_{PR,r,n} \tag{10.63b}$$

$$K = PF_{PK,r,n} = C_d \left[1 - tF_{SLP,r,n} \right] F_{PK,r,n} \tag{10.63c}$$

It is important to note that an expression for conversion factor from straight-line depreciation to the present value with tax is given by

$$F_{SLP,r,n} = \frac{1}{n} \cdot \frac{(1+r)^n - 1}{r(1+r)^n} \tag{10.64}$$

EXAMPLE 10.35

Calculate the conversion factor from straight-line depreciation to the present value with and without tax for 15% rate of return and 25% tax for a period of 10 years.

Solution

The conversion factor from straight-line depreciation to the present value is given by
(a) Without tax

$$F_{SLP,i,n} = \frac{1}{10} \frac{(1+0.15)^{10} - 1}{0.15(1+0.15)^{10}} = 0.5018$$

(b) With tax, here replacing $r = i(1 - t) = 0.15(1 - 0.25) = 0.1125$ we have

$$F_{SLP,r,n} = \frac{1}{10} \frac{(1+0.1125)^{10} - 1}{0.1125(1+0.1125)^{10}} = 0.5828$$

EXAMPLE 10.36

Derive an expression for the present value (P) for a uniform end-of-year cost (R) occurring simultaneously with the tax instant t.

Solution

The uniform end-of-year cost after taxes at the end of each year = $R(1 - t)$
The $R(1 - t)$ will be the same up to n years at the end of each year.
In order to obtain the expression for P, convert uniform annualized cost $R(1 - t)$ into present value as

$$P = R(1-t)F_{RP,r,n} \Rightarrow P = R(1-t)\frac{(1+r)^n - 1}{r(1+r)^n}$$

> ### EXAMPLE 10.37
>
> Derive an expression for the present value (P) for a given salvage value (C_{sal}) at end of nth year by treating as a non-depreciable first cost, an expense.
>
> #### Solution
>
> The C_{sal}, a non-depreciable expense is invested now and fully recovered at the end of n years with no tax consideration. The present value is
>
> $$P = C_{sal} - \frac{C_{sal}}{(1+r)^n} = C_{sal}\frac{(1+r)^n - 1}{(1+r)^n}$$

10.13 ESTIMATING COST OF A PROJECT [1]

To evaluate the life cycle cost (LCC), it is important to reasonably assess the various associated costs and revenues with the project.

10.13.1 CAPITAL COST

This is the cost incurred to bring a project to the operable status. These are one-time expenses and are independent of the level of output. Labor costs are not a part of capital cost except when labor is used in construction activities.

For a BiPVT system, the cost of purchasing the land over which the building will be built, permitting and legal costs, the cost of equipment required to bring the system in operation, construction costs, the cost of financing and the cost of commissioning incurred prior to operation of the system comes under the category of capital costs.

10.13.2 VARIABLE COST

This is the cost that may vary or change in proportion to the activity of a system. The costs associated with production and manufacture constitutes the variable cost.

For a BiPVT system, the cost of labor for periodic cleaning of the surface of the PV array, which is exposed to the sun rays, and maintenance and repair during the operational phase is the variable cost.

10.13.3 STEP-VARIABLE COST

These remain the same up to a certain level of activity, and the cost rises when the activity crosses this certain level. The jump in costs after a certain level of activity will form steps.

For a BiPVT system, the replacement of components and batteries calls for step-variable costs, where the number and timing of replacements depend on the lifespan of the components.

10.13.4 NON-PRODUCT COST

These costs have nothing to do with manufacturing, selling, marketing and administrative purposes.

Life Cycle Analysis

10.14 A CASE STUDY OF BUILDING INTEGRATED PHOTOVOLTAIC THERMAL (BiPVT) SYSTEMS

There are multiple costs associated with acquisition, operation, maintenance and disposition of a PV system. A cash flow diagram at different time intervals of BiPVT system is given in Figure 10.27. We have already discussed LCC assessment in Chapter 9. All the associated present and future costs of the system are clubbed in the present value for the LCC assessment. The purpose is to estimate the overall cost of project alternatives and to select the design that ensures the facility will provide the lowest overall cost.

10.14.1 Cost Estimation [1]

The following have been taken into consideration for LCC assessment of the BiPVT system.

- Initial cost (P_I): This is the total of all the costs associated with the system technology (like BiPVT system, charge controller, inverter, battery bank, etc.), utility interconnection amount and other associated costs for building permits.
- Maintenance and repair costs: Sometimes the quotes from the suppliers and published estimating guides provide the information on maintenance and repair costs. But in the case of BiPVT systems, there is a huge variation even for the same type and lifetime of the system in consideration.

  For the present analysis, let us consider the annual maintenance and repair cost of the BiPVT system considering as M = USD$150 (Rs. 10,866). Then the maintenance and repair costs in terms of the present value is given by

$$P_{MR} = M \times \left[\frac{(i+1)^n - 1}{i(i+1)^n} \right] \qquad (10.65)$$

- Replacement costs: This is the cost associated with replacement of batteries and other components. The number and time of any replacement is dependent on the life of the particular component and system.

  Let us assume the service life of a battery is 5 years, and replacements are made at the end of 5 years.

  If $R_5, R_{10}, R_{15}, \ldots, R_n$, are the replacement costs incurred in batteries and other components made every 5 years then the net replacement costs in terms of present values is

$$P_R = R_5 \times \left[\frac{1}{(i+1)^5} \right] + R_{10} \times \left[\frac{1}{(i+1)^{10}} \right] + R_{15} \times \left[\frac{1}{(i+1)^{15}} \right] + \ldots + R_n \times \left[\frac{1}{(i+1)^n} \right] \qquad (10.66)$$

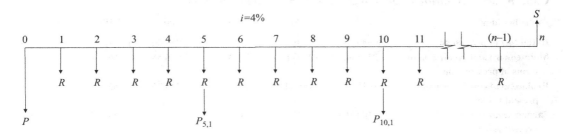

FIGURE 10.27 Cash flow diagram of BiPV and BiPVT systems.

344 Photovoltaic Thermal Passive House System

- Salvage value: This is the cost associated with demolition and disposal of the system. If S is the salvage value at the end of the system then the net salvage value in terms of present value is

$$P_S = S\left[\frac{1}{(i+1)^n}\right] \tag{10.67}$$

Table 10.10 gives the initial capital costs, maintenance and repair costs, replacement costs and salvage value in terms of present value for different technology of BiPVT system.

10.14.2 MODELING OF ANNUALIZED UNIFORM COST

The LCC assessment of the BiPVT system in terms of present value is given as:

$$LCC = P_I + P_{MR} + P_R - P_S$$
$$= P_I + R\left[\frac{(i+1)^n - 1}{i(i+1)^n}\right] + \frac{R_5}{(i+1)^5} + \frac{R_{10}}{(i+1)^{10}} + \ldots + \frac{R_n}{(i+1)^n} - \frac{S}{(i+1)^n} \tag{10.68}$$

Using the capital recovery factor as given in Equation (10.11c), the annualized uniform cost (unacost) is given by

$$Unacost = LCC \times CRF \tag{10.69}$$

The cost per unit electricity generated by the BiPVT system is determined as the ratio of annualized uniform cost and the electrical energy consumed by the load in a year.

10.14.3 METHODOLOGY [1]

The energy, exergy and the outlet air temperature has been computed using basic thermal modeling and heat transfer relations. Annual exergy output for BiPV and BiPVT systems using different solar cells is given in Figure 10.28. The velocity of air and net mass flow rate of air inside the duct is 3.2 m/s and 1 kg/s, respectively. The annual interest rate usually offered by government sectors in India to promote the use of renewable energy applications is 4%.

TABLE 10.10

Capital Costs of Various Items of the BiPVT System in USD\$ [1]

PV Technology	c-Si	p-Si	r-Si	a-Si	CdTe	CIGS
Initial cost of the system	28,989.49	24,208.84	19,796.79	8,251.36	10,640.37	13,015.86
Maintenance and repair cost in terms of present value	2593.80	2593.80	2343.31	2,038.55	1,667.76	667.77
Replacement cost in term of present value	1913.01	1661.44	1229.60	545.85	522.41	0
Salvage value in terms of present value	104.86	87.72	134.16	158.39	279.72	644.52
Net present value	33,391.45	28,376.37	23,235.53	10,677.37	12,550.82	13,039.12

Life Cycle Analysis   345

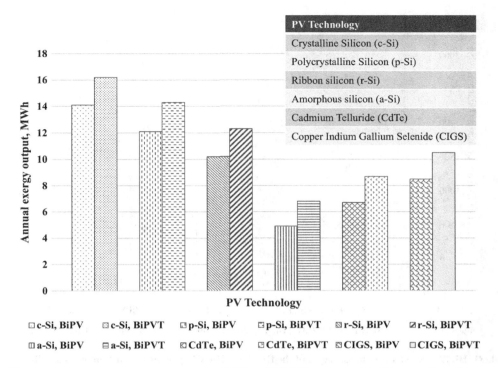

FIGURE 10.28 Annual exergy output from the BiPV and BiPVT systems with different solar cells.

Using Equations (10.11c) and (10.69), annualized uniform cost has be calculated. Figure 10.29 gives the unacost of the BiPV and BiPVT system with different solar cells. By dividing unacost with net exergy output, we arrive at the cost per unit generation as given in Figure 10.30 for different solar cells.

Table 10.11 gives the annual electrical output, thermal outputs, overall thermal and exergy efficiencies and annualized uniform cost for the BiPVT systems with different solar cell technologies.

10.14.4 Results and Discussions [1]

From Figure 10.28, maximum annual exergy output has been found in crystalline silicon (c-Si) BiPVT system with 16,225 kWh and minimum in the case of amorphous silicon (a-Si) BiPVT system with 7790 kWh. It has also been found that the annual exergy output of BiPVT system is about 15%–30% higher than that of a BiPV system of same type. The reason is that the solar cells of the BiPVT system are cooled, which helps in producing higher electrical energy than the BiPV system. Another additional advantage of BiPVT systems over BiPVT systems is the thermal gains, which can be used for space heating.

From Figure 10.29, we can see that the copper indium gallium selenide (CIGS) BiPVT system has a relatively higher annualized cost (USD$2928.94 or Rs. 212,172.4) owing to higher initial investment made that for a short life span, whereas the a-Si BiPVT system has relatively lower annualized cost (USD$785.66). It has also been found that the unacost of BiPVT system is about 2%–7 % higher than that of a BiPV system of same type.

BiPVT system reduces the unit power generation cost by 12%–25% versus that of the similar BiPV systems as given in Figure 10.30. The figure also shows that the unit power generation cost of the CIGS BiPVT system is highest (USD$0.2654/kW) and is lowest for a-Si (USD$0.1009/kW). Thus, from an economic point of view, the a-Si BiPVT are more suitable for the rooftop. Also, the cost of unit power generation from the a-Si BIPVT system is quite closer to the cost of unit power

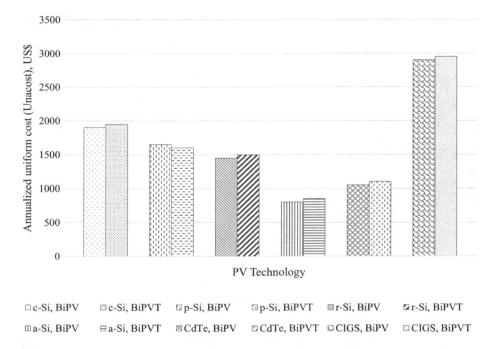

FIGURE 10.29 Annualized uniform cost of the BiPV and BiPVT systems with different solar cells.

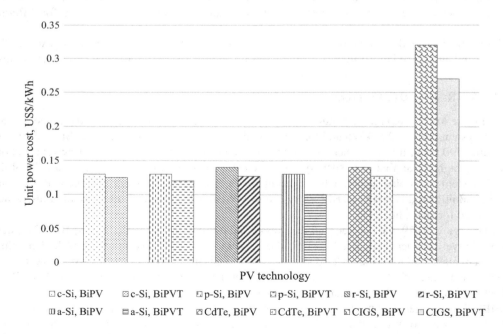

FIGURE 10.30 Unit power generation cost of the BiPV and BiPVT systems using different solar cells.

Life Cycle Analysis

TABLE 10.11

The Annual Outputs, Overall Efficiencies and Annualized Cost of Power Generation through BiPVT Systems [1]

PV Technology		c-Si Crystalline Silicon	p-Si Polycrystalline Silicon	r-Si Ribbon silicon	a-Si Amorphous silicon	CdTe Cadmium Telluride	CIGS Copper Indium Gallium Selenide
Thermal output from BIPVT	kWh	16,764	17,535	18,306	20,615	19,845	19,074
Electrical output from BiPVT	kWh	15,131	13,141	11,179	6066	7958	9578
Overall thermal efficiency of BiPVT system	%	51.99	47.88	43.84	33.54	37.41	40.65
Overall exergy efficiency of BiPVT system	%	14.91	13.19	11.50	7.13	8.75	10.13
Annualized cost of BiPVT	USD$	1931.03	1641.01	1487.35	785.66	1128.83	2928.94

generation through conventional grid. Therefore, it may be concluded that the application of such systems in residential and commercial buildings will help reduce greenhouse gas emissions, which is necessary for sustainable development.

OBJECTIVE QUESTIONS

10.1 Future value is dependent on
   (a) F_{PS}
   (b) F_{RS}
   (c) F_{RP}
   (d) F_{PK}
10.2 Present value is dependent on
   (a) F_{PS}
   (b) F_{RS}
   (c) F_{RP}
   (d) F_{PK}
10.3 The project is economically accepted when B/C
   (a) < 1
   (b) equals 0
   (c) equals 1
   (d) > 1
10.4 A uniform end of year annual amount (say, a) is related to a uniform beginning of year
   (a) $a = b\,(i + i)$
   (b) $a = b\,(1 + i)^n$
   (c) $a = b\,[(1 + i)^n - 1]$
   (d) $a = b\dfrac{\left[(1+i)^n - 1\right]}{i}$

348 Photovoltaic Thermal Passive House System

10.5 Capital recovery factor can be calculated as

(a) $\dfrac{(1+i)^n - 1}{(1+i)^n}$

(b) $\dfrac{(1+i)^n}{(1+i)^n - 1}$

(c) $\dfrac{(1+i)^n - 1}{i(1+i)^n}$

(d) $\dfrac{i(1+i)^n}{(1+i)^n - 1}$

10.6 Capitalized cost factor can be calculated as

(a) $\dfrac{(1+i)^n - 1}{(1+i)^n}$

(b) $\dfrac{(1+i)^n}{(1+i)^n - 1}$

(c) $\dfrac{(1+i)^n - 1}{i(1+i)^n}$

(d) $\dfrac{i(1+i)^n}{(1+i)^n - 1}$

10.7 The payback time period depends on
(a) Initial investment
(b) Cash flow
(c) Use of these systems
(d) Both (a) and (b)

10.8 The expression for future depreciation at end of mth year is
(a) $C_i + C_d \sum D_{fi}$
(b) $C_i - C_d \sum D_{fi}$
(c) $C_i \times C_d \sum D_{fi}$
(d) $C_i / C_d \sum D_{fi}$

10.9 Depreciation rate for straight-line depreciation is expressed as

(a) $\dfrac{C_i - C_{sal}}{n}$

(b) $\dfrac{C_i + C_{sal}}{n}$

(c) $C_i + nDt$

(d) $-nDt$

10.10 F_{SLP} can be expressed as:

(a) $n \times \dfrac{(1+r)^n - 1}{r(1+r)^n}$

(b) $n \times \dfrac{(1+r)^n - 1}{(1+r)^n}$

Life Cycle Analysis

(c) $\dfrac{1}{n} \times \dfrac{(1+r)^n - 1}{(1+r)^n}$

(d) $\dfrac{1}{n} \times \dfrac{(1+r)^n - 1}{r(1+r)^n}$

ANSWERS

10.1 (a)
10.2 (c)
10.3 (d)
10.4 (a)
10.5 (c)
10.6 (b)
10.7 (d)
10.8 (b)
10.9 (a)
10.10 (d)

PROBLEMS

10.1 Draw the cash flow diagrams for both the methods given in Example 10.14.

10.2 Calculate future (F_{ps}) and present (F_{sp}) value factor for a given number of years for a 15% rate of interest and show that $F_{ps} F_{sp} = 1$ for each case.
Hint: Use Equations (10.2c) and (10.6) for $n = 0, 2, 4, 6, 8, 10$ and 12

10.3 Compute the effective rate of return for different value of p for 20% rate of interest.
Hint: Use Equation (10.3d) for $p = 1, 2, 3, 4, 6$

10.4 Calculate capital recovery (F_{PR}) and sinking fund (F_{SR}) factors for different number of years ($n = 1, 5, 10, 15$ and 20) for a given rate of interest ($r = 0.05, 0.10, 0.15$ and 0.20 percentages)
Hint: Use Equation (10.9c)

10.5 Prove that $F_{SR} \times F_{RS} = 1$
Hint: Use Equations (10.14a) and (10.14c), respectively.

10.6 A solar cooker purchased for USD$1500 is expected to generate annual revenues of USD$250 and have salvage value of USD$500 at the end of 15 years. If a 12% per year required return is imposed on the purchase, compute the payback period.
Hint: Solve the problem with a cash flow diagram.

10.7 Find the equivalent present value (p) at the end of fifth year if money is worth 10% per year for the following series:

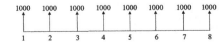

Hint: See Example 10.16

350 Photovoltaic Thermal Passive House System

10.8 Two swimming pool have been heated by solar water heating systems, which have the following cost comparison. Find out which system is more economical if the money is worth 15% per year. Consider USD$1= Rs. 72.44

Economic Components	System A	System B
First cost (Rs.)	50,000	25,000
Uniform end-of-year maintenance cost per annum (Rs.)	3000	5000
Overall, end of fifth year (Rs.)	2500	7000
Salvage value (Rs.)	11,000	3500
Life of the system (years)	10	10

Draw cash flow diagram of both the systems.
Hint: See Example 10.20

10.9 Two solar distillation plant of capacity 500 l/d have been constructed by using concrete/brick/cement and fiber reinforced plastic (FRP) materials. These distillation plants have the following cost comparison:

Economic Components	System A	System B
First cost (Rs.)	150,000	250,000
Uniform end-of-year maintenance cost (Rs.)	15,000	20,000
Salvage value (Rs.)	25,000	-
Life of the system (years)	10	20

By using a cash flow diagram, find out which system is more economical if the money is worth 15%/year.

(Consider USD$1= Rs. 72.44)
Hint: See Example 10.22

10.10 Solve Problem 10.9 by using the capitalized cost method.
Hint: Refer to Section 10.7.2

10.11 Draw the curve between FPR and n for different value of r of Problem 10.4.

REFERENCES

[1] B. Agrawal and G. N. Tiwari, *Building Integrated Photovoltaic Thermal Systems: For Sustainable Developments*, United Kingdom: RSC Publishing, 2010.
[2] G. N. Tiwari, *"Solar Energy–Fundamentals, Design, Modelling and Applications,"* Delhi: Naros, 2012.
[3] G. N. Tiwari and R. Mishra, *Advanced Renewable Energy Sources*, UK: RSC publishing, 2012.
[4] G. N. Tiwari, A. Tiwari and Shyam, *Handbook of Solar Energy*, New Delhi: Springer, 2016.
[5] K. Humphreys, *Jelen's Cost and Optimization Engineering*, 3rd Ed. New York: McGraw-Hill, Inc., 1991.
[6] L. Blank, T. Tarquin and J. Antony, *Engineering Economy*, 3rd Ed. New York: McGraw-Hill Inc. Editions, 1989.

11 Photovoltaic Application in Architecture

11.1 INTRODUCTION

Building integrated photovoltaic (BiPV) systems forms a part of the building's envelope, and various research has been conducted to understand the feasibility of the system and analyze the performance of BiPV systems, a brief description of which is given in Chapter 8. BiPV technologies are capable of achieving noteworthy reductions in cost since they constitute a part of the building and offset the cost of the building materials by replacing them. BiPV on the other hand are the systems where photovoltaics (PV) are mounted on the building to extract electrical power and do not form an integral part of the building. This chapter illustrates few examples of applications of photovoltaics in architecture around the world. Also, case studies have been conducted for further clarity.

11.2 IMPLEMENTATION OF PV SYSTEMS AROUND THE WORLD

The global photovoltaic market grew by more than 60% in terms of production in 2007 reaching approximately 4 GW. For installed PV systems, the market reached 2825 MW, which is again more than 60% [1].

Germany or Italy have exchanged their first two places in the PV market from 2010–2012. China, Japan and the United States ranked themselves in the top three places in the years 2013 and 2014. In 2014, only major markets reached the top ten, which marked the end of a long-term trend seeing small European markets booming during one year before collapsing. Globally, the centralized PV market was governed by China, the United States and emerging PV markets and represented more than a 50% share of the market during 2014 [2].

The prices continued to fall from 2017 with a boom in deployment in China, which led to an expansion in the renewable power generation. A fall of 70% in prices has been administered for new solar photovoltaic large utility-scale systems with a growth in generation of power by over a third in 2017, up to 460 TWh, representing almost 2% of total world electricity production. China continues to remain the main driver for PV deployment worldwide [3]. In 2017, India took the third spot from Japan. Seven of the top 10 leaders in 2012 were still in the top 10 in 2017, while the others have varied from one year to another. Turkey and Brazil joined for the first time in 2017. The UK entered the top 10 in 2013 and moved out in 2017, Korea entered in 2014 and is still there, and Thailand came in 2016 to leave in 2017. Greece left in 2013 and Canada in 2016. Romania entered the top 10 in 2013 and left in 2014. France came back in 2014 and confirmed its position in 2015 before leaving in 2016. South Africa entered briefly in 2014 and left in 2015. Globally, centralized PV represented more than 60% of the market in 2017, mainly driven by China, the United States, and emerging PV markets. However, distributed PV increased significantly in 2017, with more than 32 GW installed; with 14 GW from China alone.

With around 45 GW installed in China in 2018, the global PV market showed a stable situation with 99.8 GW as compared to 98.9 GW in 2017. Behind China, India topped the charts by securing second position with around 10.8 GW of annual installations in 2018, followed by the US market with 10.6 GW. The European Union followed with some growth and achieved 8.3 GW, and Japan achieved 6.5 GW. The top five countries for 2018 were similar to the chart of 2017, but the major development in 2018 was the growth of several other countries behind top five listings: Australia installed 3.8 GW, Mexico 2.7 GW, Korea 2.0 GW and Turkey 1.6 GW [4].

DOI: 10.1201/9780429445903-11

351

Some established Asian markets like Taiwan or Malaysia experienced growth in 2018. The Asian market represented slightly less than 70% of the global PV market, which was a decline in the year 2018 when compared to 2017. Asian countries continue to dominate the global PV market. The decline in the American share (10.6 GW) was somehow balanced by Mexico, which contributed about 2.7 GW installations in 2018. The Americas represented around 15% of the global PV market in 2018. In Europe, Germany took the leading position on the continent and installed nearly 3.0 GW in 2018. The Netherlands, on the other hand, installed 1.3 GW, which was one of the major measure followed by France. Europe represented slightly more than 9% of the global PV market in 2018. African and the Middle East represented around 6% of global PV installations in 2018 with South Africa becoming first African country to install close to 1 GW of PV in 2014 but in 2018 only 60 MW were installed [4].

In the year 2018, about 10 countries crossed the GW mark for annual PV installations. Eight countries have more than 10 GW of total capacity, 4 nos. more than 40 GW, and China single-handedly had 176.1 GW as tabulated in Table 11.1.

Thirty-two countries had at least 1 GW of cumulative PV capacity at the end of 2018. Total installed capacity at the end of 2018 globally has been quantified to be at least 500 GW. About 2.58% of the world's electrical power is generated from PV technology. One hundred GW were installed all over the world by the end of 2018 [4].

11.2.1 CHINA

Solar cell research was started in China in 1958. In 1973, the terrestrial application of solar cells began. During 2003 and 2005, rapid development of China's production capacity started under the influence of the European PV market, especially the German market. In 2002, a 10 KW integrated grid system was built in Fengxian, Shanghai, which realized automation. In 2003, an ecological

TABLE 11.1

Top-10 Countries for Installation and Total Installed Capacity in 2018 [4]

	Annual Installed Capacity		Cumulative Capacity	
Rank	Country	Annual Installed Capacity (GW)	Country	Cumulative Capacity (GW)
1	China	45.0	China	176.1
2	India	10.8	United States	62.2
3	United States	10.6	Japan	56.0
4	Japan	6.5	Germany	45.4
5	Australia	3.8	India	32.9
6	Germany	3.0	Italy	20.1
7	Mexico	2.7	UK	13.0
8	Korea	2.0	Australia	11.3
9	Turkey	1.6	France	9.0
10	Netherlands	1.3	Korea	7.9
*	EU	8.3	EU	115

* The European Union should come in fourth place for the capacity installed in 2018 and in second place for the cumulative capacity.

Photovoltaic Application in Architecture

demonstration project was established in Shanghai and the 5 kW integrated grid system combined well with construction with international top technologies [5]. Total production capacity reached 400 MWp by the end of 2005, out of which 140 MWp was manufactured in the year 2005 alone. Only 5 MWp and 10 MWp were installed in China in the years 2005 and 2006, respectively, the rest were exported to the European market. In 2008, 40 MWp of new PV installations were made, making the cumulative installed capacity to reach 145 MWp [6].

According to the National Energy Administration, in 2014 China installed 10.6 GW, which was slightly less than its own record of 10.95 GW for the year 2013 and was ranked number one with regard to all-time PV installations in 2013. Total PV annual installed capacity was quantified to be 10,640 MW and cumulative installed capacity to be 28,330 MW for the year 2014 [2].

PV annual installed capacity has been quantified to be 53,068 MW and PV cumulative installed capacity was calculated to be 131,141 MW for the year 2017. Rapid deployment in China with 53.1 GW capacity was supported by feed-in-tariffs that were economically attractive; this led to a record growth of PV in year 2017 with 98 GW. For the fifth year in a row, China ranked the first position in the year 2017 with significantly higher values than those in the year 2016 (34.6 GW). Since 2012, China's share of the global PV demand has grown from 10% to more than 55% [3].

Various schemes were formulated to promote the development of utility-scale PV in China, rooftop PV in city areas and micro grids and off-grid applications in unelectrified areas in the country. Few of the schemes came into existence in year 2014 like FiT scheme for utility-scale PV and rooftop photovoltaic which were completely financed by a renewable energy surcharge paid by electricity consumers. The National Energy Agency (NEA), in September 2014, issued a "Notice on Further Implementation of Policies of Distributed PV Power" promoting the rooftop PV systems for both large-scale industrial and commercial enterprises having large roof areas, high electrical demand and high retail electrical prices. NEA and the State Council Leading Group Office of Poverty Alleviation and Development (LGOP) issued a "Work Program on PV Poverty Alleviation Construction Implementation" in October 2014. The implementation plan of 2015 targeted 1.5 GW of installations. A Joint Announcement on Climate Change was issued in November 2014 by the United States and China, which stated that non-fossil energy will reach 20% of the primary energy consumption by 2030. Solar power (including PV and solar thermal) should reach at least 400 GW in 2030. This translated into about 20–25 GW of new installations from 2016 onwards [2].

Some of the regulations that were put in place in 2017 included NEA issuing a new document during the "Thirteenth Five-year Plan" period framing the implementation of renewable energies and PV in the future in July 2017. In December 2017, with drop in price of components of photovoltaics, the national development and reform commission issued the "notice on 2018 PV power project price policy". The PV Poverty alleviation program states that 15 GW of PV will be installed before 2020 for poverty alleviation. Since 2016, various competitive projects were established to accelerate the competitiveness of PV projects. In March 2018, China completed the bidding for 7 PV "front runner" plants in the third batch of projects [7].

China has been in first position in the global PV market for the fifth time in a row in 2017. This was made possible due to NEA's Guiding Opinions on the Implementation of the 13th five-year plan for renewable energy development. A target of achieving 60 TWh of PV power by 2020 has been established out of 210 TWh in total. According to the statistics of the NEA, PV contributed to 1.87% of the total electricity consumption [7].

The following are a few examples of implementation of Photovoltaic in China:

1. Jiangsu plant

   Amongst the world's largest floating farm is the China's Jiangsu plant, a 40 MW power plant having about 120,000 solar panels. It was built on a top of a coal mine, which was converted into a lake after being flooded with groundwater. It is estimated this plant could power about 15,000 homes. Floating farms are comparatively costlier than the farms built on the land, but they do have additional advantages. The water helps to reduce the solar cell temperature; thus the efficiency of the system increases [8].

2. Shanghai World Expo 2010

BiPV applications have been adopted at the Shanghai World Expo. A roughly 2.8 MW integrated solar system was installed in the theme pavilion comprised of custom-designed triangular sections covering an area of about 30,000 m^2, which is also considered the largest single solar roof, which is half of the total roof area. Installation of 16,250 polycrystalline silicon panels was done with estimated gross power output to be 2.83 MW. Solar panels were cut into patterns of 18 rhomboids (i.e., diamond shape), each measuring 36 × 72 m and 12 triangles instead of conventional solar panels. The China Pavilion is comprised of 0.3 MWp photovoltaics integrated with traditional Chinese architectural characteristics consisting of 1260 solar panels integrated with the roof and outer walls. The solar system is installed on the China Pavilion's 68-meter platform and 60-meter sightseeing platform. The project completes 3.12 MW BiPV installation in 2010 with annual production of 2.8 GWh of electrical power. Along with large-scale BiPV integration, the pavilion also displays other applications like solar streetlights, solar garden lights, solar lawn lamps, solar electronic displays, solar sculptures, solar fountains, solar kiosks, solar ice bars, solar kiosks, solar mobile toilets, solar bus shelters, solar boats and so on. The pavilion reflects the "Better City, Better Life" theme, through the promotion of solar and other new energy technologies and expansion of these applications [9, 10].

3. Hongqiao Railway Station, Shanghai

Hongqiao Railway Station was the world's largest single stand-alone BiPVT project pumping out power into the grid at that time and was built by the Ministry of Railways (Figure 11.1). About 23,710 panels have been integrated on a 61,000 m^2 roof, with installed nominal power of 6.68 MWp, and it is quantified to generate about 6.3 GW of electrical power annually, which is sufficient to power 12,000 Shanghai households [11, 12].

4. Electrical and Mechanical Services Headquarter, Hong Kong

The roof of the headquarters of the Electrical and Mechanical Services Department (EMSD) of the Hong Kong Special Administrative Region located in Kowloon Bay has been mounted with PV panels. It is one of the largest installations within the territory and Far East. The PV system consists 2357 standard PV modules mounted on the rooftop with each module to be rated at 150 W. Every PV module consists of 72 nos. Monocrystalline

FIGURE 11.1 Shanghai Hongqiao Station [13].

Photovoltaic Application in Architecture

silicon connected in series facing southwards with an inclination angle of 22°. In addition to this, 20 sets of SPV modules have also been integrated with the roof of the viewing gallery for daylighting along with electrical power production. Each set consists of monocrystalline PV cells sandwiched between two glass laminates connected in series. The total installed capacity of both PV systems is approximately 350 kWp with annual power production estimated to be 400 MW covering a total area of about 3100 m² and resulting in savings of 280 tonnes of carbon dioxide emissions [14].

11.2.2 UNITED STATES OF AMERICA

With 205 MW of PV installations and 152 MW grid connected, the United States secured its rank in fourth position in 2007 [1]. In 2012, the U.S. Department of Commerce issued orders to commence enforcing duties to be levied on products with Chinese-made PV cells. The majority of the traffic were in the range of 23%–34% of the product price. In 2013, new anti-dumping and countervailing petitions were filled with the US Department of Commerce (DoC) and the United States International Trade Commission (ITC) against Chinese and Taiwanese manufacturers of photovoltaic cells and modules.

In 2014, the United States installed 6.2 GW of photovoltaic systems with PV cumulative installed capacity in the year 2014 to be estimated as 18.3 GW. The PV market of the United States has mostly been driven by tax credits that were granted by the federal US government with net metering offered in 44 states as a complementary measure. Meanwhile at least 6 states and 17 utilities were offering power purchase agreements that were similar to FiTs. Also, 6 state public utility commissions and utilities were in the process of developing a Value-of-Solar Tariff (VOST) as an alternative to net metering at the end of year 2014. Under the Californian Solar Initiative being financed by a third party, about 60% of residential systems were installed. Power purchase agreements (PPA) were existing with regard to utility-scale PV projects. Property Assessed Clean Energy (PACE) programs were introduced in more than 30 states, which were means of financing renewable energy systems, energy-efficiency measures, and they also prevented significant up-front investments and eases the inclusion of PV system costs in property sales [2].

China, Japan and the United States secured the top three positions from 2013 to 2016, with the United States jumping to second place in 2016. In 2017, the United States was in second place with 10.7 GW installed out of which 6.2 GW (against 10.8 GW in 2016) were installed as utility-scale plants. The 2017 PV annual installed capacity and PV cumulative installed capacity for the United States has be quantified to be 10.68 GW and 51.63 GW, respectively. These are still concentrated in a small number of states, such as California, North Carolina, Arizona, Nevada, Texas and New Jersey, that cover roughly two-thirds of the market. By the end of 2017, there were more than 1.6 million distributed PV systems interconnected across the United States.

Net metering (with specifics) remained the most significant measure for distributed PV with some states having FiTs. Some states have net billing instead of net metering, and a few states have virtual net metering for community solar policies. Several electricity utilities began to engage with PV development, either through direct ownership of centralized and distributed PV assets, community solar programs, partial ownership in PV development companies, or via joint marketing agreements. In 2017, North Carolina passed a bill allowing investor-owned utilities to lease PV systems to their customers. The United States' PV market has been mainly driven by the Investment Tax Credit (ITC) and an accelerated five-year tax depreciation. The ITC was set initially to expire in 2016, however it was finally extended to 2020. Beginning in 2020, the credits will step down (from 30% today) gradually until they reach 10% in 2022 for commercial entities and expire for individuals. An expected market boom caused by the ITC cliff didn't happen but a part of the expected installations will take place in the coming years in any case [7].

The following are a few examples of implementation of photovoltaic in United States:

1. Topaz Solar Farm
   Topaz Solar Farm (Figure 11.2) is one of the world's largest solar energy plant in Carrizo Plain with an expected capacity of 1100 GWh. Photovoltaic modules based on thin-film technology having 9 million panels have been installed in an area of 260 hectares of land [15]. This facility will lead to 377,000 tonnes of annual reduction of carbon dioxide. The site was selected because of the availability of solar resources, close proximity of existing Moro Bay to Midway transmission lines, land uses and environmental factors.
2. Whirlpool Corporation Regional Distribution Center, Perris, California
   The 2011 installation of solar panels on the Whirlpool Corporation Regional Distribution Center is estimated to generate 10 MW of electricity, which is sufficient to power about 5000 homes in Perris. The project has been constructed to continue the city's commitment to "going green" through the use of renewable energy sources. About 29,600 solar panels have been installed on the Whirlpool Center's 1.7 million square foot roof [16].
3. Mandalay Bay Resort Convention Center, Las Vegas
   In 2014, photovoltaics were installed on the rooftop of Mandalay Bay Resort Convention Center. This was the first utility-scale rooftop with 21,325 arrays that provided a solution to the energy demand on the southern Nevada electrical grid covering an area of 11 acres. The photovoltaic arrays have been designed to produce electrical energy equivalent to 1000 homes annually at peak production levels. It has the capacity to accommodate more than 20% of the hotel and casino's power requirement generating more than 6.4 MWp of clean energy utilizing a state-of-the-art monitoring and control system. It has been quantified that the project can displace approximately 6300 metric tons of carbon dioxide, which is the equivalent of taking more than 1300 cars off the road [17].
4. IKEA Distribution Center, Perryville, Maryland
   The solar array installed in the IKEA center is one of the largest rooftop solar projects in North America spread over an area of 467,618 ft^2 having 2.2 MW system built with 7337 modules. The project produces and estimated 2,695,355 kWh of electrical power annually. Including the existing system, this distribution center's total 4.9 MW solar installation of 25,913 panels is capable of generating 6,092,533 kWh of clean electricity annually, which is the equivalent of reducing 4299 tons of carbon dioxide, thus eliminating the emissions of 896 cars or powering 591 homes [18].

FIGURE 11.2 Topaz Solar Farm, California, US [19].

Photovoltaic Application in Architecture

11.2.3 JAPAN

In 2006, 88.5% or 254 MW_p of the new installations were grid-connected residential systems, bringing the accumulated power of solar systems under the Japanese PV Residential Program to 1617 MW_p (out of 1709 MW_p total installed PV capacity). However, in the year 2007, the declination to 210 MWp was noted in the Japanese market, which was slightly recovered to 230 MWp in 2008 By the end of 2008, total cumulative installed capacity was 2.15 GW_p, which was less than half of the original 4.8 GW_p goal set for 2010 [1].

In 2013, Italy's record of 9.3 GW of the annual installation of power was beaten by China, which achieved 10.95 GW. Japan holds its second position for both the year 2013 and 2014 by achieving 6.9 GW and 9.7 GW of power, respectively, with a 40% increase in the year 2014 from 2013. The 2014 PV cumulative installed capacity for Japan is quantified to be 23,409 MW. The development of the PV market in Japan was in the form of the traditional rooftop, which at the end of 2014 represented almost 5 GW of the cumulative capacity; 2013 and 2014 saw the development of large-scale centralized PV systems [2].

Japan's solar PV annual capacity for the years 2015, 2016 and 2017 has been quantified to be 11 GW, 8 GW and 7 GW.

In order to encourage installation of PV systems in residential areas, the New Energy Foundation managed the first program "Monitoring Program for Residential PV Systems" from 1994 to 1996. As per which, 50% of the installation cost was paid resulting in an increase from 539 (year 1994) to 1986 (year 1996) in the participation of households [6]. With declination in the unit cost, the total annual cost of subsidies doubled from 2 to 4 billion yen, which is as per expectation. A bigger program to develop the infrastructure for introduction of residential PV systems was launched with a larger budget. The budget rose from 11.1 billion yen in 1997 to 23.5 billion yen in 2001. For the same period, a decline from 340,000 yen/kW_p to 120,000 yen/kW_p in the individual subsidies was noted. With average yearly electricity generation levels of 950 kWh/kW_p, the average savings in electrical power amounted to 23,400 yen/kW_p [20]. The savings of approximately 70,000 yen/annum for a 3 kW_p system were modest compared to the investment costs, so the lowering of initial capital costs was a critical feature to successfully stimulate the market.

In 2009, a subsidy program was initiated, which aimed to promote distribution of high-efficiency and low-price PV systems below 10 kW. This was terminated at the end of fiscal year 2013. In 2012, the existing scheme of purchasing excess production of PV power was replaced by new feed-in tariff (FiT). Its cost is shared among electricity consumers with some exceptions from electricity-intensive industries. This scheme resulted in fast growth of the PV market in Japan in 2013. The market achieved a balance between residential (below 10 kW), commercial, industrial and large-scale centralized plants in 2014. The FiT program was used to compensate excess electrical power from PV not self-consumed for systems below 10 kW. However, with tariffs above the retail electricity prices, self-consumption was not incentivized. Apart from these programs, various other support schemes were also enforced in Japan. In 2011, the Ministry of Economy, Trade and Industry (METI) launched a project that supported acceleration for introduction of renewable energy and supports, among other technologies, PV in the regions damaged by the great eastern Japan earthquake of 2011. The Ministry of Environment proposed another subsidy that supports the climatic changes and enables the technologies for local authorities' facilities, industrial facilities, schools, local communities and cities. They also helped promote the use of local storage (batteries) to favour the development of nonconventional sources of energy [2]. The Federation of Electric Power Companies of Japan (FEPC) announced that they aim to install PV plants with a cumulative installed capacity of 10 GW_p by 2020 [6].

The market in Japan slightly decreased to 7.5 GW installed capacity in 2017 from the record-high level of 10.8 GW in 2015. In the year 2017, Japan installed 7.5 GW of new PV capacity and cumulative PV installed capacity was estimated to be 49.5 GW. After having reached close to 11 GW in 2015, the market stabilized at a lower level in 2017 due to change in policies and the need to better streamline PV development. In 2017, major utility-scale plants grew, and the residential market reached 730 MW in 2017 as against 766 MW in 2016. In 2017, the FiT was adjusted downwards with a certain effect on the PV market. For prosumers' PV systems below 10 kW, the FiT program is used to remunerate excess PV electricity. The self-consumed part of PV electricity is not incentivized. Self-consumed electricity is not subject to taxation and transmission and distribution charges. The fiscal year 2017, a preferential tax treatment was initiated. The market for BiPV was comparatively smaller when compared to BaPV and around 35 MW were installed in 2017. METI runs a project on "International standardization of BiPV modules" for commercialization of BiPV [7].

The following are the few examples of implementation of photovoltaic in Japan:

1. Sharp Corporation LCD Plant, Kameyama
   Sharp facility with installation of rooftop (polysilicon solar cells) and façade installation, i.e., on walls (semitransparent thin-film) and with a nominal power capacity to be 5.21 MW located in Kameyama, established in 2006 [21], with a PV array covering an area of 47,000 m². The façade integrated with semi-transparent photovoltaic panels made of a structure where in a crystalline thin-film silicon solar cell is stacked on top of an amorphous silicon solar cell on the same glass substrate. This framework ensures the effective use of natural light from a wider band of wavelengths, thus improving the conversion efficiency to 11%. The electricity generated is used in the LCD TV assembly processes in addition to providing electric power for lighting inside the manufacturing facility. This PV would make possible to cut CO_2 emissions by about 3400 tons annually [6].

2. Villa Garten Shin-Matsudo, Chida and Tiara Court Kasukabe, Saitama
   In 1999, a PV system (crystalline silicon PV cell type) was mounted on an inclined roof of Villa Garten Shin in Chiba and Tiara Court Kasukabe with total power capacity to be 123 kW and 101 kW, respectively. This was a massive and creative project where all houses are equipped with PV systems and was launched in two areas of the Tokyo metropolitan region: 41 PV houses in Matsudo, Chiba, and 35 PV houses in Kasukabe, Saitama. These projects received the "New Energy Award in FY1999" in Japan [22]. In Matsudo, the PV system produced 2.86–3.1 kW/house. It was a grid-connected system. Annual energy yield was estimated to be 2800–3050 kWh/3kW/annum [23]. In Saitama, the PV system produced 2.88 kW/house. It was also a grid-connected system. Annual energy yield was estimated to be 2840 kWh/3kW/annum [23].

3. Itoman City Government Building, Okinawa
   The Itoman city government building started its operation in the year 2002 and has been integrated with PV systems as fixed sunscreens on façade and mounted on a flat roof with a total PV power potential of 195.6 kWp. Due to the location of this building in a coastal landfill site with high wind, the PV modules had to be prepared against salt damage. This structure was designed to sweep away salt content acquired on the surface of the PV module by rain. Further, since Okinawa is a path of typhoon, the structural design was sufficient to withstand the wind pressure. Apart from the above, since Okinawa is located in the southern part of Japan, it receives a good amount of solar insolation. Keeping this in consideration, the shading is very effective to decrease the cooling demand in summer. Thus, PV modules/arrays have been installed as a part of louvers and shelters, wherein the incident light is cut down; thus energy demand for cooling in summer is saved by the shading effect provided by the installed PV system. Horizontal louvers are provided on the southern side of the hall and pre-cast concrete screens on the eastern and western sides. The initial design for utilizing a PV system on the roof was to cover the south side of the roof. However, for maintenance and ventilation of the roof, for overview of the building and for avoiding being counted as floor height, the shelter was redesigned to cover the whole

roof by 50% of coverage factor by PV modules. The PV energy yield has been estimated to be 211,400 kWh [22]. For the effective operation, the PV power generation system was designed to interact with power utility so that excess power can be reversed. The PV system consists of power conditioners, power collectors and multi-crystalline silicon solar cells. The solar cells are protected with two layers of tempered glass provided on both front and back sides to have a self-washing effect from rainwater. An auxiliary 200 Ah battery is installed in the system to operate during the power failure in case of emergency [24].

4. Bus and Taxi Terminal at Kanazawa Station

    The city of Kanazawa enacted an environmental conservation regulation in 1997. Based on the policies laid under the mentioned regulation, ensuring the civil right of a healthy and cultural life, a PV system has been installed as a shelter of a bus and taxi terminal (Figure 11.3(a)). Amorphous silicon PV has been integrated as roof tile. The electrical energy produced by the system is mainly used for lighting of underground passages, etc. The project was subsidized as the next-generation city improvement project, and a "Green Electricity fund" was also applied to the PV system. The shelter of the terminal is approximately 4100 m^2 in area with PV modules installed on the overall shelter. Transparent PV modules were integrated with the glass roof in order to harmonize with the glass dome, station house and to develop a well-illuminated pathway with appropriate shadow areas. Total PV power output for the project has been estimated to be 110 kWp [22]. Annual power output is estimated to be 84,400 kWh.

(a)

FIGURE 11.3 (a) Bus and Taxi Terminal at Kanazawa Station, Japan [25]. (*Continued*)

(b)

FIGURE 11.3 (Continued) (b) The Solar Ark Building, Anpachi, Gifu Prefecture, Japan [26].

5. The Solar Ark Building, Anpachi, Gifu Prefecture
    The solar ark (Figure 11.3(b)) is one of the best examples of BiPV designs, a 630 kW solar-collecting building that boasts over 5000 solar panels and kicks off over 500,000 kWh of energy per annum. Most of the monocrystalline modules used were factory rejected and were headed to the scrap pile [27]. Solar Ark is 315 m wide and 37 m tall and weighs about 5000 tons. At its center is stationed Solar Lab, which helps raise children's awareness of global environmental issues, making science more interesting and appealing [28].

11.2.4 Germany

In 2008, Germany dominated the installations of photovoltaic worldwide with a share of 35%. Due to the dynamism of the German market, the growth in the overall European market energized, from 195 MW of cumulative solar PV installations in 2001 to 5337 MW in 2008 [29]. With three consecutive years above 7 GW of PV systems connected to the grid, Germany installed at least 32 GW of PV systems until the end of 2012. In the year 2014, PV annual installed capacity was 1900 MW with PV cumulative installed capacity of 38,250 MW. In 2013 and 2014, the market saw a degradation and went down to 3.3 GW and 1.9 GW, respectively, below the political aim to frame the development of PV within a range of 2.4–2.6 GW per annum [2].

In the year 2017, the PV power plant capacity was noted to be 1.75 GW, which is equivalent to 2% of total new PV capacity worldwide [30]. In order to meet the total power requirement of Germany by renewables by 2050, ca. 150–200 GW photovoltaic capacity is required [31, 32]. According to this, about 4–5 GW of PV power must be installed annually up to 2050. Germany has been able to position itself for huge investments in this sector because of the strong government support. Availability of a highly-qualified workforce, scientific research centers and universities (like Fraunhofer and Max-Planck-Institutes) have acted as a huge support facility to ensure rapid and smooth implementation and development of PV. As per [33], about 40 TWh of electrical power is generated by PV technologies, which covers approximately 7.2% of Germany's net electricity consumption including the grid losses.

The Renewable Energy Sources Act or EEG (Erneuerbare-Energien-Gesetz) law introduced the FiT for electrical power from the PV system that is mutualized in the electricity bill of the consumers with exemption given to energy-intensive industries in 2000. Due to decline in PV, Germany had to introduce the concept of "Breathing FiT" in the year 2009. This allowed the level of FiTs to decline as per the market evolution. Then "corridor" concept was introduced in 2013 due to a fast drop in PV prices. With this, the FiT levels declined with growth in the market during a defined

Photovoltaic Application in Architecture   361

period of time. Initially, the period defined between the two updates of the tariffs was a long period, i.e., up to 6 months, and this triggered some exceptional booms in the market. The biggest one was recorded for the month of December 2011 with 3 GW in one single month. Later, in September 2012, the update period was reduced to 1 month, with an additional announcement of updates every three months in order to control the market evolution efficiently. The latest change has been in place since August 2014. In September 2012, Germany abandoned FiT for installations of systems above 10 MW in size and further continued to reduce FiT levels in 2013 and 2014 [2].

A scheme named Scambio sul posto came to the surface as an alternative support scheme which favored self-consumption through an economic compensation of production of PV and consumption of electrical power for systems up to 200 kW (increased to 500 kW for the plants that were commissioned in 2015). The net-billing policy was revised in August 2012, wherein new PV systems can benefit from a self-consumption premium in complement to the FiT for the injected electricity, pushing PV systems to be progressively adjusted to the consumption pattern of users. The Sistemi Efficienti di Utenza (SEU) scheme got properly regulated in the year 2014, which enabled the management of one or more PV systems directly connected from the producer to the final consumer. In December 2014, new rules pertaining to storage of electricity connected to the grid were published [2].

Since 2014, yearly installations are constantly increasing from 1.2 GW and for the year 2017 annual PV installed capacity was recorded as 1776 MW. By the end of 2017, cumulative PV installed capacity for Germany was recorded as 42,491 MW which is about 7.6% of the electrical need of the overall electrical power production [7].

Following are the few examples of implementation of photovoltaic in Germany:

1. Solar parks/ large rooftops
   Waldpolenz Solar Park (Figure 11.4(a)) is located on 110 ha with rated power to be 40 MWp. It was the world's largest thin-film solar park using a cadmium telluride (CdTe) module when it was completed by the end of year 2008 [34]. The state-of-the-art Waldpolenz Solar Park has an installed capacity of 40 MW. Over 550,000 thin-film solar modules were used in the construction of this plant [15]. Some examples of largest photovoltaic roofs installed in Germany are at Rüsselsheim with 9.2 MWp (OPEL facility, 2011), Heddesheim with 8.1 MWp (logistics facility, 2013), Philippsburg with 7.4 MWp (Goodyear Dunlop facility, 2010, Hassleben with 6.96 MWp (Feedstock Farm, 2008–2010) of power etc. [21].

2. Mont-Cenis Academy, Herne
   Mont-Cenis Academy located in Herne was completed in the year 1999 and demonstrates a spectacular example of glazed shell, accommodating a building within a building. BiPV is integrated as rooftop and façade at the academy. An area of 10,500 m^2 are installed with solar cells for a peak power generation of 1 MW. The configuration allows the interior spaces to be optimally lit and shaded. For this purpose, solar modules with lighting densities between 53% and 93% (output 190–420 Wp) were use. Monocrystalline, multicrystalline and thin-film photovoltaics have been installed with estimated energy production to be 650,000 kWh/annum. The solar field also supplies the necessary shade for the hall. Light reflectors in front of the windows of the inside houses intensify the supply of daylight to the rear areas [35].

3. Logistics company at Bürstadt, Hessen
   A 5 MWp BiPV system was installed on the rooftop of a logistics company at Bürstadt, Hessen and has been in operation since 2005. BiPV on a flat roof of an area of 45,000 m^2 has been covered with about 29,182 monocrystalline PV modules. With efficiency of 10%, the setup is capable of generating about 4200 MWh of electrical power per annum accounting for about 9% of Bürstadt's annual consumption of electrical power. This is equivalent to electrical demand of about 1200 households [6].

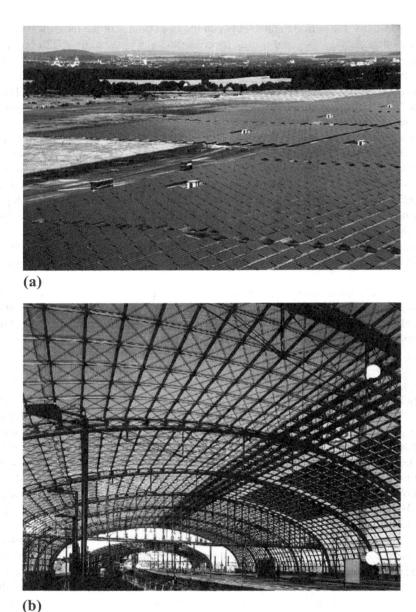

FIGURE 11.4 (a) Waldpolenz Solar Park, Germany [36]. (b) Lehrter Station, Berlin, Germany [37].

4. Logistics company Hartmann AG, Muggensturm

   The flat roof integrated PV roof system over logistics company Hartmann AG, Muggensturm, became operational in 2006. About 20,900 monocrystalline modules were installed covering an area of 80,000 m^2 with power output of 3839 kp resulting in a reduction of 2100 t of carbon dioxide per annum [22]. The annual solar power output is sufficient for the consumption of nearly 900 four-person households.

5. Lehrter Station, Berlin

   The PV modules integrated with the façade of Lehrter Station in Berlin have a capacity of 189 kWp of electrical power (Figure 11.4(b)). The PV modules consist of two disks and an

Photovoltaic Application in Architecture   363

in-between layer of synthetic resin, into which the solar cells are embedded. Because of the shape and geometry of the glass hall, all the PV modules are different in size, and none of them are identical.

The PV system is integrated with the already-constructed east-west glass roof covering an area of 1870 m². This is the biggest PV system in Berlin having an integrated performance of 189 kW and 160,000 kWh yearly electricity production. The system is made of 780 glass–glass-modules with Saturn-cells, each module manufactured fittingly precisely for the integration into the curved glass roof [22].

11.2.5 INDIA

India, as a tropical country, has been blessed with 300 sunny days amounting to sunshine equivalent to 5000 trillion kWh. Almost all parts of the country receives solar insolation of about 4–7 kWh/m² with about 2300–3200 sunshine hours per annum depending on the geographical location [38]. India has the potential to become one of the largest markets in the world, and PV can offer a solution to electricity shortages. The falling prices of solar panels are on the verge of coincidence with the growing cost of grid power in India.

India was one of the major countries that has accounted for the highest cumulative PV installations for 2014 with 3046 MW capacity, and the annual capacity for 2014 was recorded as 779 MW [2]. The Indian market jumped to 9.1 GW in 2017 from 779 MW in 2014, 2 GW in 2015 and 4.1 GW in 2016, 1.1 GW in 2013 and 1 GW in 2012 powered by various incentives in different states [7]. The PV market in India is driven by a mix of national targets and support schemes at various legislative levels. In November 2009, it was reported that India is ready to launch its Solar Mission under the National Action Plan on Climate Change, with plans to generate 1000 MW of power by 2013 [39]. The Jawaharlal Nehru National Solar Mission aims to install 20 GW of grid-connected PV system by 2022 and an additional 2 GW of off-grid systems, including 20 million solar lights [2]. Among the other major countries that accounted for the highest cumulative installations at the end of 2017 is India with more than 18 GW. India has also initiated the launch of the International Solar Alliance, aiming at accelerating the development of solar in emerging countries.

The following are a few examples of implementation of photovoltaic in India:

1. Dera Baba Jaimal Singh (Radha Soami Sect), Beas
   The first phase of rooftop solar installation at Radha Soami Satsang Beas headquarters' project was set up under the Punjab government's grid having capacity of 7.5 MW on a single rooftop and was synchronized with the grid in 2014. This has been commissioned in a single phase on multiple roofs covering an area of 1.621 million sq.ft. with varying roof profiles, roof types and truss frames [40]. In the second phase, a rooftop solar plant with 11.5 MW capacity was installed at Beas and came into operation in 2016 (Figure 11.5). The rooftop solar system is spread over an area of 42 acres. This solar installation became one of the country's largest projects with solar panels spread over an area of 82 acres on 8 rooftops and a capacity to generate 27 million units of electricity per annum, which is equivalent to catering to the electrical needs of about 8000 households [41]. The plant will offset over 19,000 tonnes of carbon emissions every year [42]. This not only highlights their green and environmental focus but also showcases the potential of rooftops in building energy independence.

2. Pata petrochemicals complex, Uttar Pradesh
   The state-owned gas utility GAIL has commissioned the one of the country's largest rooftop solar plants with a capacity of 5.76 MWp at the Pata petrochemicals complex in Uttar Pradesh. This project will reduce carbon emissions by 6300 tonnes per annum. It has been estimated that this project will save Rs. 5.5 crores annually in energy bills with payback time as 4 years. The plant is spread over an area of 65,000 m² covering rooftops of two

FIGURE 11.5 Dera Baba Jaimal Singh (Radha Soami Sect), Beas, India [43].

warehouses at the complex. More than 79 lakh units of electrical power have been estimated to be produced per annum with an expected plant load factor of 15% annually [44].

3. Mangalore refinery

The Mangalore refinery and petrochemical limited commissioned the solar power project at a refinery site with total capacity of 6.063 MWp. This is spread across 34 roof tops within the premises. The solar panels are installed to generate more than 24,000 units per day, totaling to more than 8.8 million units per annum. It has been dedicated to MRPL's commitment to sustainable development aiming at reducing carbon emissions and renewable energy generation. As estimated, annual metric tons of oil equipment savings will be 2682 with an MBN reduction of 0.1297, F&L reduction amounting to 0.0165% and a reduction in carbon emissions equal to 7000 tonnes [45].

4. Sutlej Textiles and Industries Ltd., Bhawanimandi, Rajasthan

The Rajasthan government commissions largest single rooftop solar project in the state and one of the largest in the country for Sutlej Textiles and Industries Ltd., Bhawanimandi, Rajasthan. The plant is spread over 2 lakh sq.feet of area having capacity of 2.2 MWp, generating 32 lakh units of green power per annum [46].

11.2.6 Spain

The PV market of Spain has grown tremendously over the years. In 2006, the Spanish photovoltaic industry grew by over 40% and established a worldwide production of 2520 MWp of PV modules [47] In 2008, about 2670 MW of annual installations were done. Due to the shrinking subsidies and a credit crisis, this growth was highly impacted and that impact was clearly visible in 2009. After 2009, the market went down to between 100 and 450 MW per annum. In 2014, PV cumulative installed capacity was 5376 MW with annual installed PV capacity at 22.6 MW [2]. In the long run, the Spanish Solar PV market will gradually grow and was estimated to reach total capacity of 33,738 MW by 2020 [29].

Spain's FiT policy triggered rapid development of the PV market in 2007 and 2008. This resulted in Spain establishing its number-one rank in the world PV market in 2008. In October 2008, a moratorium was put in place in order to control the growth, and the FiT was granted only after a registration process capping the installations at 500 MW a year. After a low year in 2009, due to the necessary time to put the new regulation in place, the market decreased to between 100 and 450 MW a year. In 2012 the Spanish government established a new moratorium for all the renewables projects with FiT. In 2014, 22 MW (around 22,6 MW_{DC}) were installed in Spain and the total installed capacity tops more than 4,8 GW_{AC} (5,4 GW_{DC}).

The regulation established under the Royal Decree (RD) 900/2015 in Spain employs neither a net metering (NM) nor a net billing (NB) scheme. Instead, two types of self-consumers are defined. Type 1 (considered to be a consumer) is limited to below 100 kW installation capacity, whereas type 2 is considered both consumer and producer. This regulation fails to explain residential households as prosumers since it does not contain concepts of net billing and net metering [48]. The Spanish PV market accounts for a share of about 3.1%–3.2% compared to the 7%–8% of other European countries like Italy, Germany and Greece [2].

In 2017, PV annual installed capacity was registered to be 148 MW and PV cumulative installed capacity to be 5284 MW. In October 2018, a new regulation of PV development significantly changed the PV landscape after years of constrained development that saw the market sink to extremely low levels [7].

The following are the few examples of implementation of Photovoltaic in Spain:

1. Solar parks/large rooftops
   The Olmedilla Photovoltaic Park, located in Olmedilla de Alarcon and completed in 2008, is one the world's largest PV parks (Figure 11.6). More than 270,000 solar panels have been installed to generate a total of about 87,000 MWh of electrical energy annually. Puertollano Photovoltaic park is another one of Spain's largest PV parks with an installed capacity of 50 MW power. It saves about 84,500 tonnes of carbon dioxide per annum. Arnedo Solar plant, located in La Rioja, produces about 34 MW of electrical power annually, set up on 70 hectares of land with over 172,000 solar panels [15]. Some examples of the largest photovoltaic roofs installed in Spain are at Figueruelas with 11.8 MW (GM facility) in 2008, Martorell with 11 MW (Seat al Sol, SEAT facility) in 2010–2013 and Castala with 5.2 MW (Actiu Technology Park) in 2008 [21].
2. Jaen University, Andalucía
   This integration of a medium-scale photovoltaic plant at Jaen University consists of a grid-connected PV system with a total power of 200 kW_p. PV modules are made up of

FIGURE 11.6 Olmedilla Photovoltaic Park, Spain [13].

monocrystalline silicon cells (made by Isofoton), anodized aluminum outline frames, frontal tempered glass and white transparent Tedlar on the back side, making the structure semi-transparent. PV modules were integrated into the university complex as parking canopies, facades, pergolas etc. to produce about 8% of the electrical demand of the university which accounts to about 280 MWh per annum. PV system 1 was integrated as one of the S-E-oriented parking canopies composed of a PV generator of 70 kWp and a triphasic inverter of 60 kW. System 2, with the same design, is located in a parking canopy parallel to the PV. In system 3, the PV system is integrated to a pergola. The system consists of a PV generator with 20 kWp power and nine string-oriented inverters of 2 kW. In PV system 4, the PV generator was integrated with the south façade of an existing building having 40 kWp polycrystalline modules and 15 string-oriented inverters of 2.5 kW [49].

3. Telefonica Business Park Complex, Madrid
   The Telefonica Business park complex is a gigantic business campus enclosed within a glass wall and the greatest solar panel roof in Europe and integrated with PV at roof and facade. The PV plant of 2.9 MWp integrated in the canopy of the new headquarters of Telefonica company (called "C District") in Madrid covers about 26,000 m^2 of area with solar panels having capacity of 4,398,000 kWh/year. electrical power reducing carbon dioxide by 2000 tonnes [50]. The PV modules are distributed in such a way that they form the letter "C" that constitutes the name of the building ensemble [22].

4. Torre Garena, Madrid
   Torre Garena, a skyscraper, has been integrated with PV modules at its rooftop and façade to produce total power of 85 kWp generated by installing the PV modules on the roof and façade of Torre Garena. The building's facade and roof include two different solar PV power generation plants. On the façade, 948 PV modules made from polycrystalline technology have been installed resulting in total power production of 75,840 Wp having savings of 85 tonnes of carbon dioxide emissions per annum. These modules have been installed at a 60° gradient to take full advantage of the available sunlight. The roof installation consists of 93 glass-to-glass modules of polycrystalline solar cells with nominal power of 100 Wp. They are installed on a south-facing roof with a gradient of 30° being perfectly integrated with the building [47].

11.3 CASE STUDY: BiSPVT SYSTEM INSTALLED AT SODHA BERS COMPLEX, VARANASI, INDIA [51]

The case study is based on various sections discussed in the book:

1. Various passive heating/cooling measures (Chapter 1, 6 and 7)
2. Calculation of thermal heat transfer gains (Chapter 2).
3. BiSPVT and BiOPVT systems (Chapter 8).
4. Fundamental energy matrices (Chapter 4).
5. Life cycle analysis of Sodha Bers complex (Chapter 9).
6. Carbon dioxide emission and mitigation (Chapter 9).
7. Earned carbon credits (Chapter 9).

Following are the assumptions for the calculations:

i. No degradation of PV model has been considered.
ii. All forms of energy have been considered equal.
iii. Average annual solar radiation has been considered.
iv. Steady-state model condition has been assumed.

Photovoltaic Application in Architecture

11.3.1 Introduction and Planning

Sodha Bers Complex (BERS) is located in Varanasi, India, and is a residential-cum training center for studies in solar energy for composite climatic conditions. The plot area of the complex is 233.98 m² with a total floor area of 794.75 m². It is a four-storey construction with a basement. The building is constructed based on various passive concepts as discussed in the book, namely earth shelter (Chapter 7), modified Trombe wall (Chapter 6), wind towers (Chapter 7), natural ventilation (Chapter 1 and 7), clerestory windows (Chapter 1), daylight (Chapter 3), orientation (Chapter 1), PVT solar roof, BiOPVT and BiSPVT systems (Chapter 8) and other features like a sunbathing area, rainwater harvesting making the building self-sustainable [52].

The key and floor plans shall be from Gupta and Tiwari [51].

In vernacular architecture, clear height of the space was kept at around 6–7 m in a single-story building with provision of open/glazed windows near the ceiling (i.e., clerestory windows) to provide an escape route for hot air from inside to outside for passive cooling and daylighting. In all the cases, thick Trombe walls were also constructed to keep the building cool as discussed in previous chapters. Further, all such buildings are either single- or double-story buildings due to less population density. Currently, the situation has changed to high-rise buildings because of rapid growth and development. Thus, the clear height of the building has been reduced to 3.0 m with glass windows which is economically viable. In SBC, the clear height of each floor has been designed to 3.6 m except for the basement with provisions of clerestory windows on each floor including the basement. Partition walls are constructed towards the corridor for proper daylighting and air circulation and also so that the hot air concentrated near the ceiling can be carried away through wind towers. The functional concept of windtowers is explained later. The additional cost due to increased height is about 5–10% more compared to the convention 3.0 clear height, but it saves a lot of conventional energy due to passive cooling, the provision of glazing areas in the east side of north portion of building and lower partition walls up to height of 2.4 m × 0.11 m. There is no need for artificial light throughout the year. In SBC concepts of clerestory, wind towers, SPV modules, partition walls, natural ventilation, natural daylight and energy savings have been used for passive cooling, explained in detail in next sections.

Figures 11.7 and 11.8 gives the 3-D and sectional views of the Sodha Bers Complex, respectively, all the elements and the entire arrangement of passive heating and cooling concepts, air circulation and daylighting as discussed in following sections.

11.3.1.1 Basement

The earth shelter concept has been used to keep the basement cool even in summer. Cross-ventilation and natural daylight are considered while designing the basement, wherein clerestory windows are provided at the top of the east-facing and north-facing sidewalls for cross-ventilation. Since hot air rises due to low density and hence is accumulated near the ceiling, the clerestory is an effective way to remove this hot air from the room. This window should be open during the summer months for thermal cooling and closed during winter months for thermal heating. The heat transfer coefficient (h_i) depends on variable air velocity (v) through the corridor which, can be considered as follows.

Here, the radiation term has not been considered due to small temperature differences between two walls.

Further, the rate of heat transfer can be evaluated through the clerestory window of the room by using the expression as given in Equation (1.8). where N is the number of air change from room air to outside and V is the volume of corridor/room through which hot air is withdrawn/removed.

The concept of integrated Trombe walls has been previously used by many researchers to reduce the cost of treated insulated walls. Therefore, the concept of modified Trombe walls with provision of storage (air cavity of 550 mm) has been adopted to reduce the heat lost to the outside. The purpose of constructing the storage space (almirah) with the Trombe wall on east and west walls is to create an air cavity. This arrangement reduces the U-value of envelope and also reduces the heat gains along the vertical exposed wall from outside.

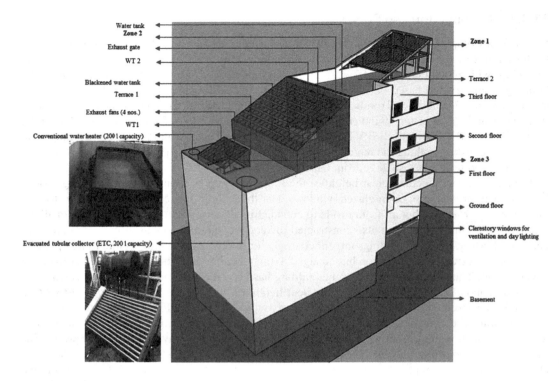

FIGURE 11.7 View defining SBC integrated with PV modules.

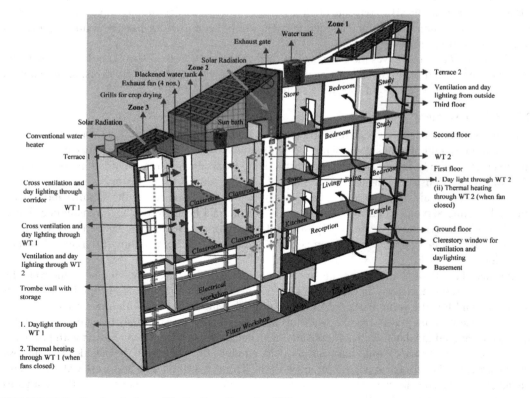

FIGURE 11.8 Sectional view of Sodha Bers Complex [51].

Photovoltaic Application in Architecture **369**

The expression for U-value for a Trombe wall with storage of 550 m depth (Figure 6.6b) can be given as [53]:

$$U = \left[\frac{1}{h_o} + \sum_{j=1}^{N} \frac{L_j}{K_j} + \sum_{j=1}^{N} \frac{1}{C_j} + \frac{1}{h_i} \right]^{-1} \tag{11.1}$$

where $C_j = 4.5$ W/m²°C is thermal air conductance of air cavity of thickness 0.550 m.

Following [53], the U-value of a solid brick wall ($L = 0.115$ m) with a wood ($K = 0.14$ W/m°C) cabinet of depth 0.550 m and thickness of 0.18 m is obtained from Equation (11.1) as.

$$U = \left[0.1754 + 0.167 + 0.222 + 1.286 + 0.3571 \right]^{-1} = \left[2.2075 \right]^{-1} = 0.4\,\text{W/m}^2\,^{\circ}\text{C} \tag{11.2}$$

If a brick wall is further waterproof insulated, then the U-value is reduced to 0.27 W. If the brick wall is further waterproof insulated, then the U-value is reduced to 0.27 W/m²°C.

Also, this concept creates a time lag and reduces the thermal load leveling thus at noon when the solar radiation is at peak, the internal heat gain is much less than the normal case. The heat stored in the Trombe wall is released with a time lag during the off-sunshine hours.

Wind Tower 1 (WT1) has been constructed in the fitter workshop towards the south end for the provision of natural daylight through the openings in the WT1 and also allows hot air to pass through it in summer months by creating wind draft to move the hot air from the basement to outside for cooling under natural mode.

11.3.1.2 Ground Floor

The ground floor has two wind towers placed from the north to south direction for proper cross-ventilation capturing the natural east wind and removing the hot air from each floor to the top of the building either forced mode or natural mode. At the same time, these wind towers (WTs) are responsible for daylighting to each connected floor. WT1 is located in the south of the building going up to the basement whereas WT2 is located at the center of the complex starting from the ground level itself. A WT functions according to the time of day and the presence or absence of wind. During the day, hot ambient air is allowed to enter the concrete tower through the openings at top. It is cooled as it comes in contact with the cool inside concrete tower by transferring its heat to the surface. The cooled air being denser than the warm air and hence it sinks down through the tower creating a downdraft. The draft is faster in the presence of wind. At night, the tower operates in reverse order like a chimney. Heat that has been stored in walls during the day warms the cool night air in the tower. As the pressure at the top of the tower is reduced, an updraft is created.

Both the towers are covered with semi-transparent photovoltaic (SPV) modules that allow sufficient daylight for the interiors of the complex.

Rainwater harvesting has also been planned on the ground floor.

11.3.1.3 First and Second Floor

Since the hot air rises, the clear height of the floors has been kept at 3.6 m to allow an easy escape for the hot air. Partition walls are constructed towards the corridor for proper daylighting and air circulation. Minimum outside exposure to the south-facing façade has been kept in kind while planning as compared to east and west façades to minimize the solar gains. Wind towers, partition walls and jali windows towards the corridor help in proper cross-ventilation. The inner walls also have the same thickness with opening 750 mm at top for ventilation as well as daylighting, and at night abundant lighting is available within the rooms via corridor lighting. The interior spaces on all floors are painted white to increase the reflection and ensure uniform distribution of natural light.

11.3.1.4 Terrace Floor Integrated with Semi-Transparent Photovoltaic (SPV) System

As discussed, wind towers are integrated by rooftop semi-transparent photovoltaic modules.

SPV modules, due to the higher electrical efficiency and daylight savings, have been considered and used in two ways in the complex. SPV modules are mounted at the rooftop in Zone 1 and integrated with the rooftop in zone 2 and zone 3, keeping in mind that none of the zones overshadow each other. The panels are south-oriented with an inclination angle of 30° to capture maximum annual solar energy. The description of zones is given in the next section.

11.3.2 ZONES [51]

The zones (Figure 11.7) depend on the type of photovoltaic arrangement and have been divided and described below.

Zone 1
48 nos. Each semi-transparent PV module (75 Wp capacity) has been integrated with the residential cum training complex at the rooftop, producing 3.6 kWp electrical power. The panels have been supported over the MS angle structure at the rooftop (terrace 2).

Zone 2
96 nos. Each semi-transparent PV modules (75 Wp capacity) have been integrated with the roof (terrace 1) of the SBC producing 7.2 kWp of electrical power on top of WT2. The SPV modules are supported on glass walls on three sides and placed over WT2, thus allowing the daylight to enter the entire building reducing the need for artificial lighting. This is an effective passive cooling concept. The enclosed area can be used for sunbathing purpose during the winter months. The thermal heat generated can be used for various purposes like space heating, domestic water heating, crop drying, etc. In winter months when there is a requirement of thermal heating, hot air is circulated within the building by closing the exhaust gate provided on the north side. Hence, hot air available in the semi-transparent greenhouse (zone 2) can be circulated to the interiors of the building. During summer months when there is a requirement of thermal cooling, hot air is thrown outside from zone 2 by opening the exhaust fan. Further, there is a 1000 L blackened water tank placed inside this zone for pre-heating, prior to re-heating by conventional and evacuated tubular collectors (ETC) water heaters for availability of hot water throughout the year. This hot water can be used for several applications in kitchens, bathrooms, etc. Figure 11.9 shows an inside view of the zone with a water drum and daylight through SPV modules.

Zone 3
On the top of WT1, there is semi-transparent photovoltaic thermal (PVT) greenhouse dryer of capacity 100 kg. The PVT dryer is used to dry all locally available medicinal/vegetables crops under forced mode to be used by residents of the complex (Figure 11.10). For forced mode of operation, there are four DC fans, each with a capacity of 20 W, which is installed at top north side of dryer. During the summer, the fan is operated for fast drying and to suck hot air from the bottom to outside. 8 nos. Each semi-transparent PV module (35 Wp capacity each) has been integrated with the roof (terrace 1) of the SBC producing 280 W of electrical power. This has been constructed above the WT1 with the space enclosed by glass walls to capture the daylight and facilitate cross-ventilation. Four exhaust fans have been installed at the north glass wall, which allows the hot air to escape creating a cooling effect for the interior spaces. The PV modules are connected to the battery through a charge controller. The same is also used to provide the light in the staircase during the night in the absence of drying. A ladder is provided for cleaning and maintenance purposes. The arrangement also produces thermal energy and can be utilized as a solar roof dryer for crop/vegetable drying, space heating and domestic water heating purposes. The heating/cooling concept is similar to what has been discussed for zone 2.

FIGURE 11.9 Inside view of Zone 2. (source: author).

FIGURE 11.10 Inside view of Zone 3. (source: author).

11.3.3 Construction Details and Materials Used [52]

The decision to build the basement was based on the ground profile of the site to avoid the landfilling cost, as well as to get bearing capacity of the natural soil as per design requirement. A minimum foundation depth for a non-load bearing wall has 0.91 m stepped footing while the loadbearing column foundation is 1.50 m trapezoidal footing below the existing ground level (GL).

Three types of columns based on the estimated load pattern of the building were designed. Decreasing the load to minimize the costing and embodied energy simultaneously, the reinforcement

372   Photovoltaic Thermal Passive House System

was curtailed, as well as size at each floor accordingly. The quantity of reinforced cement concrete (RCC) quantity is about 42 m^3.

Based on the span and load, again three types of beams were designed. Design is made as per T-beam theory, and reinforced cement concrete (RCC) work quantity is nearly 107 m^3.

Table 11.2 gives the details of construction materials used in the Sodha Bers Complex. It is clear that sand, stone ballast, bricks/tile and cement constitute the major part. However, stone and sand have minimum energy density. Savings of about 340 tonnes of brick has been done over conventional construction [52].

11.3.4 THERMAL HEAT GAINS [51]

To evaluate building's energy performance storage of storage of thermal energy in the building forms an important aspect.

$$Hourly\ heat\ gain, Q = \left(M_a \times C_a \times \Delta T\right) \times 2.77 \times 10^{-4}\ kWh \tag{11.3a}$$

$$Daily\ heat\ gain\ of\ SBC = Q \times 24\ kWh \tag{11.3b}$$

$$Annual\ heat\ gain\ of\ SBC = \left(Q \times 24\right) \times n \tag{11.3c}$$

$$M_a\ is\ the\ mass\ of\ air\ in\ kg, \quad M_a = A \times clear\ height\ of\ the\ room \times \rho_a$$

$$M_{a,basement} = 856.95\ kg$$

$$M_{a,ground\ floor} = 866.76\ kg$$

$$M_{a,first\ and\ second\ floor\ each} = 832.98\ kg$$

$$M_{a,third\ floor} = 401.71\ kg$$

n is the number of clear days = 300
ΔT is the average temperature difference between room air and the ambient in °C
ρ_a is the density of air = 1.2 kg/m^3
C_a is the specific heat of air = 1005 J/kg/K
Height of the room = 3.6 m

Table 11.3 (part A) gives the hourly, daily and annual heat gain of the building along with thermal energy efficiency of the Sodha Bers Complex for average daily solar intensity of 450 W/m^2

$$Efficiency, \eta = \frac{Annual\ thermal\ energy\left(i.e., output\ energy\right)}{Annual\ solar\ energy\left(i.e., input\ energy\right)} \times 100 \tag{11.4a}$$

or,

$$\eta = \frac{Annual\ heat\ gain}{Area\ of\ terrace \times av.\ daily\ solar\ intensity \times 10^{-3} \times n \times 24} \times 100 \tag{11.4b}$$

TABLE 11.2

Materials Used in Sodha Bers Complex [53]

Sr. No.	Material/Activity	Quantity (kg)	Percentage Breakup of Quantity (%)	Embodied Energy (MJ)	Embodied Energy (kWh)	Percentage Breakup of Embodied Energy (%)	Carbon Dioxide Emission (tonne)	Percentage Breakup of Carbon Dioxide Emission (%)
1	Cement	122,250	8.12	672,375	186,770.83	74	183.04	22.51
2	Bricks/tiles	337,935	22.45	844,837.5	234,677.08		229.98	28.28
3	Stone ballast	447,590	29.73	53,710.8	14,919.67		14.62	1.79
4	Sand	540,937	35.93	43,815.897	12,171.08		11.93	1.46
5	Lime/pop/putty	2220	0.14	11,766	3268.33		3.20	0.39
6	Reinforcement	22,574	1.49	799,119.6	221,977.67		217.54	26.75
7	Aluminum	34	0	5270	1463.89		1.43	0.17
8	Glass	1248	0.08	18,720	5200.00		5.10	0.62
9	Plywood	10,950	0.72	164,250	45,625.00		44.71	5.49
10	Wood	8200	0.54	69,700	19,361.11		18.97	2.33
11	PVC/plastic	168	0.01	12,936	3593.33		3.52	0.43
12	Paint (oil bound)	210	0.01	14,280	3966.67		3.89	0.47
13	Paint (water bound)	1125	0.07	59,625	16,562.50		16.23	1.99
14	Marvel/makrana stone	525	0.03	1050	291.67		0.29	0.03
15	Ceramic tiles/sanitary ware	4457	0.29	89,140	24,761.11		24.27	2.98
16	Iron	4458	0.29	111,450	30,958.33		30.34	3.73
17	Copper	180	0.01	7560	2100.00		2.06	0.25
18	Brass	150	0	6600	1833.33		1.80	0.22
	Total	1,505,211		2,986,205.797	829,501.6		812.91	
19	Process			867,664	241,017.78	21		
20	Transportation			180,844	50,234.44	5		
	Total			4,034,713.797	1,120,753.83			

1 kWh electrical energy generation from the thermal power plant emits 0.98 kg CO_2 [54]

Labor with T&P	Rs. 1,255,714
Electrification	Rs. 193,136

TABLE 11.3

(i) Hourly, Daily and Annual Heat Gain, (ii) Thermal Energy Efficiency of SBC [51]

Av. daily I(t)	ΔT	Basement			Ground Floor			First and Second Floor			Third Floor			SBC		
		4°C	6°C	8°C	4°C	6°C	8°C	4°C	6°C	8°C	4°C	6°C	8°C	4°C	6°C	8°C
(A) 450 W/m²	Hourly heat gain (kWh)	0.949	1.42	1.89	0.960	1.44	1.92	0.922	1.38	1.84	0.445	0.66	0.89	4.2	6.3	8.4
	Daily heat gain (kWh)	22.78	34.18	45.57	23.04	34.57	46.09	22.15	33.22	44.30	10.68	16.02	21.36	100.8	151.23	201.64
	Annual heat gain (kWh)	6836.47	10,254.70	13,672.94	6914.70	10,372.05	13,829.40	6645.20	9967.80	13,290.40	3204.73	4807.10	6409.47	**30,246.3**	**45,369.46**	**60,492.62**
	η						-							**4.65%**	**6.97%**	**9.30%**
(B) 650 W/m²	**Annual heat gain (kWh)**						-							43,663.27	65,447.96	87,326.55

Photovoltaic Application in Architecture

where

$$Area \ of \ terrace = 200.64 \, m^2$$

Using Equation 11.4b and the efficiency from Table 11.3 (part A), annual heat gain for average daily solar intensity = 650 W/m^2 has been calculated and tabulated in Table 11.3 (part B).

Table 11.3 shows that with an increase in the difference of average temperature between the room and the ambient (ΔT) from 4°C to 8°C (2-fold), there is an increase in annual heat gain by 100% for average daily solar insolation, $I(t)$, equal to 450 W/m^2. An increase of 44.35% in the annual heat gain is noticed with change in average daily solar insolation changes from 450 W/m^2 to 650 W/m^2 for $\Delta T = 8$°C. This increase in the annual heat gain is because there is an increase of 44.44% in the solar intensity, which means that ΔT and $I(t)$ have a linear effect on the thermal performance of the system.

11.3.5 Electrical Power (E_p)

Installed PV panels under standard test conditions, $I(t) = 1000$ W/m^2 in all the zones is tabulated below:

$$Therefore, total \ E_p \ produced \ by \ BiSPVT \ system$$
$$= \left[11.08 \times no. of \ hours \times n \times \left(solar \ intensity/1000 \right) \right] kWh. \tag{11.5}$$

Zones	Zone 1	Zone 2	Zone 3	Total
Electrical power (kWp)	3.6	7.2	0.28	11.08

The number of hours for summer months is 7 hours and for the winter months is 5 hours. Thus, the average number of hours taken for calculations is 6 hours.

Electrical power has also been calculated for average daily solar intensity as 450 W/m^2 and 650 W/m^2 like thermal energy storage and given below

$$E_{p,450} = 8974.80 \, kWh \, and$$

$$E_{p,650} = 12,963.60 \, kWh$$

From these results, an increase in electrical power of about 44.44% is seen with change in average daily solar insolation from 450 W/m^2 to 650 W/m^2 since the model is in steady state.

11.3.6 Daylight Energy Savings

As already discussed in previous chapters, a major advantage of semi-transparent modules over opaque PV modules is the admittance of the natural light in the building, thus reducing the building's demand for artificial lighting. Following Sudan and Tiwari [55], 12.8 kWh/m^2 is the annual energy savings in SBC due to daylighting and diffused components.

Annual energy savings due to daylight in the complex for av. daily solar intensity = 450 W/m^2 can be calculated as $A_{SBC} \times$ Annual energy savings in SBC due to daylighting and diffused components per m^2

$$A_{SBC} = 877.64\,\text{m}^2$$

Therefore, annual energy savings due to daylighting for av. daily solar intensity = $450\,\text{W/m}^2 = 877.64 \times 12.8\,\text{kWh} = 11{,}233.79\,\text{kW}$.

For $\Delta T = 4°\text{C}$ and av. daily solar intensity = 450 W/m², efficiency can be calculated as follows:

$$\eta = \frac{Annual\ energy\ savings\ due\ to\ day\ lighting\ (i.e., output\ energy)}{Annual\ solar\ energy\ (i.e., input\ energy)} \times 100 \qquad (11.6a)$$

$$\eta_{450} = 1.58\% \qquad (11.6b)$$

Now, using the above η_{450} and Equation (11.6b), annual energy savings due to daylighting of SBC for average daily solar intensity equal to 650 W/m² has been calculated as follows:

$$1.58 = \frac{Annual\ energy\ savings\ due\ to\ day\ lighting}{877.64 \times 650 \times 10^{-3} \times 6 \times 300} \times 100$$

Therefore, annual energy savings due to daylighting for average daily solar intensity is equal to $650\,\text{W/m}^2 = 16{,}224.05\,\text{kW}$.

11.3.7 Total Energy Savings

Total energy savings or total energy generated per year (E_{total}) is determined by adding electrical power and daylight savings as calculated in Table 11.3, Sections 11.3.5 and 11.3.6

Total energy generated per annum = Thermal energy gain + Electrical power + daylight savings.

For average daily solar intensity equal to 450 W/m²,

$$E_{total,\Delta T=4°\text{C}} = 30{,}246.3 + 8974.8 + 11{,}233.79 = 50{,}454.89\,\text{kWh}$$

$$E_{total,\Delta T=6°\text{C}} = 45{,}369.46 + 8974.8 + 11{,}233.79 = 65{,}578.05\,\text{kWh}$$

$$E_{total,\Delta T=8°\text{C}} = 60{,}492.62 + 8{,}974.8 + 11{,}233.79 = 80{,}701.21\,\text{kWh}$$

Similarly, for average daily solar intensity equal to 650 W/m²,

$$E_{total,\Delta T=4°\text{C}} = 43{,}663.27 + 12{,}963.0 + 16{,}224.05 = 72{,}850.92\,\text{kWh}$$

$$E_{total,\Delta T=6°\text{C}} = 65{,}447.96 + 12{,}963.0 + 16{,}224.05 = 94{,}635.61\,\text{kWh}$$

$$E_{total,\Delta T=8°\text{C}} = 87{,}326.55 + 12{,}963.0 + 16{,}224.05 = 116{,}514.2\,\text{kWh}$$

The total energy savings as calculated above gives a similar trend as found earlier in Section 11.3.5 for electrical power because of similar trends observed in thermal energy.

11.3.8 Embodied Energy

As already discussed in Chapter 4, it is necessary to know the mass of each material used for construction of building and its energy density to calculate the embodied energy. Table 11.2. Tiwari

Photovoltaic Application in Architecture 377

et al. [53] gives the total embodied energy (EE) of the materials and other activities involved which is equal to 1,120,753.83 kWh. It is clear that the embodied energy of the materials used constitutes 74% of the total embodied energy. However, the embodied energy of a PV module integrated into the building is 1271 kWh/m^2 [56].

Total number of PV modules in the complex are 48 nos in Zone 1 + 96 nos in Zone 2 + 8 nos in Zone 3 = 152 nos.

Area covered by 1 module = 0.6 m^2

Therefore, area covered by 152 PV modules = 91.2 m^2

Embodied energy of PV modules used in SBC complex (EE) = area covered by the PV modules × 1271 kWh.

Therefore, EE of PV modules = 91.2 × 1271 = 115,915.20 kWh.

Referring to Table 11.2, total EE of the SBC complex = 1,120,753.83 + 115,915.2 kWh =1,236,669.030 kWh.

11.3.9 Energy Payback Time (EPBT)

As discussed in Chapter 4, energy payback time (EPBT) is the total time period required to recover the total energy consumed to construct the building (embodied energy). It is the measure of energy sustainability of a building and is one of the major criteria to compare the viability of one technology against the other.

It depends on:

- The energy spent to prepare the materials used for fabrication of the system and its components. This is referred to as embodied energy.
- The annual energy yield (output) obtained from such system.

EPBT can be calculated as given in Equation (4.6) and can also be written as:

$$EPBT = \frac{Embodied\ Energy}{Energy\ generated\ per\ year} \tag{11.7}$$

EPBT can be calculated from E_{total} (energy savings or energy generated per year) as calculated in Section 11.3.7 and EE (embodied energy = 1,236,669.030 kWh) as calculated in Section 11.3.8.

For average daily solar intensity equal to 450 W/m^2,

$$EPBT_{\Delta T=4°C} = 24.51\ years$$

$$EPBT_{\Delta T=6°C} = 18.85\ years$$

$$EPBT_{\Delta T=8°C} = 15.32\ years$$

Similarly, for average daily solar intensity equal to 650 W/m^2,

$$EPBT_{\Delta T=4°C} = 16.97\ years$$

$$EPBT_{\Delta T=6°C} = 13.06\ years$$

$$EPBT_{\Delta T=8°C} = 10.61\ years$$

378   Photovoltaic Thermal Passive House System

From the above results, it can be seen that there is a fall in EPBT with increase in the difference of average temperature between the room air and ambient (ΔT) from 4°C to 8°C by 37.49% for average daily solar insolation as 450 W/m². When ΔT changes from 4°C to 8°C for insolation level of 650 W/m² and life of the building = 300 years, there is a 37.47% decrease in EPBT. Minimum EPBT with 10.61 years is found for $\Delta T = 8$°C and av. daily $I(t) = 650$ W/m² while the maximum EPBT with 24.51 years has been found for $\Delta T = 4$°C and av. daily $I(t) = 450$ W/m².

11.3.10 ENERGY PRODUCTION FACTOR (EPF)

As discussed in Chapter 4, energy production factor (EPF) helps predicts the overall performance of the building and should be greater than 1.

EPF can be calculated as given in Equation (4.11) and can also be written as:

$$EPF = \frac{Annual\ saved\ energy \times Life\ of\ the\ building}{Embodied\ energy} > 1 \tag{11.8}$$

EPF has been calculated considering the life of the building as 100, 200 and 300 years.

EPF can be calculated from E_{total} (energy savings or energy generated per year) as calculated in Section 11.3.7 and EE (embodied energy = 1,236,669.030 kWh) as calculated in Section 11.3.8 and is tabulated in Table 11.4.

From the Table 11.4, maximum EPF with 28.26 years has been found when average daily solar intensity is 650 W/m², $\Delta T = 8$°C, life of the building is 300 years and the minimum EPF of 4.08 years corresponds to average daily solar intensity of 450 W/m², $\Delta T = 4$°C and life of the building as 100 years. When ΔT changes from 4°C to 8°C for the life of the building as 300 years, there is an increase of 59.96% in EPF for the insolation level of $450\,\text{W}/\text{m}^2$, and there is an increase of a 59.93% in EPF for an insolation level of 650 W/m².

11.3.11 LIFE CYCLE CONVERSION EFFICIENCY

As discussed in Chapter 4, life cycle conversion efficiency (LCCE), is considered to be the most important from an energy point of view. Equation (4.15) can be expressed as:

$$LCCE = \varphi = \frac{\left[\left(Annual\ saved\ energy \times Life\ of\ the\ building\right) - EE\right]}{Total\ annual\ solar\ energy \times Life\ of\ the\ building} \tag{11.9a}$$

TABLE 11.4
Energy Production Factor and Life Cycle Conversion Efficiency [51]

Life of the Building (Years)	Energy Production Factor (Years)			Life Cycle Conversion Efficiency		
	100 Years	200 Years	300 Years	100 Years	200 Years	300 Years
(A) For average daily solar intensity = 450 W/m²						
$\Delta T = 4$°C	4.08 years	8.16 years	12.24 years	0.23	0.27	0.28
$\Delta T = 6$°C	5.30 years	10.61 years	15.91 years	0.32	0.36	0.37
$\Delta T = 8$°C	6.53 years	13.05 years	19.58 years	0.42	0.45	0.47
(B) For average daily solar intensity = 650 W/m²						
$\Delta T = 4$°C	5.89 years	11.78 years	17.67 years	0.25	0.28	0.44
$\Delta T = 6$°C	7.65 years	15.30 years	22.96 years	0.35	0.37	0.47
$\Delta T = 8$°C	9.42 years	18.84 years	28.26 years	0.44	0.38	0.48

Photovoltaic Application in Architecture

or,

$$LCCE = \varphi = \frac{\left[\left(Annual\ saved\ energy \times Life\ of\ the\ building\right) - EE\right]}{Av.\ daily\ solar\ radiation \times 10^{-3} \times n \times 6 \times Life\ of\ the\ building \times Area_{floor}} \quad (11.9b)$$

Like EPF, LCCE is also calculated for 100, 200 and 300 years as the life of the building and given in Table 11.4.

From Table 11.4, an increase in LCCE of 67.85% is observed with a change in ΔT from 4°C to 8°C for average daily solar radiation = 450 W/m^2 and the life of the building = 300 years. When ΔT changes from 4°C to 8°C for insolation level of 650 W/m^2 and life of the building = 300 years, there is a 65.51% increase in LCCE.

Maximum LCCE with 0.47 value and 0.48 value for av. daily solar insolation = 450 W/m^2 and 650 W/m^2, respectively, is found for $\Delta T = 8$°C and life of the building as 300 years.

It can be said that the embodied energy, annual energy savings and life of the building plays an important role in evaluating the energy matrices. The PV module is best suited when the EPBT has a lower value, whereas EPF and LCCE demonstrate higher values. This results in better efficiency, cost-effectiveness and is more environment friendly.

11.3.12 CARBON DIOXIDE EMISSION

The major contributors of CO_2 emission are brick/tiles, cement and reinforcement (77.55%) as given in Table 11.2. Here, note that the brick contributes about 28% due to use of 0.11 m brick exterior and partition wall, otherwise its contribution might have been very high.

If unit power (kWh) is used by a consumer and the losses due to poor domestic appliances is L_a (=0.20), then the transmitted power should be $\dfrac{1}{1-L_a}$ units.

Referring to Equation (9.15), annual CO_2 emission per year for

$$n_{sys} = 100\ years = \frac{1{,}236{,}669.030}{100} \times \frac{1}{1-0.20} \times \frac{1}{1-0.4} \times 0.98\,kg = 25{,}248.65\,kg = 25.24\ tonnes$$

$$n_{sys} = 200\ years = \frac{1{,}236{,}669.030}{200} \times \frac{1}{1-0.20} \times \frac{1}{1-0.4} \times 0.98\,kg = 12{,}624.32\,kg = 12.62\,tonnes$$

$$n_{sys} = 300\ years = \frac{1{,}236{,}669.030}{200} \times \frac{1}{1-0.20} \times \frac{1}{1-0.4} \times 0.98\,kg = 8416.22\,kg = 8.41\ tonnes$$

11.3.13 NET CARBON DIOXIDE MITIGATION

If the transmission and distribution losses are $L_{td} = 0.40$, then the power that has to be generated in the power plant is $\dfrac{1}{1-L_a} \times \dfrac{1}{1-L_{td}}$ units.

Referring to Section 9.7.1, for unit power consumption by the consumer the amount of CO_2 emission $= \dfrac{1}{1-L_a} \times \dfrac{1}{1-L_{td}} \times 0.98\,\dfrac{g}{kWh}$.

CO_2 emission over the lifetime is given in Equation (9.16) and can be calculated as,

$$CO_2\ emission\ over\ the\ lifetime = 1{,}236{,}669.030 \times \frac{1}{1-0.20} \times \frac{1}{1-0.4} \times 0.98\,kg$$
$$= 2{,}524{,}865.936\,kg = 2524.86\,tonnes$$

380 Photovoltaic Thermal Passive House System

TABLE 11.5

Net CO₂ Mitigation for Sodha Bers Complex

Average Temperature Difference (ΔT) between Room Air and Ambient Air Temperature (°C)	The Net CO₂ Mitigation over the Lifetime of the System (Tonnes)		
	For L = 100 Year	For L = 200 Year	For L = 300 Year
(A) For average daily solar intensity = 450 W/m²			
4°C	7776.34	18,077.55	28,378.75
6°C	10,863.99	24,252.84	37,641.69
8°C	13,951.63	30,428.13	46,904.63
(A) For average daily solar intensity = 650 W/m²			
	For L = 100 year	For L = 200 year	For L = 300 year
4°C	12,348.86	27,222.59	42,096.32
6°C	16,796.57	36,118.01	55,439.45
8°C	21,263.45	45,051.77	68,840.08

- Calculation of carbon dioxide mitigation for SBC (Usin Equation 9.17)
  Embodied energy = 1,236,669.030 kWh (Section 11.3.8) and E_{aout} as total energy savings (=Thermal gains + Electrical power + Daylight savings), refer Section 11.3.7

  The net CO₂ mitigation over the lifetime of the system, lifetime = 100, 200 and 300 years for different ΔT and av. daily solar insolation of 450 W/m² and 650 W/m² is given in Table 11.5.

11.3.14 EARNED CARBON CREDITS

As already discussed, carbon credits are defined as "a key component of national and international emissions trading schemes that have been implemented to mitigate global warming".

If CO₂ emission is being traded at 10 US\$/tonne of CO₂ mitigation, then the carbon credit earned by the system is obtained as

$$Carbon\ credit\ earned = Net\ CO_2\ mitigation \times 10US\$$$

Using the equation and Table 11.5, carbon credit has been evaluated the results are reported in Table 11.6

TABLE 11.6

Carbon Credit Earned by Sodha Bers Complex

Average Temperature Difference (ΔT) between Room Air and Ambient Air Temperature (°C)	The Net CO₂ Mitigation over the Lifetime of the System (Tonnes)		
	For L = 100 Year	For L = 200 Year	For L = 300 Year
(A) For average daily solar intensity = 450 W/m²			
4°C	77,763.4	180,775.5	283,787.5
6°C	108,639.9	242,528.4	376,416.9
8°C	139,516.3	304,281.3	469,046.3
(A) For average daily solar intensity = 650 W/m²			
	For L = 100 year	For L = 200 year	For L = 300 year
4°C	123,488.6	272,225.9	420,963.2
6°C	167,965.7	361,180.1	554,394.5
8°C	212,634.5	450,517.7	688,400.8

OBJECTIVE QUESTIONS

11.1 Embodied energy of a PV module integrated into the building can be considered as:
- (a) 1271 kWh/m^2
- (b) 1271 Wh/m^2
- (c) 1271 kWh/ft^2
- (d) 1271 Wh/ft^2

11.2 Thermal heat gains can be measured by:
- (a) $M_a \times C_a \times \Delta T$
- (b) $M_a \times C_a \times T$
- (c) $\dot{M}_a \times C_a \times \Delta T$
- (d) $\dot{M}_a \times C_a \times T$

11.3 Energy payback time is the ratio of
- (a) $\dfrac{Embodied\ Energy}{Life\ of\ the\ building}$
- (b) $\dfrac{Embodied\ Energy}{Annual\ energy\ output\ from\ the\ system}$
- (c) $\dfrac{Annual\ saved\ energy \times Life\ of\ the\ building}{Embodied\ energy}$
- (d) $\dfrac{Annual\ saved\ energy}{Embodied\ energy \times Life\ of\ the\ building}$

11.4 Energy production factor should always be
- (a) Less than one
- (b) Zero
- (c) Equal to one
- (d) Greater than one

11.5 Carbon dioxide emissions over the lifetime can be calculated as:
- (a) $E_{in} \times \dfrac{1}{L_a - 1} \times \dfrac{1}{L_{td} - 1} \times 0.98\,\text{kg}$
- (b) $E_{in} \times \dfrac{1}{1 - L_a} \times \dfrac{1}{1 - L_{td}} \times 0.98\,\text{kg}$
- (c) $\dfrac{1}{1 - E_{in}} \times \dfrac{1}{1 - L_a} \times \dfrac{1}{1 - L_{td}} \times 0.98\,\text{kg}$
- (d) $\dfrac{1}{E_{in}} \times \dfrac{1}{1 - L_a} \times \dfrac{1}{1 - L_{td}} \times 0.98\,\text{kg}$

ANSWERS

11.1 (a)
11.2 (a)
11.3 (b)
11.4 (d)
11.5 (b)

REFERENCES

[1] IoE, "PV status report 2008," JRC European Commission, 2008.

[2] IEA, "Trends 2015 in photovoltaic applications. Survey report of selected IEA countries between 1992 and 2014. Photovoltaics power system program," IEA PVPS, 2015.

[3] IEA, 2018a. [Online]. Available: https://www.iea.org/tcep/power/renewables/solar/ [Accessed April 22, 2019].

[4] IEA, "Snapshot of Global PV MArket," IEA PVPS, 2019.

[5] C. R. Intelligence, "Research Report of China's Building Integrated Photovoltaic (BIPV) Industry," 2009. [Online]. Available: https://www.prlog.org/10250896-research-report-of-chinas-building-integrated-photovoltaic-bipv-industry-2009.html [Accessed April 22 2019].

[6] B. Agrawal and G. N. Tiwari, *Building Integrated Photovoltaic Thermal Systems: For Sustainable Developments*, United Kingdom: Royal Society of Chemistry 2010.

[7] IEA, "Trends 2018 in photovoltaic applications," IEA PVPS T1-34:2018, 2018.

[8] A. Rathi, *Quartz*, 2017. [Online]. Available: https://qz.com/1056019/satisfyingly-the-worlds-largest-floating-solar-farm-is-producing-energy-atop-a-former-coal-mine/ [Accessed April 29, 2019].

[9] E. Hughes, *Pvtech*, 2010. [Online]. Available: https://www.pv-tech.org/news/project_focus_suntech_completes_3.12mw_bipv_installation_shanghai [Accessed April 22, 2019].

[10] Dricus, *Sinovoltaics*, 2010. [Online]. Available: https://sinovoltaics.com/building-integrated-photovoltaics/shanghai-expo-2010-solar-bipv/ [Accessed April 22, 2019].

[11] B. Meinhold, *inhabitat*, 2010. [Online]. Available: https://inhabitat.com/worlds-largest-integrated-photovoltaic-bipv-project-online/ [Accessed April 22, 2019].

[12] C. B. M. S. A. P. B. A. Committee, *bipvcn*. [Online]. Available: http://www.bipvcn.org/project/case-domestic/15202.html [Accessed April 22, 2019].

[13] Google Earth, 2020.

[14] S. K. Ho, J. K. Chan and I. P. L. Lau, "Performance evaluation of a large building integrated photovoltaic system," *ICEE*, July 2007.

[15] J. Josh, *Solarelectrician*, 2017. [Online]. Available: https://solarelectrician.com/largest-solar-power-plants/ [Accessed April 15, 2019].

[16] J. Vargo, "Perris press release," [Online]. Available: http://www.cityofperris.org/news/2011_stories/09-19-11_solarpanels.html [Accessed April 29, 2019].

[17] DLR Group. [Online]. Available: http://www.dlrgroup.com/work/mandalay-bay-solar-array/ [Accessed April 29, 2019].

[18] IKEA, *CSR*, 2014. [Online]. Available: http://www.csrwire.com/press_releases/37232-IKEA-Completes-Near-Doubling-of-Maryland-s-Largest-Rooftop-Array-on-Distribution-Center-in-Perryville-Making-it-One-of-The-Largest-Such-Installations-in-U-S- [Accessed April 29, 2019].

[19] Allen, J. 2015. *Wikipedia*. Retrieved June 2021 from https://en.wikipedia.org/wiki/Topaz_Solar_Farm#/media/File:Topaz_Solar_Farm,_California_Valley.jpg.

[20] P. Parker, "Residential solar photovoltaic market simulation: Japanese and Australian lessons for Canada," *Renewable and Sustainable Energy Reviews*, vol. 12 no. 7, pp. 1944–1958 2008.

[21] Pvresources. [Online]. Available: http://www.pvresources.com/en/pvpowerplants/top50pvroofs.php#notes [Accessed April 13, 2019].

[22] pvdatabase, "Building integrated and urban photovoltaic solar projects and products," 2006. [Online]. Available: http://www.pvdatabase.org/index.php [Accessed April 15, 2019].

[23] "Community-scale PV: Real examples of PV based housing and public developments." [Online]. Available: http://www.pvdatabase.org/pdf/VillaGartenTiaraCourt.pdf [Accessed April 20, 2019].

[24] Japan Sustainable Building Database, "Itoman City Hall," 2008. [Online]. Available: http://www.ibec.or.jp/jsbd/H/tech.htm [Accessed April 22, 2019].

[25] Wikipedia, "JR Kanazawa Station East entrance." *Wikipedia*. Retrieved June 2021 from https://upload.wikimedia.org/wikipedia/commons/3/3a/Kanazawa_eki.jpg.

[26] Monami, 2005, *Wikipedia*. Retrieved June 2021 from https://en.wikipedia.org/wiki/Solar_Ark#/media/File:Solar_Ark.jpg.

[27] Inhabitat, "SOLAR ARK: World's Most Stunning Solar Building," 2008. [Online]. Available: https://inhabitat.com/solar-ark-worlds-most-stunning-solar-building/ [Accessed April 22, 2019].

[28] TFOT. [Online]. Available: https://thefutureofthings.com/5823-solar-ark/ [Accessed April 22, 2019].

[29] PRLOG, "Global Solar Photovoltaic Market Analysis and Forecasts to 2020," [Online]. Available: https://www.prlog.org/10198293-global-solar-photovoltaic-market-analysis-and-forecasts-to-2020.html [Accessed April 13, 2019].

[30] Fraunhofer, "Recent trends about photovoltaics in Germany," Fraunhofer Institute for Solar Energy Systems ISE Germany, 2016.

[31] H.-M. Henning and A. Palzer, "100% Erneuerbare Energien für Strom und Wärme in Deutschland," Tech. Rep., Fraunhofer Institute for Solar Energy Systems ISE, 2012. https://www.ise.fraunhofer.de/de/veroeffentlichungen/ veroeffentlichungen-pdf-dateien/studien-und-konzeptpapiere/ studie-100-erneuerbare-energien-in-deutschland.pdf.

[32] EE-Strom, Interaktion; Wärme; Verkehr, "Studie im Auftrag des Bundesmi-nisterium für Wirtschaft und Energie," Projektleitung Fraunhofer-Institut für Windenergie und Energiesystemtechnik (IWES), 2015.

[33] B. P. Conference, 2017.

[34] SMA Solar Technology AG, "System manager for solar power stations." [Online]. Available: http://files.sma.de/dl/4988/REFWALDPOLENZ-AEN100411-web.pdf [Accessed April 13, 2019].

[35] Institute of Applied Sustainability to the Built Environment (ISAAC), "SUPSI." [Online]. Available: http://www.bipv.ch/index.php/en/administration-s-en/item/590-montcenis [Accessed April 13, 2019].

[36] JUWI Group, 2008, *Wikipedia*. Retrieved June 2021 from https://upload.wikimedia.org/wikipedia/commons/f/fd/Juwi_PV_Field.jpg.

[37] M. Cwyll, R. Michalczyk, N. Grzegorzewska and A. Garbacz, "Predicting performance of aluminum–glass composite facade systems based on mechanical properties of the connection," *Periodica Polytechnica Civil Engineering*, vol. 62, issue 1, pp. 259–266, 2018.

[38] N. Sasikumar and P. Jayasubramaniam, 2013, "Solar energy system in India," *IOSR Journal of Business and Management*, vol. 7 no. 1, pp. 61–68.

[39] N. Sethi, *Times of India*, 2009. [Online]. Available: https://timesofindia.indiatimes.com/india/India-targets-1000mw-solar-power-in-2013/articleshow/5240907.cms [Accessed May 2, 2019].

[40] T. P. Solar, "Tata power solar commissions world's largest rooftop solar for RSSB- EES," 2015.

[41] A. Bassi, *Hindustan Times*, 2016. [Online]. Available: https://www.hindustantimes.com/punjab/world-s-biggest-rooftop-solar-power-plant-inaugurated-at-radha-soami-dera-in-beas-by-cm-badal/story-VgCHxP3LbG5qCak5eFpNTP.html [Accessed April 30, 2019].

[42] The Hindu, *The Hindu-BusinessLine*, 2015. [Online]. Available: https://www.thehindubusinessline.com/companies/tata-power-solar-commissions-12-mw-rooftop-project-in-amritsar/article7971544.ece [Accessed April 30, 2019].

[43] Kuvampuri, 2020, *Wikipedia*. Retrieved June 2021 from https://en.wikipedia.org/wiki/Radha_Soami_Satsang_Beas#/media/File:Satsang_shed.jpg.

[44] S. Dutta *Times of India*, 2018. [Online]. Available: https://timesofindia.indiatimes.com/business/india-business/gail-sets-up-indias-2nd-largest-rooftop-solar-plant-in-up/articleshow/62327725.cms [Accessed April 30, 2019].

[45] MRPL, "Mangalore Refinery and Petrochemical Limited," 2018. [Online]. Available: https://mrpl.co.in/solar_project [Accessed April 30, 2019].

[46] *Business World*, 2018. [Online]. Available: http://www.businessworld.in/article/Largest-Single-Rooftop-Solar-Project-in-Rajasthan-Commissioned-By-Mahindra-Susten-/06-02-2018-139705/ [Accessed April 30, 2019].

[47] V. Salas and E. Olias, "Overview of the photovoltaic technology status and perspective in Spain," *Renewable and Sustainable Energy Reviews*, vol. 13, pp. 1049–1057, 2009.

[48] J. L. Prol and K. W. Steininger, "Photovoltaic self-consumption regulation in Spain: Profitability analysisand alternative regulation schemes," *Energy Policy,* vol. 108, pp. 742–754, 2017.

[49] M. Drif, P. J. Perez-Higueras, J. Aguilera, G. Almonacid, P. Gomez, J. d. l. Casa and J. D. Aguilar, "Univer Project. A grid connected photovoltaic system of 200 kW p at Jaén University. Overview and performance analysis," *Solar Energy Materials and Solar Cells*, vol. 91, no. 8, pp. 670–683, 2007.

[50] Syncronia, 2018. [Online]. Available: https://www.syncronia.com/en/magazine/facades/telefonicas-headquarters-multilayer-envelope [Accessed April 15, 2019].

[51] N. Gupta and G. N. Tiwari,"Energy matrices of building integrated photovoltaic thermal systems: Case study," *Journal of Architectural Engineering*, vol. 23, no. 4, pp. 05017006-1–05017006-14, 2017.

[52] A. Deo and G. N. Tiwari, "Performance analysis of 1.8 kWp rooftop photovoltaic system in India," in *Proceedings of 2nd International Conference on Green Energy and Technology (ICGET)*, Dhaka, September 2014, pp. 87–90.

[53] G. N. Tiwari, A. Deo, V. Singh and A. Tiwari, 2016, "Energy efficient passive building: A case study of SODHA BERS COMPLEX," *Foundations and Trends in Renewable Energy*, vol. 1, no. 3, pp 109–183. http://dx.doi.org/10.1561/2700000003.

[54] M. E. Watt, A. Johnson, M. Ellis and H. Outhred, "Life-cycle air emissions from PV power plants," *Progress in Photovoltaics Research and Applications,* vol. 6, no. 2, pp. 127–136, 1998.

[55] M. Sudan and G. N. Tiwari, "Energy matrices of the building by incorporating daylight concept for composite climate: An experimental study," *Journal of Renewable and Sustainable Energy,* vol. 6, p. 053122, 2014.

[56] G. N. Tiwari and R. Mishra, *Advanced Renewable Energy Sources*, United Kingdom: RSC, 2012, p. 463.

Appendix A

Parameters on Horizontal Surface for Sunshine Hours =10 for All Four Weather Type of Days for Different Indian Climates [1]

Type of day	Month▶ Parameters▼	January	February	March	April	May	June	July	August	September	October	November	December
New Delhi													
a	T_R	2.25	2.79	2.85	2.72	3.54	2.47	2.73	2.58	2.53	1.38	0.62	0.72
	α	0.07	0.10	0.17	0.23	0.16	0.28	0.37	0.41	0.29	0.47	0.59	0.54
	K_1	0.47	0.39	0.33	0.28	0.20	0.27	0.41	0.40	0.23	0.21	0.21	0.28
	K_2	−13.17	−6.25	5.61	38.32	65.04	31.86	−40.57	−55.08	39.92	32.77	30.62	9.73
b	T_R	2.28	2.78	2.89	3.15	5.44	4.72	5.58	5.43	3.23	4.56	0.19	1.83
	α	0.15	0.13	0.14	0.17	0.16	0.20	0.24	0.18	0.31	0.22	1.14	0.42
	K_1	0.51	0.54	0.49	0.46	0.45	0.45	0.53	0.39	0.37	0.42	0.35	0.40
	K_2	−21.77	−28.26	−9.22	−11.55	1.54	23.99	−51.61	9.46	14.07	−9.50	17.47	−0.07
c	T_R	5.88	6.36	6.11	7.77	9.20	10.54	7.13	7.97	5.51	5.01	4.93	3.23
	α	0.27	0.37	0.37	0.31	0.07	0.06	0.41	0.51	0.49	1.26	1.06	0.64
	K_1	0.39	0.36	0.33	0.35	0.56	0.48	0.47	0.35	0.39	0.36	0.31	0.43
	K_2	−14.73	−7.97	10.87	20.45	−56.00	−0.37	−52.27	47.70	35.64	−0.68	13.06	−7.04
d	T_R	7.47	8.97	10.77	11.18	13.69	12.47	8.21	8.58	9.40	7.24	4.30	4.02
	α	0.96	1.04	0.24	0.07	0.07	0.61	1.26	1.10	0.84	1.29	1.43	1.70
	K_1	0.35	0.30	0.43	0.49	0.48	0.46	0.43	0.43	0.41	0.36	0.31	0.38
	K_2	−25.89	−6.48	−36.46	−44.07	−42.58	−62.66	−56.75	−61.08	−27.09	3.90	20.10	−11.78
Bangalore													
a	T_R	3.36	3.27	3.63	5.05	4.24	4.32	5.18	4.75	4.10	2.28	1.66	1.65
	α	0.07	0.13	0.06	−0.06	0.10	0.19	0.10	0.18	0.13	0.33	0.35	0.36
	K_1	0.33	0.35	0.33	0.29	0.21	0.25	0.32	0.23	0.20	0.05	0.03	0.12
	K_2	−18.05	−22.11	−5.44	14.54	47.81	22.40	−26.04	10.14	38.54	107.04	103.64	47.70
b	T_R	3.24	5.25	6.21	5.72	5.90	7.35	4.12	5.27	4.83	2.43	1.89	3.68
	α	0.31	0.24	0.21	0.19	0.25	0.17	0.51	0.44	0.62	0.56	0.78	0.39
	K_1	0.50	0.45	0.48	0.50	0.41	0.50	0.46	0.50	0.33	0.26	0.37	0.41
	K_2	−60.12	−60.50	−80.04	−75.59	−28.55	−103.35	−90.54	−115.27	13.80	69.14	9.08	−33.76

c	T_R	3.70	4.51	7.74	5.83	4.95	4.39	5.68	2.67	6.64	4.71	5.68	2.02
	α	0.96	0.94	0.63	0.98	0.96	1.12	1.07	1.35	0.78	1.03	0.93	1.44
	K_1	0.46	0.57	0.36	0.50	0.53	0.58	0.50	0.55	0.48	0.43	0.36	0.43
	K_2	−63.02	−129.68	−20.76	−61.13	−103.14	−156.14	−108.34	−161.61	−52.93	−26.53	−15.95	−47.21
d	T_R	6.13	7.49	7.35	6.86	6.33	4.84	4.45	6.68	3.94	3.91	3.84	2.80
	α	1.61	1.31	1.41	1.48	1.59	2.00	2.32	1.69	2.16	2.00	2.04	2.58
	K_1	0.29	0.30	0.40	0.45	0.53	0.61	0.41	0.50	0.38	0.42	0.55	0.27
	K_2	36.80	83.73	−39.85	−72.22	−99.52	−213.29	−79.79	−146.94	−88.62	−125.35	−177.28	−12.29
Jodhpur													
a	T_R	1.26	1.33	1.59	2.82	3.72	3.87	3.25	3.39	3.20	2.26	1.56	1.54
	α	0.37	0.38	0.37	0.27	0.21	0.21	0.27	0.28	0.27	0.33	0.39	0.31
	K_1	0.22	0.14	0.18	0.21	0.20	0.13	0.10	0.17	0.26	0.24	0.23	0.26
	K_2	30.67	63.90	56.40	47.66	50.84	87.88	105.23	59.41	14.42	27.40	22.71	9.48
b	T_R	2.34	2.03	3.00	4.07	5.21	5.50	5.07	4.73	3.81	2.90	2.28	3.43
	α	0.46	0.55	0.42	0.31	0.23	0.28	0.37	0.40	0.35	0.38	0.46	0.24
	K_1	0.33	0.29	0.31	0.34	0.33	0.33	0.34	0.33	0.34	0.30	0.33	0.40
	K_2	12.89	43.13	42.22	23.50	31.22	33.40	35.81	29.57	8.71	24.12	12.35	−11.64
c	T_R	3.81	4.78	4.04	4.97	6.87	5.58	4.90	5.10	3.40	3.71	3.28	4.23
	α	0.93	1.32	0.98	0.64	0.61	0.67	1.02	0.88	0.97	2.05	1.31	1.06
	K_1	0.43	0.40	0.42	0.47	0.47	0.46	0.41	0.50	0.48	0.53	0.44	0.44
	K_2	−33.72	12.44	−19.11	−26.93	−44.76	−35.15	2.06	−60.42	−26.96	−62.06	−35.85	−32.84
d	T_R	2.25	5.20	7.09	9.33	8.01	3.52	9.62	3.17	1.63	7.67	1.71	1.94
	α	1.89	1.64	2.03	1.59	1.66	2.37	2.37	2.77	3.24	0.86	2.89	2.03
	K_1	0.44	0.46	0.42	0.44	0.43	0.28	0.52	0.44	0.44	0.52	0.36	0.39
	K_2	−19.31	−45.44	−89.92	−149.27	−117.01	60.69	−221.29	−87.34	−77.55	−26.47	−15.46	−14.88

Type of day	Month ▶ Parameters ▼	January	February	March	April	May	June	July	August	September	October	November	December
Mumbai													
a	T_R	1.95	1.80	2.88	3.95	5.40	3.20	3.31	4.25	4.22	3.16	2.97	3.27
	α	0.34	0.37	0.23	0.14	-0.02	0.16	0.61	0.33	0.15	0.30	0.23	0.18
	K_1	0.26	0.19	0.28	0.34	0.28	0.25	0.09	0.12	0.24	0.24	0.26	0.30
	K_2	19.77	53.96	27.13	-0.75	30.06	4.55	27.28	47.27	30.02	15.87	9.11	-4.81
b	T_R	2.96	2.68	3.57	4.98	6.25	6.08	7.74	6.70	4.78	3.93	3.40	4.21
	α	0.43	0.49	0.37	0.25	0.15	0.19	0.20	0.37	0.47	0.47	0.45	0.24
	K_1	0.35	0.31	0.35	0.40	0.42	0.44	0.31	0.39	0.41	0.36	0.34	0.37
	K_2	-0.14	24.17	11.73	-13.57	-13.69	-19.52	61.35	22.16	-14.71	5.99	0.60	-14.17
c	T_R	3.06	2.26	3.24	4.39	5.91	5.97	8.17	4.24	5.36	3.16	2.97	3.75
	α	1.14	1.18	1.10	1.00	0.79	0.86	0.62	1.26	0.98	1.13	1.10	0.91
	K_1	0.59	0.58	0.52	0.54	0.60	0.52	0.54	0.43	0.44	0.47	0.57	0.54
	K_2	-59.86	-47.12	-58.09	-78.37	-111.97	-81.79	-95.21	-34.40	-39.31	-28.02	-48.41	-52.45
d	T_R	3.38	7.42	4.45	2.30	4.71	4.71	6.41	7.40	7.46	3.22	5.13	3.05
	α	1.71	1.73	2.29	2.08	2.95	2.66	2.68	1.81	2.14	2.15	1.53	1.51
	K_1	0.52	0.56	0.50	0.35	0.41	0.38	0.32	0.47	0.34	0.42	0.57	0.53
	K_2	-59.78	-26.16	-82.34	63.52	-101.81	-87.19	-61.50	-108.37	-38.68	-25.89	-78.03	-40.51

Srinagar

a	T_R	1.45	5.37	3.31	4.25	5.41	3.63	5.77	6.45	4.06	2.61	4.03	0.72
	α	0.33	-0.36	-0.03	-0.03	-0.12	0.08	-0.09	-0.23	0.03	0.20	-0.37	0.53
	K_1	0.37	0.63	0.69	0.37	0.51	0.33	0.17	0.37	0.46	0.43	0.66	0.33
	K_2	-6.14	-82.86	-94.01	-10.95	-79.57	-13.73	68.06	-42.79	-60.27	-47.83	-37.00	-6.60
b	T_R	3.09	6.98	4.65	6.92	5.86	6.82	7.40	7.58	6.41	4.04	0.04	0.35
	α	0.38	-0.48	0.23	0.06	0.29	0.11	0.00	-0.13	-0.04	0.19	1.16	1.00
	K_1	0.39	0.83	0.59	0.42	0.32	0.63	0.48	0.38	0.48	0.52	0.37	0.41
	K_2	-23.08	-110.23	-107.74	-49.61	0.26	-167.86	-80.06	-13.91	-66.64	-62.52	-14.63	-12.20
c	T_R	2.35	6.59	6.31	7.57	8.69	8.00	9.72	8.23	7.36	5.02	1.86	0.76
	α	1.64	0.86	1.35	0.57	0.61	0.81	0.69	0.90	0.99	1.49	1.47	1.98
	K_1	0.41	0.42	0.48	0.54	0.50	0.39	0.56	0.49	0.44	0.52	0.41	0.31
	K_2	-37.87	-85.68	-180.45	-120.38	-146.97	-87.44	-228.91	-147.96	-62.10	-93.64	-40.07	-12.15
d	T_R	1.69	1.36	7.52	9.09	9.48	10.79	10.93	8.54	8.16	7.75	3.78	2.44
	α	2.63	2.97	1.87	1.35	1.13	1.56	3.08	1.71	3.15	1.70	1.74	2.04
	K_1	0.43	0.36	0.35	0.62	0.92	0.80	0.45	0.75	0.67	0.55	0.48	0.63
	K_2	-41.27	-44.68	-65.17	-254.24	-467.30	-421.63	-129.49	-356.92	-261.85	-119.53	-49.16	-64.02

REFERENCE

[1] G. N. Tiwari and R. K. Mishra, *Advanced Renewable Energy Sources*, Royal Society of Chemistry, 2011.

Appendix B

The Turbidity Factor (T_R) for Different Months [1]

Month → Region ↓	1	2	3	4	5	6	7	8	9	10	11	12
Mountain	1.8	1.9	2.1	2.2	2.4	2.7	2.7	2.7	2.5	2.1	1.9	1.8
Flatland	2.2	2.2	2.5	2.9	3.2	3.4	3.5	3.3	2.9	2.6	2.3	2.2
City	3.1	3.2	3.5	3.9	4.1	4.2	4.3	4.2	3.9	3.6	3.3	3.1

For cloudy conditions, the value of T_R will be more than 10.0.

REFERENCE

[1] G. N. Tiwari and R. K. Mishra, *Advanced Renewable Energy Sources*, London: Royal Society of Chemistry, 2011.

Appendix C

The hourly variation of ambient air temperature and solar intensity for different months for various stations in India have been given below in tables [1].

I. KOLKATA

	Kolkata: Monthly Solar Radiation (kWh/m²)												
Hour/ Months	6	7	8	9	10	11	12	13	14	15	16	17	18
Jan/Nov total	0.1	0.124	0.31	0.49	0.629	0.716	0.746	0.716	0.629	0.49	0.31	0.124	0.1
Jan/Nov diffuse	0.1	0.117	0.141	0.156	0.165	0.168	0.17	0.168	0.165	0.156	0.148	0.111	0.1
Feb/Oct total	0.1	0.165	0.374	0.563	0.707	0.798	0.829	0.798	0.707	0.563	0.374	0.165	0.1
Feb/Oct diffuse	0.1	0.122	0.147	0.161	0.168	0.172	0.174	0.172	0.168	0.161	0.147	0.122	0.1
March/Sep total	0.1	0.237	0.458	0.653	0.801	0.896	0.928	0.896	0.801	0.653	0.458	0.237	0.1
March/Sep diffuse	0.1	0.133	0.154	0.166	0.173	0.177	0.178	0.177	0.173	0.166	0.154	0.133	0.1
April/Aug total	0.115	0.314	0.534	0.727	0.875	0.967	0.997	0.967	0.875	0.727	0.536	0.314	0.115
April/Aug diffuse	0.108	0.142	0.159	0.169	0.178	0.179	0.18	0.179	0.176	0.169	0.159	0.142	0.108
May/July total	0.15	0.366	0.58	0.762	0.904	0.991	1.022	0.991	0.904	0.762	0.58	0.366	0.15
May/July diffuse	0.118	0.147	0.162	0.171	0.177	0.18	0.181	0.18	0.177	0.171	0.162	0.147	0.118
June total	0.173	0.385	0.594	0.771	0.91	0.994	1.025	0.994	0.91	0.771	0.594	0.385	0.173
June diffuse	0.123	0.148	0.163	0.171	0.177	0.18	0.181	0.18	0.177	0.171	0.163	0.148	0.12
Dec total	0.1	0.117	0.289	0.465	0.603	0.688	0.717	0.688	0.603	0.465	0.289	0.117	0.1
Dec diffuse	0.1	0.109	0.139	0.154	0.163	0.167	0.169	0.167	0.163	0.154	0.139	0.109	0.1

Kolkata: Monthly Hourly Ambient Temperature (°C)

Hour/Months	1	2	3	4	5	6	7	8	9	10	11	12	13	14	15	16	17	18	19	20	21	22	23	24
Jan	15.3	14.7	14.1	13.7	13.6	13.9	14.5	15.7	17.4	19.4	21.7	23.8	25.3	26.4	26.8	26.4	25.5	24.0	22.3	20.6	19.1	17.8	16.8	16.0
Feb	18.2	17.5	17.0	16.6	16.5	16.8	17.4	18.6	20.3	22.2	24.4	26.5	28.1	29.1	29.5	29.1	28.2	26.8	25.1	23.4	22.0	20.7	19.6	18.1
March	23.2	22.5	22.0	21.6	21.5	21.8	22.4	23.5	25.2	27.1	29.3	31.4	32.9	33.9	34.3	33.9	33.0	31.6	29.9	28.3	26.9	25.6	24.6	23.8
April	26.5	25.9	25.5	25.1	25.0	25.2	25.8	26.8	28.3	30.0	31.9	33.7	35.1	36.0	36.3	36.0	35.2	33.9	32.5	31.0	29.7	28.6	27.7	27.0
May	27.7	27.2	26.9	26.6	26.5	26.7	27.2	28.0	29.2	30.6	32.2	33.7	34.8	35.5	35.8	35.5	34.9	33.8	32.6	31.4	30.4	29.5	28.7	28.2
June	27.7	27.3	27.0	26.8	26.7	26.8	27.2	27.9	28.8	30.0	31.2	32.4	33.3	33.9	34.1	33.9	33.4	32.5	31.6	30.6	29.8	29.1	28.5	28.0
July	27.0	26.8	26.5	26.4	26.3	26.4	26.7	27.2	28.0	28.8	29.8	30.7	31.4	31.8	32.0	31.8	31.4	30.8	30.1	29.3	28.7	28.1	27.7	27.3
Aug	27.0	26.8	26.5	26.4	26.3	26.4	26.7	27.2	28.0	28.8	29.8	30.7	31.4	31.8	32.0	31.8	31.4	30.8	30.1	29.3	28.7	28.1	27.6	27.2
Sep	26.9	26.6	26.3	26.1	26.1	26.2	26.5	27.1	27.9	28.8	29.9	30.9	31.6	32.1	32.3	32.1	31.7	31.0	30.2	29.4	28.7	28.1	27.6	27.2
Oct	29.4	24.5	24.2	24.0	23.9	24.1	24.5	25.2	26.2	27.4	28.7	30.0	30.9	31.6	31.8	31.6	31.0	30.1	29.1	28.1	27.2	26.4	25.8	25.3
Nov	19.8	19.3	18.8	18.5	18.4	18.6	19.2	20.2	21.6	23.3	25.2	26.9	28.3	29.2	29.5	29.2	28.4	27.2	25.7	24.3	23.1	22.0	21.1	20.4
Dec	15.9	15.2	14.7	14.3	14.2	14.5	15.1	16.2	17.9	19.8	22.0	24.1	25.6	26.6	27.0	26.6	25.7	24.3	22.6	21.0	19.6	18.3	17.3	16.5

II. JODHPUR

Hour/Months	Jodhpur: Monthly Solar Radiation (kWh/m²)												
	6	7	8	9	10	11	12	13	14	15	16	17	18
Jan/Nov total	0.100	0.114	0.278	0.450	0.586	0.670	0.698	0.670	0.586	0.450	0.218	0.114	0.100
Jan/Nov diffuse	0.100	0.107	0.138	0.153	0.163	0.166	0.168	0.166	0.163	0.153	0.138	0.107	0.100
Feb/Oct total	0.100	0.148	0.347	0.530	0.670	0.757	0.788	0.757	0.670	0.530	0.347	0.148	0.100
Feb/Oct diffuse	0.100	0.118	0.145	0.159	0.166	0.170	0.172	0.170	0.166	0.159	0.145	0.118	0.100
March/Sep total	0.100	0.227	0.442	0.630	0.774	0.866	0.892	0.866	0.774	0.630	0.442	0.227	0.100
March/Sep diffuse	0.100	0.132	0.152	0.165	0.171	0.175	0.177	0.175	0.171	0.165	0.152	0.132	0.100
April/Aug total	0.119	0.316	0.533	0.717	0.861	0.951	0.980	0.951	0.861	0.717	0.533	0.316	0.119
April/Aug diffuse	0.109	0.142	0.159	0.169	0.175	0.179	0.180	0.179	0.175	0.169	0.159	0.142	0.109
May/July total	0.166	0.378	0.586	0.763	0.901	0.985	1.016	0.985	0.901	0.763	0.586	0.378	0.166
May/July diffuse	0.122	0.147	0.163	0.171	0.177	0.180	0.181	0.180	0.177	0.171	0.163	0.147	0.122
June total	0.196	0.402	0.605	0.777	0.911	0.993	1.023	0.993	0.911	0.777	0.605	0.402	0.196
June diffuse	0.128	0.149	0.164	0.172	0.178	0.180	0.181	0.180	0.178	0.172	0.164	0.149	0.128
Dec total	0.100	0.108	0.253	0.423	0.557	0.640	0.668	0.640	0.557	0.423	0.253	0.108	0.100
Dec diffuse	0.100	0.105	0.135	0.151	0.161	0.165	0.166	0.165	0.161	0.151	0.135	0.105	0.100

Jodhpur: Monthly Hourly Ambient Temperature (°C)

Hour/Months	1	2	3	4	5	6	7	8	9	10	11	12	13	14	15	16	17	18	19	20	21	22	23	24
Jan	11.5	10.7	10.1	9.7	9.5	9.8	10.6	11.9	13.9	16.1	18.7	21.1	22.9	24.1	24.6	24.1	23.1	21.4	19.5	17.5	15.8	14.3	13.1	12.2
Feb	14.1	13.3	12.6	12.2	12	12.3	13.1	14.5	16.6	19	21.7	24.2	26.2	27.4	27.9	27.4	26.3	24.6	22.5	20.4	18.7	17.1	15.8	14.9
March	19.2	18.4	17.7	17.3	17.1	17.4	18.2	19.7	21.8	24.2	27	29.6	31.5	32.8	33.3	32.8	31.7	29.9	27.8	25.7	23.9	22.3	21	20
April	24.5	23.7	23	22.6	22.4	22.7	23.5	24.9	27	29.4	31.1	34.6	36.6	37.8	38.3	37.8	36.7	35	32.9	30.8	29.1	27.5	26.2	25.3
May	29.2	28.4	27.9	27.4	27.3	27.6	28.3	29.6	31.4	33.6	36	38.3	40	41.2	41.6	41.2	40.2	38.6	36.7	34.9	33.3	31.9	30.7	29.9
June	30	29.4	29	28.6	28.5	28.7	29.3	30.4	31.9	33.6	35.6	37.4	38.8	39.8	40.1	39.8	38.7	37.7	36.2	34.6	33.4	32.2	31.3	30.6
July	27.7	27.4	27.1	26.9	26.8	26.9	27.3	27.9	28.8	29.8	31	32.1	32.9	33.5	33.7	33.5	33	32.3	31.4	30.5	29.7	29	28.5	28
Aug	26.2	25.8	25.5	25.3	25.2	25.4	25.8	26.5	27.5	28.7	30.1	31.4	32.3	33	33.2	33	32.4	31.5	30.5	29.4	28.6	27.8	27.1	26.6
Sep	25.5	24.9	24.5	24.2	24.1	24.3	24.8	25.8	27.2	28.8	30.6	32.3	33.5	34.4	34.7	33.4	33.6	32.5	31.1	29.7	28.6	27.5	26.6	26
Oct	21.7	20.9	20.2	19.8	19.6	19.9	20.7	22.2	24.3	26.7	29.4	32	33.9	35.2	35.7	35.2	34.1	32.3	30.2	28.1	26.4	24.8	23.5	22.5
Nov	16.2	15.3	14.6	14.1	13.9	14.2	15.1	16.7	19	21.6	24.6	27.4	29.5	30.9	31.4	30.9	29.6	27.7	25.4	23.2	21.2	19.5	18.1	17
Dec	12.8	12	11.3	10.9	10.7	11	11.8	13.3	15.3	17.7	20.5	23	24.9	26.2	26.7	26.2	25.1	23.3	21.3	19.2	17.4	15.8	14.5	13.6

Appendix C

III. CHENNAI

Chennai : Monthly Solar Radiation (kWh/m²)													
Hour/ Months	6	7	8	9	10	11	12	13	14	15	16	17	18
Jan/Nov total	0.100	0.178	0.39	0.582	0.728	0.82	0.852	0.82	0.728	0.582	0.39	0.178	0.1
Jan/Nov diffuse	0.100	0.124	0.148	0.162	0.169	0.173	0.175	0.173	0.169	0.162	0.148	0.124	0.1
Feb/Oct total	0.100	0.212	0.436	0.636	0.788	0.885	0.918	0.885	0.788	0.635	0.436	0.212	0.1
Feb/Oct diffuse	0.100	0.13	0.152	0.165	0.172	0.177	0.178	0.177	0.172	0.165	0.152	0.13	0.1
March/Sep Total	0.100	0.257	0.492	0.696	0.854	0.953	0.983	0.953	0.854	0.696	0.492	0.257	0.1
March/Sep diffuse	0.100	0.136	0.156	0.168	0.175	0.179	0.18	0.179	0.175	0.168	0.156	0.136	0.1
April/Aug total	0.107	0.301	0.536	0.736	0.893	0.988	1.023	0.988	0.893	0.736	0.535	0.301	0.107
April/Aug diffuse	0.105	0.141	0.159	0.169	0.177	0.18	0.181	0.18	0.177	0.169	0.159	0.141	0.105
May/July total	0.199	0.326	0.552	0.745	0.893	0.986	1.020	0.986	0.895	0.745	0.552	0.326	0.119
May/July diffuse	0.109	0.143	0.160	0.170	0.177	0.180	0.181	0.18	0.177	0.17	0.16	0.143	0.109
June total	0.125	0.335	0.554	0.742	0.888	0.978	1.01	0.979	0.888	0.742	0.554	0.335	0.125
June diffuse	0.111	0.144	0.160	0.17	0.177	0.180	0.181	0.180	0.177	0.170	0.16	0.144	0.111
Dec total	0.1	0.166	0.374	0.563	0.706	0.797	0.828	0.797	0.706	0.563	0.374	0.166	0.1
Dec diffuse	0.1	0.122	0.147	0.161	0.168	0.172	0.174	0.172	0.168	0.161	0.147	0.122	0.1

Chennai: Monthly Hourly Ambient Temperature (°C)

Hour/Months	1	2	3	4	5	6	7	8	9	10	11	12	13	14	15	16	17	18	19	20	21	22	23	24
Jan	21.4	21	20.6	20.4	20.3	20.5	20.9	21.7	22.8	24	25.5	26.8	27.9	28.5	28.8	28.5	27.9	27	25.9	24.8	23.9	23	22.3	21.8
Feb	22.3	21.9	21.5	21.2	21.1	21.3	21.8	22.6	23.9	25.3	26.9	28.4	29.6	30.3	30.6	30.3	29.6	28.6	27.4	26.1	25.1	24.1	23.4	22.8
March	24.3	23.9	23.5	23.2	23.1	23.3	23.8	24.6	25.9	27.3	29	30.5	31.6	32.4	32.7	32.4	31.7	30.7	29.4	28.2	27.1	26.2	25.4	24.8
April	27.2	26.7	26.4	26.1	26	26.2	26.6	27.4	28.6	29.9	31.4	32.9	33.9	34.6	34.9	34.6	34	33	31.9	30.7	29.7	28.8	28.1	27.6
May	29.1	28.6	28.2	27.9	27.8	28	28.5	29.4	30.6	32.1	33.8	35.3	36.5	37.3	37.6	37.3	36.6	35.5	34.3	33	31.9	30.9	30.2	29.6
June	28.9	28.4	28	27.7	27.6	27.8	28.3	29.2	30.4	31.9	33.5	35.1	36.2	37	37.3	37	36.3	35.3	34	32.7	31.7	30.7	29.9	29.3
July	27.5	27	26.7	26.4	26.3	26.5	26.9	27.7	28.9	30.2	31.7	33.2	34.2	34.9	35.2	34.9	34.3	33.3	32.2	31	30	29.1	28.4	27.9
Aug	26.9	26.5	26.1	25.9	25.8	26	26.4	27.2	28.3	29.6	31.1	32.5	33.5	34.2	34.5	34.2	33.6	32.7	31.5	30.4	29.5	28.6	27.9	27.4
Sep	26.5	26.1	25.7	25.5	25.4	25.6	26	26.8	27.9	29.1	30.6	31.9	33	33.6	33.9	33.6	33.1	32.1	31	29.9	29	28.1	25.2	26.9
Oct	25.4	25	24.7	24.5	24.4	24.5	24.9	25.6	26.5	27.7	28.9	30.1	31	31.6	31.8	31.6	31.1	30.2	29.3	28.3	27.5	26.8	25.2	25.7
Nov	23.4	23	22.8	22.6	22.5	22.6	23	23.6	24.4	25.4	26.6	27.7	28.5	29	29.2	29	28.5	27.8	26.9	26.1	25.3	24.6	24.1	23.7
Dec	21.9	21.6	21.3	21.1	21	21.1	21.5	22.2	23.1	24.2	25.4	26.5	27.4	28	28.2	28	27.5	26.7	25.8	24.8	24	23.3	22.7	22.3

Appendix C

IV. NEW DELHI

Hour/Months	New Delhi: Monthly Solar Radiation (kWh/m^2)												
	6	7	8	9	10	11	12	13	14	15	16	17	18
Jan/Nov total	0.1	0.108	0.256	0.424	0.558	0.641	0.669	0.641	0.558	0.424	0.256	0.108	0.1
Jan/Nov diffuse	0.1	0.105	0.136	0.151	0.161	0.165	0.166	0.165	0.161	0.151	0.136	0.105	0.1
Feb/Oct total	0.1	0.138	0.329	0.508	0.646	0.732	0.761	0.731	0.646	0.508	0.329	0.138	0.1
Feb/Oct diffuse	0.1	0.115	0.143	0.157	0.165	0.169	0.171	0.169	0.165	0.157	0.143	0.115	0.1
March/Sep total	0.1	0.222	0.431	0.616	0.756	0.846	0.877	0.846	0.756	0.616	0.431	0.222	0.1
March/Sep diffuse	0.1	0.131	0.151	0.164	0.17	0.174	0.176	0.174	0.17	0.164	0.151	0.131	0.1
April/Aug total	0.121	0.317	0.529	0.71	0.851	0.939	0.693	0.939	0.851	0.71	0.529	0.317	0.121
April/Aug diffuse	0.11	0.142	0.159	0.168	0.175	0.178	0.179	0.178	0.175	0.168	0.159	0.142	0.11
May/July total	0.177	0.382	0.589	0.762	0.898	0.98	1.011	0.98	0.898	0.762	0.589	0.384	0.177
May/July diffuse	0.124	0.148	0.163	0.171	0.177	0.18	0.181	0.18	0.177	0.171	0.163	0.118	0.124
June total	0.207	0.411	0.61	0.779	0.911	0.99	1.02	0.99	0.911	0.779	0.61	0.411	0.207
June diffuse	0.13	0.15	0.164	0.172	0.177	0.18	0.181	0.18	0.177	0.172	0.164	0.158	0.13
Dec total	0.1	0.103	0.232	0.397	0.528	0.61	0.637	0.61	0.528	0.397	0.232	0.103	0.1
Dec diffuse	0.1	0.103	0.133	0.149	0.159	0.164	0.165	0.164	0.159	0.149	0.133	0.103	0.1

New Delhi: Monthly Hourly Ambient Temperature (°C)

Hour/Months	1	2	3	4	5	6	7	8	9	10	11	12	13	14	15	16	17	18	19	20	21	22	23	24
Jan	9.1	8.4	7.9	7.4	7.3	7.6	8.3	9.5	11.4	13.5	15.8	18.1	19.8	20.9	21.3	20.9	19.9	18.4	16.5	14.7	13.2	11.8	10.7	9.8
Feb	11.9	11.2	10.6	10.2	10.1	10.4	11	12.3	14	16	18.3	20.5	22.1	23.2	23.6	23.2	22.2	20.8	19	17.3	15.8	14.4	13.3	12.5
March	17.1	16.3	15.7	15.3	15.1	15.4	16.2	17.5	19.5	21.7	24.3	26.7	28.5	29.7	30.2	29.7	28.7	27	25.1	23.1	21.4	19.9	18.7	17.8
April	23	22.2	21.6	21.2	21	21.3	22.1	23.4	25.4	27.7	30.3	32.7	34.5	35.7	36.2	35.7	34.7	33	31	29.1	27.4	25.9	24.6	23.7
May	28.4	27.7	27.2	26.7	26.6	26.9	27.6	28.8	30.6	32.7	35.1	37.3	39	40.1	40.5	40.1	39.1	37.6	35.8	34	32.4	31	29.9	29.1
June	30.2	29.6	29.1	28.8	28.7	28.9	29.5	30.5	31.9	33.6	35.5	37.3	38.7	39.6	39.9	39.6	38.8	37.5	36.1	34.6	33.4	32.3	31.4	30.7
July	28.3	27.8	27.5	27.3	27.2	27.4	27.8	28.5	29.5	30.8	32.1	33.4	34.4	35.1	35.3	35.1	34.5	33.6	32.5	31.5	30.6	29.8	29.1	28.7
Aug	27.1	26.7	26.4	26.2	26.1	26.3	26.6	27.3	28.3	29.4	30.7	32	32.9	33.5	33.7	33.5	32.9	32.1	31.1	30.1	29.3	28.5	27.9	27.5
Sep	25.8	25.4	25	24.7	24.6	24.8	25.3	26.1	27.4	28.8	30.4	31.9	33.1	33.8	34.1	33.8	33.1	32.1	30.9	29.6	28.6	27.6	26.9	26.3
Oct	20.6	19.9	19.3	18.8	18.7	19	19.7	21	22.9	25	27.5	29.8	31.5	32.7	33.1	32.7	31.7	30.1	28.2	26.3	24.7	23.3	22.2	21.3
Nov	14	13.2	12.5	12	11.8	12.1	13	14.5	16.7	19.2	22.1	24.8	26.8	28.2	28.7	28.2	27	25.2	23	20.8	18.9	17.2	15.9	14.8
Dec	11.3	10	9	8.3	8	8.5	9.8	12.1	15.4	19.2	23.5	27.6	30.6	32.6	33.4	32.6	30.9	28.1	24.8	21.5	18.7	16.1	14.1	12.4

Appendix C

Surface	Hour/Months	6	7	8	9	10	11	12	13	14	15	16	17	18
Horizontal	Jan total	0.1	0.108	0.256	0.424	0.558	0.641	0.669	0.641	0.558	0.424	0.256	0.108	0.1
	Jan diffuse	0.1	0.105	0.136	0.151	0.161	0.165	0.166	0.165	0.161	0.151	0.136	0.105	0.1
South	Jan total	0.06	0.096	0.363	0.539	0.652	0.717	0.738	0.717	0.652	0.539	0.363	0.096	0.06
	Jan diffuse	0.05	0.053	0.068	0.075	0.08	0.083	0.083	0.083	0.08	0.075	0.068	0.053	0.005
Horizontal	June total	0.207	0.411	0.61	0.779	0.911	0.99	1.02	0.99	0.911	0.779	0.61	0.411	0.207
	June diffuse	0.13	0.15	0.164	0.172	0.177	0.18	0.181	0.18	0.177	0.172	0.164	0.158	0.13
South	June total	0.065	0.075	0.082	0.086	0.211	0.258	0.274	0.258	0.211	0.086	0.082	0.075	0.065
	June diffuse	0.065	0.075	0.082	0.086	0.089	0.09	0.09	0.09	0.089	0.086	0.082	0.075	0.065

V. SRINAGAR

Srinagar: Monthly Solar Radiation (kWh/m²)													
Hour/Months	6	7	8	9	10	11	12	13	14	15	16	17	18
Jan/Nov total	0.1	0.1	0.207	0.363	0.486	0.566	0.593	0.566	0.486	0.363	0.207	0.1	0.1
Jan/Nov diffuse	0.1	0.1	0.13	0.146	0.156	0.161	0.163	0.161	0.156	0.146	0.13	0.1	0.1
Feb/Oct total	0.1	0.12	0.287	0.453	0.584	0.665	0.692	0.665	0.584	0.453	0.287	0.12	0.1
Feb/Oct diffuse	0.1	0.11	0.139	0.153	0.162	0.166	0.167	0.166	0.162	0.153	0.139	0.11	0.1
March/Sep total	0.1	0.207	0.401	0.576	0.708	0.793	0.821	0.793	0.708	0.576	0.401	0.207	0.1
March/Sep diffuse	0.1	0.13	0.149	0.162	0.168	0.172	0.173	0.172	0.168	0.162	0.149	0.13	0.1
April/Aug total	0.127	0.318	0.518	0.689	0.821	0.904	0.933	0.904	0.821	0.689	0.518	0.318	0.127
April/Aug diffuse	0.112	0.142	0.158	0.167	0.173	0.177	0.178	0.177	0.173	0.167	0.158	0.142	0.112
May/July total	0.202	0.398	0.592	0.755	0.883	0.963	0.987	0.963	0.883	0.755	0.592	0.398	0.202
May/July diffuse	0.129	0.149	0.163	0.17	0.177	0.179	0.18	0.179	0.177	0.17	0.163	0.149	0.129
June total	0.236	0.432	0.619	0.779	0.903	0.979	1.006	0.979	0.903	0.779	0.619	0.432	0.236
June diffuse	0.133	0.152	0.164	0.172	0.177	0.188	0.181	0.18	1.77	0.172	0.164	0.152	0.133
Dec total	0.100	0.1	1.82	0.332	0.454	0.532	0.558	0.532	0.454	0.332	0.182	0.1	0.1
Dec diffuse	0.100	0.1	0.125	0.144	0.153	0.159	0.161	0.159	0..153	0.144	0.125	0.1	0.1

Srinagar: Monthly Hourly Ambient Temperature (°C)

Hour/Months	1	2	3	4	5	6	7	8	9	10	11	12	13	14	15	16	17	18	19	20	21	22	23	24
Jan	-1.4	-1.8	-2	-2.2	-2.3	-2.2	-1.8	-1.2	-0.4	0.6	1.8	2.9	3.7	4.2	4.4	4.2	3.7	3	2.1	1.3	0.5	-0.2	-0.7	-1.1
Feb	0.3	-0.1	-0.5	-0.7	-0.8	-0.6	-0.2	0.6	1.7	3	4.5	5.9	6.9	7.6	7.9	7.6	7	6.1	4.9	3.8	2.9	2	1.3	0.8
March	4.8	4.3	3.9	3.6	3.5	3.7	4.2	5.1	6.4	7.9	9.5	11.1	12.3	13.1	13.4	13.1	12.4	11.3	10	8.7	7.7	6.7	5.9	5.3
April	8.9	8.4	7.9	7.5	7.4	7.6	8.2	9.3	10.9	12.6	14.7	16.6	18	18.9	19.3	18.9	18.1	16.8	15.3	13.7	12.4	11.2	10.3	9.5
May	12.9	12.3	11.7	11.3	11.2	11.5	12.1	13.3	15.1	17.1	19.4	21.5	23.1	24.2	24.6	24.2	23.3	21.8	20	18.3	16.8	15.5	14.4	13.6
June	16.3	15.6	15	14.5	14.4	14.7	15.4	16.7	18.6	20.8	23.3	25.6	27.4	28.6	29	28.6	27.5	25.9	24	22.1	20.5	19.1	17.9	17
July	20	19.4	18.9	18.5	18.4	18.6	19.3	20.4	22	23.9	26	27.9	29.4	30.4	30.8	30.4	29.6	28.2	26.6	25	23.6	22.4	21.4	20.6
Aug	19.5	18.9	18.4	18	17.9	18.1	18.7	19.8	21.4	23.2	25.2	27.1	28.6	29.5	29.9	29.5	28.7	27.4	25.8	24.3	22.9	21.7	20.8	20.1
Sep	14.7	13.9	13.3	12.9	12.7	13	13.8	15.2	17.2	19.6	22.2	24.7	26.6	27.8	28.3	27.8	26.7	25	23	21	19.3	17.7	16.4	15.5
Oct	7.9	7.1	6.4	5.9	5.7	6	6.9	8.4	10.6	13.1	16	18.7	20.7	22.1	22.6	22.1	20.9	19.1	16.9	14.7	12.8	11.1	9.8	8.7
Nov	1.9	-1	-1.4	-1.7	-1.8	-1.6	-1.1	-0.1	1.3	2.9	4.7	6.4	7.6	8.5	8.8	8.5	7.7	6.6	5.2	3.8	2.7	1.6	0.7	0.1
Dec	-0.4	-1	-1.4	-1.7	-1.8	-1.6	-1.1	-0.1	1.3	2.9	4.7	6.4	7.6	8.5	8.8	8.5	7.7	6.6	5.2	3.8	2.7	1.6	0.7	0.1

REFERENCE

[1] G. N. Tiwari, *Solar Energy: Fundamentals, Design, Modelling and Applications*, New Delhi: Narosa, 2002.

Appendix D

ESTIMATION OF SOLAR RADIATION AT INCLINED ANGLE

```
% IMD Pune data for January Type a, New Delhi
It = [85.80
264.25
411.27
517.48
579.85
576.18
520.45
400.43
256.38
60.29];0

Id = [35.81
85.52
104.56
121.08
116.58
122.58
112.07
102.08
76.97
27.16];

Ib = It - Id;

phi = degtorad(26.3); % phi = latitude
beta = phi;
gamma = 0;
n = 1; % n = day of the year
delta = degtorad(23.45*sin((360/365)*(284 + n)));
ST = [8 9 10 11 12 13 14 15 16 17];
omega = degtorad((ST-12)*15);

p = 0.2;
num= ((cos(phi)*cos(beta) + sin(phi)*sin(beta)*cos(gamma))*cos(d
elta)*cos(omega) + cos(delta)*sin(omega)*sin(beta)*sin(gamma) +
sin(delta)*(sin(phi)*cos(beta) - cos(phi)*sin(beta)*cos(gamma)));
%cos(thetai)
denom= cos(phi)*cos(delta)*cos(omega) + sin(delta)*sin(phi); %cos(thetaz)
Rb = num./denom;
Rd = (1+cos(beta))/2;
Rr = (1-cos(beta))/2;

It_new = Ib.*Rb' + Id*Rd + p*Rr*(Ib+Id);
```

Appendix E [1]

TABLE E.1
Properties of Air at Atmospheric Pressure

T (K)	P (kg/m³)	C_p (kJ/kgK)	μ (kg/m-s) × 10⁻⁵	ν (m²/s) × 10⁻⁶	K (W/m²K) × 10⁻³	α (m²/s) × 10⁻⁵	Pr
100	3.6010	1.0259	0.6924	1.923	9.239	0.2501	0.770
150	2.3675	1.0092	1.0283	4.343	13.726	0.5745	0.753
200	1.7684	1.0054	1.3289	7.490	18.074	1.017	0.739
250	1.4128	1.0046	1.488	9.49	22.26	1.3161	0.722
300	1.1774	1.0050	1.983	15.68	26.22	2.216	0.708
350	0.9980	1.0083	2.075	20.76	30.00	2.983	0.697
400	0.8826	1.0134	2.286	25.90	33.62	3.760	0.689

The value of μ, K, C_p and Pr are not strongly pressure-dependent and may be used over a fairly wide range of pressures.

TABLE E.2
Properties of Water (Saturated Liquid)

Temperature		C_p (kJ/kgK)	ρ (kg/m³)	μ_k (kg/ms)	K (W/mK)	Pr	$\dfrac{g\ \beta P^2 C_p}{\mu_k}$ (1/m³K)
°F	°C						
32	0.00	4.225	999.8	1.79×10^3	0.566	13.25	1.91×10^9
40	4.44	4.208	999.8	1.55	0.575	11.35	6.34×10^9
50	10.00	4.195	999.2	1.31	0.585	9.40	1.08×10^{10}
60	15.56	4.186	998.6	1.12	0.595	7.88	1.46×10^{10}
70	21.11	4.179	997.4	9.8×10^4	0.604	6.78	1.46×10^{10}
80	26.67	4.179	995.8	8.6	0.614	5.85	1.91×10^{10}
90	32.22	4.174	994.9	7.65	0.623	5.12	2.48×10^{10}
100	37.78	4.174	993.0	6.82	0.630	4.53	3.3×10^{10}
110	43.33	4.174	990.6	6.16	0.637	4.04	4.19×10^{10}
120	48.89	4.174	988.8	5.62	0.644	3.64	4.89×10^{10}
130	54.44	4.179	985.7	5.13	0.649	3.30	5.66×10^{10}
140	60.00	4.179	983.3	4.71	0.654	3.01	6.48×10^{10}
150	65.55	4.183	980.3	4.3	0.659	2.73	7.62×10^{10}
160	71.11	4.186	977.3	4.01	0.665	2.53	8.84×10^{10}
170	76.67	4.191	973.7	3.72	0.668	2.33	9.85×10^{10}
180	82.22	4.195	970.2	3.47	0.673	2.16	1.09×10^{10}
190	87.78	4.199	966.7	3.27	0.675	2.03	
200	93.33	4.204	963.2	3.06	0.678	1.90	
210	104.40	4.216	955.1	2.67	0.684	1.66	

Appendix E

TABLE E.3
Properties of Metals

Metal		Properties at 20°C			
		P (kg/m³)	C_p (kJ/kgK)	K (W/mK)	α (m²/s × 10⁻⁵)
Aluminum	Pure	2707	0.896	204	8.418
	Al-Si (Silumin, copper bearing) 86% Al, 1% Cu	2659	0.867	137	5.933
Lead	Pure	11,400	0.1298	34.87	7.311
Iron	Pure	7897	0.452	73	2.034
	Steel (Carbon steel)	7753	0.486	63	0.970
Copper	Pure	8954	0.3831	386	11.234
	Aluminum bronze (95% Cu, 5% Al)	8666	0.410	383	2.330
Bronze	75% Cu, 25% Sn	8666	0.343	326	0.859
Red Brass	85% Cu, 9% Sn 6% Zn	8714	0.385	61	1.804
Brass	70% Cu, 30% Zn	8600	0.877	85	3.412
German Silver	62% Cu, 15% Ni, 22% Zn	8618	0.394	24.9	0.733
Constantan	60% Cu, 40% Ni	8922	0.410	22.7	0.612
Magnesium	Pure	1746	1.013	171	9.708
Nickel	Pure	8906	0.4459	90	2.266
Silver	Purest	10,524	0.2340	419	17.004
	Pure (99.9%)	10,524	0.2340	407	16.563
Tin	Pure	7304	0.2265	64	3.884
Tungsten	Pure	19,350	0.1344	163	6.271
Zinc	Pure	7144	0.3843	112.2	4.106

TABLE E.4
Properties of Non-Metals

Material	Temperature (°C)	K (W/mK)	ρ (kg/m³)	C (kJ/kgK)	α (m²/s) × 10⁻⁷
Asbestos	50	0.08	470	-	-
Building brick	20	0.69	1600	0.84	5.2
Common face	-	1.32	2000	-	-
Concrete, Cinder	23	0.76	-	-	-
Stone 1-2-4 mix	20	1.37	1900–2300	0.88	8.2–6.8
Glass, window	20	0.78 (avg)	2700	0.84	3.4
Borosilicate	30–75	1.09	2200	-	-
Plaster, Gypsum	20	0.48	1440	0.84	4.0
Granite	-	1.73–3.98	2640	0.82	8–18
Limestone	100–300	1.26–1.33	2500	0.90	5.6–5.9
Marble	-	2.07–2.94	2500–2700	0.80	10–13.6
Sandstone	40	1.83	2160–2300	0.71	11.2–11.9
Fir	23	0.11	420	2.72	0.96
Maple or Oak	30	0.166	540	2.4	1.28
Yellow Pine	23	0.147	640	2.8	0.82
Cord, board	30	0.043	160	1.88	2–5.3
Cork, regranulated	32	0.045	45–120	1.88	2–5.3
Ground	32	0.043	150	-	-
Sawdust	23	0.059	-	-	-
Wood shaving	23	0.059	-	-	-

TABLE E.5
Physical Properties of Some Other Materials

S. No.	Material	Density (kg/m³)	Thermal Conductivity (W/mK)	Specific Heat (J/kgK)
1.	Air	1.117	40.026	1006
2.	Alumina	3800	29.0	800
3.	Aluminum	41–45	211	0.946
4.	Asphalt	1700	0.50	1000
5.	Brick	1700	0.84	800
6.	Carbon dioxide	1.979	0.145	871
7.	Cement	1700	0.80	670
8.	Clay	1458	11.28	879
9.	Concrete	2400	1.279	1130
10.	Copper	8795	385	-
11.	Cork	240	0.04	2050
12.	Cotton Wool	1522	-	1335
13.	Fiberboard	300	0.057	1000
14.	Glass-crown	2600	1.0	670
15.	Glass-window	2350	0.816	712
16.	Glass-wool	50	0.042	670
17.	Ice	920	2.21	1930
18.	Iron	7870	80	106
19.	Limestone	2180	1.5	-
20.	Mudphuska	-	-	-
21.	Oxygen	1.301	0.027	920
22.	Plasterboard	950	0.16	840
23.	Polyesterene-expanded	25	0.033	1380
24.	P.V.C. —rigid foam	25–80	0.035–0.041	-
25.	P.V.C. —rigid sheet	1350	0.16	-
26.	Sawdust	188	0.57	-
27.	Thermocole	22	0.03	-
28.	Timber	600	0.14	1210
29.	Turpentine	870	0.136	1760
30.	Water (H_2O)	998	0.591	4190
31.	Seawater	1025	-	3900
32.	Water vapor	0.586	0.025	2060
33.	Wood wool	500	0.10	1000

TABLE E.6
Absorptivity of Various Surfaces for the Sun's Ray

Surface	Absorptivity	Surface	Absorptivity
White paint	0.12–0.26	**Walls**	
Whitewash/glossy white	0.21	White/yellow brick tiles	0.30
Bright aluminum	0.30	White stone	0.40
Flat white	0.25	Cream brick tile	0.50
Yellow	0.48	Burl brick tile	0.60
Bronze	0.50	Concrete/red brick tile	0.70
Silver	0.52	Red sand line brick	0.72
Dark aluminum	0.63	White sandstone	0.76
Bright red	0.65	Stone rubble	0.80
Brown	0.70	Blue brick tile	0.88
Light green	0.73	**Surroundings**	
Medium red	0.74	Sea/lake water	0.29
Medium green	0.85	Snow	0.30
Dark green	0.95	Grass	0.80
Blue/black	0.97	Light-colored grass	0.55
Roof		Sand gray	0.82
Asphalt	0.89	Rock	0.84
White asbestos cement	0.59	Green leaves	0.85
Copper sheeting	0.64	Earth (black ploughed field)	0.92
Uncolored roofing tile	0.67	White leaves	0.20
Red roofing tiles	0.72	Yellow leaves	0.58
Galvanized iron, clean	0.77	Aluminum foil	0.39
Brown roofing tile	0.87	Unpainted wood	0.60
Galvanized iron, dirty	0.89		
Black roofing tile	0.92		
Metals			
Polished aluminum/copper	0.26		
New galvanized iron	0.66		
Old galvanized iron	0.89		
Polished iron	0.45		
Oxidized rusty iron	0.38		

Appendix E

TABLE E.7
Theoretical Model for the Thermal Conductivity of Nanofluids Found in Literature

Models	Thermal Conductivity (k) (W/mK)	Physical Models
Maxwell-Eucken [2]	$$k_{nf} = k_{bf} \times \frac{\left[k_p + 2k_{bf} + 2\varphi_p \left(k_p - k_{bf} \right) \right]}{\left[k_p + 2k_{bf} - \varphi_p \left(k_p - k_{bf} \right) \right]}$$ **Remarks:** Spherical particles	Based on the conduction solution through a stationary random suspension of spheres
Bruggeman [3]	$$k_{nf} = \frac{\left[(3\varphi - 1) k_p + (2 - 3\varphi) k_{bf} + k_{bf} \sqrt{\Delta} \right]}{4}$$ $$\Delta = \left[(3\varphi_p - 1) \frac{k_p}{k_{bf}} + (2 - 3\varphi) \right]^2 + 8 \frac{k_p}{k_{bf}}$$ **Remarks:** – Applicable to high-volume fraction of spherical particles – Suspension with spherical inclusions	Based on the differential effective medium (DEM) theory to estimate the effective thermal conductivity of composites at high particle concentrations
Hamilton-Crosser [4]	$$k_{nf} = k_{bf} \frac{\left[k_p + (n-1) k_{bf} + \varphi (n-1) \left(k_p - k_{bf} \right) \right]}{\left[k_p + (n-1) k_{bf} - \varphi \left(k_p - k_{bf} \right) \right]}$$ **Remarks:** Spherical and non-spherical particles, $n = 3$ (spheres), $n = 6$ (cylinders)	Based on the effective thermal conductivity of a two-component mixture when the ratio of thermal conductivity is more than 100
Wasp [5]	**Remarks:** Special case of Hamilton and Crosser's model with $n = 3$	Based on effective thermal conductivity of a two-component mixture

TABLE E.8
Correlation Developed for Thermal Conductivity of Nanofluids

Khanafer and Vafai [6]

$$k_{nf} = k_{bf} \times \left(1 + 1.0112\varphi_p + 2.4375\varphi_p \times \left(\frac{47}{d_p(\text{nm})} \right) - 0.0248\varphi_p \left(\frac{k_p}{0.613} \right) \right)$$

Al_2O_3-H_2O
CuO-H_2O

Remarks: At ambient temperature

$$k_{nf} = k_{bf} \times \left[\begin{array}{l} 0.9843 + 0.398\varphi_p^{0.7383} \left(\dfrac{1}{d_p(\text{nm})} \right)^{0.2246} \left(\dfrac{\mu_{nf}(T)}{\mu_{bf}(T)} \right)^{0.0235} \\ - (3.9517) \times \left(\dfrac{\varphi_p}{T} \right) + (34.034) \times \left(\dfrac{\varphi_p^2}{T^3} \right) + (32.509) \times \left(\dfrac{\varphi_p}{T^2} \right) \end{array} \right]$$

$$\mu_{bf} = 2.414 \times 10^{-5} \times 10^{247.8/(T-140)}$$

$0 \leq \varphi_p \leq 10 \ \%$, $11 \leq d_p \leq 150$ nm, $20 \leq T \leq 70°C$

$$\mu_{nf} = -0.4491 + \frac{28.837}{T} + 0.574\varphi_p - 0.1634\varphi_p^2 + 23.053 \times \left(\frac{\varphi_p^2}{T^2} \right)$$
$$+ 0.0132\varphi_p^3 - 2354.735 \times \left(\frac{\varphi_p}{T^3} \right) + 23.498 \times \left(\frac{\varphi_p^2}{d_p^2} \right) - 3.0185 \times \left(\frac{\varphi_p^3}{d_p^2} \right)$$

$1 \leq \varphi_p \leq 9 \ \%$, $13 \leq d_p \leq 131$ nm, $20 \leq T \leq 70°C$

TABLE E.9
Models of Viscosity of Nanofluids

Models	Dynamic Viscosity (μ) (kg/m-s)	Physical Model
Einstein [7]	$\mu_{nf} = \mu_{bf}(1 + 2.5\phi_p)$ **Remarks:** – Infinitely dilute suspension of spheres (no interaction between the spheres) – Valid for relatively low particle volume fraction $\phi_p < 5\%$	– Based on the phenomenological hydrodynamic equations – Considered a suspension containing n-solute particles in a total volume, V
Brinkman [8]	$\mu_{nf} = \dfrac{\mu_{bf}}{\left(1 - \phi_p\right)^{2.5}}$ **Remarks:** – Spherical particles – Valid for high moderate particle concentrations – Used Einstein's factor: $(1 + 2.5\phi_p)$	– Based on Einstein model – Derived by considering the effect of the addition of one solute-molecule to an existing solution
Batchelor [9]	$\mu_{nf} = \mu_{bf}\left(1 + 2.5\phi_p + 6.2\phi_p^2\right)$ $\quad = \mu_{bf}\left(1 + \eta\phi_p + k_H\phi_p^2\right)$ Here, Huggins coefficient, $k_H = 6.2$ (5.2 from hydrodynamic effect and 1.0 from Brownian motion) **Remarks:** Brownian motion Isotropic structure	– Based on reciprocal theorem in Stokes flow problem to obtain an expression for the bulk stress due to the thermodynamic forces – Incorporated both effects: hydrodynamic effects and Brownian motion
Lundgren [10]	$\mu_{nf} = \dfrac{\mu_{bf}}{\left(1 - 2.5\phi_p\right)}$ **Remarks:** – Dilute concentration of spheres – Random bed of spheres	– Based on a Taylor series expansion in terms of ϕ_p

TABLE E.10
Viscosity Models at Room Temperature Based on Experimental Data (TiO$_2$-H$_2$O, CuO-H$_2$O and Al$_2$O$_3$-H$_2$O)

Reference	Viscosity, μ	Remarks
Maïga et al. [11]	$\mu_{nf} = \mu_{bf}\left(1 + 7.3\varphi_p + 123\varphi_p^2\right)$	Least-square curve fitting of Wang et al. [12] data, Al$_2$O$_3$-H$_2$O, $d_p = 28$ nm
Khanafer and Vafai [6]	$\mu_{nf} = \mu_{bf}\left(1 + 0.164\varphi_p + 302.34\varphi_p^2\right)$	Least-square curve fitting of experimental data (1993, 1999), Al$_2$O$_3$-ethylene glycol, $d_p = 28$ nm
Buongiorno [13]	$\mu_{nf} = \mu_{bf}\left(1 + 5.45\varphi_p + 108.2\varphi_p^2\right)$	Curve fitting of Pak and Cho [14] data, TiO$_2$-H$_2$O, $d_p = 27$ nm

Appendix E

413

TABLE E.10 (Continued)
Viscosity Models at Room Temperature Based on Experimental Data (TiO$_2$-H$_2$O, CuO-H$_2$O and Al$_2$O$_3$-H$_2$O)

Reference	Viscosity, μ	Remarks
Khanafer and Vafai [6]	$\mu_{nf} = \mu_{bf}\left(1 + 23.09\varphi_p + 1525.3\varphi_p^2\right)$	Curve fitting of Pak and Cho [14] data, Al$_2$O$_3$-H$_2$O, $d_p = 13$ nm, $0 \leq \varphi_P \leq 0.04\%$
Nguyen et al. [15]	$\mu_{nf} = \mu_{bf}[0.904 \times \exp(0.148\varphi_p)]$, d_p = **47 nm** $\mu_{nf} = \mu_{bf}\left(1 + 0.0025\varphi_p + 0.00156\varphi_p^2\right)$, d_p = **37 nm**	Curve fitting of the experimental data, Al$_2$O$_3$-H$_2$O
Nguyen et al. [15]	$\mu_{nf} = \mu_{bf}\left(1.475 - 0.319\varphi_p + 0.051\varphi_p^2 + 0.009\varphi_p^3\right)$	Curve fitting of the experimental data, CuO-water, $d_p = 29$ nm
Tseng and Lin [16]	$\mu_{nf} = \mu_{bf}[13.47 \times \exp(35.98\varphi_p)]$	TiO$_2$-H$_2$O, $0.05 \leq \varphi_P \leq 0.12\%$

TABLE E.11
Effect of Temperature and Volume Fraction on Dynamic Viscosity of Al$_2$O$_3$-H$_2$O Nanofluid

Reference	Viscosity, μ	Remarks
Khanafer and Vafai [6]	$\mu_{nf} = 0.44 - 0.254\varphi_p^2 + 0.0368\varphi_p^2 + 26.33\dfrac{\varphi_p}{T} - 59.311\dfrac{\varphi_p^2}{T^2}$	Curve fitting of Pak and Cho [14] data, and $d_p = 13$ nm, $20 < T\,(°C) < 70$, $\varphi_p = 1.34\%$ and 2.78%; Units: **mPa-s**
Nguyen et al. [15]	$\mu_{nf} = \mu_{bf}(1.125 - 0.0007 \times T\,(°C))$, $\varphi_p = 1\%$ $\mu_{nf} = \mu_{bf}(2.1275 - 0.0215 \times T\,(°C) + 0.0002 \times T^2\,(°C))$, $\varphi_p = 4\%$	Units: **mPa-s**
Namburu et al. [17, 18]	$\mu_{nf} = \exp[A\exp(-BT)]$ Here $A = -0.2995\varphi_p^3 + 6.7388\varphi_p^2 - 55.44\varphi_p + 236.11$ $B = \left(-6.4745\varphi_P^3 + 140.03\varphi_p^2 - 1478.5\varphi_P + 20,341\right)\times 10^{-6}$	Experimental Al$_2$O$_3$-ethylene glycol and water mixture; $1\% < \varphi_p < 10\%$, $d_p = 53$ nm, and $238 < T(K) < 323$, Units: **mmPa-s**

TABLE E.12
Effect of Temperature and Volume Fraction on the Dynamic Viscosity of TiO$_2$-H$_2$O and CuO-H$_2$O Nanofluids

Reference	Viscosity, μ	Remarks
Duangthongsuk and Wongwises [19]	$\mu_{nf} = \mu_{bf}\left(1.0226 + 0.0477\varphi_p - 0.0112\varphi_p^2\right); T = 15°C$ $\mu_{nf} = \mu_{bf}\left(1.013 + 0.092\varphi_p - 0.015\varphi_p^2\right); T = 25°C$ $\mu_{nf} = \mu_{bf}\left(1.018 + 0.112j_p - 0.0177j_p^2\right); T = 35°C$	– Experimental data, TiO$_2$-H$_2$O – $d_p = 21$ nm, $0.2\% \leq \varphi_p \leq 2\%$ – Units: **mPa-s**
Khanafer and Vafai [6]	$\mu_{nf} = 0.6002 - 0.569\varphi_p + 0.0823\varphi_p^2 + 28.8763\dfrac{\varphi_p}{T}$ $\qquad - 204.2202\dfrac{\varphi_P^2}{T^2} + 561.3175\dfrac{\varphi_P^3}{T^3}$	– Curve fitting of Pak and Cho [14] data, TiO$_2$-H$_2$O – $d_p = 27$ nm, $20 < T\,(°C) < 70$, $\varphi_p = 0.99\%$, 2.04%, 3.16% – Units: **mPa-s**

(Continued)

TABLE E.12 (Continued)
Effect of Temperature and Volume Fraction on the Dynamic Viscosity of TiO$_2$-H$_2$O and CuO-H$_2$O Nanofluids

Reference	Viscosity, μ	Remarks
Namburu et al. [17, 18]	$\mu_{nf} = \exp[A\exp(-BT)]$ Here $A = 1.8375\phi_p^2 + 29.643\phi_p + 165.56$ $B = \left(4\times10^{-6}\phi_P^2 - 0.001\phi_P + 0.0186\right)$	– CuO–ethylene glycol and water mixture – $1 \leq \phi_p \leq 6\%$, $d_p = 29$ nm, $238 < T\,(K) < 323$ – Units: **mmPa-s**
Kulkarni et al. [20, 21]	$\mu_{nf} = \exp\left(\dfrac{A}{T} - B\right)$ Here $A = 20,587\phi_p^2 + 15,857\phi_p + 1078.3$, $B = -107.12\phi_p^2 + 53.54\phi_p + 2.8715$	– CuO-H$_2$O, $0.5\% \leq \phi_P \leq 0.15\%$ – $d_p = 29$ nm – $238 < T\,(K) < 323$, – Units: **mmPa-s**

TABLE E.13
Theoretical Models and Correlations for Thermo-Physical Properties

Properties	References	Theoretical Formulae	Correlations
Density (kg/m^3)	Pak and Cho [14]	$\rho_{nf} = \rho_p\phi_p + \rho_{bf}(1-\phi_p)$	– $\rho_{nf} = 1001.064 + 2738.6191\phi_p - 0.2095T$, $0 \leq \phi_P \leq 0.4\%$, $5 < T\,(°C) < 40$ – Curve fitting of Ho et al. [22] measured the density of Al$_2$O$_3$-water nanofluid at different temperatures and nanoparticle volume fraction
Specific heat (J/kgK)	Pak and Cho [14] Xuan and Roetzel [23]	$C_{nf} = \dfrac{\phi_p\rho_p C_{pp} + \left(1-\phi_p\right)\rho_{bf}C_{pbf}}{\rho_{bf}}$ $(\rho C_p)_{nf} = (1-\phi_p)$ $(\rho C_p)_{bf} + \phi_p(\rho C_p)_p$	– Vajjha and Das [24] for Al$_2$O$_3$ and SiO$_2$, ZnO nanofluids $C_{pnf} = C_{pbf}\left[\dfrac{A\left(\dfrac{T}{T_0}\right) + B\left(\dfrac{C_{pp}}{C_{pbf}}\right)}{\left(C+\phi\right)}\right]$

Nanofluids	A	B	C	Max error (%)	Avg. absolute error (%)
Al$_2$O$_3$	0.00089	0.5179	0.4250	5	2.28
SiO$_2$	0.00176	1.1937	0.8021	3.1	1.5
ZnO	0.00046	0.9855	0.299	4.4	2.7

– $0 \leq \phi_p \leq 0.1\%$ for Al$_2$O$_3$ and SiO$_2$-ethylene glycol and water mixture (60:40 by weight)
– $0 \leq \phi_p \leq 0.07\%$ for ZnO-ethylene glycol and water mixture (60:40 by weight)
– $315 < T\,(K) < 363$

(Continued)

Appendix E

TABLE E.13 (Continued)
Theoretical Models and Correlations for Thermo-Physical Properties

Properties	References	Theoretical Formulae	Correlations
Thermal expansion coefficient (K^{-1})	Khanfer et al. [25] Wang et al. [26], Ho et al. [27]	$$\beta_{nf} = \frac{(1-\varphi_p)(\rho\beta)_{bf} + \varphi_p(\rho\beta)_p}{\rho_{nf}}$$ $$\beta_{nf} = (1-\varphi_p)\beta_{bf} + \varphi_p\beta_p$$	– A correlation for the thermal expansion coefficient of Al_2O_3-water nanofluid as a function of temperature and volume fraction of nanoparticles based on the data presented in Ho et al. [22] $$\beta_{nf} = \left(-0.479\phi_p + 9.3149 \times 10^{-3}T - \frac{4.7211}{T^2}\right) \times 10^{-3},$$ $0 \leq \varphi_P \leq 0.04\ \%\ ,\ 10 < T\ (^\circ C) < 40$
Thermal diffusivity (mm^2/s)		$$\alpha_{nf} = \frac{k_{nf}}{\rho_{nf}C_{pnf}}$$	

$\phi_p = V_p/V_{bf} + V_p$ = Volume fraction of nanoparticles, k_{nf} = thermal Conductivity of nanofluids, k_{bf} = Thermal conductivity of base fluid, k_p = Thermal conductivity of nanoparticles, k_H = 6.2, Huggins coefficient, n = Empirical shape factor, rp = Particle radius, h = Inter-particle spacing, t = thickness of the nano-layer, T = Temperature, $\phi_{p,\max}$ = maximum volume fraction of nanoparticles, k_{layer} = thermal conductivity of the nano-layer.

Subscript f = fluid, bf = base fluid, p = nanoparticle, nf = nanofuid.

TABLE E.14
Heat Transfer Coefficient of Nanofluids

References	Nusselt Number and Heat Transfer Coefficient	Remarks
Seider–Tate Equation [28]	$$\left(Nu_{nf}\right)_{th} = 1.86 \times \left(Re_{nf}\ Pr_{nf}\ \frac{D}{L}\right)^{\frac{1}{3}} \left(\frac{\mu_{nf}}{\mu_{bf}}\right)^{0.14}$$ $$Re_{nf} = \frac{u_m D}{\mu_{nf}},\ Pr_{nf} = \frac{(C_p)_{nf}\mu_{nf}}{K_{nf}},\ Nu_{nf} = \frac{h_{nf}D}{K_{nf}}$$	– Circular tube
Pak and Cho [14]	$Nu = 0.021Re^{0.8}Pr^{0.5}$	– Circular tube – Limited to dilute concentration up to 3% – ultrafine metallic oxide particles suspended in water (γ-Al_2O_3, TiO_2) – mean diameter 13 nm (γ-Al_2O_3) and 17 nm (TiO_2)
Xuan and Li [29]	$$Nu_{nf} = c_1\left(1 + c_2\varphi^{m_1}Pe_d^{m_2}\right)Re_{nf}^{m_3}\ Pr_{nf}^{m_4}$$ $$Pe_p = \frac{u_m d_p}{\alpha_{nf}},\ Re_{nf} = \frac{u_m D}{\mu_{nf}}$$ $$Pr_{nf} = \frac{\mu_{nf}}{\alpha_{nf}},\ \alpha_{nf} = \frac{k_{nf}}{(\rho C_p)_{nf}}$$	(see table below)

	c_1	c_2	m_1	m_2	m_3	m_4
Laminar flow	0.4328	11.285	0.754	0.218	0.333	0.4
Turbulent flow	0.0059	7.6286	0.6886	0.001	0.9238	0.4

– The case $C_2 = 0$ refers to zero thermal dispersion, which just corresponds to the case of the pure base fluid
– limited to dilute up to 2%

(Continued)

416 Appendix E

TABLE E.14 (Continued)
Heat Transfer Coefficient of Nanofluids

References	Nusselt Number and Heat Transfer Coefficient	Remarks
Maïga et al. [30]	Laminar flow $Nu = 0.086Re^{0.55}Pr^{0.5}$ (For constant wall flux) $Nu_{nf} = 0.28Re^{0.35}Pr^{0.36}$ (For constant wall temperature) Turbulent flow $(Nu_{nf})_{fd} = 0.085 \times Re^{0.71}Pr^{0.35}$	– Circular tube – Al_2O_3 nanoparticles suspension in water

Fotukian and Esfahany [31]

Turbulent flow

$$h_{nf} = \frac{C_{pnf}\,\rho_{nf}\,uA\left(T_{b2} - T_{b1}\right)}{\pi DL\left(T_w - T_b\right)_{LM}}$$

– Circular tube
– CuO-H_2O
– Turbulent convective heat transfer performance and pressure drop of very dilute (less than 0.24% volume)

Nano-sized particle	Mean diameter (nm)	Density (kg/m³)	Thermal conductivity (J/kg-K)	Specific heat (W/mK)
CuO	30–50	6350	69	535.6

Qiang and Yimin [32]

Mouromtseff numbers

Laminar Flow

$$Mo = \left[1 + \left(11.285\varphi_p^{0.754}\left(\frac{d_p}{\alpha_{nf}}\right)^{0.218}\right)\frac{\rho_{nf}^{0.33}C_{pnf}^{0.4}K_{nf}^{0.6}}{\mu_{nf}^{-0.07}}\right]$$

Turbulent Flow

$$Mo = \left[1 + \left(7.6286\varphi_p^{0.6886}\left(\frac{d_p}{\alpha_{nf}}\right)^{0.001}\right)\frac{\rho_{nf}^{0.9238}C_{pnf}^{0.4}K_{nf}^{0.6}}{\mu_{nf}^{0.5232}}\right]$$

– Friction factor: $f_{nf} = \dfrac{2P_{nf}Dg}{Lu_m^2}$

– Derived from equation of Xuan and Li [29] for fully developed internal laminar and turbulent flow at a specific velocity of 1 m/s.

Gnielinski [33]

For single phase flow

$f = (1.58 \times \ln Re - 3.82)^{-2}$

$$Nu_{nf} = \frac{\left(0.125f\right)\left(Re - 1000\right)Pr}{1 + 12.7\left(0.125f\right)^{0.5}\left(Pr^{2/3} - 1\right)}$$

Heat Exchanger
Al_2O_3-H_2O
TiO_2-H_2O

Duangthongsuk and Wongwises [19]

For each flow rate

$f = 0.961 \times \left(Re^{-0.375}\,\varphi_p^{0.052}\right)$

$Nu_{nf} = 0.074 \times Re_{nf}^{0.707}\,Pr_{nf}^{0.385}\,\varphi_p^{0.074}$

$Re = \dfrac{VD}{\mu}, Pe = \dfrac{VD}{\alpha_{nf}}, Pr = \dfrac{\mu_{nf}}{\alpha_{nf}}, \alpha_{nf} = \dfrac{k_{nf}}{\rho_{nf}\left(C_p\right)_{nf}}$

Heat Exchanger
Al_2O_3-H_2O
TiO_2-H_2O

$(Pe)_p$ = Particle Péclet number of the nanoparticle, Re_{nf} = Reynolds number of nanofluids, Pr_{nf} = Prandtl number of nanofluids, α_{nf} = Thermal diffusivity of nanofluids, K_{nf} = thermal conductivity of nanofluids, D = tube diameter, d_p = particle diameter, u = fluid velocity, A = cross-section area of the tube, L = tube length, T_{b1} = inlet bulk temperature (K), T_{b2} = exit bulk temperature (K), T_w = wall temperature of the tube (K), $(T_w - T_b)_{LM}$ = logarithmic mean temperature difference in which T_w is the wall temperature that is the average of ten measured temperatures on tube wall at different positions. P_{nf} = Pressure drop of the pressure drop test section, L = Length of the pressure drop test section, g = Acceleration gravity, f = friction factor, volume concentration is $\varphi_v = [1/(100/\varphi_m)((\rho_p/\rho_w) + 1)] \times 100\%$.

REFERENCES

[1] G. N. Tiwari, A. Tiwari and Shyam, *Handbook of Solar Energy: Theory, Analysis and Applications*, New Delhi: Springer, 2016.

[2] J. Maxwell, *A Treatise on Electricity and Magnetism*, Oxford: Clarendon Press, vol. 1, second edition, 1881.

[3] V. D. A. G. Bruggeman, "Betechnung vershiedener physikalischer konstanten von heterogenen substanzen," *Annalen der Physik*, vol. 24, pp. 636–664, 1935.

[4] R. Hamilton and O. Crosser, "Thermal conductivity of heterogeneous two component systems," *Industrial & Engineering Chemistry Fundamentals*, vol. 1, no. 3, pp. 17–191, 1962.

[5] E. Wasp, J. Kenny and R. Gandhi, *Solid–Liquid Flow, Slurry Pipeline TransportationA*, Clausthal, Germany: Trans Tech Publications, 1977.

[6] K. Khanafer and K. Vafai, "A critical synthesis of thermophysical characteristics of nanofluids," *International Journal of Heat and Mass Transfer*, vol. 54, no. 19, pp. 4410–4428, 2011.

[7] A. Einstein, "Eine neue Bestimmung der Molekul-dimensionen," *Annalen der Physik*, vol. 324, no. 2, pp. 289–306, 1906.

[8] H. Brinkman, "The viscosity of concentrated suspensions and solutions," *The Journal of Chemical Physics*, vol. 20, pp. 571–581, 1952.

[9] G. Batchelor, "The effect of Brownian motion on the bulk stress in a suspension of spherical particles," *Journal of Fluid Mechanics*, vol. 83, pp. 97–117, 1977.

[10] T. Lundgren, "Slow flow through stationary random beds and suspensions of spheres," *Journal of Fluid Mechanics*, vol. 51, pp. 273–299, 1972.

[11] S. Maïga, S. J. Palm, C. T. Nguyen, G. Roy and N. Galanis, "Heat transfer enhancement by using nanofluids in forced convection flows," *International Journal of Heat and Fluid Flow*, vol. 26, no. 4, pp. 530–546, 2005.

[12] X. Wang, X. Xu and S. Choi, "Thermal conductivity of nanoparticle – fluid mixture," *Journal of Thermophysics and Heat Transfer*, vol. 13, no. 4, pp. 474–480, 1999.

[13] J. Buongiorno, "Convective transport in nanofluids," *Journal of Heat Transfer*, vol. 128, no. 3, pp. 240–250, 2006.

[14] B. Pak and Y. Cho, "Hydrodynamic and heat transfer study of dispersed fluids with submicron metallic oxide particles," *Experimental Heat Transfer an International Journal*, vol. 11, no. 2, pp. 151–170, 1998.

[15] C. T. Nguyen, G. Roy, C. Gauthier and N. Galanis, "Heat transfer enhancement using Al_2O_3–water nanofluid for an electronic liquid cooling system," *Applied Thermal Engineering*, vol. 27, no. 8–9, pp. 1501–1506, 2007.

[16] W. J. Tseng and K.-C. Lin, "Rheology and colloidal structure of aqueous TiO_2 nanoparticle suspensions," *Materials Science and Engineering: A,* vol. 355, no. 1–2, pp. 186–192, 2003.

[17] P. K. Namburu, D. Kulkarni, A. Dandekar and D. K. Das, "Experimental investigation of viscosity and specific heat of silicon dioxide nanofluids," *IET Micro & Nano Letters*, vol. 2, no. 3, pp. 67–71, 2007.

[18] P. K. Namburu, D. K. Das, K. M. Tanguturi and R. S. Vajjha, "Numerical study of turbulent flow and heat transfer characteristics of nanofluids considering variable properties," *International Journal of Thermal Sciences*, vol. 48, no. 2, pp. 290–302, 2009.

[19] W. Duangthongsuk and S. Wongwises, "Measurement of temperature dependent thermal conductivity and viscosity of TiO_2–water nanofluids.," *Experimental Thermal and Fluid Science*, vol. 33, no. 4, pp. 706–714, 2009.

[20] D. Kulkarni, D. Das and G. Chukwu, "Temperature dependent rheological property of copper oxide nanoparticles suspension (nanofluid)," *Journal of Nanoscience and Nanotechnology*, vol. 6, pp. 1150–1154, 2006.

[21] D. Kulkarni, D. K. Das and S. Patil, "Effect of temperature on rheological properties of copper oxide nanoparticles dispersed in propylene glycol and water mixture," *Journal of Nanoscience and Nanotechnology*, vol. 7, no. 7, pp. 2318–2322, 2007.

[22] C. Ho, W. Liu, Y. Chang and C. Lin, "Natural convection heat transfer of alumina-water nanofluid in vertical square enclosures: An experimental study.," *International Journal of Thermal Sciences*, vol. 49, pp. 1345–1353, 2010.

[23] Y. Xuan and L. Qiang, "Heat transfer enhancement of nanofluid," *International Journal of Heat and Fluid Flow*, vol. 21, no. 1, pp. 58–64, 2000.

418 Appendix E

[24] R. S. Vajjha and D. K. Das, *Measurements of Nanofluids Properties and Heat Transfer Computation: Correlations for Nanofluids Properties*, Saarbrücken: LAP LAMBERT Academic Publishing, 2010.

[25] K. Khanafer, K. Vafai and M. F. Lightstone, "Buoyancy-driven heat transfer enhancement in a two-dimensional enclosure utilizing nanofluids," *International Journal of Heat and Mass Transfer*, vol. 46, no. 19, pp. 3639–3653, 2003.

[26] Z. Wang, Q. Zhang, D. Sun and S. Yin, "Study on thermal expansion of Nd^{3+}: $Gd_3Ga_5O_{12}$ laser crystal," *Journal of Rare Earths*, vol. 25, pp. 244–246, 2007.

[27] C. Ho, M. W. Chen and Z. W. Li, "Numerical simulation of natural convection of nanofluid in a square enclosure: Effects due to uncertainties of viscosity and thermal conductivity," *International Journal of Heat and Mass Transfer*, vol. 51, no. 17–18, pp. 4506–4516, 2008.

[28] E. Sieder and G. Tate, "Heat transfer and pressure drop of liquids in tubes," *Industrial & Engineering Chemistry Research*, vol. 28, no. 12, pp. 1429–1435, 1936.

[29] Y. Xuan and Q. Li, "Investigation on convective heat transfer and flow features of nanofluids," *Journal of Heat Transfer*, vol. 125, no. 1, pp. 151–155, 2003.

[30] S. E. B. Maïga, C. T. Nguyen, N. Galanis, N. Galanis, G. Roy, T. Mare and M. Coqueux, "Heat transfer enhancement in turbulent tube flow using Al_2O_3 nanoparticle suspension," *International Journal of Numerical Methods for Heat and Fluid Flow*, vol. 16, no. 3, pp. 275–292, 2006.

[31] S. Fotukian and M. N. Esfahany, "Experimental investigation of turbulent convective heat transfer of dilute γ-Al_2O_3/water nanofluid inside a circular tube," *International Journal of Heat and Fluid Flow*, vol. 31, no. 4, pp. 606–612, 2010.

[32] L. Qiang and X. Yimin, "Convective heat transfer and flow characteristics of Cu-water nanofluid," *Science in China Series E: Technological Science volume*, vol. 45, pp. 408–416, 2002.

[33] V. Gnielinski, "New equations for heat and mass transfer in turbulent pipe and channel flow," *Scholarly Articles for Int Chem Eng*, vol. 16, pp. 359–368, 1976.

Appendix F

List of Embodied Energy Coefficients [1]

Material	MJ/kg	MJ/m³
Aggregate, general	0.10	150
virgin rock	0.04	63
river	0.02	36
Aluminum, virgin	191	515,700
extruded	201	542,700
extruded, anodized	227	612,900
extruded, factory-painted	218	588,600
foil	204	550,800
sheet	199	537,300
Aluminum, recycled	8.1	21,870
extruded	17.3	46,710
extruded, anodized	42.9	115,830
extruded, factory-painted	34.3	92,610
foil	20.1	54,270
sheet	14.8	39,960
Asphalt (paving)	3.4	7140
Bitumen	44.1	45,420
Brass	62.0	519,560
Carpet	72.4	-
felt underlay	18.6	-
nylon	148	-
polyester	53.7	-
e Polyethylene terephthalate (PET)	107	-
polypropylene	95.4	-
wool	106	-
Cement	7.8	15,210
cement mortar	2.0	3200
fiber cement board	9.5	13,550
soil-cement	0.42	819
Ceramic		-
brick	2.5	5170
brick, glazed	7.2	14,760
pipe	6.3	-
tile	2.5	5250
Concrete		-
block	0.94	-
brick	0.97	-
GRC	7.6	14,820
paver	1.2	-
pre-cast	2.0	-

(*Continued*)

420   Appendix F

Material	MJ/kg	MJ/m³
ready mix, 17.5 MPa	1.0	2350
30 MPa	1.3	3180
40 MPa	1.6	3890
roofing tile	0.81	-
Copper	70.6	631,160
Earth, raw		-
adobe block, straw stabilized	0.47	750
adobe, bitumen stabilized	0.29	-
adobe, cement stabilized	0.42	-
rammed soil cement	0.80	-
pressed block	0.42	-
Fabric		-
cotton	143	-
polyester	53.7	-
Glass	66.2	-
float	15.9	40,060
toughened	26.2	66,020
laminated	16.3	41,080
tinted	14.9	375,450
Insulation		-
cellulose	3.3	112
fiberglass	30.3	970
polyester	53.7	430
polystyrene	117	2340
wool (recycled)	14.6	139
Lead	35.1	398,030
Linoleum	116	150,930
Paint	90.4	118 per liter
solvent-based	98.1	128 per liter
water-based	88.5	115 per liter
Paper	36.4	33,670
building	25.5	-
kraft	12.6	-
recycled	23.4	-
wall	36.4	-
Plaster, gypsum	4.5	6460
Plasterboard	6.1	5890
Plastics		-
ABS	111	-
high density polyethylene (HDPE)	103	97,340
low density polyethylene (LDPE)	103	91,800
polyester	53.7	7710
polypropylene	64.0	57,600
polystyrene, expanded	117	2340
polyurethane	74.0	44,400
PVC	70.0	93,620
Rubber		-
natural latex	67.5	62,100
synthetic	110	-
Sand	0.10	232
Sealants and adhesives		-

(*Continued*)

Appendix F   **421**

phenol formaldehyde	87.0	-
urea formaldehyde	78.2	-
Steel, recycled	10.1	37,210
reinforcing, sections	8.9	-
wire rod	12.5	-
Steel, virgin, general	32.0	251,200
galvanized	34.8	273,180
imported, structural	35.0	274,570
Stone, dimension		-
local	0.79	1890
imported	6.8	1890
Straw, baled	0.24	30.5
Timber, softwood		-
air-dried, rough-sawn	0.3	165
kiln-dried, rough-sawn	1.6	880
air-dried, dressed	1.16	638
kiln-dried, dressed	2.5	1380
moldings, etc.	3.1	1710
hardboard	24.2	13,310
MDF	11.9	8330
glulam	4.6	2530
particleboard	8.0	-
plywood	10.4	-
shingles	9.0	-
Timber, hardwood		-
air-dried, rough-sawn	0.50	388
kiln-dried, rough-sawn	2.0	1550
Vinyl flooring	79.1	105,990
Zinc	51.0	364,140
galvanizing, per kg steel	2.8	-

REFERENCE

[1] G. N. Tiwari, V. Singh, P. Joshi Shyam, and A. Deo Prabhakant and A. Gupta, "Design of an earth air heat exchanger (EAHE) for climatic condition of Chennai, India" *The Open Environmental Sciences*, vol. 8, no. 18, pp. 24–34, 2014.

Appendix G

Conversion of units [1]

i) Length, m

1 yd (yard) = 3 ft = 36 in (inches) = 0.9144 m

1 m = 39.3701 in = 3.280839 ft = 1.093613 yd = 1,650,763.73 wavelength

1 ft =12 in = 0.3048 m

1 in = 2.54 cm = 25.4 mm

1 mil = 2.54×10^{-3} cm

1 μm = 10^{-6} m

1 nm = 10^{-9} m = 10^{-3} μm

ii) Area, m²

1 ft² = 0.0929 m²

1 in² = 6.452 cm² = 0.00064516 m²

1 cm² = 10^{-4} m² = 10.764×10^{-4} ft² = 0.1550 in²

1 ha =10,000 m²

iii) Volume, m³

1 ft³ = 0.02832 m³ = 28.3168 l (liter)

1 in³ = 16.39 cm³ = 1.639×10^2 l

1 yd³ = 0.764555 m³ = 7.646×10^2 l

1 UK gallon = 4.54609 l

1 US gallon = 3.785 l = 0.1337 ft³

1 m³ = 1.000×10^6 cm³ = 2.642×10^{12} US gallons = 109 l

1 l = 10^{-3} m³

1 fluid ounce = 28.41 cm³

iv) Mass, kg

1 kg = 2.20462 lb = 0.068522 slug

1 tonne (short) = 2000 lb (pounds) = 907.184 kg

1 tonne (long) = 1016.05 kg.

1 lb = 16 oz (ounces) = 0.4536 kg

1 oz = 28.3495 g

1 quintal = 100 kg

1 kg = 1000 g = 10,000 mg

1 μg = 10^{-6} g

1 ng = 10^{-9} g

v) Density and specific volumes, kg/m³, m³/kg

1 lb/ft³ = 16.0185 kg/m³ = 5.787×10^{-4} lb/in³

1 g/cm³ = 10^3 kg/m³ = 62.43 lb/ft³

1 lb/ft³ = 0.016 g/cm³ = 16 kg/m³

1 ft³ (air) = 0.08009 lb = 36.5 g at N.T.P.

1 gallon/lb = 0.010 cm³/kg

1 μg/m³ = 10^{-6} g/m³

vi) Pressure, Pa (Pascal)

1 lb/ft² = 4.88 kg/m² = 47.88 Pa

1 lb/in² = 702.7 kg/m² = 51.71 mm Hg = 6.894757×10^3 Pa = 6.894757×10^3 N/m²

1 atm = 1.013×10^5 N/m² = 760 mm Hg = 101.325 kPa

1 in H_2O = 2.491×10^2 N/m² = 248.8 Pa = 0.036 lb/in²

1 bar = 0.987 atm = 1.000×10^6 dynes/cm² = 1.020 kgf/cm² = 14.50 lbf/in² = 10^5 N/m² = 100 kPa

1 torr (mm Hg 0°C) = 133 Pa

1 Pa = 1 N/m² = 1.89476 kg.

1 in of Hg = 3.377 kPa = 0.489 lb/in²

vii) Velocity, m/s

1 ft/s = 0.3041 m/s

1 mile/h = 0.447 m/s = 1.4667 ft/s = 0.8690 knots

1 km/h = 0.2778 m/s

1 ft/min = 0.00508 m/s

viii) Force, N

1 N (Newton) = 10^5 dynes = 0.22481 lb wt = 0.224 lb f

1 pdl (poundal) = 0.138255 N (Newton) = 13.83 dynes = 14.10 gf

1 lbf (i.e., wt of 1 lb mass) = 4.448222 N = 444.8222 dynes

1 tonne = 9.964×10^3 N

1 bar = 10^5 Pa (Pascal)

1 ft of H_2O = 2.950×10^{-2} atm = 9.807×10^3 N/m²

1 in H_2O = 249.089 Pa

1 mm H_2O = 9.80665 Pa

1 dyne = 1.020×10^{-6} kg f = 2.2481×10^{-6} lb f = 7.2330×10^{-5} pdl = 10^{-5} N

1 mm of Hg = 133.3 Pa

1 atm = 1 kg f/cm² = 98.0665 kPa

1 Pa (Pascal) = 1 N/m²

ix) Mass flow rate and discharge, kg/s, m³/s

1 lb/s = 0.4536 kg/s

1 ft³/min = 0.4720 1/s = 4.179×10^{-4} m³/s

1 m³/s = 3.6×10^6 l/h

1 g/cm³ = 10^3 kg/m³

1 lb/h ft² = 0.001356 kg/s m²

1 lb/ft³ = 16.2 kg/m²

1 liter/s (l/s) = 10^{-3} m³/s

x) Energy, J

1 cal = 4.187 J (joules)

1 kcal = 3.97 Btu = 12×10^{-4} kWh = 4.187×10^3 J

1 Watt = 1.0 J/s

1 Btu = 0.252 kcal = 2.93×10^{-4} kWh = 1.022×10^3 J

1 hp = 632.34 kcal = 0.736 kWh

1 kWh = 3.6×10^6 J = 1 unit

1 J = 2.390×10^{-4} kcal = 2.778×10^{-4} Wh

1 kWh = 860 kcal = 3413 Btu

1 erg = 1.0×10^{-7} J = 1.0×10^{-7} Nm = 1.0 dyne cm

Appendix G

1 J = 1 Ws = 1 Nm

1 eV = 1.602×10^{-19} J

1 GJ = 10^9 J

1 MJ = 10^6 J

1 TJ (terajoules) = 10^{12} J

1EJ (exajoules) = 10^{18} J

xi) Power, Watt (J/s)

1 Btu/h = 0.293071 W = 0.252 kcal/h

1 Btu/h = 1.163 W = 3.97 Btu/h

1 W = 1.0 J/s = 1.341×10^{-3} hp = 0.0569 Btu/min = 0.01433 kcal/min

1 hp (F.P.S.) = 550 ft lb f/s = 746 W = 596 kcal/h = 1.015 hp (M.K.S.)

1 hp (M.K.S.) = 75 mm kg f/s = 0.17569 kcal/s = 735.3 W

1 W/ft^2 = 10.76 W/m^2

1 tonne (refrigeration) = 3.5 kW

1 kW = 1000 W

1 GW = 10^9 W

1 W/m^2 = 100 lux

xii) Specific Heat, J/kg°C

1 Btu/lb°F = 1.0 kcal/kg°C = 4.187×10^3 J/kg°C

1 Btu/lb = 2.326 kJ/kg

xiii) Temperature, °C and K used in SI

$T_{(Celsius, \,°C)} = (5/9)\,[T_{(Fahrenheit, \,°F)} + 40] - 40$

$T_{(°F)} = (9/5)\,[T_{(°C)} + 40] - 40$

$T_{(Rankine, °R)} = 460 + T_{(°F)}$

$T_{(Kelvin, K)} = (5/9)\,T_{(°R)}$

$T_{(Kelvin, K)} = 273.15 + T_{(°C)}$

$T_{(°C)} = T_{(°F)}/1.8 = (5/9)\,T_{(°F)}$

xiv) Rate of heat flow per unit area or heat flux, W/m^2

1 Btu/ft^2 h = 2.713 kcal/m^2 h = 3.1552 W/m^2

1 kcal/m^2 h = 0.3690 Btu/ft^2 h = 1.163 W/m^2 = 27.78×10^{-6} cal/s cm^2

1 cal/cm^2 min = 221.4 Btu/ft^2 h

1 W/ft^2 = 10.76 W/m^2

1 W/m^2 = 0.86 kcal/hm^2 = 0.23901×10^{-4} cal/s cm^2 = 0.137 Btu/h ft^2

1 Btu/h ft = 0.96128 W/m

xv) Heat transfer coefficient, W/m^2°C

1 Btu/ft^{2}h°F = 4.882 kcal/m^{2}h°C = 1.3571×10^{-4} cal/cm^2 s°C

1 Btu/ft^{2}h°F = 5.678 W/m^2°C

1 kcal/m^{2}h°C = 0.2048 Btu/ft^2 h°F = 1.163 W/m^2°C

1 W/m^{2}K = 2.3901×10^{-5} cal/cm^{2}sK = 1.7611×10^{-1} Btu/ft^2°F = 0.86 kcal/m^{2}h°C

xvi) Thermal conductivity, W/m°C

1 Btu/ft h°F = 1.488 kcal/m h°C = 1.73073 W/m°C

1 kcal/m h°C = 0.6720 Btu/ft h°F = 1.1631 W/m°C

1 Btu in/ft^2 h°F = 0.124 kcal/mh°C = 0.144228 W/m°C

1 Btu/in h°F = 17.88 kcal/mh°C

1 cal/cm s°F = 4.187×10^2 W/m°C = 242 Btu/h ft°F

1 W/cm°C = 57.79 Btu/h ft°F

xvii) Angle, rad

2π rad (radian) = 360° (degree)

1° (degree) = 0.0174533 rad = 60′ (minutes)

$1' = 0.290888 \times 10^{-3}$ rad = 60′ (seconds)

$1' = 4.84814 \times 10^{-6}$ rad

1° (hour angle) = 4 minute (time)

xviii) Illumination

1 lx (lux) = 1.0 lm (lumen)/m²

1 lm/ft² = 1.0 foot candle

1 foot candle = 10.7639 lx

100 lx = 1 W/m²

xix) Time, h

1 week = 7 days = 168 h = 10,080 minutes = 604,800 s

1 mean solar day = 1440 minute = 86,400 s

1 calendar year = 365 days = 8760 h = 5.256×10^5 minutes

1 tropical mean solar year = 365.2422 days

1 sidereal year = 365.2564 days (mean solar)

1 s (second) = 9.192631770×10^9 hertz (Hz)

1 day = 24 hour = 360° (hour angle)

xx) Concentration, kg/m³ and g/m³

1 g/l = 1 kg/m³

1 lb/ft³ = 6.236 kg/m³

xxi) Diffusivity, m²/s

1 ft²/h = 25.81×10^{-6} m²/s

REFERENCE

[1] G. N. Tiwari, A. Tiwari and Shyam, *Handbook of Solar Energy: Theory, Analysis and Applications*, New Delhi: Springer, 2016.

Index

Note: Page numbers in *italics* refer figures and **bold** refer tables.

A

acoustics, 90
air conductance/gap, 167, 171, 191–192
air electricity, 90
air ingredients, 90
air mass, 9, 86
air movement, 87–88
air pressure, 90
air temperature, 86
air vent, 198, *199*
altitude, 7
angle of incidence, 7
atrium, 30

B

beam radiation, 10
benefits, 321
bioclimatic design, 23–26
BiOPV, 223
    with façade, 224
    on rooftop, 227
    with rooftop, 223–224
Biot number, 44–45
BiPVT, 92–93, *93*
BiSPVT, 219–220, 225
    façade, 211, 226–227
    on rooftop, 227
    with rooftop, 176, 225–226
    with roof and vent, 177–180, 220–221
book value, 332
building mass, 30
building shape, 18–19, 30

C

capital cost, 342
capitalized cost, 308
Caratheodory principle, 113
carbon dioxide emissions, 271–272, *273*, **273**
    for PVT systems, *274*
carbon dioxide mitigation, 274–275
    with use of photovoltaics, 278–279
cash flow diagrams, 284
Clausius's principle, 113
climatic conditions, 13, 25–26
    criteria for the classification of climates, **13**
    dry climates, 14
    moist mid-latitude climates with cold winters, 14
    moist mid-latitude climates with mild winters, 14
    polar climates, 14–15
    tropical moist climates, 13
clothing insulation, 96, **97**
conduction, 41, 99
convection, 46, 100
    convective heat transfer coefficient, 59, 101

forced convection, 56, 90, 101
free/natural convection, 56, 90, 101
mixed mode convection, 59, 101
rate of convection, 100
conversion factor, 286
cool roof, 31, 192–193
cost analysis, 285
costs, 321
    conductive heat transfer coefficient, 44
    rate of conduction through clothing, 99–100

D

daylighting, 90–91
declination, 5, *6*
decrement factor, 106
depreciation, 332
    accelerated, 336
    rate, 333
diffuse radiation, 10
direct radiation, 10
disbenefits, 321
disbursements, 284
doping, 139
double roof, 31
double skin facades, 29
dry bulb temperature, 86

E

earned carbon credit, 274–275
earth-air heat exchanger, 199–203
Earth coupling, 199
Earth tube, 30
economic analysis, 283
effective rate of return, 291
embodied carbon, 270
embodied energy, 115, 261–262
    for BiPV, 115
    different materials, 262–263
    in floor/roofing systems, 263–264
    for PV module, **121**, 264
    renewable energy technologies, 267
    transportation, 264
emissive power, 12
energy analysis, 114
    BiSPVT, 228–229
    high grade, 124
    low grade, 124
energy balance, 75
    cloudy, 75, *77*
    intermediate season, 78, *79*
    PV modules, 153–156
    summer, 77, *78*
    winter, 75, *76*
energy conservation, 26
energy consumption, 26

428

energy density, 115
    for PV module, **121**
energy efficiency, 26–28
energy matrices, 115
    for PV module, 121, **123**
energy payback time, 116–118
    for BiPV, 117
energy production factor, 118
energy storage, 32
entropy, 114
evaporation, 66–68, 102–103
evaporative cooling, 31, 193–195
exergy, 114, 123
exergy analysis, 122
    BiSPVT, 228–229, 236–237
    of PVT systems, 129

F

façade air collector, 30
fill factor, 148
Fourier number, 45
Fourier's heat conduction equation, 41
future value, 286

G

glazing, 29
global radiation, 10
Grashof number, 48
greenhouse, 30–31
greenhouse gas, 259, 272
green roof, 31, 192–193

H

hour angle, 5, *6*

I

infiltration, 22, 27, 184–186
initial cost, 332, 343
insulation, 20–21, *20*, 27, 29
    exterior, 197
    interior, 197
    movable, 197
    Trombe wall, 210–211
integrated optical-thickness, 10
interior design, 22–23
internal rate of return, 327–328
irradiance, 11
irradiation, 11

K

Kelvin's principle, 113
Kirchhoff's law, 61
Kyoto Protocol, 276–278

L

latitude, 5, **5**
life cycle analysis, 283
life cycle assessment, 259–261

life cycle conversion efficiency, 118, 121
light shelves, 93–94
light tube, 30

M

macroclimate, 15
market value, 333
mass transfer, 66
mean radiant temperature, 88–90
metabolism, 96, **98**
microclimate, 15–16

N

net cash flow, 284
net present value, 311
net zero energy building, 2
night ventilation, 30
non-packing area of SPV, 165–166
non-product cost, 342
normal solar radiation/irradiance, 9–10
Nusselt number, 47

O

optimum load resistance, 148
overall heat transfer coefficient, 69
    parallel slabs, 69
    parallel slabs with cavity, 71
owner, 321

P

packing area of SPV, 152, 165, 248
passive cooling, 30–31
passive houses, 16, 27
    architectural design, 17, 27–28
    building envelope, 18
    building location and orientation, 17–18, 29
    building shape, 18–19, 30
    clustering, 18
    entrances and windows, 19, 27
    infiltration, 22, 27
    insulation, 20–21, *20*, 27
    interior design, 22–23
    solar access, 16
    solar shading techniques, 19–20, 29
    site planning, 17
    wind control, 16–17
payback time, 316
phase change material, 183
photo current, 143
photovoltaic cell, 139
photovoltaic effect, 143–144
photovoltaic material, 144–147
photovoltaic system, 139
    photovoltaic solarium, *175*, 180
Planck's law, 61
    first radiation constant, 61
    second radiation constant, 61
Prandtl number, 47
    photovoltaic thermal Trombe walls, 176, 205–207
predicted mean vote (PMV) index, 103–105, *105*, **105**

Index

predicted percentage dissatisfied (PPD) index, 104–105
present value, 285
psychometric chart, *87*
pure semiconductors, 141

R

radiant
    existence, 12
    exposure, 11
    self-existence, 12
radiation, 60, 101
    involving real surfaces, 60
    radiative heat transfer coefficient, 64
radiosity, 12
Rayleigh atmosphere, 10
Rayleigh number, 48
receipts, 284
recovery period, 333
relative humidity, 86–87, 248
replacement costs, 342
respiration, 103
Reynolds number, 47

S

salvage value, 299, 332, 343
short circuit current, 143
sinking fund factor, 299
skylight, 92
    PV skylight, 222
sky radiation, 62
slope, 7
smart glass windows, 94
sol-air temperature, 161
    bare surface, 161, 163
    blackened and glazed surface, 164
    direct gain through (single and double) glazed
        windows, 167
    evaporative cooling, 193
    wetted surface, 163
solar altitude angle, 7
solar azimuth angle, 7, **7**
solar cell current, 144
solar cell efficiency, 148–150
solar cell (photovoltaic) materials, 144–147
solar cooling, 211
solarium, 174
solar shading techniques, 19–20, 29, 186, **187**,
    188–190
    PV, 223
solar tube, 92
specific heat capacity, 183
specific humidity, 86
Stefan–Boltzmann law, 62
    Stefan-Boltzmann constant, 3, 62, 101
step-variable cost, 342
surface azimuth angle, 7

T

thermal circuit analysis, 73
    composite roof, 74
    composite wall, 73
    thermal resistance, 73–74
thermal comfort, 85
    behavioral, 96–98
    comfort equation, 99
    comfort indices, 103, **104**
    physical, 86–95
    physiological, 95
    standards, 107
thermal conductivity, 42–43
thermal cooling, 114, 183, 192, 203–204
thermal diffusivity, 43
thermal heating, 114, 161, 164, *165*, 167, 177
thermal load levelling, 106, 171
thermal storage, 171, 174
thermodynamics, 111
total heat loss, 8
    latent heat loss, 88
    sensible heat loss, 88
total heat transfer coefficient, 68
total radiation, 10
trans wall, 174, *175*
Trombe wall, 30, 171, 204–205
    PCM, 205–207
    rate of heat transfer, 209
    U value, 207–209
    vented, 205, *206*, 209
turbidity factor, 10

U

unacost, 297
uniform annual cost, 297–298
U value, 29, 69, 168
    Trombe wall, 207–209

V

variable cost, 342
ventilation and infiltration, 184–186

W

water wall, 174
weather conditions, **12**, 15
wet bulb temperature, 86
Wien's displacement law, 62, 64
windows and fenestration, 19, 27, 91–92, 187–188
wind towers, 197–198

Z

zenith, 6
zero energy buildings, 1–2
    design approach, 28–32

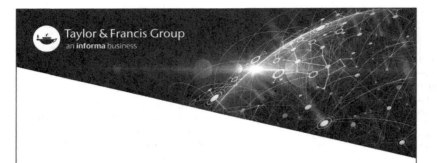

9781138333550